Elements of Crustal Geomechanics

This key textbook considers the mechanics of geomaterials at a wide range of scales, both in time and space. It provides detailed introduction to the study of crustal geomechanics, focusing specifically on the seismogenic crust.

Following an introduction to the necessary fundamentals of structural geology and material science, the book demonstrates how the application of continuum mechanics principles can provide efficient solutions to geomechanics problems at various scales, taking into account the multiphase characteristics of the geomaterials as well as discontinuities such as fractures and faults. It shows how field and laboratory observations can be combined with basic mathematical theory to build solutions with known levels of uncertainty. Particular consideration is given to the use of microseismicity in constraining geomechanical models – especially those involving fluid–rock interactions. Case studies are provided that illustrate how *in situ* stress determinations at very different scales provide unique constraints on the rheological characteristics of the seismogenic crust, and practical results from numerical modeling are used to illustrate the applicability and limitations of current theories.

Elements of Crustal Geomechanics introduces students to the common basic principles used in solving geomechanics problems ranging from exploitation of geothermal energy and long-term storage of nuclear waste to mitigating the impacts of volcanic eruptions. Accessible explanations of the mathematical formulations, convenient summaries of the key equations, and exercises that encourage students to put their learning into practice make this a valuable reference for students and researchers in geomechanics, geophysics, structural geology and engineering.

François Henri Cornet is a Professor at the Institut de Physique du Globe de Strasbourg. Prior to this he worked in the Department of Seismology at the Institut de Physique du Globe de Paris, and was also Visiting Scientist at Stanford University and at The Lawrence Berkeley National Laboratory. His main research interests are in rock mechanics, specializing in the measurement and modeling of stress fields; in rock–fluid interactions, including induced seismicity and applications to geothermal energy development; and in the development of large-scale, *in situ*, geophysical laboratories. Professor Cornet has extensive experience of teaching geomechanics courses at undergraduate and graduate levels and has also consulted internationally on stress field evaluations.

Elements of Crustal Geomechanics

FRANÇOIS HENRI CORNET
Institut de Physique du Globe de Strasbourg, Université de Strasbourg

CAMBRIDGE
UNIVERSITY PRESS

CAMBRIDGE
UNIVERSITY PRESS

University Printing House, Cambridge CB2 8BS, United Kingdom

Cambridge University Press is part of the University of Cambridge.

It furthers the University's mission by disseminating knowledge in the pursuit of education, learning and research at the highest international levels of excellence.

www.cambridge.org
Information on this title: www.cambridge.org/9780521875783

First published 2015

Printed in the United Kingdom by TJ International Ltd., Padstow, Cornwall

A catalogue record for this publication is available from the British Library

Library of Congress Cataloguing in Publication data
Cornet, F. H.
Elements of crustal geomechanics / François Henri Cornet.
pages cm
ISBN 978-0-521-87578-3
1. Geodynamics. 2. Geology, Structural. I. Title.
QE501.C634 2015
551.1–dc23
2014037526

ISBN 978-0-52187578-3 Hardback

Additional resources for this publication at www.cambridge.org/cornet

L'observation scientifique est toujours une observation polémique; elle confirme ou infirme une thèse antérieure, un schéma préalable.

Gaston Bachelard, *Le nouvel esprit scientifique*

(Scientific observation is always polemical; it confirms or contradicts a previous thesis, an earlier sketch.)

Contents

Online resources available at **www.cambridge.org/cornet**:

- worked solutions to the exercises
- field-based datasets
- MATLAB codes

Preface

Geomechanics refers to the mechanics of geomaterials, i.e. to the deformation and flow processes that affect the materials which make up the planet earth.

Geomechanics issues are encountered in a great variety of situations with very different scales, both in space and time. Generally, in engineering applications, time scales vary from a few days to a few tens of years and the volumes under consideration vary from a few hundreds of cubic meters to a few cubic kilometers. In earth science, however, time scales range from seconds to tens of millions of years and volumes vary from a few cubic kilometers to that of the entire planet. Accordingly, each domain of application has developed its own appropriation of the geomechanics concept, given that engineers have to deal mostly with perturbations of an existing system, with particular concern for safety issues and production or construction efficiency, while earth scientists are trying to understand natural phenomena such as fault motion, mountain building and sedimentary basin evolution.

For the last 30 years engineers have been confronted with much longer time scales and much greater volumes. For example the development of a repository for nuclear waste must be proved to be safe for up to a million years. The exploitation of geothermal energy or the filling of dams must not reactivate large faults and so trigger destructive earthquakes. Similarly, earth scientists must come up with precise seismic risk analysis, which requires an accurate description of the expected ground motion at specific locations. They must analyze, in real time, deformation fields on volcanoes in order to mitigate the hazards associated with eruption.

Today, geoengineers and geoscientists dealing with the mechanics of earth materials need to speak the same language. The objective of this text book is to introduce the basic principles of mechanics that earth scientists and mining, petroleum, civil and environmental engineers need to apply for solving problems in geomechanics. The only materials which are considered here are crustal geomaterials. The only paradigm considered for describing the deformation and flow processes of these geomaterials is that of continuum mechanics, but the limits of this paradigm are pointed out occasionally.

The aim of this book is to introduce the material for a two-semester class on geomechanics for upper undergraduate and first-year graduate students in earth sciences. It is based on notes prepared for my classes and inspired by notes from P. R. Fosdick's continuum mechanics classes at the University of Minnesota.

In the first part of the book (chapters 1 to 7) the basic concepts of solid and fluid mechanics necessary for understanding the mechanical behavior of geomaterials are introduced. The second part of the book (chapters 8 to 12) discusses various specificities of geomechanics that result from the complexity of geomaterials. Special attention is given to dynamic phenomena (such as microseismicity) as well as to solid–fluid interactions. In the last part of the book (chapters 13 and 14) various *in situ* stress determination methods are introduced and practical examples at various scales illustrate how a sound evaluation of the stress field helps a better understanding of the various mechanical processes at work in the seismogenic crust.

The first chapter introduces the concept of equivalent geomaterials and a description of their discontinuities (fractures and faults). The second chapter presents various unidirectional rheological models that help one to understand the basic concepts of elasticity, viscosity, plasticity and friction. The third and fourth chapters discuss the concepts of stress, strain and deformation. In the fifth chapter the behavior of linearly elastic solids is discussed and problems frequently encountered in geomechanics are solved. The sixth chapter introduces some basic elements of continuum mechanics with application to the laminar flow of incompressible materials. The seventh chapter presents basic principles of linear fracture mechanics. With chapter 8, our attention turns more specifically to geomaterials, and the results of laboratory investigations are presented. Chapter 9 addresses the application of continuum mechanics principles to geomechanics, and chapter 10 introduces specific characteristics of fractures and faults. In chapter 11 we describe the various types of wave observed in seismology and then we discuss more specifically seismic sources. Chapter 12 addresses various aspects of solid–fluid interactions, including linear poroelasticity, thermoelasticity and the nonlinear effects associated with failure processes (hydraulic fracturing and fluid induced shear fractures). Chapter 13, on *in situ* stress determination methods, gives practical applications of the various concepts that have been introduced throughout the book. In the final chapter these methods are illustrated through examples that concern the design of an underground hydroelectric power scheme (km^3 scale), the design of a nuclear waste repository (100 km^3 scale) and the stress fields in the upper Rhine graben (1000 km^3 scale) and the west-central European lithosphere (10^6 km^3 scale).

I would like to thank very sincerely Susan Francis from Cambridge University Press, who suggested that I should take the time to write up my lecture notes. She did not anticipate that I would be so slow in doing so, however! I also thank her two assistants, Laura Clark and Zoe Pruce, for their help during the various preparatory phases, as well as Susan Parkinson for her thorough copyediting of the manuscript.

My sincere gratitude goes to Marco Calo, and to my son Jan, for their help in preparing most of the figures. The manuscript has also greatly benefitted from the help of my colleagues Patrick Baud, Daniel Billaux, Dominique Bruel, Michel Cara, Mai Linh Doan, Emmanuel Detrournay, Emmanuel Gaucher, Georges Jobert, Sophie Lambotte, Olivier Langline, Vincent Magnenet, Romain Prioul, Daniel Quesada and Jean Schmittbuhl for reading early versions of some chapters. They pointed out a multitude of typing errors and contributed significant improvements. But I bear the entire responsibility for all the errors that are still left in the present document.

Finally my sincere gratitude to my wife, Basia, who has helped me through all these years and kept my morale up especially during the last, never-ending, phase of this project.

F. H. Cornet

1 Geomaterials and crustal geomechanics

Geomechanics is concerned with the deformation and flow of geomaterials. A specific aspect of geomaterials, i.e. the materials that make up the planet earth, is their complex combination of solid and fluid phases.

In engineering geomechanics it is customary to talk of rock masses in a way that refers to both the rocks and the fracture systems that affect the volumes of concern. This concept, however, is often too vague for efficient mechanical modeling, and specific attention must be given to both the geomaterials and their discontinuities.

In the two first sections of this chapter we define more precisely the notion of a geomaterial and the related concept of a representative elementary volume (REV), with a brief reference to the various methods available for identifying geomaterials. In the third section we discuss the concepts of fracture sets and faults, with special attention to scaling laws. Finally, in the fourth section our attention turns to the various loading processes that may be encountered in geomechanics, whether of human or natural origin.

1.1 Rocks, soils and other geomaterials

Three kinds of rock can be identified: igneous, metamorphic and sedimentary. This characterization refers to the origin as well as to the past thermal and loading history of the rock. It implies strong consequences for the rock fabric, namely, the structure of its constitutive (solid) grains and of the complementary pore space, which is generally filled with fluid. And this introduces immediately the fact that materials of geological origin are most often multiphasic in their natural environment, i.e. they include solid, liquid and gas phases.

First, we introduce definitions that are used to describe the relative volumes occupied by the various phases. Then we introduce Goodman's classification of rocks according to their texture (Goodman, 1989), since in this book attention is given to the behavioral rather than to the genetic attributes of rocks. Such a nomenclature is helpful as a starting point for defining the material properties of import for a given mechanical problem.

1.1.1 Porosity, phase relationships, density

The representative elementary volume (REV) concept

Geomaterials always include some solid parts and some voids, the voids usually being filled with fluids, whether liquid or gas or both. The porosity of a geomaterial describes

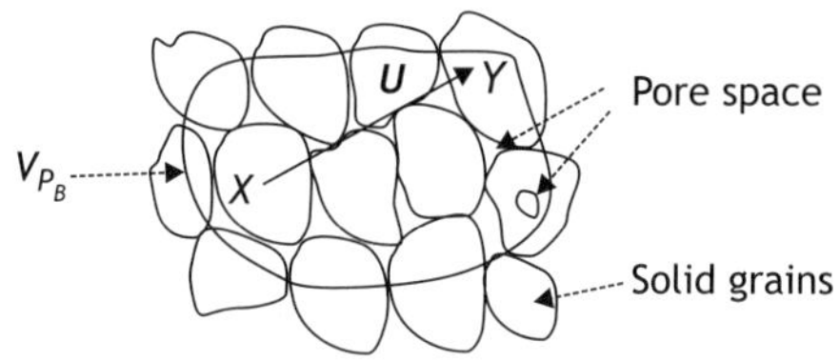

Fig. 1.1 Linear porosity.

the relative percentage of solids and voids. Let us consider a body B made up of two components, a solid component M and a fluid component F. Component M is made up of many grains that touch each other (fig. 1.1) and fluid F fills completely the voids in between the solid grains, i.e. the pore space is fully interconnected. The pore space is said to be fully interconnected when any point of it may be related to any other point of it by a continuous line, all the points of which remain within the pore space.

A small part P_B of body B, with volume V_{P_B}, includes a part P_S consisting of solid S with volume V_{P_S} and a part P_F consisting of fluid F with volume V_{P_F}, such that $V_{P_B} = V_{P_S} + V_{P_F}$. The volume V_{P_F} defines the pore space of part P_B. The *volume porosity* n of P_B is defined as $n = V_{P_F}/V_{P_B}$ while the ratio $e = V_{P_F}/V_{P_S}$ is called the *void ratio*.

The void ratio and the porosity are interrelated:

$$e = \frac{n}{1-n}, \qquad n = \frac{e}{1+e} \tag{1.1}$$

In soil mechanics, the specific volume v is defined as the total volume of soil that contains a unit volume of solid ($v = 1 + e$).

If the pore space is not fully interconnected, the volume V_{P_F} includes only the interconnected part of the total pore volume V_{P_P}. Then the fluid within the non-interconnected pore space may be different from that in the interconnected pore space, as e.g. in volcanic rocks. Furthermore, as will be discussed in section 9.3, the physical properties of the fluid that fills up the pores depend on the distance to the contact with the solid phase. For example, for water, for very small distances to the solid interface (in the micrometer range), a thin film exists that cannot flow and that can be removed only by heating up the material. It is called adsorbed water. For the purpose of defining the interconnected pore space, the adsorbed water is "assimilated" to the solid.

Let us now consider two points X and Y that define the vector $\boldsymbol{U}$ with origin at X and extremity at Y (fig. 1.1). The linear porosity $l(X, \boldsymbol{U})$ associated with the vector $\boldsymbol{U}$ at X is defined by the ratio

$$l(X, \boldsymbol{U}) = \frac{|\boldsymbol{U}_P|}{|\boldsymbol{U}|} \tag{1.2}$$

where $|\boldsymbol{U}|$ is the modulus of vector $\boldsymbol{U}$ while $|\boldsymbol{U}_P|$ is the modulus of that part of $|\boldsymbol{U}|$ that intersects the pore space. When point Y is the same as point X, so that $|\boldsymbol{U}| = 0$, the linear porosity is set equal to 1 if point X is in a void and equal to 0 if it is located in a grain. As the modulus $|\boldsymbol{U}|$ gets larger and larger, the variation in the linear porosity gets smaller and smaller (fig. 1.2) so that, for a length $|\boldsymbol{U}|$ larger than, say, U^T, variations in linear porosity with increasing reference length $|\boldsymbol{U}|$ may be neglected. The definition of this critical length

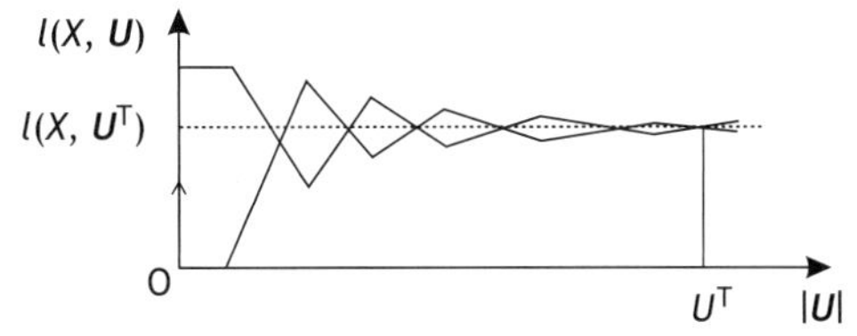

Fig. 1.2 Variation of linear porosity with the length of the defining vector ***U***.

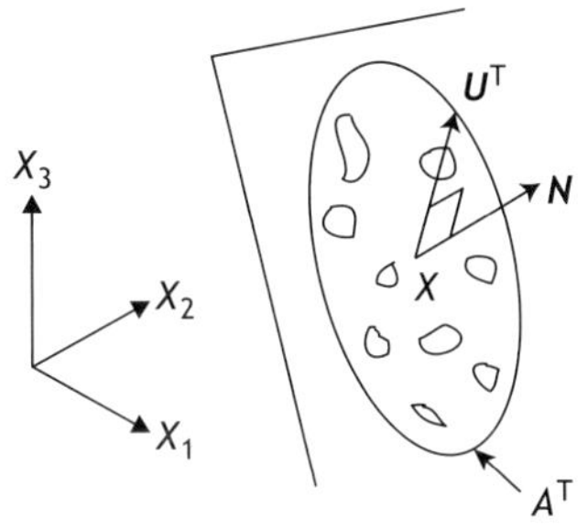

Fig. 1.3 Areal porosity at point *X* in a plane with normal ***N***.

U^T is somewhat arbitrary and may be chosen so that the variations in linear porosity are smaller than, say, 0.005 or 0.001 when $|\boldsymbol{U}|$ is larger than U^T.

If the linear porosity does not vary with the orientation of $\boldsymbol{U}^T$ then it is isotropic and if it does not vary with the spatial position of X in the body, it is homogeneous.

A similar approach may be followed to define an areal porosity. Consider a planar surface with normal $\boldsymbol{N}$. The envelope of the extremities of all vectors $\boldsymbol{U}^T$ in the plane normal to $\boldsymbol{N}$ defines a closed planar contour with area A^T centered at X (fig. 1.3).

Let A_P be that part of A^T that passes through the pores. The areal porosity $f(X, \boldsymbol{N})$ at the point X in the plane normal to $\boldsymbol{N}$ is defined by the ratio

$$f(X, \boldsymbol{N}) = \frac{A_P}{A^T} \tag{1.3}$$

If $f(X, \boldsymbol{N})$ does not vary with the orientation of $\boldsymbol{N}$, the areal porosity is isotropic and equal to $l(X,\ U^T)$.

Finally, when the dip (see fig. 1.9) of the normal $\boldsymbol{N}$ varies from 0 to π, the envelope of all surfaces with area A^T normal to $\boldsymbol{N}$ defines a volume. This volume corresponds to the smallest part P of B for which the volume porosity $n(X)$ may be defined. It is called the *representative elementary volume* (REV). For bodies with isotropic porosity,

$$l(X,\ U^T) = f(X,\ A^T) = n(X) \tag{1.4}$$

As already mentioned, when the porosity does not depend on X it is said to be homogeneous. If, however, the porosity varies in space, it is said to be heterogeneous. In general, a geomaterial is said to be heterogeneous with respect to a given property when this property varies with position in the volume under consideration.

For a body made up of various constitutive minerals or grains, the concept of the REV may be extended to define the smallest part of B that is statistically representative of the body, i.e. the smallest volume for which the characteristic properties of the body under consideration may be defined. As a rule of thumb, the REV is often considered to range somewhere between 50 and 100 times the mean dimension of the grains that make up the geomaterial.

When applicable, the REV provides a means to define an equivalent body that behaves as a continuum, in which the theory of continuous functions may be applied. Expressed in terms of probability, the concept of the REV implies a stationarity in the statistical properties of the material. This concept of a statistically representative volume is discussed further in section 1.3.3 and in chapter 10, where fracture fields are discussed and are found *not* to satisfy the concept of the REV.

Various techniques have been developed for measuring the porosity, whether based on the percolation of fluids (the mercury porosity) or by direct visual inspection of planar surfaces (Bourbié *et al.*, 1986; Guéguen and Palciauskas, 1992; Zinszner and Pellerin, 2007).

Phase relationships and density

In what follows we refer only to a representative elementary volume of material. For a completely dry geomaterial, only two phases are involved: solid and air. A fully saturated geomaterial also involves only two phases. For example, saturated soils involve solid grains and liquid water. Hence the concept of saturation refers to the idea that the pore space is filled with only one liquid. A material in which the pore space is filled with liquid and gas is described as being partially saturated. For example, a partially saturated soil has liquid water and air in the pore space. A partially saturated deep geothermal system has liquid water and vapor. However, the concept of saturation may also refer to geomaterials in which the pore space is filled with diverse immiscible liquid materials, such as brine and hydrocarbons, without any gas phase.

The moisture or water content, w, of a soil is defined as the ratio of the mass of liquid water, M_w, and the mass of solid, M_S, for an REV:

$$w = \frac{M_w}{M_S} \tag{1.5}$$

while the degree of water saturation, S_r, is defined as the ratio of the volume of liquid water, V_w, and the total pore volume, V_p, which is assumed to be fully interconnected:

$$S_r = \frac{V_w}{V_p} \tag{1.6}$$

The bulk density of a body B, ϱ_B, made up of solid S and water, is simply the ratio of the total mass of B and the total volume (see fig. 1.4):

$$\varrho_B = \frac{M_B}{V_B} = \frac{M_S + M_w}{V_B} \tag{1.7}$$

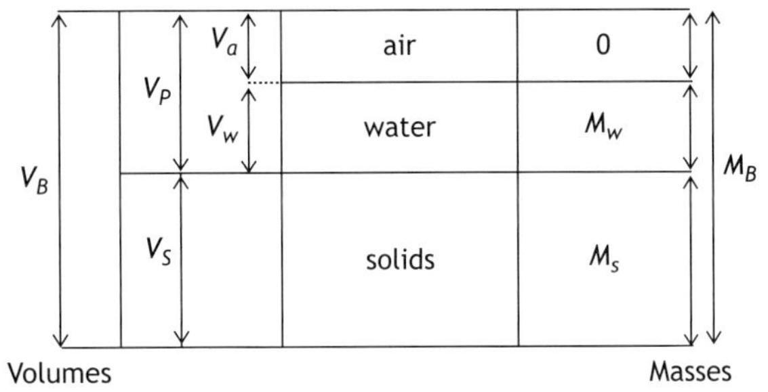

Fig. 1.4 The various phase relationships in masses (M) and volumes (V) for a representative elementary volume (REV) (adapted from Craig, 1987, fig. 1.10, with permission from Taylor & Francis).

If ϱ_w is the density of liquid water and ϱ_S is the density of solid, the bulk density of a saturated body B is

$$\varrho_B = (\varrho_S V_S + \varrho_w V_P)/V_B = (1-n)\varrho_S + n\varrho_w \tag{1.8}$$

For a soil with water content w, liquid water density ϱ_w, specific gravity of solid particles $G_s = M_s/(V_s\varrho_w)$ and void ratio e, its bulk density ϱ_B is given by

$$\varrho_B = \frac{G_s(1+w)}{1+e}\varrho_w = \frac{G_s + S_r e}{1+e}\varrho_w \tag{1.9}$$

When the soil is saturated, $S_r = 1$, the bulk density is

$$\varrho_B = \frac{G_s + e}{1+e}\varrho_w \tag{1.10}$$

The mechanical properties of soils depend in particular on their particle size distribution and on their water content. For example, a dry clay may behave as a solid; yet when immersed in water it will become progressively more liquid. When the clay particles are dispersed, the density of the fluid part may be larger than that of some solid parts (e.g. those with a high unconnected pore space). These solid parts may start to float when the liquid part gets dense enough, giving rise to mud slides. Hence an uncemented clastic geomaterial may behave as a solid when its water content is low, but may behave as a dense fluid when saturated with water.

Similarly lava, which is made up of solid crystals, melted elements and gas, may behave as a fluid when the solid grains are completely dispersed in the melted part but as a very viscous solid when the solid portion dominates.

Thus the mechanical behavior of geomaterials ranges in a continuous manner from that of a solid-type material to that of a fluid-type material. For example, for soils, it depends on water saturation as well as on the void ratio and attention must be given to both these variables. Standardized tests have been devised in soil mechanics in order to define a liquid limit and a plastic limit for such soils (e.g. Craig, 1987).

More generally, geomaterials always involve diverse constituents and phases. The objective of geomechanics is to describe the deformation and/or flow of these complex systems, which cover the complete spectrum from solids to fluids. Equivalent materials are defined

in order to investigate the mechanical behavior of these systems at scales much larger than those of their representative elementary volume.

Hence, while in fact rocks are always heterogeneous because they involve various minerals and phases, the equivalent geomaterials are assumed to be homogeneous and their mechanical behavior is assumed to represent that of volumes much larger than the REV.

1.1.2 Rock and soil classification

Crystalline texture

Crystalline rocks consist of tightly interlocked crystals of silicate minerals (e.g. quartz, feldspar, micas etc.), or carbonates, sulfates or other salts. Their mechanical properties depend on those of the constitutive minerals, on those of the contact interfaces between the minerals and also on the geometry of voids between the minerals and the fluids filling up the pore space. Following Goodman, six crystalline textures may be distinguished.

Soluble carbonates and salts

These include limestone, dolomite, marble (a metamorphic limestone), rock salt and gypsum. Their mechanical properties depend strongly on the mean pressure which is applied to them. According to terminology to be introduced in the next chapter they may exhibit either an elastic or a plastic behavior, depending on their environment. Further, they, or parts of them, may dissolve in water at relatively fast rates depending on environmental conditions (time constants ranging from days to years).

Rocks with continuous bands of planar minerals

These include schists with alignments of planar minerals such as mica, chlorite or graphite schists. Mica and other sheet minerals strongly influence the mechanical properties of the rock. They are strongly anisotropic, i.e. many of their properties depend strongly on orientation.

Banded silicate minerals without mica sheets

These include gneiss. Their properties generally show some anisotropy. This anisotropy is not as strong as that of rocks with continuous bands of planar minerals yet is easily detectable by various petrophysical means.

Randomly oriented and distributed silicate minerals

Examples are granite, chlorite, gabbro, syenite etc. Their properties usually do not depend on orientation, so that they are isotropic. Chemical interaction with water is generally rather slow at low temperatures (it happens over time scales usually in thousands of years).

Randomly oriented and distributed silicate minerals in a background of very fine grains and/or glass, with vugs (small cavities)

These correspond to lava-type rocks (basalt, rhyolite etc.). The density of these materials depends strongly on the pore space, which may be very large, yet they may include many unconnected pores (i.e. each pore is completely isolated from its neighbor) that are filled with gas.

Sheared rocks

Examples are mylonites or serpentinite. Their mechanical properties are markedly anisotropic.

Clastic texture: from rocks to soils

Clastic rocks are composed of pieces of various rock types and assorted mineral grains. Their mechanical properties depend strongly on those of the cement binding the various fragments together. Five different clastic textures may be identified.

Stably cemented

Examples are silica-cemented sandstone or limonite sandstone. Examples may also be found in volcanic breccia.

With slightly soluble cement

These rocks include calcite-cemented sandstones or conglomerates. The percolation of acidic fluids rapidly modifies their properties.

With highly soluble cement

With gypsum-cemented sandstones and conglomerates, the effect of water may be very rapid (a matter of days or months).

Incompletely or weakly cemented

The difference from the two previous classes rests mostly on the quantity of the cement and is qualitative.

Uncemented: soils and granular materials

This class is described in the soil mechanics literature (e.g. Craig, 1987). It involves particles with a very wide spectrum of sizes and shapes. Particle-size analysis determines the percentage by weight of the particles within the different size ranges. The classification distinguishes the following: clays, with particles smaller than 2 microns; silts, with particle sizes ranging from 2 microns to 60 microns; sands, with particle sizes ranging from 60 microns to 2 mm; gravels, with sizes ranging from 2 mm to 60 mm; and cobbles and boulders, containing blocks larger than 6 cm. The properties of uncemented clastic materials (often called "soils") depend strongly on their particle-size distribution. An example of grain-size analysis is shown in fig. 1.19 for the gouge of Aigion fault in the Corinth rift (western Greece).

Very fine grained rocks

These rocks include, for example, shales, slates, flagstones and marls and are made up of clays and silts with various degrees and kinds of cementation.They also include chalk and other very fine-grained materials, such as radiolarites, which result from the accumulation of plankton skeleton.

Organic rocks

These include coals and bituminous coals, oil shale and tar sand. Sometimes oil and gas are also called rock to emphasize their geological origins. According to such a definition, water would also be a rock. However, we prefer to keep the word *rock* to refer to solid material, and we introduce the more generic term *geomaterial* to refer to all materials of geological origin encountered on earth.

Hence a geomaterial may be a rock, a soil, a liquid or a gas or any mixture of these, provided that each component is of geological origin.

1.2 Rock masses and crustal geomechanics

In the previous section the concept of pore space was introduced without any consideration of the morphology of the pores. However, when discussing rock textures we mentioned that laminar minerals and their relative alignments are of great importance for the mechanical properties of the equivalent geomaterials. Similarly *fractures*, which correspond to structures for which one dimension is much smaller than either of the others, by orders of magnitude, are the most significant structuring elements of rock masses. A key question when tackling geomechanics problems is determining how to integrate fractures into the solution. Various approaches are presented in chapters 9 and 10.

We introduce in this section, and in the next, various types of rock-mass reconnaissance for identifying equivalent geomaterials and the correlated fracture field, i.e. those fractures that are dealt with explicitly as opposed to those which are included implicitly through the definition of equivalent homogeneous geomaterials.

In order to define the equivalent geomaterials filling the volumes under consideration, these volumes must be probed with various geophysical means and/or directly sampled.

1.2.1 Geophysical exploration

Engineering applications

Essentially three classes of geophysical techniques are applied for distant rock mass reconnaissance: seismic methods, electrical and electromagnetic methods and gravimetric methods. See e.g. Milsom and Eriksen (2011) for a review of geophysical reconnaissance techniques commonly used in geotechnical and environmental engineering or Sheriff and Geldart (1995) for deeper investigations as conducted for example in hydrocarbon exploration.

In seismic methods, seismic waves (mechanical vibrations) are artificially generated, and then specific receivers positioned at suitable locations record some of these seismic waves after they have traveled through the domain of interest. Exploitation of the results is based on seismic wave propagation theory, which rests on the theory of elasticity. This technique helps to identify the domains in which homogeneous materials may be defined, with particular attention to their fluid content. It also helps to identify whether the main reflecting structures are linked to the interfaces between different materials or to major fracture zones. This is briefly discussed in chapter 11.

Because hydrocarbons and most rocks are much more electrically resistant than water, various electrical and electromagnetic methods have been developed for identifying zones with high water content. These methods are also efficient for determining the clay content, given the high electrical conductivity of clays. However, such methods are not discussed any further in this book, with the exception of the borehole electrical imaging technique, specifically developed for fracture identification in boreholes.

Gravimetric techniques help to identify contrasts in density at various depths, whether these are linked to saturation or, more generally, to fluid content or to density contrasts between different geomaterials. Such techniques have also turned out to be efficient for identifying underground cavities, both man-made (mines and quarries) and natural (e.g. karstic structures).

Crustal geomaterials

Through the combined analysis of various geophysical and geochemical observations, earth scientists have progressively refined their understanding of the structure of the earth. Their analysis has identified four successive layers starting from the earth surface, namely, the lithosphere, the asthenosphere, which starts just below the lithosphere and extends down to 3480 km, the liquid core between 3480 and 1220 km from the earth's center and finally the solid inner core (e.g. Mussett and Khan, 2000); recall that the earth's radius is equal to about 6380 km). The contact between the lithosphere and the asthenosphere is characterized by a change in mechanical behavior and occurs at variable depths depending on the location on earth.

The lithosphere itself is subdivided into the crust, which extends from the earth's surface down to the Moho discontinuity, and the mantellic lithosphere, which exhibits seismic wave velocities much faster than those of the crust. The Moho discontinuity is found at depths ranging from 5 to 10 km in the oceanic lithosphere. However, its depth in the continental lithosphere is much more variable and may be found from 20 km in thin crust down to 60 km in subduction zones (see e.g. Stein and Wysession, 2003; for further reading on the earth's structure determined from seismic wave analysis).

The crust itself is often separated into an upper crust and a lower crust, the mechanics of which is still being debated (Burov and Diament, 1995; Burov, 2010).

The object of this text book is to introduce concepts that may be applied to analyze the mechanical behavior of the upper crust, i.e. from the ground surface down to the lower limit of the zone where earthquakes occur (generally between 15 and 25 km, except in subducting zones where it can go down to 80 km). It is often referred to as the brittle crust or the seismogenic crust.

1.2.2 Borehole reconnaissance

A major difference between engineering and earth science applications of geomechanics is that in the former direct access to the rock mass is required.

Until recently the deepest borehole ever drilled was undertaken by Russian earth scientists who were hoping to reach the Moho discontinuity (Kozlovski, 1984). After a depth of 12 262 km was reached in 1989 the project was stopped because of numerous technical difficulties linked in particular to higher than anticipated temperatures (around 180 °C at a depth of 11 500 m). At the time of writing (2014), the deepest well in the world, off the Sakalin Island coast in eastern Russia reaches 12 345 m. However, most attempts at deep drilling for scientific research purposes have been to depths shallower than 9 km, and this constitutes a major difference between engineering and earth sciences applications of geomechanics.

Indeed, in all engineering applications direct access to the rock mass is possible, and this helps to obtain *in situ* characterization of the geomaterials and of the fracture field. But in many earth science applications, as in fault mechanics, information on deep geomaterials comes only through indirect geophysical investigations.

Coring

As discussed above, direct reconnaissance of a rock mass may be made through boreholes of various sizes and lengths as well as through excavation from the ground surface or through underground access (shafts or adits). In the latter case, the underground access is often exploited further by the drilling of shallow boreholes.

Boreholes may be drilled either with core recovery, i.e. the drilling operation produces in theory a continuous cylindrical core made of the various materials and fractures encountered by the drill bit, or in a destructive manner so that only cuttings are produced. Cuttings are rock particles the size of which is in the millimeter to centimeter range. They help to identify the various geological materials intersected by the borehole.

Cores, in addition to producing samples that may be tested in the laboratory, may also be used to retrieve important information on the fracture field, as will be discussed in section 1.3.1. Various indices characteristic of the fracture density are defined for engineering

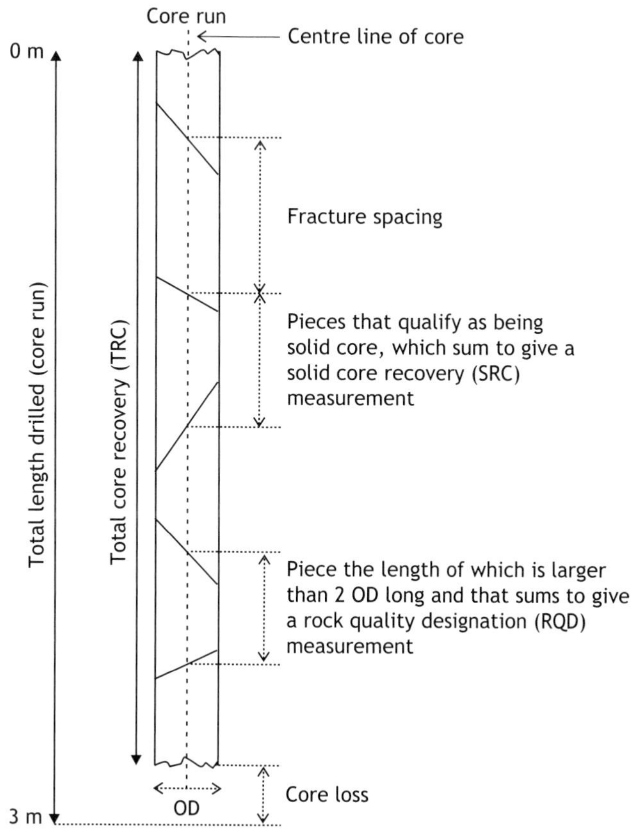

Fig. 1.5 Fracture spacing and rock quality designation (RQD) index for a 3 m long coring operation.

purposes (e.g. Price, 2009). The total core recovery (TCR) is the ratio of the total drilled length and the length of core recovered in the core barrel. The rock quality designation (RQD) index of a rock mass (Deere, 1963) is evaluated by determining the percentage recovery of core in lengths greater than twice the diameter of the core (fig. 1.5):

$$RQD = \sum_{i=1}^{n} \frac{x_i}{L} \tag{1.11}$$

where x_i are the spacings between fractures that are greater than twice the core diameter and L is the total drilling length for which coring has been conducted.

The concept of indexing rock masses for engineering investigation probably represents one of the first attempts at defining equivalent geomaterials. The goal was to define equivalent continuum materials for conducting the large-scale stability analysis required for construction purposes while introducing for each continuum a single ad hoc parameter characteristic of the fracture field within the whole volume of the equivalent continuum.

It has since been realized that the drilling operation itself generates some breakage of rock cores and results in some loss of material as well as alteration of the rock material properties. Hence estimates on fracture fields based only on core analysis are frequently misleading. It is generally much preferable to investigate fracture fields in boreholes directly rather than through a core analysis, as will be discussed below.

Similarly, the properties of rock samples are often perturbed by various mechanisms (microcracking, swelling, loss of water content, etc.), so that direct *in situ* investigation through geophysical logging gives very useful information, the deeper the boreholes, the more so.

Geophysical logging

In some civil engineering applications boreholes are empty or are filled with clear water so that video cameras may be used to observe the rock formation directly, especially for fracture orientation analysis. Generally, however, boreholes are filled with mud or opaque water so that optical investigations are ruled out. Instead, various efficient techniques have been developed for investigating *in situ* the properties of rock masses, their fluid content and the ability of these fluids to flow through the rock mass. The description of the variation with depth of these characteristics, or of related physical rock mass properties, is referred to as a geophysical log. This is the domain of rock physics (e.g. Zinszner and Pellerin, 2007).

Hence, different logging techniques are used depending on whether attention is being given to the fluid content (hydrocarbon or water), the pore space and its inter-connectivity, the bulk material (solid plus fluid) and its density, the clay content, the sequence of sedimentary beds or the thickness of the various formations, etc. See e.g. Ellis and Singer (2008) or Hearst (2000) for geophysical logs used in the oil industry and in hydrogeology or other deep applications.

1.3 Fractures and faults as structural discontinuities

As already mentioned, a fracture is defined as a structure with one dimension that is orders of magnitude smaller than the two other dimensions. A fracture is defined as a *structural discontinuity* if it is taken into consideration explicitly in the definition of the geomechanical problem under consideration. Fractures smaller than a critical dimension, or fractures that do not correspond to significant discontinuities in geomaterial properties, are either dealt with implicitly as specific properties of the equivalent geomaterial or are simply ignored.

1.3.1 Morphology and orientation of fractures

Seen from a certain distance, a fracture is essentially a surface, which often appears to be planar but is not necessarily so. As one looks more closely, the fracture is found to correspond to a very thin volume, with a very small and complex thickness (fig. 1.6).

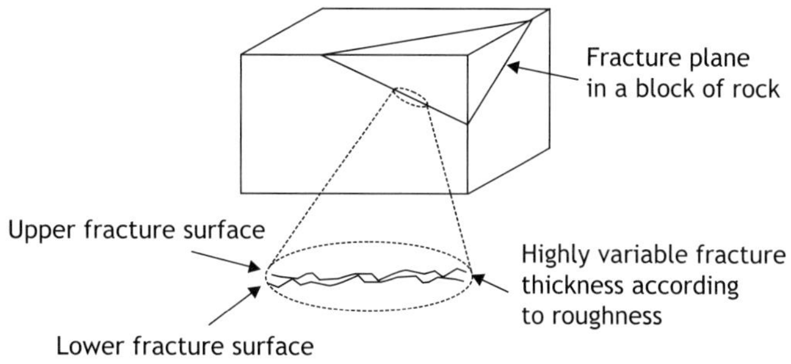

Fig. 1.6 Fracture plane, fracture surfaces and fracture roughness.

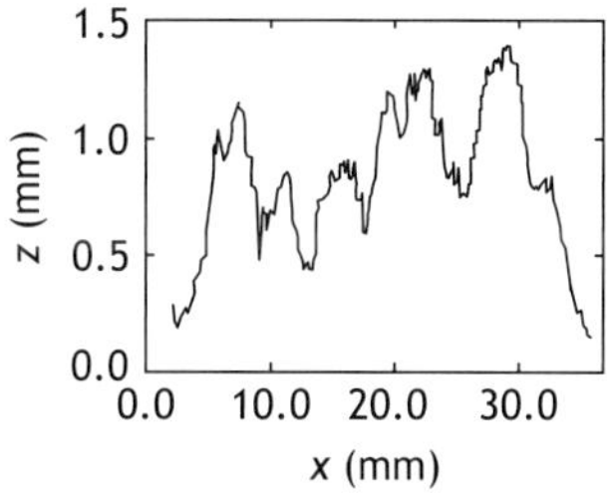

Fig. 1.7 Profile of a fracture surface (reproduced from Schmittbuhl *et al.*, 1995, fig. 2, with permission from Wiley).

Fracture roughness and fracture filling

A fracture always involves two irregular surfaces, the roughness of which renders the fracture thickness spatially highly variable. Adler and Thovert (1999) compiled an extensive review of the means to measure experimentally and to characterize the roughness of fracture surfaces. Measurements are conducted with profilometers to generate one-dimensional profiles of the surface height along parallel, regularly spaced, straight lines (fig. 1.7). An arbitrary planar surface reference, $z = 0$, is defined so that each point measured on the profile is characterized by its distance z to this reference plane.

A common practice in engineering applications, is to characterize the roughness of fracture surfaces by statistical properties of z such as the central line average (CLA), the mean squares value (MSV) and the root mean square average (RMS):

$$CLA = \langle |z_i| \rangle, \quad MSV = RMS^2 = \langle z_i^2 \rangle \tag{1.12}$$

where the notation $\langle a \rangle = (\sum_{i=1}^{N} a_i)/N$ refers to the mean value of a_i and z_i refers to the z component of the ith point on the profile.

When the reference plane $z = 0$ is made to correspond with the mean fracture surface $CLA = 0$, then the RMS is simply the standard deviation of the surface height measurements. More elaborate statistical characterizations have been proposed (e.g. by Pigott and Elsworth, 1995).

However, as pointed out by Brown and Scholz (1985), the definition of the reference plane depends on the size of the sampled surface, i.e. individual measurements are not independent of each other but some spatial correlation exists between them, at least for wave lengths between 10^{-5} and 1 m.

Schmittbuhl *et al.* (1995) showed that fracture surfaces are self-affine structures, i.e. they are statistically invariant under an affine transformation. For a horizontal direction amplitude Δx in the mean fracture plane and a vertical direction amplitude Δz, an affine transformation is defined by

$$\Delta x \rightarrow \lambda \Delta x \tag{1.13}$$

$$\Delta z \rightarrow \beta \Delta z \tag{1.14}$$

where $\beta = \lambda^{\zeta}$ and ζ is the Hurst exponent.

Schmittbuhl *et al.* showed that, for very different rock types and different fracture origins, the Hurst exponent is constant and equal to 0.8, with a dispersion of 0.05. Hence, measuring a fracture surface profile for a given situation helps to generate other statistically representative fracture surfaces for similar conditions.

Theoretically, once the topology of the two rough surfaces that bound a fracture volume are known, it may be used to determine the mechanical and hydraulic characteristics of the fracture.

Very often, though, this void space is partially filled by deposits associated with previous fluid circulations. Hence, a simple description of the topology of the fracture surface is not enough to characterize the pore space that may be open to fluid flow for this fracture.

In practice, fractures are generally made up of solid bridges and open spaces. The solid bridges control the mechanical behavior of the fracture and are often called asperities, while the interconnected pore space is of major significance when analyzing fluid flow; it is often taken into account through an equivalent so-called *fracture hydraulic aperture* (see section 10.2). These features, the asperities and fracture hydraulic apertures, are self-affine fractals within a large domain, which remains bounded nevertheless. The upper bound of this domain defines the scale for which a mean fracture orientation may be defined.

Fracture orientation and reference systems

Depending on the context of the investigation, different reference systems may be used to define orientation. With the geographical system, the reference angles are the longitude and latitude with positive rotations counterclockwise (fig. 1.8).

For many geomechanical problems a direct cartesian frame of reference is chosen in which the horizontal axis X_1 is oriented north, the horizontal axis X_2 is oriented east and the X_3 axis is vertical and oriented positive downward, toward the center of the earth. Hence, in such a geological reference system, an azimuth which is defined as N57°E implies that a 57° angle is measured with respect to the geographic north, turning positively toward the east direction.

An alternative direct geological frame of reference may be defined by choosing to rotate in the counterclockwise direction (i.e. toward the west) with the X_3 axis positive upward.

Experience shows that most fractures are planar surfaces when looked at from a certain distance, a feature consistent with the self-affine geometrical model of fractures.

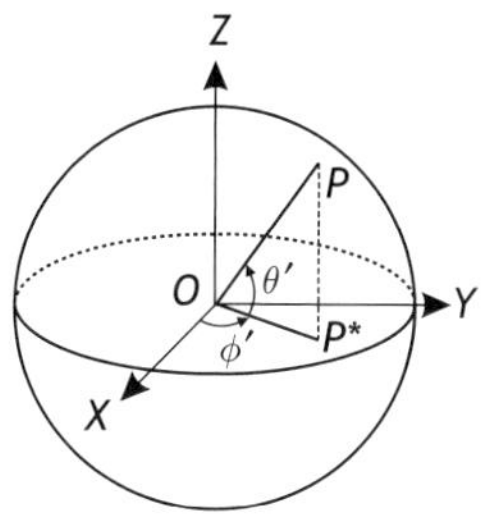

Fig. 1.8 Geographical reference system; ϕ' is longitude and θ' is latitude. In spherical coordinates, the radius OP is variable and often called ρ; the angular coordinate $\theta = 90° - \theta'$ is complementary (see fig. 3.12).

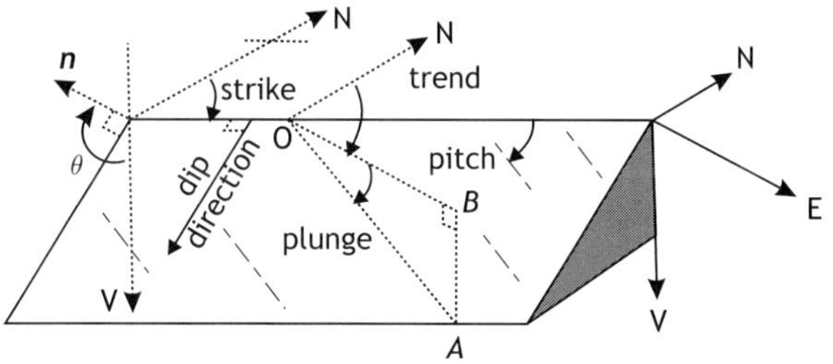

Fig. 1.9 Definition of the angles commonly used to characterize the orientation of a planar fracture or a straight line within the fracture. The line OB lies in a horizontal plane.

Although it does not need to be so, we will assume here that, indeed, fractures are roughly planar structures. The orientation of the mean fracture plane is defined by two angles (see fig. 1.9). These can be the *strike s*, which is the orientation of a horizontal straight line in the fracture plane with respect to north, and the *dip angle d*, which is the angle between the steepest straight line in the plane, oriented downward, and the horizontal plane. The strike is usually defined within an 180° angle (from north to south), and then the dip direction must also be specified. Alternatively, reference may be made to the *azimuth* of the plane, a, which is defined within a 360° angle, while the dip is still defined within a 90° angle when considering the downward oriented steepest straight line. Often, fracture orientations are defined by their *dip direction dd* and their dip d.

In order to avoid 180° ambiguities, it is common practice to consider that the line of steepest descent is oriented downward so that dip directions are defined between 0 and 360°. The direct geological frame of reference mentioned earlier is shown on the right in fig. 1.9. Note that many data on fracture orientation are produced by borehole geophysical logs, which use magnetic north as reference rather than geographical north.

It is also very common to characterize the fracture plane orientation by that of its normal $\boldsymbol{n}$. In this case it is necessary to specify whether the unit normal is oriented upward or downward. The characteristic angles for the normal to the fracture plane are the dip direction dd and the angle θ of the normal ($\theta = 90° + d$) with respect to the vertical direction, when the normal is oriented upward (fig. 1.9) and the vertical direction is positive

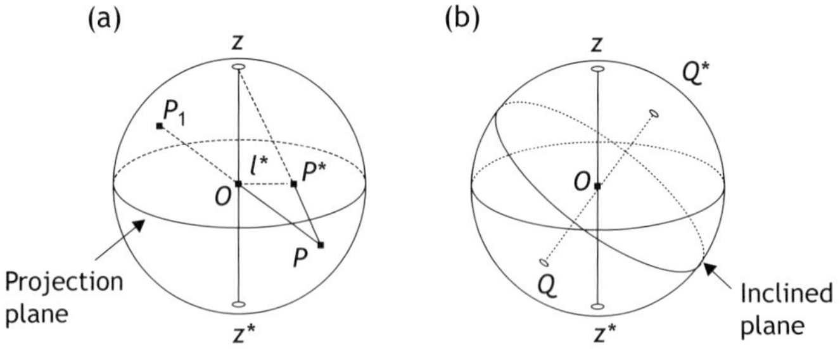

Fig. 1.10 Two-dimensional plots of orientation data: (a) equal-angle (Wulff) stereographic projection; (b) the poles Q and Q^* of an inclined plane.

downward. Accordingly, the direction cosines n_i $(i = 1, 2, 3)$ of the normal to the fracture plane as expressed in the direct geological system are

$$n_1 = \sin\theta \cos dd, \quad n_2 = \sin\theta \sin dd, \quad n_3 = \cos\theta \tag{1.15}$$

when expressed in terms of the dip direction dd and the inclination θ of the normal. These direction cosines may also be written in terms of the strike and the dip of the plane, on observing that $s = dd - \pi/2$ and $d = \pi - \theta$, with the vertical axis positive downward:

$$n_1 = -\cos s \sin d, \quad n_2 = \sin s \sin d, \quad n_3 = -\cos d \tag{1.16}$$

A straight line in a plane, such as a striation in a fault plane, is characterized by its *trend* (the angle between the horizontal projection OB of the straight line and the north direction) and its *plunge* (the angle defined in a vertical plane between the straight line and the horizontal projection OB of the straight line; see fig. 1.9). The pitch of such a straight line in a given plane is the angle between the straight line and a horizontal line in the given plane.

Stereographic and equal-area projections

Orientation data, whether they concern planar structures or straight lines, always involve only two angles. When only orientations are considered for an analysis, without reference to the exact location in space of the objects under consideration, orientation data are most conveniently plotted on stereographic projections or on equal-area projections, which project the orientation data onto a plane, i.e. a two-dimensional reference system, as shown in fig. 1.10.

Figure 1.10a shows the stereographic projection of a plunge line. The line passes through the center O of a reference sphere and a point P in the lower hemisphere of the reference sphere. The horizontal plane passing through O is termed the projection plane. The vertical axis, perpendicular to the projection plane at O, pierces the top of the reference sphere at Z, which for lower hemisphere stereographic projection is termed the focus. A stereographic projection consists of projections of lines and points located on the surface of the reference sphere from a single perspective point (the focus) to corresponding lines and points in the projection plane.

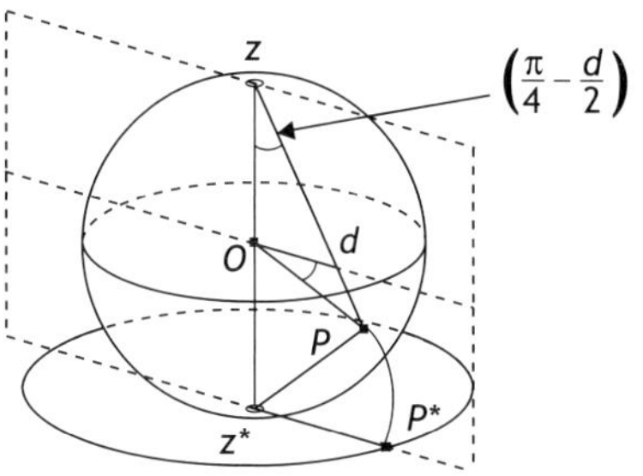

Fig. 1.11 Equal-area (Lambert or Schmidt) projection.

For example, the stereographic projection of point P in the lower hemisphere is point P^*, which is defined by the intersection between the line ZP and the projection plane (fig. 1.10(a)). The distance $l^* = OP^*$ is equal to $\tan(\pi/4 - d/2)$ if d is the dip of OP, i.e. the angle between OP and OP^*. The azimuth of P^* is the same as that of P. The point P_1 shown in fig. 1.10(a) is diametrically opposed to P. It may be used for characterizing the direction PP_1, provided that the point Z^*, which is diametrically opposed to Z, is used as the focus for the upper hemisphere stereographic projection.

Consider now an inclined plane passing through the center of the reference sphere. Its descending normal intersects the lower hemisphere at a point Q (see fig. 1.10(b)); Q is called the pole of the plane. Hence the plane's orientation may be characterized on a stereographic projection by that of its pole. Alternatively, one may consider the ascending normal, which defines the pole Q^* in the upper hemisphere. The stereographic projection of points in the upper hemisphere is performed by choosing the focus of the projection to be at Z^*.

Stereographic or Wulff projection is much in use in engineering and structural geology, as discussed for example by Goodman (1989) or Hoek and Brown (1980), because it conserves angles.

Equal-area projection, also called the Lambert or Schmidt projection, is very helpful for analyzing the statistics of orientations (see section 1.3.3). If a and d are the azimuth and the dip of a given direction, the equal-area projection transforms point P, which corresponds to the values a and d on the reference sphere, into point P^* on a horizontal circle with center Z^*, where the horizontal plane is tangent to the reference sphere (fig. 1.11). Choosing polar coordinates to describe the position of P^* in the reference circle, the distance ρ between Z^* and P^* is $\rho = 2\sin(\pi/4 - d/2)$ and the azimuth of P^* is simply a.

Because areas on the surface of the sphere are preserved in this transformation, the density of points on the sphere is the same as the density of points in the reference circle. Hence equal-area projection is convenient for the statistical analysis of orientation data (Fisher *et al.*, 1987), in particular for the statistical analysis of the distribution of fracture poles.

1.3.2 *In situ* reconnaissance of fractures

Fractures are observed either at their intersection with free surfaces (outcrops or underground openings) or at their intersection with boreholes.

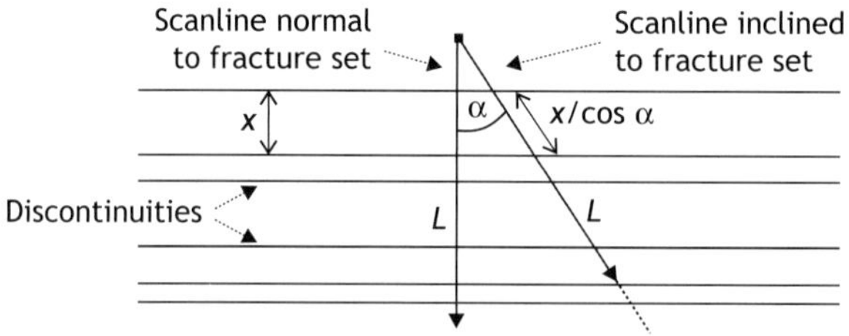

Fig. 1.12 Fracture frequency and the bias induced by the scanline's orientation.

Direct observation on rock mass exposures

Fractures are three dimensional in nature but will be assimilated to sub-planar structures here. These sub-planar structures are observed only at their intersection with free surfaces and therefore observations concern mostly linear segments within the fracture surface.

As can been seen from fig. 1.6, the direction of a line of intersection of a fracture plane and a free surface cannot be used to characterize the fracture plane orientation if the free surface is planar. Only if the free surface is curved can the fracture plane orientation be determined, and this is discussed below in the case of cylindrical borehole investigations.

Planar rock mass exposures are very useful for investigating the fracture density, i.e. the number of times a straight scanline is intersected by fractures. The distance between two successive fractures, along the scanline, is called the fracture spacing x (fig. 1.5).

The fracture frequency is defined as the number of fractures per unit length of scanline. Clearly the fracture frequency depends on the orientation of the scanline within the planar rock mass exposure. In fig. 1.12 we consider a set of parallel fractures and two scanlines, one normal to the fracture set and the other inclined at an angle α to this perpendicular direction. For the same fracture set, the frequency sampled by the inclined scanline, λ_s, is related to the frequency sampled by the scanline normal to the fracture set, λ, as follows:

$$\lambda_s = \frac{N}{L/\cos\alpha} = \frac{N}{L}\cos\alpha = \lambda\cos\alpha \tag{1.17}$$

This clearly shows that scanline orientation introduces some bias in the spacing distribution and therefore in the RQD index already mentioned. Further, the fracture length observed at the exposure is generally very different from the maximum fracture length of the fracture. Identifying the real shape and the dimensions of fractures *in situ* remains an open question.

Borehole fracture imaging

When a planar structure intersects a cylinder, it defines an ellipse on the cylinder wall (fig. 1.13). If the cylindrical surface is cut along a generatrix and is opened up so as to be transformed into a rectangular area, the ellipse transforms into a sinusoid. The imaging tool's orientation is known thanks to an orienting device fixed on the tool, and this information is used to define the orientation of the leftmost generatrix of the rectangular area where the fracture is mapped.

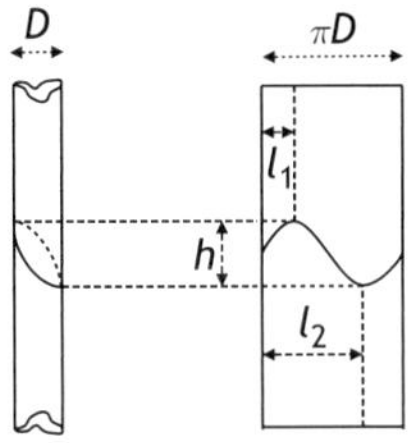

Fig. 1.13 Determination of the dip direction and dip of a fracture with respect to the borehole axis.

When the borehole is vertical, orientations are generally defined with respect to magnetic North. For inclined boreholes, however, reference is often made to "top of hole (ToH)", which is the uppermost generatrix of the cylindrical surface.

The upper and lower points of the sinusoid correspond to the top and bottom of the ellipse. Determining their position at the borehole wall provides a means to identify the orientation of the dip direction *dd*. The angle θ between the upward oriented normal to the fracture plane and the borehole axis is determined from the difference in depth h of these two extreme points (fig. 1.13):

$$dd = \frac{1}{2}\left[\frac{l_1}{\pi D} + \frac{l_2 - (\pi D/2)}{\pi D}\right], \qquad \theta = \arctan\left(\frac{h}{D}\right) + \frac{\pi}{2} \tag{1.18}$$

where *dd* and θ are expressed in radians. For better accuracy, a least squares method may be applied for evaluating the dip direction by considering the intersections between the ellipse and various horizontal planes.

Two borehole-wall imaging techniques, ultrasonic borehole imaging (Zemanek *et al.*, 1970) and electrical resistivity imaging (Mosnier, 1982), have been shown to be efficient for identifying fracture orientation and fracture spacing along a borehole direction. Today, the practice of deviated and horizontal drilling has become very common, so that fracture imaging techniques provide a unique means for obtaining data on the fracture field at depth and in three dimensions. Complete fracture orientation determination can be completed, so as to define fracture sets. Then the fracture frequency for the various sets is determined along scanlines with diverse orientations, so that bias can be eliminated.

Ultrasonic borehole televiewer

With the ultrasonic borehole-wall imaging tool, a very high frequency mechanical pulse is focused on a small part of the borehole wall and the reflected signal is recorded (fig. 1.14). If the wave propagation velocity in the fluid filling the borehole is known, the time elapsed between wave emission and wave reception equals the travel time for twice the distance between the sensor and the borehole wall.

The measurements also gives the ratio of the maximum amplitude of the emitted wave and that of the reflected wave. These measurements are conducted at azimuth increments of 1° or 2° to cover the complete borehole wall at a given depth. During the logging process, the tool is moved up very slowly (generally at a velocity equal to about 1 m/min) whilst

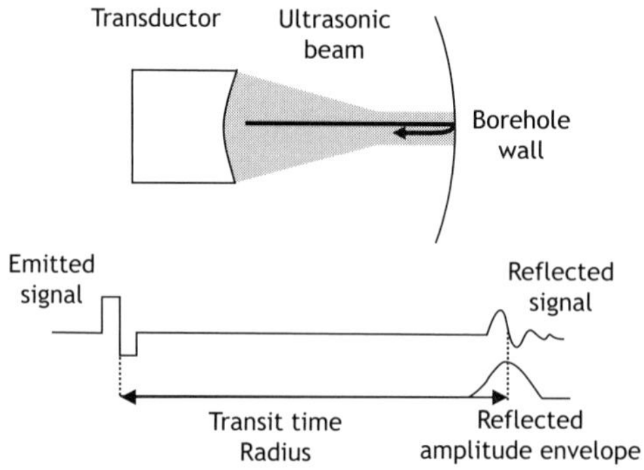

Fig. 1.14 The acoustic televiewer.

the acoustic sensor covers twice the 360° azimuth range in 2 seconds. An example of the fracture images produced by an ultrasonic borehole televiewer and the amplitude ratios are shown on fig. 1.16.

Borehole-wall electrical imaging

With an electrical imaging tool, an alternating electrical current is applied between the metallic armor of the logging cable and the body of the tool. Focusing electrodes ensure that electrical current lines converge normally to the tool at the 24 measuring electrodes. These are evenly distributed in a row, so as to provide measurements of the azimuthal variations in electrical resistivity at a given depth (fig. 1.15). When the tool is moved along the borehole, it provides a complete log of the electrical resistivity variations at the borehole wall.

Sometimes the tool is combined with a double inflatable packer element which allows the injection of water at a given depth in a borehole. Thanks to electrical imaging, the "straddle packer" system can be positioned precisely on an existing fracture; then hydraulic testing provides information on the hydraulic characteristics of the fracture (see chapter 10 and section 13.2).

Other versions of the tool include four pads applied to the borehole wall, with 36 electrodes on each pad. With such high electrode-density tools the resolution is much improved, but azimuthal coverage of the borehole wall may not be complete when the borehole diameter gets too large.

Some electrical and acoustic fracture images are shown for comparison in fig. 1.16.

1.3.3 Fracture fields and scaling laws

The RQD index defined in section 1.2.2 is intended to characterize in one single parameter the fracture field. It is defined for each particular volume for which an equivalent geomaterial is considered. Clearly, in many situations this simple approach is not satisfactory, and the fracture field must be described in more detail.

An important concept, when considering a fracture field, is that of fracture sets, i.e. sets of fractures that are all approximately parallel to a given direction. Our objective now is to formalize this concept of "approximately parallel fractures".

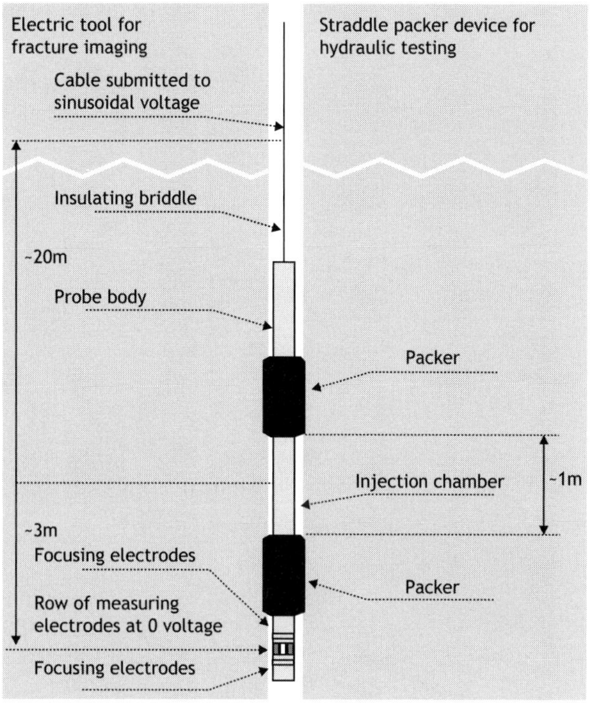

Fig. 1.15 Electrical imaging combined with hydraulic testing equipment for deep boreholes.

Fracture set identification

When fracture orientation data are retrieved from borehole imaging logs conducted in boreholes drilled in various directions, the poles of the fractures may be plotted on an equal-area projection net. On such projections, *fracture sets* appear as clusters of points (fig. 1.17) and the objective is to analyze these clusters in a rigorous statistical manner.

In fig. 1.17, sets 1 and 2 are close to being vertical and are well defined. Set 3 is also close to vertical but no clear direction can be simply identified. Various techniques have been developed for conducting the quantitative analysis of such plots. Of particular interest is the paper by Vollmer (1995), which proposes a numerical code for the automatic contouring of spherical orientation data plotted on equal-area projections.

Consider a set of S fractures with the orientation of the jth fracture, $j = 1, \ldots, S$, characterized by its normal $\boldsymbol{N}^j$, with direction cosines n_i^j, $i = 1, 2, 3$, defined in a direct geological frame of reference (see section 1.3.1). The mean orientation of this set of fractures is defined by Goodman (1989) as

$$n_1^m = \frac{\sum_{j=1}^S n_1^j}{|\boldsymbol{R}|}, \qquad n_2^m = \frac{\sum_{j=1}^S n_2^j}{|\boldsymbol{R}|}, \qquad n_3^m = \frac{\sum_{j=1}^S n_3^j}{|\boldsymbol{R}|} \tag{1.19}$$

with $|\boldsymbol{R}| = [(\sum_{j=1}^S n_1^j)^2 + (\sum_{j=1}^S n_2^j)^2 + (\sum_{j=1}^S n_3^j)^2]^{1/2}$; $|\boldsymbol{R}|$ is the modulus of the vector $\boldsymbol{R}$ defined as the sum of the vectors $\boldsymbol{N}^j$. If the vectors $\boldsymbol{N}^j$ were all parallel to one another, the

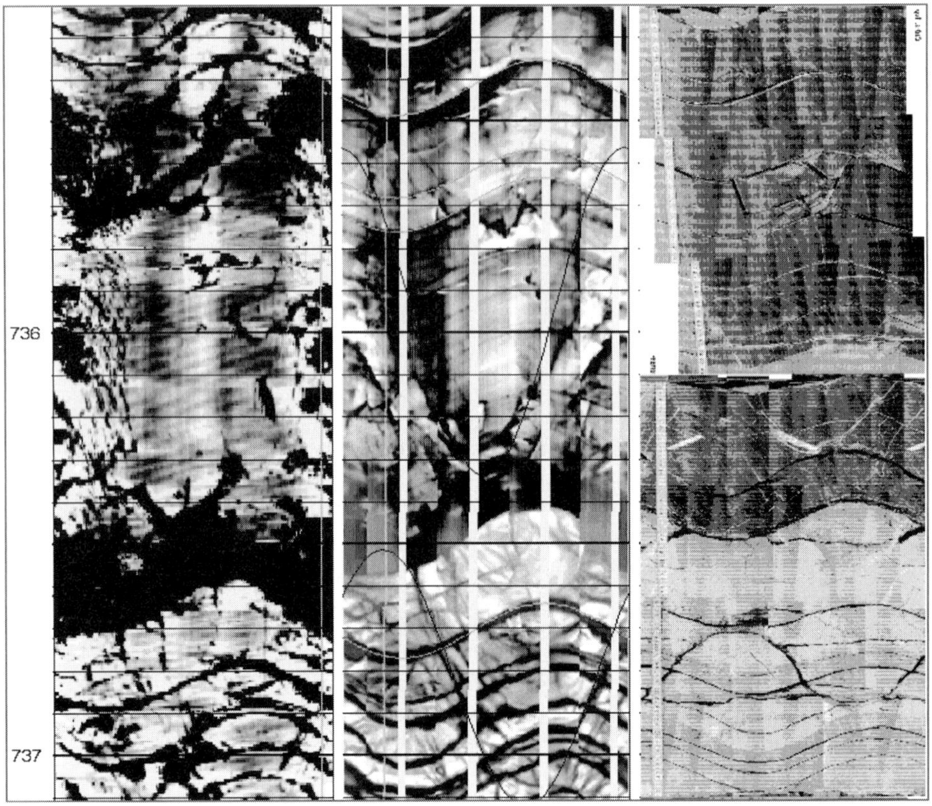

Fig. 1.16 Acoustic televiewer (left), electrical borehole images (central) and those for a scanned core (right). Observe on the electrical image the four axial white straps, which correspond to an absence of data specific to this tool. The numbers on the left of the figure correspond to the height in meters. The borehole diameter is 16.0 cm (6.5").

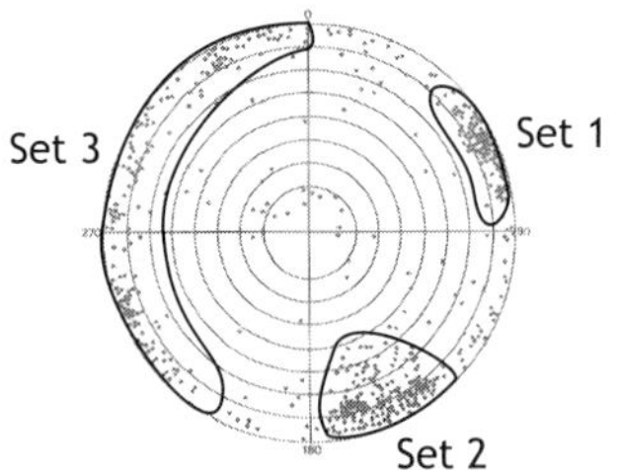

Fig. 1.17 Identification of fracture sets from equal-area projection plots of fracture orientation data. Only the poles of the fracture surfaces are considered. Note that some poles have not been included in the definition of the fracture sets. This reflects the effects of the selection criteria.

modulus $|\boldsymbol{R}|$ would be equal to S. Because not all the fracture planes are parallel, however, $|\boldsymbol{R}|$ is smaller than S, and its value depends on the dispersion of the fracture orientations. Fisher *et al.* (1987) discussed the statistical analysis of spherical data; they defined the dispersion of a fracture set by the quantity

$$K_F = \frac{S}{S - |\boldsymbol{R}|} \tag{1.20}$$

It may be shown (Goodman, 1989) that, for a hemispherical normal distribution, the probability P that the normal to a fracture makes an angle that is equal to or smaller than ψ to the mean set's orientation is defined implicitly by

$$\cos\psi = 1 + \frac{1}{K_F}\ln(1 - P) \tag{1.21}$$

where ln is the natural logarithm. For such a normal distribution the standard deviation is $\bar{\psi} = 1/K_F$.

Fracture set characterization and scaling laws

Fracture set characterizations are important in many different problems. Designing hazardous waste repositories, overcoming the problem of multiphase flow in hydrocarbon reservoirs and understanding the scaling laws for earthquakes are but a few examples of the fields where fracture sets are encountered. Excellent reviews of the work done in this field over the past 25 years are provided by Bonnet *et al.* (2001) as well as by Adler and Thovert (1999).

As mentioned above, a fracture set is made up of fractures that are roughly parallel to each other. In order to describe such sets, for each fracture one must describe its shape, its size, the location of its center, possibly its thickness in relation to the displacement that has occurred between its two faces and many other attributes, which are discussed further in chapter 10. Usually fractures are assimilated to flat disks, also called penny-shaped fractures, so that the size can be simply related to the radius.

We saw in section 1.3.2 that a fracture set may be characterized by the number of times, M, that fractures from this set are intersected by a scanline of length L, divided by L; however, this ratio depends strongly on the size of the sampling length L. If a particular fracture spacing is x then the mean fracture spacing $\bar{x} = L/M$. Figure 1.18 shows how the frequency of fracture spacings with length x may be plotted as a function of x, given that the number of spacing measurements is large enough (say more than 200).

In the example shown in fig. 1.18 the histogram may be approximated by a negative-exponential probability density function.

More generally fracture-set characteristics are described so as to obey representative statistical requirements, and all the work that has been conducted in this respect has concluded that these characteristics have fractal dimensions: it is not possible to define, for characteristics related to fractures, a representative elementary volume as was introduced in section 1.1.1.

Consequently, the approach that is applied today is to generate statistically representative simulations, called realizations of the discrete fracture network (DFN), of the given fracture population and then to analyze many such realizations so as to conduct a statistical analysis of the results. In chapter 10 this is discussed further, in the context of flow through fractured material.

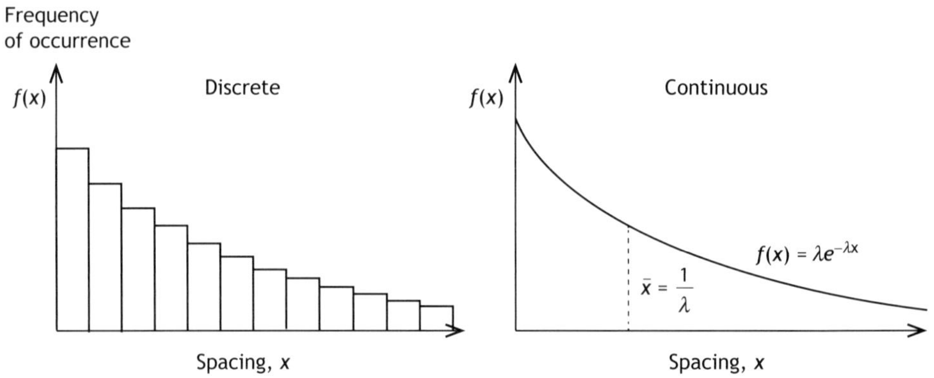

Fig. 1.18 Identifying probability density functions for spacing values. The frequency of occurrence of spacing x is the number of times x has been measured, divided by the total number of spacing measurements. This example assumes a negative exponential distribution.

Four different frequency distributions $n(w)$ have been investigated for characterizing a fracture property w, where w is length or displacement, etc.

Log normal distribution

The log normal distribution is given by

$$n(w) = \frac{1}{[w\sigma(2\pi)^{0.5}]} \exp\left[-\frac{(\log w - \langle \log w \rangle)^2}{2\sigma^2}\right] \tag{1.22}$$

where $\langle \log(w) \rangle$ and σ are respectively the logarithmic mean and the variance of w (the fracture radius, in this case). It has been used to describe the fracture length distribution. Arguments were put forward that truncated power laws, because of sampling difficulties, may result in only apparent log normal distributions. Today, though, the power law distribution seems to be favored because of the accompanying absence of bounds.

Exponential law

Exponential laws are used to describe the size of the discontinuities in continental rocks:

$$n(w) = A_2 \exp\left(-\frac{w}{w_0}\right) \tag{1.23}$$

where A_2 is a constant. The exponential law incorporates a characteristic scale w_0, which may reflect either a physical length, such as the thickness of a sedimentary layer, or some effects related to the fracturing process.

Gamma law

The gamma distribution is a power law with an exponential tail. It is commonly used to analyze fault populations or earthquake statistics. It is characterized by a power law exponent a and a characteristic scale w_0:

$$n(w) = A_3 w^{-a} \exp\left(-\frac{w}{w_0}\right) \tag{1.24}$$

where A_3 is a constant.

The characteristic scale w_0 may be related to a correlation length in the spatial pattern. It implies an upper bound to fractal behavior.

Power law

Numerous studies at various scales and in different tectonic settings have shown that the distribution of many fracture properties (length, displacement, etc.) often obeys a power law:

$$n(w) = A_4 w^{-a} \tag{1.25}$$

where A_4 is a constant and a is the fractal dimension of w. Power law distributions have the important property that they contain no characteristic length scale. It is now generally recognized that the effects of resolution and finite size on a power law population can result in distributions that appear exponential or log normal (Bonnet *et al.*, 2001). In nature, power laws are limited by physical length scales: these form the upper and lower limits to the scale range over which they are valid. Indeed, at some scale bounds must exist; for example the thickness of the sedimentary layers become significant when one is investigating fracture fields in sedimentary formations. The advantage of such truncated power laws over the previous three laws is that the upper and lower bounds may be defined arbitrarily, independently of the *shape* of the law. For the other three laws, their shape is linked to the bounds.

1.3.4 Faults and their morphology

Faults are *not* large fractures. They always involve a more complex texture, with crushed rocks, which gives them their specificity. We will discuss faults in chapter 10 and only consider here some issues concerning the geological materials involved.

The texture of faults has been often discussed. We follow here Sibson's analysis (Sibson, 1977) which identifies for all faults an incohesive part and a cohesive part. The incohesive part includes fault brecciae and fault gouge, the definition of which depends on the particle size distribution in a manner somewhat reminiscent of the soil mechanics approach for uncemented clastic textures (fig. 1.19). The cohesive part may include pseudotachylite (with some glass elements) and cataclasite series, the structures of which are related to grain size.

An example of structure for a presently active fault is provided by the Aigion Fault (the Corinth Rift) in western Greece and is shown in fig. 1.20. This fault is about 15 km long and exhibits a roughly 200 m offset. It dips at about 60 degrees to the North so that the apparent 1 m long gouge core corresponds in reality to a zone 50 cm thick.

The important feature is that the fine to very fine, granular, part of a fault is always embedded within a cohesive part. The characteristics of the incohesive part depend on many features: the nature of the material that has been faulted, the amplitude of the fault

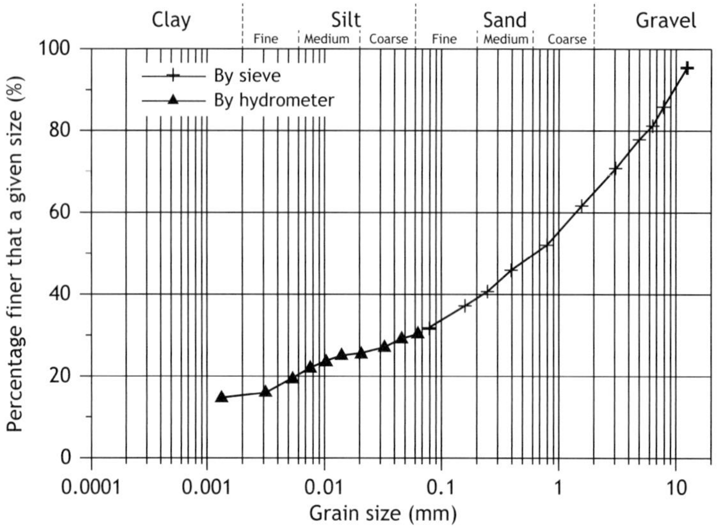

Fig. 1.19 Particle size distribution of Aigion fault gouge. The percentage of clay (particles smaller than 2 μm) is 15%. The gouge is shown in fig. 1.20 as the fault core (reproduced from Sulem *et al.*, 2004, fig. 2, with permission from Elsevier Masson).

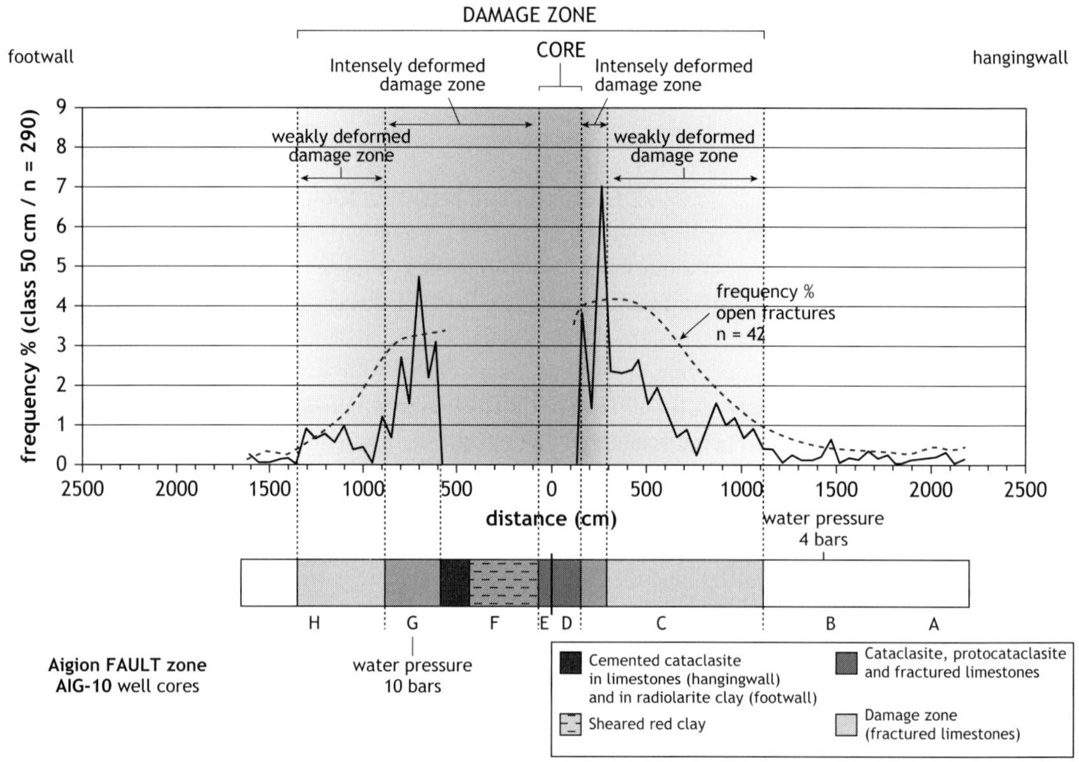

Fig. 1.20 Structure of the Aigion fault in the Corinth rift, as seen on a vertical core. The fault gouge is about 1 m thick in the core. The extensions of the cataclastic zones on both sides of the gouge are about 15 m thick (reproduced from Micarelli *et al.*, 2006, fig. 8, with permission from Elsevier).

motion and the depth at which the fault motion occurs. In the particular case of the Aigion fault, the fault affects a sedimentary series with thick radiolarite deposits. These deposits have been stretched by the faulting process but have not broken up, hence producing a continuous 50 cm thick film all over the fault surface. This is known as clay smearing, a deformation process with very strong consequences for both the mechanical and the hydraulic properties of the fault, as will be discussed in chapter 10.

As the depth increases, higher and higher temperatures and pressures are encountered, so that uncohesive characteristics are progressively lost to the milonite textures mentioned in the first section of this chapter. Various factors control fault growth and the geometry of faults. These are discussed in chapter 10.

Another important feature of fault zones is their apparent segmentation when observed on two-dimensional outcrops. When viewed in three-dimensional perspective, however, the various branches may or may not coalesce at depth; faults are always three dimensional in nature (Ben-Zion and Sammis, 2003). Recent developments point out to the fractal characteristics of fault segmentation (Bonnet *et al.*, 2001).

Whatever the scale of the problem under investigation, homogeneous geomaterials may always be defined in such a way as to describe the behavior of the bulk materials that fill up the volume under consideration. However, because of their fractal characteristics, discontinuities, whether fractures or faults, must always be introduced explicitly in models. Because they exhibit different morphologies, fractures and faults must be dealt with separately. Hence bounds must be defined to describe domains in which the respective fractal characterizations of faults and of fractures are valid.

In practice most geomechanics problems involve various geomaterials and different fracture fields but only a few (say, less than five) fault zones. However, because fracture fields can be approached only through their statistical properties, no unique solution exists to any given problem and a solution must be formulated in terms of an expected value together with an associated confidence level (e.g. Einstein and Baecher, 1983).

1.4 Loading processes

Man-made loading processes may involve mostly the solid phase of geomaterials, as in an excavation, or just the fluid phase, as in oil extraction. They may also involve both solid and fluid phases, and they always correspond to a perturbation of an already existing system. Therefore an important step in analyzing man-made loading processes is the characterization of the initial and boundary conditions for both solid and fluid phases, which themselves result from natural loading processes.

1.4.1 Natural loading processes

Natural loading processes may have various origins. They may be related to plate tectonics, i.e. the motion and deformation of the lithospheric plates that are distributed over

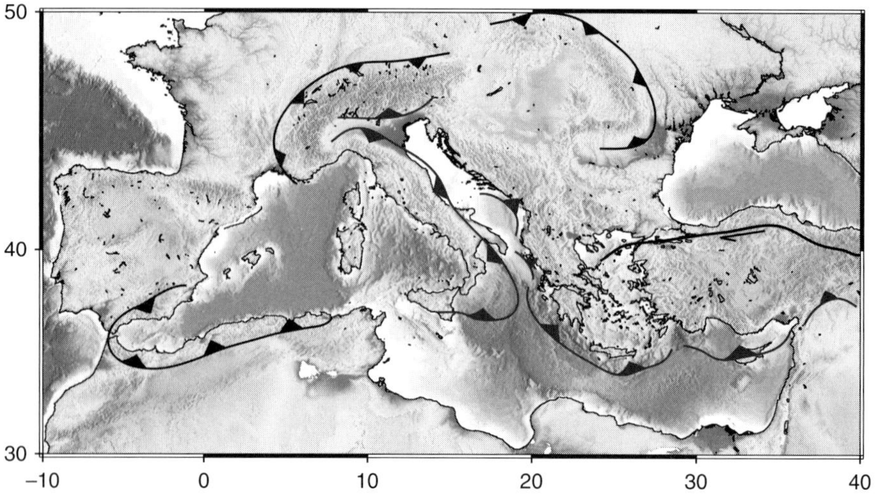

Fig. 1.21 Present day tectonics in the Mediterranean area. The subduction zones are indicated as well as the North Anatolian fault (line with half-arrow). The different shades of gray in the sea areas indicate the varying sea depth. Although the Europe–Africa convergence rate is only mm per year, the velocities at the subducting plates may reach a few centimeters per year where rollback takes place (courtesy of L. Husson).

the surface of the earth (e.g. Mussett and Khan, 2000). They may also be related to the evolution over time of their phase relationships, i.e. the evolution of the pore space and the correlated phase content, as well as to the various fluxes that concern the fluid phases.

Further, plate tectonics involves the contacts between the plates within the lithosphere but also the rate of penetration by the lithospheric plates of the asthenosphere. This is particularly well illustrated by the present-day geodynamic context of the Mediterranean area.

Figure 1.21 (after Royden, 1993) illustrates the present-day distribution of subduction zones in the Mediterranean area and the location of the most active lithospheric fault (the North Anatolian fault to the east of Greece). For many of these subduction zones rollback is taking place (fig. 1.22, right), i.e. the subducting part of the plate does not move forward any longer but is falling within the asthenospheric mantle, thus generating an extension zone in the other lithospheric plate (Royden, 1993; Jolivet and Faccenna, 2000).

Clearly, the loading processes that affect the northern shore as well as the southern shore of the Mediterranean sea are significantly dependent on the relative motions that occur at the various subducting zones in between these shores. It would be erroneous simply to consider the rate of convergence between southern Europe and northern Africa when trying to understand the deformation process at any of these shores. Further, given the scale of the phenomenon, the planar frame of reference suggested by the figure is misleading for the effect of the sphericity of the earth is an important consideration for such a geomechanical problem.

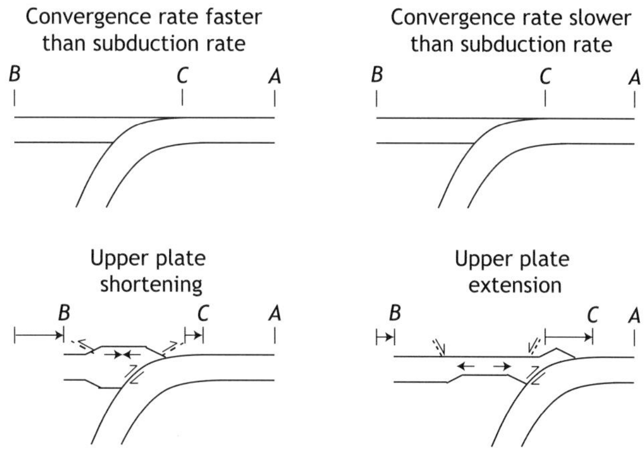

Fig. 1.22 Schematic diagram illustrating regional deformation depending on the relative velocity of the advancing plate *BC* and the retreating plate *CA*. When the retreating velocity of *CA* is faster than the advancing velocity of *BC* (on the right of the figure), a zone of extension exists in the advancing plate, *BC*, close to the contact zone; in the converse situation there is a zone of shortening in *BC* (reproduced from Royden, 1993, fig. 1, with permission from Wiley).

No consideration is given in this text book to the geomechanics of subduction zones. Here, we just want to stress that definition of the initial and boundary conditions which characterize a given crustal geomechanics problem often requires a preliminary three-dimensional analysis at a larger scale.

For engineering applications an alternative solution is to measure, directly in the field, prior to any perturbation, the boundary values required for the analysis. This is described in the two last chapters, where methods for measuring stress fields are presented and some results are discussed.

1.4.2 Coupling or no coupling

Although all crustal geomechanics problems involve both solid and fluid phases, it is often possible to consider only one phase (e.g. in the case of flow through porous materials when the fluid pressure remains smaller than a critical threshold or in the case of wave propagation in a solid elastic earth).

However, in many problems interactions between the solid and fluid phases cannot be ignored. We define as *coupled* a geomechanics problem in which the solution of the problem involves the description of fluid and solid interactions. When only fluid flow and solid deformation are concerned, we speak of hydromechanical coupling even though the fluid may not be water. When the fluid flow component is not significant but temperature effects are of importance, we speak of thermomechanical coupling. When fluid flow, heat transfer and solid deformation are all of importance, we speak of hydrothermomechanical coupling. Finally, the fluid may interact with the solid phase by partial dissolution or by scale deposits, and this introduces the concept of chemical coupling.

These various coupling issues are addressed in chapter 12. While hydrothermomechanical couplings may be addressed with a continuum mechanics approach, couplings that involve chemical interactions imply thermodynamics and thermo-statistics considerations. Chemical couplings are very briefly mentioned in chapter 12 for the sake of completeness, but thermo-statistics and thermodynamics are beyond the scope of this book.

1.5 Exercises

1. Observe that the intersection of a plane (say, a planar fracture F) with an infinite cylindrical surface (say, the inner surface of a cylindrical borehole B) defines an ellipse on the cylindrical surface (fig. 1.13). Assume that the cylindrical surface is cut along one of its generatrixes, G (the left-hand side of the rectangle in fig. 1.13), and that it is opened up so as to become a planar surface. Show that the ellipse is transformed into a sinusoid on this planar surface.
2. Assume that the cylinder discussed in the previous exercise has a vertical axis and radius $R = 5$ cm and that the generatrix G used to cut the cylindrical surface is oriented N50°E with respect to the cylinder axis. Let $a_1 = 30°$ be the angular coordinate of the uppermost point of the ellipse and assume that the plane F makes an angle equal to 45° with respect to the cylinder's axis. In the planar surface, X is the axis perpendicular to G that intersects the sinusoid halfway from its top. Give the coordinates of the lowest point of the ellipse in the XG plane and its angular coordinate on the cylindrical surface.
3. What are the coordinates in the XG plane of a unit vector normal to the fracture F at its intersection with the borehole axis?
4. Consider the reference sphere used for the Schmidt equal-area stereographic projection and write a *Matlab* program for plotting the pole of the fracture F.
5. Using the real electrical images of a borehole wall that are provided on the internet site associated with this book, plot on a Schmidt stereographic projection all the poles of fractures identified between 30 and 78 m in this borehole.
6. Considering the plot of the previous exercise, how many fracture sets can you define?
7. Given that coring operations do not provide any information on the orientation of the collected cores, how would you proceed to find this orientation after the cores have been collected ?

2 Elements of rheology

When a load is applied to a material the material always deforms, possibly in an almost infinitesimal manner. The science that describes the way in which materials deform under load is called rheology.

In nature, a load may have very diverse origins: it may be associated with gravity, heat, humidity or any other source. Often, only one load source is of interest. In specific conditions, when many sources are acting simultaneously, the response of the material to the various loading processes may be estimated by adding the responses of the material to each loading taken individually. This application of the superposition principle is valid when the response to each loading is linear within the complete range of deformation and when the responses to the various different loadings are not coupled, e.g. when the parameters that characterize the linear response to a loading do not depend on any of the loadings and remain constant within the domain of interest.

In this chapter we introduce some simple materials described by elementary rheological models and show how combinations of these elementary models lead to more elaborate, realistic, model materials. For all these models, only a one-dimensional space is considered so as to keep the mathematical formulation as simple as possible and yet introduce some essential physical aspects of rock-mass deformation.

Because we consider only a one-dimensional space, the forces F and displacements δ are represented by scalars. Their time derivatives are indicated by dots on the corresponding letter (e.g. $\dot{F} = dF/dt$, $\ddot{\delta} = d^2\delta/dt^2$). The response of a given rheological model to a load is characterized by a relationship between the force, the displacement and their respective time derivatives. This relationship is called the model constitutive equation. It characterizes a specific rheological behavior.

2.1 The elastic or Hookean solid

An elastic, or Hookean, solid can be represented by a spring (fig. 2.1). When a force F is applied to the spring with original length l_0, it gives rise to a displacement δ proportional to F:

$$\delta = F/k \tag{2.1}$$

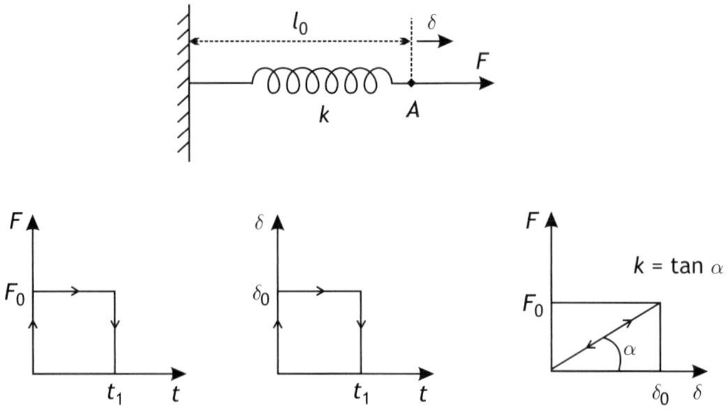

Fig. 2.1 Elastic or Hookean solid.

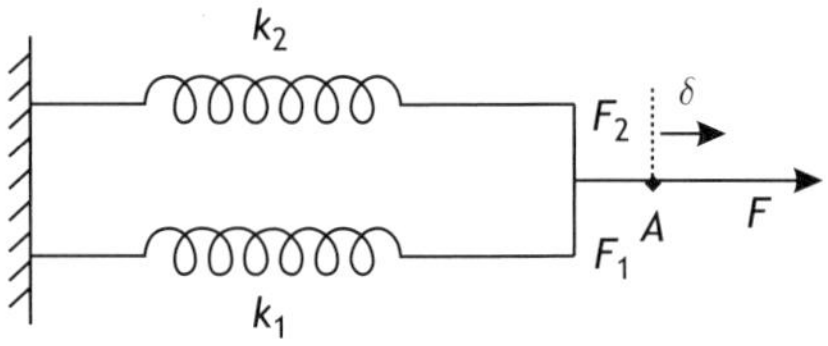

Fig. 2.2 Two springs loaded in parallel.

where $\delta = l - l_0$ if l is the final spring length; k is called the material stiffness. The material response to a loading is instantaneous for an elastic body, i.e. the displacement occurs at the same time as the force is applied. When the load is dropped back to zero, the spring returns instantaneously to its original length l_0. A typical example of an elastic response for geological materials is the transmission of teleseismic waves, i.e. the propagation at very long distances, of motions at the 1/10 to 1/100 mm scale, generated by very distant earthquakes (more than a thousand kilometers away).

Let us consider now two springs with respective stiffnesses k_1 and k_2 loaded in parallel by the same force F (fig. 2.2).

The force F induces in each spring forces F_1 and F_2 respectively, while the displacement for each spring is equal to δ:

$$F_1 = k_1\delta, \qquad F_2 = k_2\delta, \qquad F = F_1 + F_2 = (k_1 + k_2)\delta \tag{2.2}$$

Hence, the load in the spring with stiffness k_1 is $F_1 = [k_1/(k_1 + k_2)]F$. The stiffer spring supports the larger load and conversely the softer spring supports the smaller load.

This has an important application to the stress fields in layered sedimentary formations of intercalated soft (clay) and stiff (limestone or sandstone) beds (sections 1.1.2 and chapter 14). Stiff beds support the largest deviatoric stress (the deviatoric stress is defined in chapter 3), while soft beds support little, if any, deviatoric stress. Sometimes combinations of stiff and soft materials are assimilated to equivalent composite materials. This is typically the case for materials consisting of stiff (carbon or glass) fibers embedded in soft resins. The stiffness of the composite material is controlled by that of the fibers and

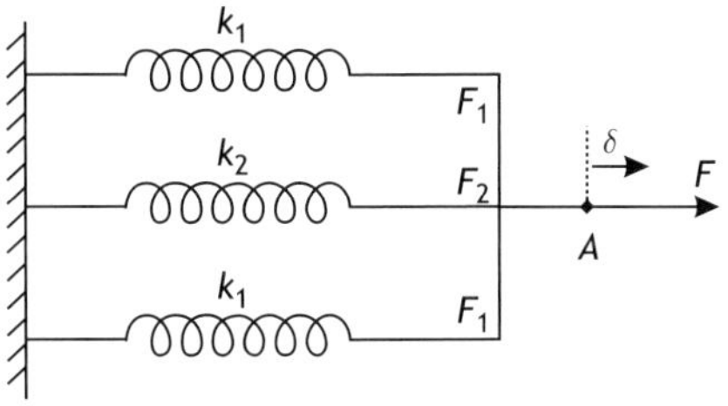

Fig. 2.3 Three springs loaded in parallel.

by the geometry of the fiber-layering process. Such a composite material approach may be proposed for modeling evaporite deposits with intercalations of very stiff but thin layers of anhydrite within softer, thicker, layers of halite or potash (sections 1.1.2 and chapter 14). If we consider now three springs in parallel (fig. 2.3), with respective stiffnesses k_1, k_2 and k_1, such that $k_2 < k_1$, then the force F_2 supported by the central spring is smaller than that supported by either of the other two springs ($F_2 = k_2/(2k_1 + k_2)F$). Such a mechanism may be invoked to explain the lateral variations of the vertical stress component observed in some horizontal sedimentary beds that exhibit a lateral stiffness variation (chapter 14).

2.2 The viscous Newtonian fluid

A viscous Newtonian fluid can support a load only during its deformation process. During deformation, the force F required to displace point A is proportional to the velocity $\dot{\delta}$ of A (fig. 2.4); the fluid's viscosity is represented by a dashpot (damping device). Thus we have

$$F = \eta\dot{\delta} \tag{2.3}$$

where η is the viscosity of the fluid. In viscous fluids the instantaneous displacement is zero at the moment when the load is applied and displacement stops only when the load drops to zero. When a constant load F_0 has been applied from time t_0 to time t_1, a permanent displacement $\delta_1 = F_0(t_1 - t_0)/\eta$ is observed after the load has returned to zero.

2.3 Viscoelastic materials

Viscoelastic materials exhibit some memory of past loading. They are modeled by combinations of springs and dashpots and may be either solid-type viscoelastic materials or liquid-type visco-elastic materials, depending on their ability to sustain a load under static conditions ($\dot{\delta} = 0$).

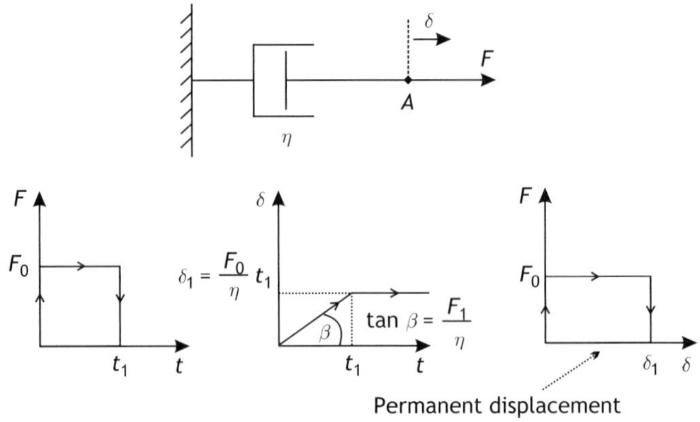

Fig. 2.4 Newtonian viscous fluid.

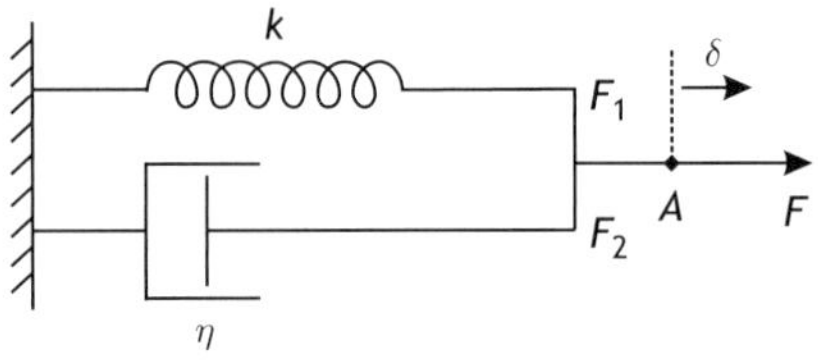

Fig. 2.5 The Kelvin–Voigt material.

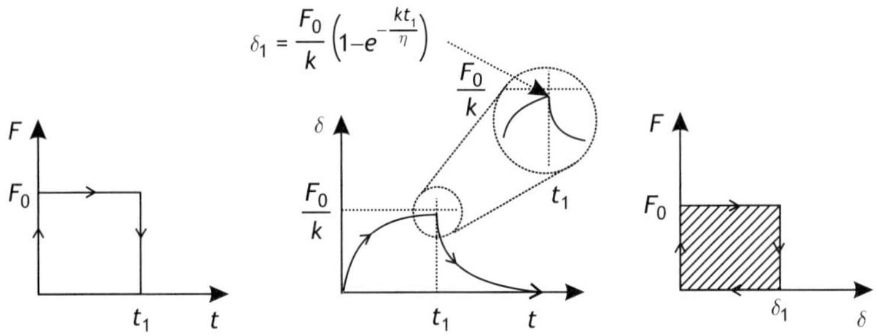

Fig. 2.6 The Kelvin–Voigt material has some memory of past loading. The shaded area in the rightmost diagram represents the work done during a loading cycle.

2.3.1 A solid-type viscoelastic material: the Kelvin–Voigt material

The Kelvin–Voigt material is modeled by a spring set in parallel with a dashpot (fig. 2.5). The spring supports a force F_1, and the dashpot a force F_2, such that

$$F_1 = k\delta, \qquad F_2 = \eta\dot{\delta}, \qquad F = F_1 + F_2 = k\delta + \eta\dot{\delta} \tag{2.4}$$

Equation (2.4) represents the constitutive equation for the Kelvin–Voigt material. Let us find the solution of equation (2.4) for the case in which the load F is equal to F_0 for $t > 0$ (fig. 2.6), with the condition that $\delta = 0$ at $t = 0$.

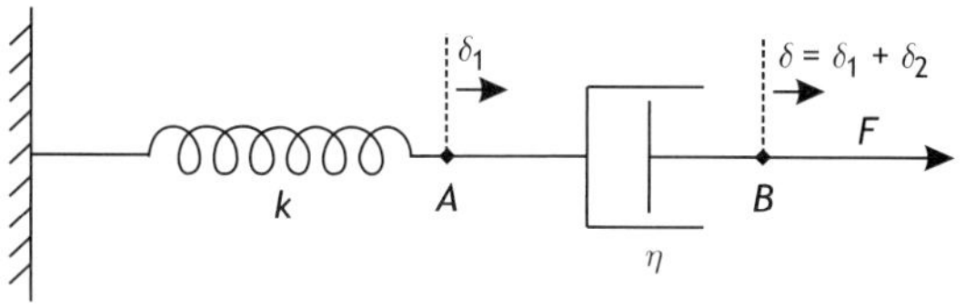

Fig. 2.7 The Maxwell material.

The solution of the equation $k\delta + \eta\dot{\delta} = 0$ is

$$\delta = Ae^{-kt/\eta} \tag{2.5}$$

in which A is a constant. A particular solution of equation (2.4) is $\delta = F_0/k$, so that the general solution of equation (2.4) is $\delta = (Ae^{-kt/\eta} + F_0/k)$. This implies that $A = -F_0/k$ if $\delta = 0$ when $t = 0$, and so the behavior of the material is described by

$$\delta = \frac{F_0}{k}(1 - e^{-kt/\eta}) \tag{2.6}$$

Laboratory tests in which the loading condition is kept constant, $F = F_0$, are called creep tests. More generally, the observation of motion under such quasistatic loading conditions is referred to as creep. An important problems in earth sciences is to understand why faults sometimes creep and sometimes slip in an unstable manner leading to catastrophic earthquakes.

When the load F_0 drops back to 0 after time t_1, the material behavior is characterized by equation (2.5) with the condition that (2.6) is obeyed at the instant $t = t_1$ when the load is dropped to zero. The material keeps deforming after time t_1 even though no load is applied to it: i.e. for a certain time it keeps some memory of the loads that were applied to it in the past. However, no residual deformation is observed after a sufficiently long time has elapsed, just as for an elastic Hookean material, but the loading process has dissipated some energy as opposed to the case of a purely elastic solid.

Questions of energy and energy dissipation are basic to many geomechanics problems; such questions are involved in the attenuation of elastic wave propagation, the stability of rock mass failure, the generation of heat during earthquakes and so on.

2.3.2 A fluid-type viscoelastic material: the Maxwell material

When a dashpot and a spring are placed in series (fig. 2.7) the equivalent material exhibits an instantaneous response because of the spring, but then deforms continuously as long as the load is applied because of the dashpot.

In such a model, the force is the same in each element of the system but the displacements at A and B add up:

$$F = k\delta_1 = \eta\dot{\delta}_2, \qquad \delta = \delta_1 + \delta_2, \tag{2.7}$$

$$\dot{\delta} = \frac{\dot{F}}{k} + \frac{F}{\eta} \tag{2.8}$$

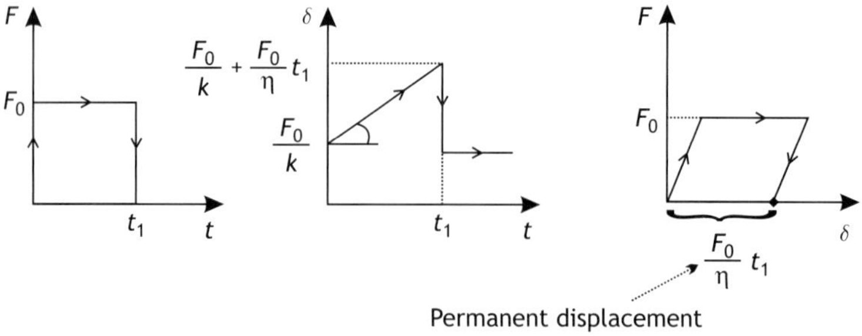

Fig. 2.8 The Maxwell material exhibits both an instantaneous response and a permanent deformation when the load is dropped back to zero. The slope angle of the diagonal line is $\arctan(F_0/\eta)$.

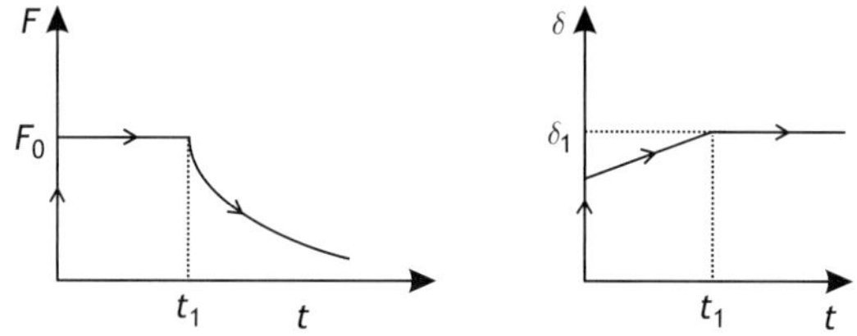

Fig. 2.9 During relaxation the force required to keep the displacement constant drops back to zero.

As previously, a simple solution may be found for the condition $F = F_0$ for all $t > 0$:

$$\delta = \frac{F_0}{\eta}t + \frac{F_0}{k} \tag{2.9}$$

When the force F_0 drops back to zero at time t_1, the spring recovers its initial length while the dashpot keeps its permanent deformation (see fig. 2.8). Another characteristic feature of the Maxwell material is the load variation required to maintain a constant displacement once it has reached a given value. We have $F = F_0$, for $0 < t < t_1$ and then δ is constant for $t > t_1$. The solution is given by

$$\delta = \frac{F_0}{\eta}t + \frac{F_0}{k}, \qquad 0 < t \le t_1 \tag{2.10}$$

$$\delta = \frac{F_0}{\eta}t_1 + \frac{F_0}{k} \Rightarrow \frac{\dot{F}}{k} + \frac{F}{\eta} = 0, \qquad t > t_1, \tag{2.11}$$

The solution of equation (2.11) is

$$F = Ae^{-kt/\eta} \tag{2.12}$$

where A is such that, for $t = t_1$, $F = F_0 \Rightarrow A = F_0 e^{kt_1/\eta}$, so that we can write (2.12) as

$$F = F_0 e^{k(t_1 - t)/\eta} \tag{2.13}$$

This is illustrated in fig. 2.9.

Hence, in viscoelastic materials two important loading conditions often occur: the relaxation configuration, in which the displacement is maintained constant with time while the load varies, and the creep configuration, in which the load is maintained constant with time

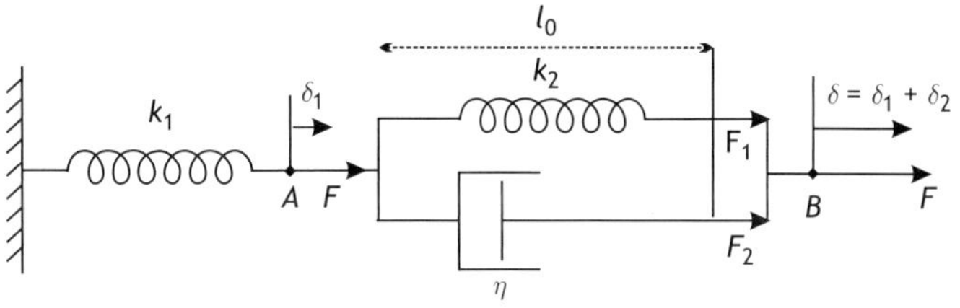

Fig. 2.10 Generalized Kelvin–Voigt model.

while deformation occurs. Typical examples of geological relaxation are given by the folds that still exist today when the load that created them has completely vanished.

2.3.3 Generalized viscoelastic materials

None of the previously discussed viscoelastic model materials describes rock materials properly. Indeed, rocks may sustain loads for a very long time, as demonstrated by the age of some deep river canyons (e.g. 30 million years for the Grand Canyon in the western USA), a feature not compatible with the Maxwell material. The loads supported by rocks at the bottom of canyons are very high and would have induced a deformation progressively filling up the empty space had the rock been simply viscous. Further all rocks transmit elastic waves, a feature not compatible with the simple Kelvin–Voigt model. Hence, slightly more complex models must be introduced to capture the main features of rock deformation.

Solid-type behavior: the Generalized Kelvin–Voigt model

A simple model used to describe the rapid response of rocks is the generalized Kelvin–Voigt model, which transmits elastic waves (with an immediate response) and yet also exhibits a time-dependent response. It is illustrated in fig. 2.10.

The forces are defined by the elementary models

$$F_1 = k_2\delta_2, \qquad F_2 = \eta\dot{\delta}_2$$
$$F = F_1 + F_2 = k_1\delta_1$$

If δ_2 is the displacement of point B with respect to point A and if δ_1 is the displacement of point A then the total displacement of point B is $\delta = \delta_1 + \delta_2$, so that $\dot{\delta} = \dot{\delta}_1 + \dot{\delta}_2$ with the condition that $\dot{\delta}_1 = \dot{F}/k_1$.

Given that $\dot{\delta}_2 = F_2/\eta = (F - F_1)/\eta = F/\eta - k_2/\eta(\delta - \delta_1)$, we conclude that

$$\dot{\delta}_2 = \frac{F}{\eta} - \frac{k_2}{\eta}\delta + \frac{k_2}{\eta k_1}F \tag{2.14}$$

The constitutive equation of the material is given by the relationship between the total displacement, the force and their time derivatives:

$$\dot{\delta} = \frac{\dot{F}}{k_1} + \frac{F}{\eta} - \frac{k_2}{\eta}\delta + \frac{k_2}{\eta k_1}F \tag{2.15}$$

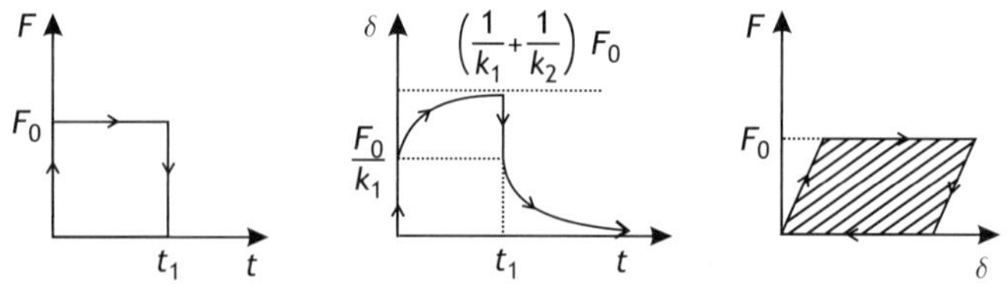

Fig. 2.11 Response of generalized Kelvin–Voigt model to a constant load.

This may be conveniently rewritten as

$$\dot{F} + \frac{k_1 + k_2}{\eta} F = k_1 \dot{\delta} + \frac{k_1 k_2}{\eta} \delta \tag{2.16}$$

We will solve equation (2.16) for the conditions $F = F_0$ for $0 < t < t_1$, with $\delta = F_0/k_1$ for $t = 0^+$. Equation (2.16) becomes

$$\frac{k_1 + k_2}{\eta} F_0 = k_1 \dot{\delta} + \frac{k_1 k_2}{\eta} \delta \tag{2.17}$$

A particular solution to equation (2.17) is $\delta = (1/k_1 + 1/k_2)F_0$, while the solution to $k_1\dot{\delta} + k_1, k_2\delta/\eta = 0$ is $\delta = Ae^{-k_2 t/\eta}$. Hence the constant A is $-F_0/k_2$ and the general solution to equation (2.17) is

$$\delta = \frac{F_0}{k_2}(1 - e^{-k_2 t/\eta}) + \frac{F_0}{k_1}. \tag{2.18}$$

The result (2.18) is shown in fig. 2.11.

Let us assume now that $F = 0$ for $t > t_1$. The displacement is given by $\delta = Ae^{-k_2 t/\eta}$, where the constant A is defined by the value of the displacement at time t_1^+,

$$\delta|_{t_1^+} = \delta_2|_{t_1} = \frac{F_0}{k_2}(1 - e^{-k_2 t_1/\eta}) \tag{2.19}$$

so that $A = (F_0/k_2)(e^{k_2 t_1/\eta} - 1)$ and the solution is

$$\delta = \frac{F_0}{k_2}\left(e^{\frac{k_2}{\eta}(t_1 - t)} - e^{-\frac{k_2}{\eta}t}\right) \tag{2.20}$$

This model fits quite well with observations for rocks that exhibit different stiffnesses when loaded either dynamically or in a quasistatic manner. The quasistatic stiffness measured in the laboratory is always much smaller than the dynamic stiffness measured through wave propagation. This raises real difficulties for evaluating the elastic moduli of large-scale rock masses. These are touched upon in chapters 8 and 9, in which we discuss the influence of loading rates on the mechanical behavior of rocks. In practice, it is customary to adopt different so-called elastic stiffness values for dynamic and for quasistatic loading conditions.

Another application of the generalized Kelvin–Voigt model concerns the determination of stress fields. We will see, for example in chapter 13, that rock cores that have been collected at great depth exhibit some delayed deformation when they are brought rapidly to the surface. Understanding this delayed deformation process is of advantage when one is characterizing the stress field at the depth where the core has been collected.

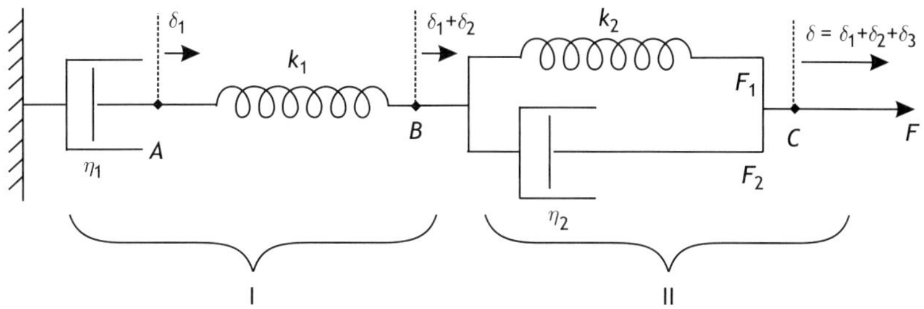

Fig. 2.12 The Burger material.

Fluid-type behavior: the Burger material

While most rock materials exhibit a solid-type behavior at time scales of engineering interest (from a few months to hundreds of years), they may exhibit a fluid-type behavior at time scales of thousands, millions or hundreds of million of years, depending on the geomaterial. A simple way to model a fluid-type behavior for very long time scales is simply to introduce a dashpot with high viscosity in series with the spring of stiffness k_1 in the generalized Kelvin–Voigt model (fig. 2.10). Such a model is called a Burger material (fig. 2.12). In fig. 2.12, elements I correspond to the Maxwell model (with displacement $\delta_I = \delta_1 + \delta_2$) while elements II correspond to the Kelvin–Voigt model (with displacement $\delta_{II} = \delta_3$). The constitutive equation is simply the sum of the constitutive equations for both models:

for I, we have $\dot{\delta}_I = \dot{F}/k_1 + F/\eta_1$;
for II, we have $F = k_2\delta_{II} + \eta_2\dot{\delta}_{II}$.

Here $\delta = \delta_I + \delta_{II}$ and therefore $\ddot{\delta} = \ddot{\delta}_I + \ddot{\delta}_{II}$. By successive substitution, we have

$$\dot{F} = k_2\dot{\delta}_{II} + \eta_2\ddot{\delta}_{II} \tag{2.21}$$

$$= k_2(\dot{\delta} - \dot{\delta}_I) + \eta_2(\ddot{\delta} - \ddot{\delta}_I) \tag{2.22}$$

$$= k_2\left[\dot{\delta} - \left(\frac{\dot{F}}{k_1} + \frac{F}{\eta_1}\right)\right] + \eta_2\left[\ddot{\delta} - \left(\frac{\ddot{F}}{k_1} + \frac{\dot{F}}{\eta_1}\right)\right] \tag{2.23}$$

so that, finally, the constitutive equation for the Burger model is given by

$$\eta_2\ddot{\delta} + k_2\dot{\delta} = \frac{k_2}{\eta_1}F + \left(1 + \frac{k_2}{k_1} + \frac{\eta_2}{\eta_1}\right)\dot{F} + \frac{\eta_2}{k_1}\ddot{F} \tag{2.24}$$

We will solve equation (2.24) for the conditions $F = F_0$ for $0 < t < t_1$ with $\delta = F_0/k_1$ for $t = 0^+$. The displacement at time t is given by

$$\eta_2\ddot{\delta} + k_2\dot{\delta} = \frac{k_2}{\eta_1}F_0 \tag{2.25}$$

A particular solution is $\dot{\delta} = F_0/\eta_1$, so that the general solution is

$$\delta = \frac{F_0}{k_1} + \frac{F_0}{\eta_1}t + \frac{F_0}{k_2}(1 - e^{-k_2 t/\eta_2}) \tag{2.26}$$

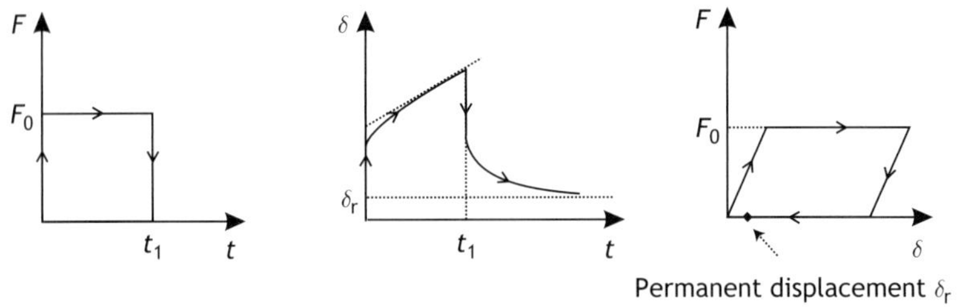

Fig. 2.13 Response of a Burger material to a constant load.

In (2.26), the first term on the right-hand side is the instantaneous elastic response (the wave propagation due to spring in I, in fig. 2.12), the second term is the very long-term creep (a million year response due to the dashpot in I, in fig. 2.12), while the third term (component II in fig. 2.12) corresponds to the so-called "quasistatic" elastic response for time scales ranging from hours to hundreds if not thousands of years. Presently, a particularly active topic of research is identifying the response of sedimentary rocks at time scales of hundreds of thousands of years, to determine safe nuclear waste repositories.

Long-term creep term controls the deformation observed at a geological time scale (the tectonic phase). It is the term to be considered when the load drops to zero at time t_1 at the end of the tectonic phase, in fig. 2.13. The dashpot in II (fig. 2.12) controls the time scale during which the geomaterial maintains a memory of past loads.

2.4 Limits to linear elasticity: friction and ductility

In all the configurations we have investigated so far, the load may be increased indefinitely without any limit. In reality, the mechanical response of a material to a load depends strongly on the magnitudes of the load and the displacement. For example, springs will eventually break, i.e. they will lose their ability to sustain any load. Other materials may simply deform faster once the load gets larger than a certain threshold and then may fail completely if the deformation gets too large.

Depending on the material, the large-deformation domain may be controlled by friction or by structural changes in the solid matrix (the ductility). In the latter case this involves the extension of faults within the material, whether microcracks (generally at the submillimeter scale) or dislocations (at the atomic scale). This issue of failure is addressed in two different chapters (7 and 8). We introduce here only the load limit concept using the Saint-Venant element, which covers both friction and ductility for the purposes of phenomenological modeling.

2.4.1 The Saint-Venant material

The concept of a load threshold is introduced by the Saint-Venant element, which is modeled by a sliding pad. The pad's motion is controlled by friction. If the load applied to

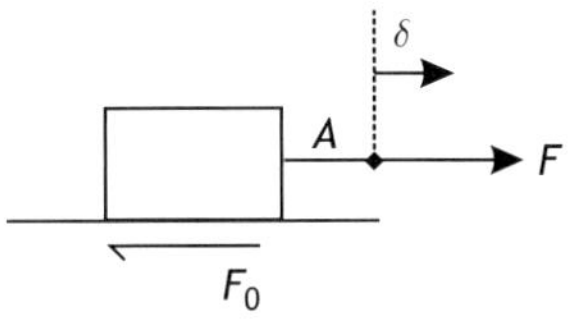

Fig. 2.14 The Saint-Venant element.

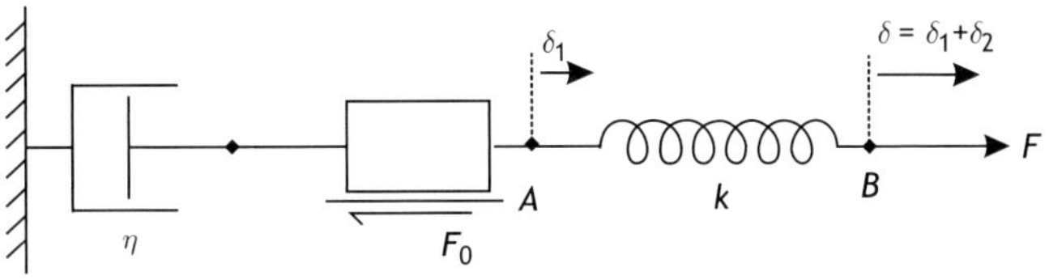

Fig. 2.15 The Bingham material.

the pad is lower than a value controlled by friction, the pad does not move. If the load exceeds the threshold, then the pad slips and its motion is determined by that of the force application point A (fig. 2.14). Hence a Saint-Venant element is never considered singly but always with additional elements, whether springs or dashpots. With the Saint-Venant element, the displacement δ of the point A where the load is applied is zero if $F < F_0$; it increases from zero if $F \geq F_0$.

2.4.2 The Bingham material

A Bingham material is elastic as long as the load remains lower than a given threshold F_0. Once the load becomes larger than this threshold the displacement of the Saint-Venant pad is modeled by a dashpot, and the material behaves like a Maxwell material. Thus, such a material exhibits a bilinear behavior and may be referred to as elastoviscoplastic:

$$\text{if } F < F_0, \quad \delta = \frac{F}{k}; \tag{2.27}$$

$$\text{if } F \geq F_0; \quad \delta = \frac{F}{k} + \frac{F - F_0}{\eta} t \tag{2.28}$$

The concept of plasticity refers to the situation above the critical load, when some permanent deformation takes place, whilst that of viscosity refers to the noninstanteous rate of deformation that occurs after the elastic. Plasticity refers to a solid-type behavior.

2.4.3 The concept of a "residual" load

Let us assume now that the above-mentioned sliding pad is introduced into the dash-pot branch of a generalized Kelvin model (fig. 2.16). The model exhibits a bilinear behavior. If the force F is smaller than a threshold value F_0, the material is linearly elastic. If the force

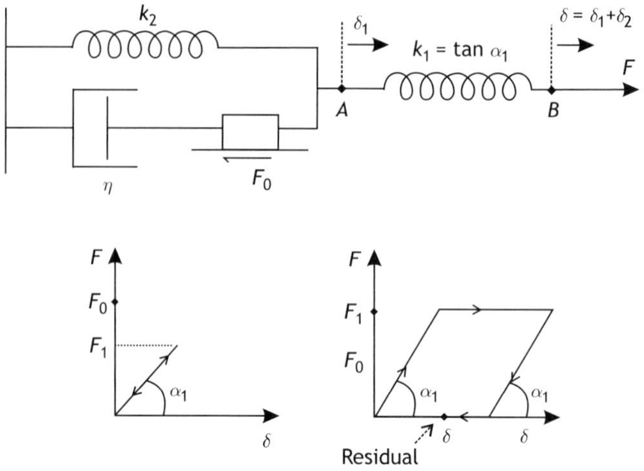

Fig. 2.16 The pad blocks the load to a value F_0 in the Kelvin–Voigt element when the force F is released.

F is set equal to a value F_1 larger than the threshold F_0 of the pad, its behavior is described by (2.30):

$$\text{if } F_1 < F_0, \quad \delta = \frac{F_1}{k_1}; \tag{2.29}$$

$$\text{if } F_1 \geq F_0, \quad \delta = \frac{F_1}{k_1} + \frac{F_1 - F_0}{k_2}(1 - e^{-k_2 t/\eta}) \tag{2.30}$$

In the long term, when the load is larger than F_0 the model's apparent stiffness changes and, as time passes, it behaves as two springs in series: the load F_1 acts on the spring with stiffness k_1 and the load $F_1 - F_0$ acts on the spring with stiffness k_2. Let us assume now that the load is released at time t_1. The spring with stiffness k_1 instantaneously returns to its original length, while the load in the spring with stiffness k_2 drops progressively until it is equal to the resisting force F_0 in the pad. At this point the system is not loaded externally and does not deform any further, but it supports an internal load and exhibits the residual deformation F_0/k_2. Internal loads are often called "residual" loads in rock mechanics. They have been observed in both sedimentary and crystalline rocks and may have very diverse origins (sections 9.1 and 14).

We observe that the load–displacement curves for the Burger model (fig. 2.13) and the "residual load" model (fig. 2.16) are identical although they represent somewhat different materials. The loading history becomes significant, and it is impossible to evaluate it from only the final deformation state.

2.5 Nonlinear models

Once a load threshold has been reached, the material behavior becomes nonlinear. This may be controlled simply by a change in friction coefficient but this can lead to recurrent instabilities if the dynamic friction coefficient is smaller than the static value.

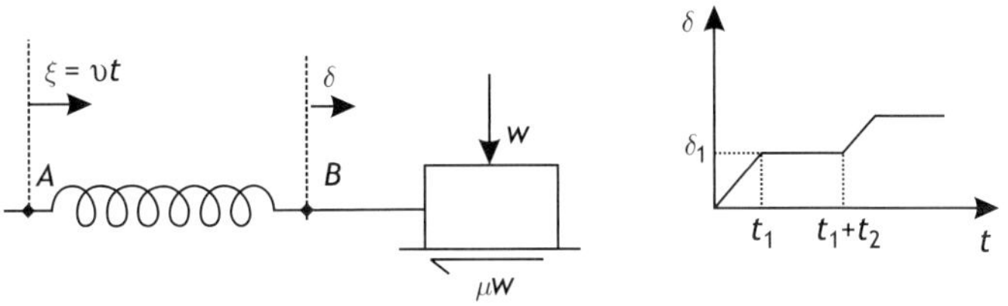

Fig. 2.17 The stick–slip principle.

2.5.1 Dynamic friction and earthquakes

The sliding pad in the Saint-Venant element introduces the concept of a force threshold through frictional resistance. Friction is a very important property of rocks and rock masses and is discussed in chapter 10. It will be shown that the friction coefficient μ exhibits different values in static and dynamic conditions. We explore here the instability that occurs when the dynamic friction coefficient is smaller than the static value.

Following Jaeger and Cook (1979), let us consider a pad of mass M pushed by a spring, the extremity A of which has a constant velocity v (fig. 2.17). The position ξ of A at time t is given by $\xi = vt$. The motion of the pad is described by its displacement δ. A weight W is applied to the pad, so that its frictional resistance to motion is μW; μ is the friction coefficient and is equal to μ^s when the pad is at rest and to μ^d when the pad moves, with the condition $\mu^d \leq \mu^s$. At time $t = 0$, the force exerted by the spring on the pad is assumed to be equal to the static frictional resistance and the position of A is noted as ξ_0, so that $\xi = \xi_0 = \mu^s W/k$.

At time t, the equation that describes the pad motion is

$$k(\xi_0 + vt - \delta) - \mu^d W = M\ddot{\delta} \tag{2.31}$$

The first term on the left-hand side of equation (2.31) is the force on the pad due to the spring, the second term is the frictional resistance and the right-hand side expresses Newton's law. Let us define $k/M \equiv n^2$. The motion equation (2.31) becomes

$$\ddot{\delta} + n^2\delta = n^2 vt + (\mu^s - \mu^d)\frac{W}{M} \tag{2.32}$$

With the condition $\delta = \dot{\delta} = 0$ at $t = 0$, the solution is

$$\delta = vt + (\mu^s - \mu^d)\frac{W}{Mn^2} - \frac{V}{n}\sin\, nt - \left[(\mu^s - \mu^d)\frac{W}{Mn^2}\right]\cos\, nt \tag{2.33}$$

The velocity $\dot{\delta}$ of the pad is given by

$$\dot{\delta} = v - v\cos\, nt + (\mu^s - \mu^d)\frac{W}{Mn}\sin\, nt \tag{2.34}$$

Since $\cos\, nt = [1 - \tan(nt/2)^2]/[1 + \tan(nt/2)^2]$ and $\sin\, nt = 2\tan(nt/2)/[1 + \tan(nt/2)^2]$, it is found that $\dot{\delta} = 0$ at time t_1, when $\tan(nt_1/2) = -(\mu^s - \mu^d)W/(Mnv)$. The solution is

$$t_1 = 2\pi/n - \frac{2}{n}\arctan\frac{(\mu^s - \mu^d)w}{Mnv}$$

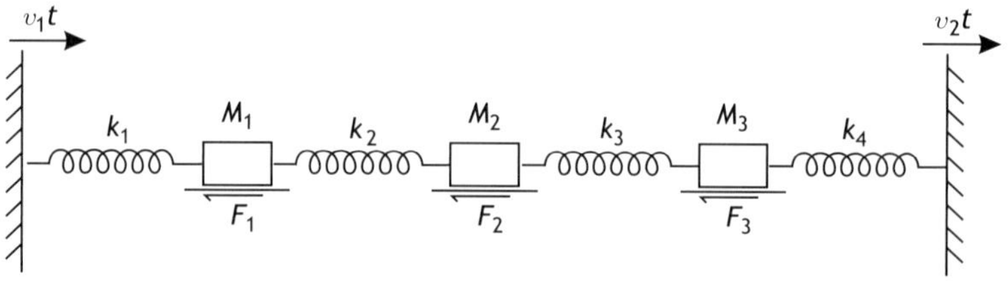

Fig. 2.18 A proposed mechanism for the nonperiodic recurrences of earthquakes.

from which we can determine the position of the pad when it stops:

$$\delta_1 = vt_1 + 2(\mu^s - \mu^d)\frac{W}{Mn^2}.$$

The force exerted by the spring on the pad at this time is given by

$$F_1 = k(vt_1 + \xi_0 - \delta_1) = k\left[\xi_0 - 2(\mu^s - \mu^d)\frac{W}{Mn^2}\right] \tag{2.35}$$

$$= \mu_s W - 2(\mu^s - \mu^d)W = (2\mu^d - \mu^s)W \tag{2.36}$$

and the time when the pad starts sliding again is $t_2 = 2(\mu^s - \mu^d)W/(kv)$.

This stop-and-go displacement process is often referred to as "stick–slip" and is somewhat illustrative of the way earthquakes occur along existing faults that are loaded more or less regularly by tectonic forces.

If instead of considering a single spring and pad we consider a set of pads linked to each other by springs with various stiffnesses (fig. 2.18) then the motion of the pads becomes much more complex and may be made to resemble somewhat the earthquake history along tectonic-plate-scale faults.

While materials such as metals exhibit a resistance to failure in tension somewhat similar to that in compression, it is customary to assume that rock masses exhibit zero tensile strength because of the many faults or fractures that are always present. Consequently, rocks are always under compression. In fig. 2.18, if $V_2 \geq V_1$ then a failure develops on the right of the spring with stiffness k_4 (corresponding to the phenomenon of rifting in tectonics) and the right-hand boundary condition for the system becomes equivalent to a free-boundary condition. In such conditions the displacements in all the springs depend only on the boundary condition V_1 on the left of the system and on the material properties of the pads and of the springs.

2.5.2 General nonlinear models

With the Bingham material, once the threshold F_0 has been reached the dashpot is mobilized and this corresponds to a transition from an elastic solid to a viscoelastic fluid with permanent deformation when the load is relaxed. However, the overall behavior of the Bingham material is that of a plastic solid, i.e. the flow process occurs only when the load become larger than a certain threshold, and a residual deformation is observed when the load drops back to zero after it has been raised above a certain threshold. The difference between a viscoelastic fluid and an elastoviscoplastic solid is the absence of a load

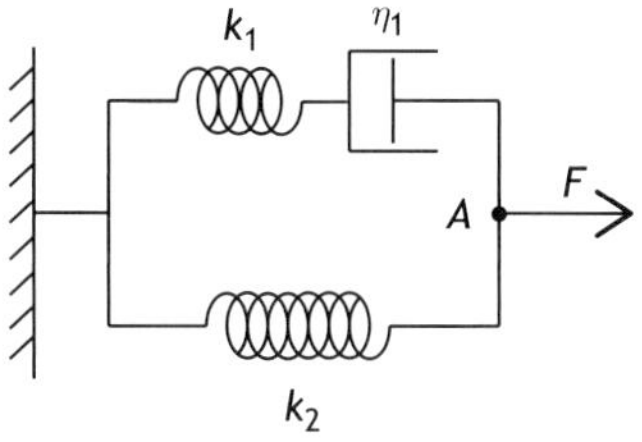

Fig. 2.19 Elasto-viscosity and more.

threshold for the viscous fluid. Similarly, the difference between a viscoelastic solid and a elastoviscoplastic solid is the absence of permanent deformation when the load is dropped back to zero for the viscoelastic solid.

Another common situation occurs when the constants that describe the material (the stiffness and viscosity) become dependent on the load or on the deformation or on their time derivatives. For such materials the response of the equivalent system cannot be determined by a simple superposition of the elementary models, and attention must be given to the loading history, i.e. to the way in which the load or displacement has been progressively increased.

Understanding loading history and its influence on present-day deformation processes is one of the goals of crustal geomechanics.

2.6 Exercises

1. Consider the elastic material schematized in fig. 2.3.
   1.1. Assume first that all three springs have the same stiffness. What must be this stiffness K if the displacement δ is equal to 1 mm when the force F reaches 28 000 newtons?
   1.2. What must be the stiffness K_2 of the central spring if the load supported by this spring is 0.8 times that supported by the two other springs?
2. Propose a rheological model that exhibits an instantaneous elastic response and a solid-type viscoelastic response as time passes.
3. Consider the rheological model shown in fig. 2.19.
   3.1. Write down the constitutive equation for this model.
   3.2. What are the units of viscosity η?
   3.3. Assume that both springs have a stiffness equal to 1.5 times the value of K found in exercise 1.1. What must be the viscosity η_1 if the displacement δ of point A reaches 1 mm after 1 minute, when the applied constant load is equal to 28 000 newtons? What is the viscosity if it takes 2 hours to reach this displacement? Call this last value η_2.
4. We consider now the case in which a Saint-Venant pad is introduced in series with the spring k_2 in the model shown in fig. 2.19.

4.1. What must be the threshold F_0 if the pad starts moving when the total force acting on A reaches 5000 newtons ?

4.2. Write down the constitutive equation for this material.

4.3. Find the displacement reached after two hours of loading if the applied constant load F at time 0 is equal to 7800 newtons.

4.4. Describe what happens when the load drops back to zero after one hour of loading. Draw a force–displacement diagram for this loading pattern.

4.5. What happens if the force remains constant indefinitely? What is the velocity of this motion?

4.6. What is the viscosity if the velocity is 1 cm/yr?

3 Forces and stresses

In this chapter we explore the concept of a force acting on a solid continuous geomaterial, as defined in chapter 1 and we investigate how forces are transmitted through the material. The ideas presented in this chapter were mostly developed during the nineteenth century (see e.g. Cauchy, 1829, 1864) and are discussed in many modern books on mechanics. The chapter draws heavily on books written in the last 50 years or so (Fung, 1965; Malvern, 1969; Jaeger *et al.*, 2007)). Applications of the concept of stress to geomaterials are discussed further in chapters 8, 12 and 13.

3.1 Forces and moments

3.1.1 Body forces, surface forces

Let us consider a body B of volume V_B enclosed within a closed surface S_B (fig. 3.1) with a cartesian frame of reference $\hat{\boldsymbol{I}}_i, i = 1, 2, 3$. The $\hat{\boldsymbol{I}}_3$ axis is assumed to be vertically upwards. In many geomechanics problems, as mentioned earlier $\hat{\boldsymbol{I}}_1$ is chosen oriented positive to the north, while $\hat{\boldsymbol{I}}_2$ is oriented positive to the east; then, in order to keep a "direct" frame of reference, $\hat{\boldsymbol{I}}_3$ is taken to be positive downward. However, in this chapter $\hat{\boldsymbol{I}}_3$ is taken to be positive upward (fig. 3.1).

We will consider a small volume dv with uniform material density ϱ and a body force $\boldsymbol{b}$ per unit mass, which acts on the matter that occupies the volume dv. In many static geomechanics problems this body force is simply gravity ($\boldsymbol{b} = b_i\hat{\boldsymbol{I}}_i = -g\delta_{i3}\hat{\boldsymbol{I}}_3$, with $\delta_{i3} = 0$ when $i \neq 3$ and $\delta_{i3} = 1$ when $i = 3$). More generally, in dynamic problems, the body forces are associated with the acceleration of the matter that occupies volume dv.

The body force that acts on the volume dv is $d\boldsymbol{f}_b = \varrho\boldsymbol{b}\, dv$, so that the total body force that acts on the body B is

$$\boldsymbol{f}_{V_B} = \int_{V_B} \varrho\boldsymbol{b}\, dv = \int_{V_B} \varrho b_i\hat{\boldsymbol{I}}_i\, dv = -\int_{V_B} \varrho g\, dv\, \hat{\boldsymbol{I}}_3 \tag{3.1}$$

Let us now consider a small surface element $\boldsymbol{ds}$ defined by $\boldsymbol{ds} = \boldsymbol{n}\, da$, where $\boldsymbol{n}$ is the normal oriented positively toward the exterior of body B and da is its area. If $\boldsymbol{t}$ is the surface force per unit area, the surface force $\boldsymbol{dt}$ acting on the surface element $\boldsymbol{ds}$ is $\boldsymbol{dt} = \boldsymbol{t}\, da$, so that the total surface force acting on the surface S_B of body B is

$$\boldsymbol{f}_{S_B} = \int_{S_B} \boldsymbol{t}\, da \tag{3.2}$$

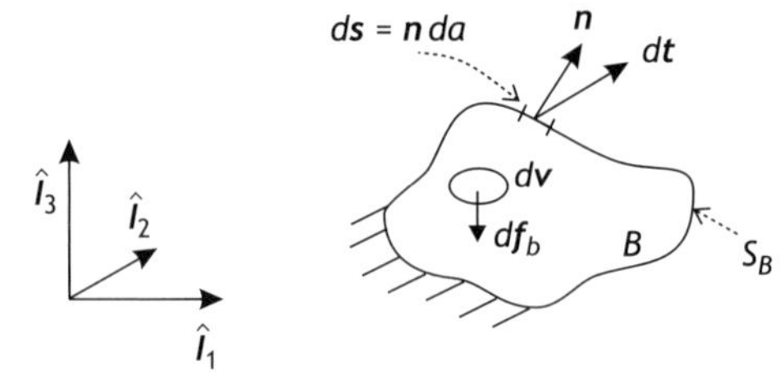

Fig. 3.1 Body forces and surface forces.

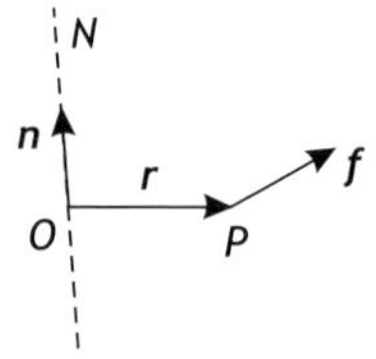

Fig. 3.2 The moment of the force $\boldsymbol{f}$.

The resulting force R_B on the body B is the sum of the total body forces and the total surface forces: $\boldsymbol{R}_B = \boldsymbol{f}_{V_B} + \boldsymbol{f}_{S_B}$.

3.1.2 Moments, momentum and Euler's laws

Let $\boldsymbol{r}$ be the vector that joins the origin O of the frame of reference $(\hat{\boldsymbol{I}}_i, i = 1, 2, 3)$ to point P, and let $\boldsymbol{f}$ be a force acting at point P (fig. 3.2). The moment $\boldsymbol{m}(\boldsymbol{f}, O)$ of force $\boldsymbol{f}$ with respect to the point O is defined by

$$\boldsymbol{m}(\boldsymbol{f}, O) \equiv \boldsymbol{r} \wedge \boldsymbol{f} \tag{3.3}$$

where the notation $\wedge$ indicates a vector product. Consider now the axis $\boldsymbol{N}$ passing through O and defined by the unit vector $\boldsymbol{n}$. The moment of the force $\boldsymbol{f}$ with respect to the axis $\boldsymbol{N}$ is defined by the scalar product $m_N \equiv \boldsymbol{m}(\boldsymbol{f}, O) \cdot \boldsymbol{n} = (\boldsymbol{r} \wedge \boldsymbol{f}) \cdot \boldsymbol{n}$.

Let G be the center of gravity of the body B and choose it as the origin for the frame of reference $(\hat{\boldsymbol{I}}_i, i = 1, 2, 3)$. The moment of the body forces with respect to G is given by:

$$\boldsymbol{m}(\boldsymbol{f}_{V_B}, G) = \int_{V_B} (\boldsymbol{r} \wedge \boldsymbol{b})\varrho \, dv = \int_{V_B} \varepsilon_{ijk} r_i \varrho b_j \, dv \, \hat{\boldsymbol{I}}_k, \quad i, j, k = 1, 2, 3 \tag{3.4}$$

where $\varepsilon_{ijk} = +1$ if the integers i, j and k are a direct permutation of 1, 2 and 3, $\varepsilon_{ijk} = -1$ for an indirect permutation, and $\varepsilon_{ijk} = 0$ when any two of the integers i, j and k are equal.

If $\boldsymbol{r}'$ is the vector that joins G to the point of application of the surface force $\boldsymbol{t}$ acting on ds, the moment of the surface forces is

$$\boldsymbol{m}(\boldsymbol{f}_{S_B}, G) = \int_{A_B} (\boldsymbol{r}' \wedge \boldsymbol{t}) \, da = \int_{A_B} \varepsilon_{ijk} r'_i t_j \, da \, \hat{\boldsymbol{I}}_k, \qquad i, j, k = 1, 2, 3 \tag{3.5}$$

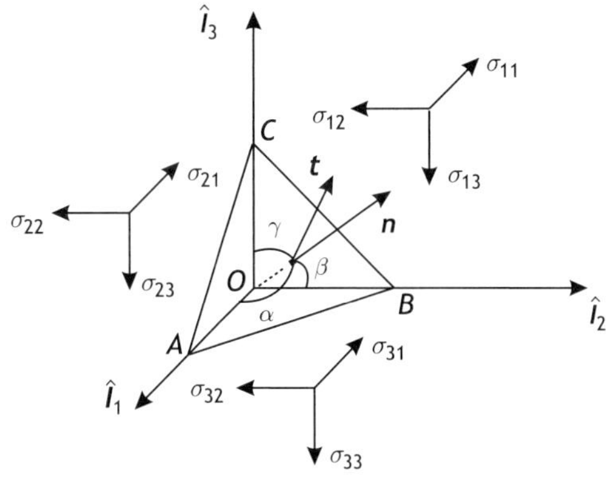

Fig. 3.5 Cauchy tetrahedron.

The sides of the square formed by the projection of the cube onto the $\hat{\boldsymbol{I}}_1\hat{\boldsymbol{I}}_2$ plane are numbered 1 to 4, with 1 and 3 perpendicular to $\hat{\boldsymbol{I}}_1$. Let us observe that the normal $\boldsymbol{n}$ to face 1 is opposite to the normal to face 3. On face 1, the components of the force dt_1 are $\sigma_{11}\,da = \sigma_{11}$, $\sigma_{12}\,da = \sigma_{12}$ and $\sigma_{13}\,da = \sigma_{13}$, since $da = 1$. When the component σ_{11} is acting in the same direction as the normal $\boldsymbol{n}$ to the face it is called a traction (some authors call it a tension).

In continuum mechanics, the traction is taken as positive. When σ_{11} is acting in the opposite direction to that of $\boldsymbol{n}$, it is referred to as a compression and is negative. Hence the signs of the components σ_{11}, σ_{22} and σ_{33} depend on their orientation with respect to the normal to the face on which they are acting and not on their orientation with respect to the unit vectors of the frame of reference.

For face 1 we observe that the normal $\boldsymbol{n}$ is oriented in the same direction as $\hat{\boldsymbol{I}}_1$ and therefore a positive σ_{11} means that the resulting force is acting in the same sense as $\hat{\boldsymbol{I}}_1$. By analogy, we assume that σ_{12} is positive if the resulting force acting in the direction of $\hat{\boldsymbol{I}}_2$ is acting in the same sense as $\hat{\boldsymbol{I}}_2$. As a result, for face 3, since the normal $\boldsymbol{n}$ is opposite to $\hat{\boldsymbol{I}}_1$, so that a traction is in the opposite direction to $\hat{\boldsymbol{I}}_1$, a positive σ_{12} implies a force in the opposite direction to that of $\hat{\boldsymbol{I}}_2$. Similar reasoning is applied to faces 2 and 4, so that positive values for $\sigma_{ij}; i,j = 1,2$, result in the orientations shown in fig. 3.4.

In many geomechanics problems only compressions are observed and it is customary, in such problems, to reckon the directions of compressions as positive. Then the sign convention for the nondiagonal components of the matrix should be changed as well.

3.2.2 The stress vector

We show in this subsection how the nine components σ_{ij} of the stress tensor may be used to determine the force $\boldsymbol{dt}$ acting on a surface element ds with any orientation $\boldsymbol{n}$. Let us consider the case in which the small volume dv introduced above is the so-called "Cauchy" tetrahedron (fig. 3.5).

One of the corners of the tetrahedron is the origin O of the frame of reference and the three other corners, A, B and C, result from the intersection of a plane normal to the unit vector $\boldsymbol{n}$ with the three axes of reference; $\boldsymbol{n}$ is characterized by its three direction cosines $n_1 = \cos\alpha$; $n_2 = \cos\beta$ and $n_3 = \cos\gamma$, where α, β and γ are respectively the angles between $\boldsymbol{n}$ and $\hat{\boldsymbol{I}}_1, \hat{\boldsymbol{I}}_2$ and $\hat{\boldsymbol{I}}_3$. Let ω be the area of the triangle ABC. Then the area of the triangle OBC is $\omega\cos\alpha = n_1\omega$. Similarly, the areas of the triangles OAB and OAC are respectively $n_3\omega$ and $n_2\omega$. If h is the height of the tetrahedron normal to ABC, the volume of the tetrahedron is simply $dv = h\omega/3$.

Now let us assume that the tetrahedron is in equilibrium, so that the sum of the surface and body forces acting on it is zero. The surface force acting on ABC is $d\boldsymbol{t} = \boldsymbol{t}\omega$, where $\boldsymbol{t}$ is the surface force per unit area on ABC. According to the definition in section 3.2.1 the surface force on the triangle OBC is

$$dt_1 = \sigma_{11}\omega n_1\hat{\boldsymbol{I}}_1 + \sigma_{12}\omega n_1\hat{\boldsymbol{I}}_2 + \sigma_{13}\omega n_1\hat{\boldsymbol{I}}_3 \tag{3.13}$$

and so on for the other faces of the tetrahedron. The body force is $d\boldsymbol{f}_b = -\rho g(1/3)h\omega\hat{\boldsymbol{I}}_3$.

We now compute the three components of the resulting force $\boldsymbol{R} = \boldsymbol{f}_s + \boldsymbol{f}_b$, starting with that along the $\hat{\boldsymbol{I}}_1$ axis, R_1. It must be null for the tetrahedron to be at rest:

$$R_1 = t_1\omega - \sigma_{11}\omega n_1 - \sigma_{21}\omega n_2 - \sigma_{31}\omega n_3 = 0 \tag{3.14}$$

Thus

$$t_1 = \sigma_{11}n_1 + \sigma_{21}n_2 + \sigma_{31}n_3 \tag{3.15}$$

From the condition $R_2 = 0$ we obtain, similarly,

$$t_2 = \sigma_{12}n_1 + \sigma_{22}n_2 + \sigma_{32}n_3 \tag{3.16}$$

In the vertical direction we must take into account the body force:

$$R_3 = t_3\omega - \sigma_{13}\omega n_1 - \sigma_{23}\omega n_2 - \sigma_{33}\omega n_3 - \rho g\frac{h\omega}{3} = 0 \tag{3.17}$$

The condition (3.17) must remain valid when the volume shrinks to zero, i.e. when h tends to zero. Hence, in this limit,

$$t_3 = \sigma_{13}n_1 + \sigma_{23}n_2 + \sigma_{33}n_3 \tag{3.18}$$

In index notation these three equations may be summarized by $t_j = \sigma_{ij}n_i = \sigma^T_{ji}n_i$, or in vectorial form $\boldsymbol{t} = \tilde{\sigma}^T\boldsymbol{n}$, where the tensor $\tilde{\sigma}^T$ is the transpose of tensor $\tilde{\sigma}$.

Symmetry of the stress tensor

Let us show now that the stress tensor is symmetrical ($\tilde{\sigma} = \tilde{\sigma}^T$) if moments must balance everywhere in the body under consideration. Let us assume that the length of a side of the cube in fig. 3.4 is l. Then the area of each face is l^2. If this body is at rest, not only must the resultant force be zero but also the resultant moment. Let us compute the moment acting on the cube. Given that the forces $\sigma_{11}l^2$ acting on faces 1 and 3 of the cube are equal and of opposite sign, the resultant moment is zero. The same is true for the forces $\sigma_{22}l^2$ acting on faces 2 and 4. The forces $\sigma_{12}l^2$ and $\sigma_{21}l^2$ acting respectively on faces 3 and 4 pass through the origin and therefore their moments with respect to it are zero. The moment of the force

$\sigma_{21}l^2$ acting on face 2 is $-\sigma_{21}l^3$ while that of $\sigma_{12}l^2$, acting on face 1, is $\sigma_{12}l^3$. Hence, the condition for the moment with respect to axis $\hat{\boldsymbol{I}}_3$ to be zero is $\sigma_{12} = \sigma_{21}$. A similar reasoning may be applied for moments with respect to $\hat{\boldsymbol{I}}_1$ and $\hat{\boldsymbol{I}}_2$, by considering volumes that shrink to zero so that the moment associated with the body force becomes insignificant. A more general demonstration of the symmetry of the stress tensor, which is valid also in dynamic conditions when the tetrahedron is not at rest, is given in section 6.1.2.

Hence, given that the stress tensor is symmetrical, the force per unit area $\boldsymbol{t}$ supported by any surface ds with normal $\boldsymbol{n}$ is given by

$$\boldsymbol{t} = \tilde{\sigma}\boldsymbol{n} \qquad \text{or, in terms of components,} \qquad t_i = \sigma_{ij}n_j, \qquad i,j = 1,2,3. \tag{3.19}$$

The "force per unit area" $\boldsymbol{t}$ is called the stress vector. The stress vector is defined at a point, just like the stress tensor, but for a surface of given orientation as characterized by its unit normal $\boldsymbol{n}$.

3.2.3 Normal stress, shear stress, principal stress components

The stress vector $\boldsymbol{t}$ may be projected onto the normal $\boldsymbol{n}$ to a surface ds:

$$\sigma_n \equiv \boldsymbol{t}\cdot\boldsymbol{n} = \tilde{\sigma}\boldsymbol{n}\cdot\boldsymbol{n} = \sigma_{ij}n_jn_i \tag{3.20}$$

$$= \sigma_{11}n_1^2 + \sigma_{22}n_2^2 + \sigma_{33}n_3^2 + 2n_1n_2\sigma_{12} + 2n_2n_3\sigma_{23} + 2n_3n_1\sigma_{31} \tag{3.21}$$

Here, σ_n is the normal component of the stress vector acting on the surface ds. It is a scalar quantity known as the normal stress acting on ds.

Similarly, $\boldsymbol{t}$ may be projected onto the surface ds:

$$\boldsymbol{\tau} = \boldsymbol{t} - \sigma_n\boldsymbol{n} = \tilde{\sigma}\boldsymbol{n} - (\tilde{\sigma}\boldsymbol{n}\cdot\boldsymbol{n})\boldsymbol{n} \tag{3.22}$$

The vector $\boldsymbol{\tau}$ is called the shear component of $\boldsymbol{t}$. For example, if we consider the triangle *OAB* of the Cauchy tetrahedron (fig. 3.5), the normal stress is σ_{33} while σ_{31} and σ_{32} are the components of the shear stress in that plane.

We observe that the diagonal elements of the matrix σ_{ij} are the normal stresses acting respectively on three surfaces normal to the axis of the frame of reference while the nondiagonal elements are the components of the shear stress supported respectively by the same planes.

Because $\boldsymbol{t}$ varies with the orientation of the surface ds, so do the normal stress and the shear stress. Let us determine the extreme values (extrema) of σ_n.

From equation (3.20) we may write

$$\sigma_n = \boldsymbol{t}\cdot\boldsymbol{n} = t_1n_1 + t_2n_2 + t_3n_3 \tag{3.23}$$

$$= (\sigma_{11}n_1 + \sigma_{12}n_2 + \sigma_{13}n_3)n_1 + (\sigma_{21}n_1 + \sigma_{22}n_2 + \sigma_{23}n_3)n_2 + (\sigma_{31}n_1 + \sigma_{32}n_2 + \sigma_{33}n_3)n_3 \tag{3.24}$$

Only two independent variables are necessary to define a direction (e.g. dip direction and dip angle). Let us choose as independent variables the components n_1 and n_2 of $\boldsymbol{n}$. This implies that n_3 is a function of n_1 and n_2. Given that $\boldsymbol{n}$ is a unit vector, so that

$$\boldsymbol{n} \cdot \boldsymbol{n} \equiv \boldsymbol{n}^2 = n_1^2 + n_2^2 + n_3^2 = 1 \tag{3.25}$$

we obtain the partial derivatives of n_3 with respect to n_1 and n_2 by differentiating equation (3.25) with respect to n_1 and to n_2:

$$\frac{\partial \boldsymbol{n}^2}{\partial n_1} = 2n_1 + 2n_3 \frac{\partial n_3}{\partial n_1} = 0 \tag{3.26}$$

$$\Rightarrow \frac{\partial n_3}{\partial n_1} = -\frac{n_1}{n_3} \tag{3.27}$$

$$\frac{\partial \boldsymbol{n}^2}{\partial n_2} = 2n_2 + 2n_3 \frac{\partial n_3}{\partial n_2} = 0 \tag{3.28}$$

$$\Rightarrow \frac{\partial n_3}{\partial n_2} = -\frac{n_2}{n_3} \tag{3.29}$$

We are looking for the extrema of σ_n with respect to n_1 and n_2 and, by differentiating equation (3.24) with respect to n_1, we obtain

$$\frac{\partial \sigma_n}{\partial n_1} = 2(\sigma_{11}n_1 + \sigma_{12}n_2 + \sigma_{13}n_3) + 2(\sigma_{31}n_1 + \sigma_{32}n_2 + \sigma_{33}n_3)\frac{\partial n_3}{\partial n_1} \tag{3.30}$$

$$= 2t_1 - 2\frac{n_1}{n_3}t_3 = 0 \tag{3.31}$$

$$\Rightarrow \frac{t_1}{n_1} = \frac{t_3}{n_3} \tag{3.32}$$

and similarly by differentiating with respect to n_2 we have :

$$\frac{\partial \sigma_n}{\partial n_2} = 2t_2 - 2\frac{n_2}{n_3}t_3 = 0 \tag{3.33}$$

$$\Rightarrow \frac{t_2}{n_2} = \frac{t_3}{n_3} \tag{3.34}$$

$$\Rightarrow \frac{t_1}{n_1} = \frac{t_2}{n_2} = \frac{t_3}{n_3} \tag{3.35}$$

$$\Rightarrow \boldsymbol{t} = \lambda \boldsymbol{n} \tag{3.36}$$

where λ in equation (3.36) is a constant scalar.

Principal stress components

The extrema of the normal stress are obtained when the stress vector is collinear to the normal to the surface, i.e. when there is no shear stress. Equation (3.36) may be rewritten as

$$(\tilde{\sigma} - \lambda \tilde{1})\boldsymbol{n} = 0 \tag{3.37}$$

which is the well-known eigenvalue and eigenvector problem for a tensor (see the appendix). Its nontrivial solution is such that $\det(\tilde{\sigma} - \lambda \tilde{1}) = 0$, which may be rewritten:

$$-\lambda^3 + I_{\tilde{\sigma}}\lambda^2 - II_{\tilde{\sigma}}\lambda + III_{\tilde{\sigma}} = 0 \tag{3.38}$$

where, if σ_1, σ_2 and σ_3 are the roots of equation (3.38), the quantities $I_{\tilde{\sigma}}, II_{\tilde{\sigma}}$ and $III_{\tilde{\sigma}}$ are known as the first, second and third invariants of $\tilde{\sigma}$. They are defined as follows :

$$I_{\tilde{\sigma}} = \sigma_1 + \sigma_2 + \sigma_3 = \text{tr}(\tilde{\sigma}^2) \tag{3.39}$$

$$II_{\tilde{\sigma}} = \sigma_1\sigma_2 + \sigma_2\sigma_3 + \sigma_3\sigma_1 = (1/2)[(\text{tr}\,\tilde{\sigma})^2 - \text{tr}(\tilde{\sigma}^2)] \tag{3.40}$$

$$III_{\tilde{\sigma}} = \sigma_1\sigma_2\sigma_3 = \det\,\tilde{\sigma} \tag{3.41}$$

where $\text{tr}\,\tilde{\sigma}$ and $\det\,\tilde{\sigma}$ are the trace and the determinant of $\tilde{\sigma}$ (see the appendix). Equation (3.38) is known as the "characteristic" equation.

Because the stress tensor is symmetrical the three eigenvalues are all real, so that three real eigenvectors exist and constitute a frame of reference within which the matrix of the tensor is diagonal. By convention $\sigma_1 \geq \sigma_2 \geq \sigma_3$.

The quantities σ_1, σ_2 and σ_3 are called the principal stress components, or simply the principal stresses, while the directions associated with the three eigenvectors $\boldsymbol{e}_i; i = 1, 2, 3$, define the principal stress directions at a point $\boldsymbol{X}$ where the stress tensor is considered.

If $\sigma_1 = \sigma_2$ then any vector in the plane normal to $\boldsymbol{e}_3$ is an eigenvector. If $\sigma_1 = \sigma_2 = \sigma_3 = P$, any vector in the space is an eigenvector and the stress tensor is known as a pressure (corresponding to a fluid at rest); it is also referred to as an isotropic stress tensor.

In a two-dimensional stress field ($\sigma_3 = 0$), the characteristic equation becomes

$$\lambda^2 - \lambda(\sigma_{11} + \sigma_{22}) + (\sigma_{11}\sigma_{22} - \sigma_{12}^2) = 0 \tag{3.42}$$

with solutions

$$\sigma_{1,2} = \frac{1}{2}(\sigma_{11} + \sigma_{22}) \pm \left[\sigma_{12}^2 + \frac{(\sigma_{11} - \sigma_{22})^2}{4}\right]^{1/2} \tag{3.43}$$

3.3 Mohr representation of the stress vector

The objective here is to examine all the possible values taken by the normal stress σ_n and by the magnitude of the shear stress τ when the surface ds rotates around the point $\boldsymbol{X}$. We wish to plot in a single diagram all the possible values taken by these two quantities and for this we consider the two-dimensional Mohr space, in which the abscissa refers to the normal stress whilst the ordinate refers to the shear stress magnitude.

Let us observe that, in the Mohr plane, the pairs (σ_n, τ) for the cases when the surface ds is normal to one eigenvector become $(\sigma_i, 0)$, so that all the principal stress components lie on the abscissa axis (fig. 3.6).

In the present discussion, we consider only the absolute value of the shear stress magnitude $|\tau|$, although the derivation is easily extended to include the sign of the shear stress

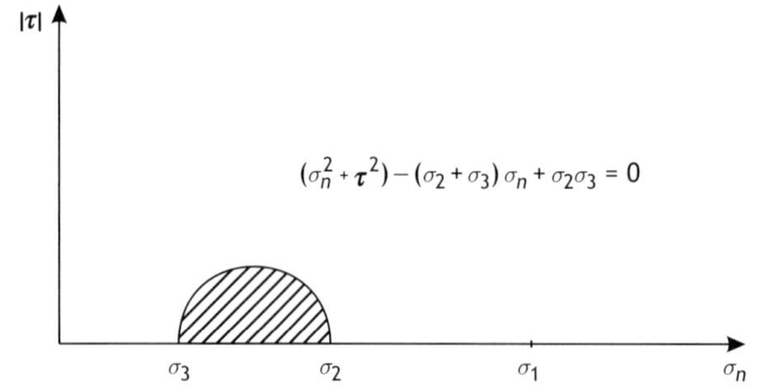

Fig. 3.6 Locus of the points $(\sigma_n, |\boldsymbol{\tau}|)$ constrained by the condition $n_1^2 \geq 0$. The semicircle shown corresponds to positive values of $\boldsymbol{\tau}$. No point may lie within the hatched area inside the semicircle. For points on the circle, $n_1^2 = 0$.

magnitude. Let us choose the frame of reference defined by the three eigenvectors $\boldsymbol{e}_i$ to rewrite equations (3.19), (3.20) and (3.22):

$$\boldsymbol{t} = \sigma_1 n_1 \boldsymbol{e}_1 + \sigma_2 n_2 \boldsymbol{e}_2 + \sigma_3 n_3 \boldsymbol{e}_3 \tag{3.44}$$

$$t^2 = \sigma_n^2 + \tau^2 = \sigma_1^2 n_1^2 + \sigma_2^2 n_2^2 + \sigma_3^2 n_3^2 \tag{3.45}$$

$$\boldsymbol{t} \cdot \boldsymbol{n} = \sigma_n = \sigma_1 n_1^2 + \sigma_2 n_2^2 + \sigma_3 n_3^2 \tag{3.46}$$

$$1 = n_1^2 + n_2^2 + n_3^2 \tag{3.47}$$

The system of equations (3.45)–(3.47) constitutes a system of three equations with three unknowns, n_1^2, n_2^2 and n_3^2. The solution for n_1^2 is

$$n_1^2 = \frac{\begin{vmatrix} (\sigma_n^2 + \tau^2) & \sigma_2^2 & \sigma_3^2 \\ \sigma_n & \sigma_2 & \sigma_3 \\ 1 & 1 & 1 \end{vmatrix}}{\begin{vmatrix} \sigma_1^2 & \sigma_2^2 & \sigma_3^2 \\ \sigma_1 & \sigma_2 & \sigma_3 \\ 1 & 1 & 1 \end{vmatrix}} \tag{3.48}$$

which may be rewritten as

$$n_1^2 = \frac{(\sigma_2 - \sigma_3)(\sigma_n^2 + \tau^2) - (\sigma_2^2 - \sigma_3^2)\sigma_n + (\sigma_2^2\sigma_3 - \sigma_3^2\sigma_2)}{(\sigma_1 - \sigma_2)(\sigma_2 - \sigma_3)(\sigma_1 - \sigma_3)} \tag{3.49}$$

$$= \frac{(\sigma_n^2 + \tau^2) - (\sigma_2 + \sigma_3)\sigma_n + \sigma_2\sigma_3}{(\sigma_1 - \sigma_2)(\sigma_1 - \sigma_3)} \tag{3.50}$$

Let us recall that, according to our convention, $\sigma_1 \geq \sigma_2 \geq \sigma_3$, so that the denominator in (3.50) is always positive.

Since n_1^2 is also always positive, it implies that the numerator in (3.50) is always positive. Let us note that the equation

$$(\sigma_n^2 + \tau^2) - (\sigma_2 + \sigma_3)\sigma_n + \sigma_2\sigma_3 = 0 \tag{3.51}$$

describes a circle passing through the points $(\sigma_3, 0)$ and $(\sigma_2, 0)$ and with radius $r = (\sigma_3 - \sigma_2)/2$, so that the condition $(\sigma_n^2 + \tau^2) - (\sigma_2 + \sigma_3)\sigma_n + \sigma_2\sigma_3 \geq 0$ implies that none of the points $(\sigma_n, |\boldsymbol{\tau}|)$ lie within this circle (in fig. 3.6 only the upper semicircle is shown).

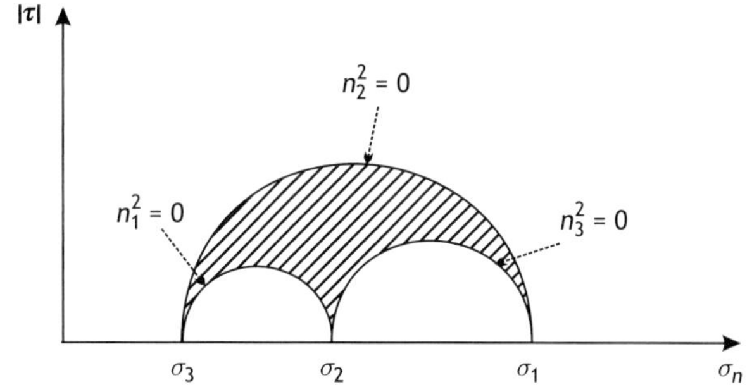

Fig. 3.7 Locus of the points $(\sigma_n, |\boldsymbol{\tau}|)$ produced by a rotation of the surface $\boldsymbol{ds}$ around the point $\boldsymbol{X}$. The diagram corresponds to positive values of $\boldsymbol{\tau}$. All points must lie within the hatched area or on any of the three semicircles.

We may solve similarly for n_2^2 and n_3^2, and we obtain

$$n_2^2 = -\frac{(\sigma_n^2 + \tau^2) - (\sigma_1 + \sigma_3)\sigma_n + \sigma_1\sigma_3}{(\sigma_1 - \sigma_2)(\sigma_2 - \sigma_3)} \tag{3.52}$$

$$n_3^2 = \frac{(\sigma_n^2 + \tau^2) - (\sigma_1 + \sigma_2)\sigma_n + \sigma_1\sigma_2}{(\sigma_2 - \sigma_3)(\sigma_1 - \sigma_3)} \tag{3.53}$$

In equation (3.52), because the denominator is always positive the pairs of values (σ_n, τ) must be such that the numerator is always negative, so that points with coordinates $(\sigma_n, |\boldsymbol{\tau}|)$ must lie either on the circle with equation $(\sigma_n^2 + \tau^2) - (\sigma_1 + \sigma_3)\sigma_n + \sigma_1\sigma_3 = 0$ or inside this circle.

From equation (3.53) we conclude that points must lie either on the circle defined by equation $(\sigma_n^2 + \tau^2) - (\sigma_1 + \sigma_2)\sigma_n + \sigma_1\sigma_2 = 0$ or outside it, so that, finally, the locus of all possible points $(\sigma_n, |\boldsymbol{\tau}|)$ must lie either on one of the circles, called *Mohr circles*, or within the shaded area between the circles (fig. 3.7).

Of particular interest are the points on the Mohr circle passing through σ_1 and σ_3. They correspond to planes such that $n_2^2 = 0$, i.e. to planes that contain $\boldsymbol{e}_2$. Thus, the surface which sustains the largest shear stress contains $\boldsymbol{e}_2$, and we show in the next section that it is oriented at an angle $\pi/4$ to either of the two other eigenvectors.

A special case occurs when $\sigma_2 = \sigma_3$. It is a frequent configuration encountered in the laboratory for testing rock under compression (chapter 8). For such conditions (called, wrongly, triaxial tests) none of the planes that contain $\boldsymbol{e}_1$ supports any shear. Finally, observe that a pressure P reduces to a single point P on the horizontal axis in the Mohr diagram.

3.4 Changing the frame of reference

As was shown in section 3.2.1, the coefficients of the matrix that characterizes the stress tensor depend on the frame of reference and it is often necessary to determine this matrix in a different frame of reference.

Let $\hat{\boldsymbol{I}}_i$ be the old frame of reference, in which the stress components σ_{ij} are known, and let $\hat{\boldsymbol{I}'}_j$ be the new frame, in which the stress tensor components are σ'_{kl}. Let Q_{ij} be the components of the axis vectors of the old frame of reference with respect to the new frame of reference: $\hat{\boldsymbol{I}}_i = Q_{ij}\hat{\boldsymbol{I}'}_j$. It is shown in the appendix that Q_{ij} are the components of an orthogonal tensor $\tilde{Q}$, so that $\tilde{Q}^T = \tilde{Q}^{-1}$ or $Q_{ij}Q_{jk} = \delta_{ik}$ where δ_{ik} is the Kronecker delta symbol.

We first determine the change of components for an arbitrary vector $\boldsymbol{u} = u_i\hat{\boldsymbol{I}}_i = u'_j\hat{\boldsymbol{I}'}_j$, where u_i are its components in the old frame while u'_j are its components in the new frame.

Expressing the components of $\hat{\boldsymbol{I}}_i$ in the new frame we obtain

$$\boldsymbol{u} = u_i\hat{\boldsymbol{I}}_i = u_iQ_{ij}\hat{\boldsymbol{I}'}_j = Q^T_{ji}u_i\hat{\boldsymbol{I}'}_j = u'_j\hat{\boldsymbol{I}'}_j \tag{3.54}$$

so that

$$u'_j = Q^T_{ji}u_i \tag{3.55}$$

Now let us consider the relation $\boldsymbol{t} = \tilde{\sigma}\boldsymbol{n}$ between the stress vector $\boldsymbol{t}$, the stress tensor $\hat{\sigma}$ and the normal $\boldsymbol{n}$ to a surface. The expressions for this relation in the old frame and in the new frame are

$$t_i = \sigma_{ij}n_j, \qquad t'_k = \sigma'_{kl}n'_l \tag{3.56}$$

Applying relation (3.55) we obtain

$$Q^T_{km}t_m = \sigma'_{kl}Q^T_{lj}n_j \tag{3.57}$$

and if we multiply from the left both terms of this equality by Q_{ik} we obtain

$$Q_{ik}Q^T_{km}t_m = Q_{ik}\sigma'_{kl}Q^T_{lj}n_j \tag{3.58}$$

Since $Q_{ik}Q^T_{km} = \delta_{im}$ and $\delta_{im}t_m = t_i$ we have

$$t_i = Q_{ik}\sigma'_{kl}Q^T_{lj}n_j \tag{3.59}$$

and therefore by analogy with (3.56)

$$\sigma_{ij} = Q_{ik}\sigma'_{kl}Q^T_{lj} \tag{3.60}$$

Finally, we obtain an important result for changing the frame of reference:

$$\sigma'_{kl} = Q^T_{ki}\sigma_{ij}Q_{jl} \tag{3.61}$$

3.4.1 Normal stress and shear stress as functions of the principal stress components

It is very often convenient to express the normal stress and shear stress magnitudes, σ_n and τ, in terms of the principal stress components for surfaces in which an eigenvector lies.

Let us consider the stress vector $\boldsymbol{t}$ supported by a plane whose normal is perpendicular to the eigenvector $\boldsymbol{e}_2$ (fig. 3.8). The corresponding σ_n and τ values describe a point that belongs to the largest Mohr circle passing through σ_1 and σ_3.

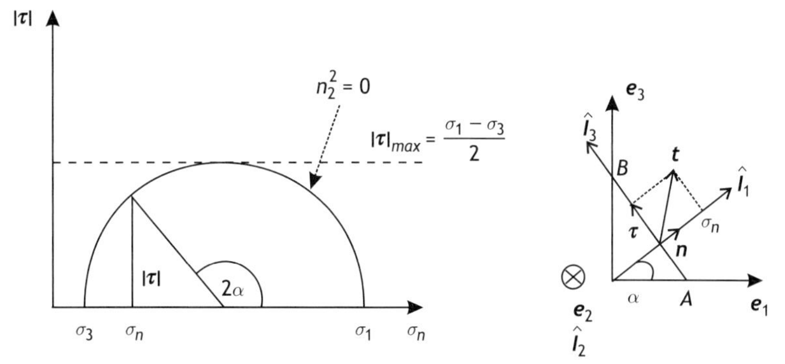

Fig. 3.8 Normal and shear stress magnitudes in terms of the principal stress components for planes whose normal is perpendicular to the $\boldsymbol{e}_2$ eigenvector.

Let AB be the projection of a surface with unit area and normal inclined by an angle α to the direction of the $\boldsymbol{e}_1$ eigenvector. Consider the direct frame of reference $\hat{\boldsymbol{I}}_i$ such that $\hat{\boldsymbol{I}}_1$ is parallel to $\boldsymbol{n}$, $\hat{\boldsymbol{I}}_2$ is parallel to $\boldsymbol{e}_2$ and $\hat{\boldsymbol{I}}_3$ is in the plane AB, oriented so that the frame $\hat{\boldsymbol{I}}_i$ is direct.

Let us observe that in the $\hat{\boldsymbol{I}}_i$ frame of reference, σ_{11} is equal to σ_n whilst $\sigma_{13} = \tau$. The components of $\boldsymbol{e}_1$ in the $\hat{\boldsymbol{I}}_i$ frame are $(\cos\alpha, 0, -\sin\alpha)$, those of $\boldsymbol{e}_2$ are $(0, 1, 0)$ and those of $\boldsymbol{e}_3$ are $(\sin\alpha, 0, \cos\alpha)$, so that the stress components σ_{ij} in the $\hat{\boldsymbol{I}}_i$ frame are given by the matrix product

$$\begin{pmatrix} \cos\alpha & 0 & \sin\alpha \\ 0 & 1 & 0 \\ -\sin\alpha & 0 & \cos\alpha \end{pmatrix} \begin{pmatrix} \sigma_1 & 0 & 0 \\ 0 & \sigma_2 & 0 \\ 0 & 0 & \sigma_3 \end{pmatrix} \begin{pmatrix} \cos\alpha & 0 & -\sin\alpha \\ 0 & 1 & 0 \\ \sin\alpha & 0 & \cos\alpha \end{pmatrix} \tag{3.62}$$

Hence we have the useful result

$$\sigma_n = \frac{\sigma_1 + \sigma_3}{2} + \frac{\sigma_1 - \sigma_3}{2}\cos 2\alpha \tag{3.63}$$

$$\tau = -\frac{\sigma_1 - \sigma_3}{2}\sin 2\alpha \tag{3.64}$$

We also observe that

$$\sigma_{11} - \sigma_{33} = (\sigma_1 - \sigma_3)\cos 2\alpha \tag{3.65}$$

so that the orientations of the principal stress directions with respect to the frame of reference $\hat{\boldsymbol{I}}_i = 1, 2, 3$, are defined by

$$\tan 2\alpha = \frac{-2\sigma_{13}}{\sigma_{11} - \sigma_{33}} \tag{3.66}$$

3.4.2 Components of the stress tensor in cylindrical and spherical coordinates

Many important problems of geomechanics involve cylindrical geometries (for example long boreholes) or spheres (such as the planet earth) and then it is necessary to consider either cylindrical or spherical coordinates.

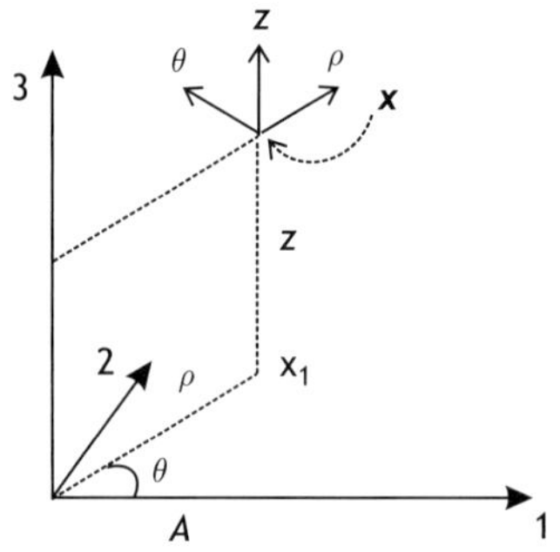

Fig. 3.9 Cylindrical coordinates ρ, θ, z for the point $\boldsymbol{x}$.

Let us consider cylindrical coordinates (fig. 3.9) defined by the three unit vectors $\hat{\boldsymbol{I}}_\rho$, $\hat{\boldsymbol{I}}_\theta$ and $\hat{\boldsymbol{I}}_z$. In the $\hat{\boldsymbol{I}}_i$ frame of reference the components of $\hat{\boldsymbol{I}}_\rho$ are $(\cos\theta, \sin\theta, 0)$, those of $\hat{\boldsymbol{I}}_\theta$ are $(-\sin\theta, \cos\theta, 0)$ and those of $\hat{\boldsymbol{I}}_z$ are $(0, 0, 1)$. Hence the components of the stress tensor in cylindrical coordinates are given by the matrix product

$$\begin{pmatrix} \cos\theta & \sin\theta & 0 \\ -\sin\theta & \cos\theta & 0 \\ 0 & 0 & 1 \end{pmatrix} \begin{pmatrix} \sigma_{11} & \sigma_{12} & \sigma_{13} \\ \sigma_{21} & \sigma_{22} & \sigma_{23} \\ \sigma_{31} & \sigma_{32} & \sigma_{33} \end{pmatrix} \begin{pmatrix} \cos\theta & -\sin\theta & 0 \\ \sin\theta & \cos\theta & 0 \\ 0 & 0 & 1 \end{pmatrix} \tag{3.67}$$

with the following results:

$$\sigma_{\rho\rho} = \sigma_{11}\cos^2\theta + \sigma_{22}\sin^2\theta + 2\sigma_{12}\cos\theta\sin\theta \tag{3.68}$$

$$\sigma_{\theta\theta} = \sigma_{11}\sin^2\theta + \sigma_{22}\cos^2\theta - 2\sigma_{12}\cos\theta\sin\theta \tag{3.69}$$

$$\sigma_{zz} = \sigma_{33} \tag{3.70}$$

$$\sigma_{\rho\theta} = \sigma_{21}(\cos^2\theta - \sin^2\theta) + (\sigma_{22} - \sigma_{11})\cos\theta\sin\theta \tag{3.71}$$

$$\sigma_{\rho z} = \sigma_{13}\cos\theta + \sigma_{23}\sin\theta \tag{3.72}$$

$$\sigma_{\theta z} = -\sigma_{13}\sin\theta + \sigma_{23}\cos\theta \tag{3.73}$$

The components of the stress tensor in spherical coordinates may be obtained following exactly the same procedure.

3.5 More definitions

Spherical and deviatoric stress components

Let $\sigma_m = \frac{1}{3}\operatorname{tr}\tilde{\sigma} = (\sigma_1 + \sigma_2 + \sigma_3)/3$. The spherical stress component $\tilde{\sigma}_s$ and the deviatoric stress component $\underset{\sim}{\tilde{\sigma}}$ of the stress tensor are defined as

$$\tilde{\sigma}_s \equiv \sigma_m \tilde{1}, \qquad \underset{\sim}{\tilde{\sigma}} \equiv \tilde{\sigma} - \sigma_m \tilde{1} \tag{3.74}$$

The spherical component is sometimes called the isotropic stress component or the hydrostatic stress component. Let us note that the trace of the deviatoric stress $\mathrm{tr}\,\tilde{\tilde{\sigma}}$ is zero. The principal directions of $\tilde{\tilde{\sigma}}$ are the same as those of $\tilde{\sigma}$ whilst the components of $\tilde{\tilde{\sigma}}$ are

$$(\tilde{\tilde{\sigma}})_{11} = \sigma_{11} - \sigma_m, \qquad (\tilde{\tilde{\sigma}})_{22} = \sigma_{22} - \sigma_m, \qquad (\tilde{\tilde{\sigma}})_{33} = \sigma_{33} - \sigma_m \tag{3.75}$$

$$(\tilde{\tilde{\sigma}})_{12} = \sigma_{12}, \qquad (\tilde{\tilde{\sigma}})_{23} = \sigma_{23}, \qquad (\tilde{\tilde{\sigma}})_{31} = \sigma_{31} \tag{3.76}$$

The invariants of the deviatoric stress are

$$J_1 = s_1 + s_2 + s_3 = 0 \tag{3.77}$$

$$J_2 = -(s_1 s_2 + s_2 s_3 + s_3 s_1) \tag{3.78}$$

$$J_3 = s_1 s_2 s_3 \tag{3.79}$$

where s_1, s_2 and s_3 are the principal components of the deviatoric stress. Through algebraic manipulation it can be shown that the second invariant may be written

$$J_2 = \frac{1}{6}[(\sigma_1 - \sigma_2)^2 + (\sigma_2 - \sigma_3)^2 + (\sigma_3 - \sigma_1)^2] \tag{3.80}$$

Octahedral stresses

The octahedral stresses are the normal and shear components of the stress vectors, which act on planes with outward unit normal vectors defined in the principal stress direction frame of reference by the direction cosines

$$(n_1,\ n_2,\ n_3) = \left(\pm\frac{1}{\sqrt{3}},\ \pm\frac{1}{\sqrt{3}},\ \pm\frac{1}{\sqrt{3}}\right) \tag{3.81}$$

Eight such planes exist; they are parallel to the faces of an octahedron whose vertices are located on the principal stress directions. For these planes the normal stress component is

$$\sigma_{oct} = \sigma_1 n_1^2 + \sigma_2 n_2^2 + \sigma_3 n_3^2 = \frac{1}{3}(\sigma_1 + \sigma_2 + \sigma_3) = \frac{I_{\tilde{\sigma}}}{3} \tag{3.82}$$

whilst the shear stress component is given by

$$|\tau_{oct}| = \frac{1}{3}[(\sigma_1 - \sigma_2)^2 + (\sigma_2 - \sigma_3)^2 + (\sigma_3 - \sigma_2)^2]^{1/2} \tag{3.83}$$

$$= \frac{\sqrt{2}}{3}(I_{\tilde{\sigma}}^2 - 3II_{\tilde{\sigma}})^{1/2} \tag{3.84}$$

$$= \left(\frac{2}{3}J_2\right)^{1/2} \tag{3.85}$$

Units for stress components

The components of the stress tensor correspond to forces per unit area, i.e. they are homogeneous to a pressure. The SI unit of pressure is the pascal (Pa); it corresponds to a force of one newton acting on a one square meter area.

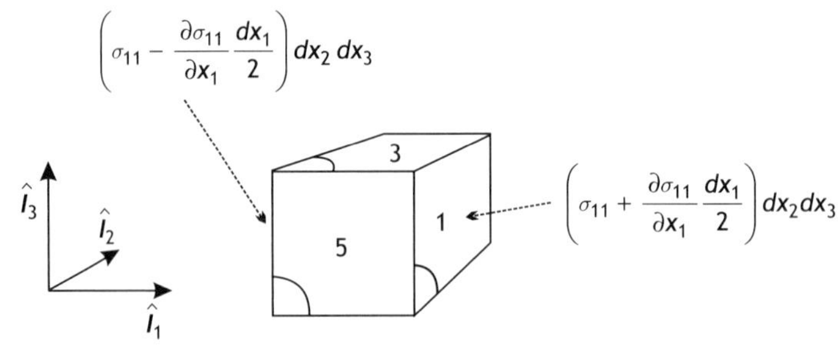

Fig. 3.10 Equilibrium conditions in cartesian coordinates.

The atmospheric pressure at ground surface is close to 10^5 pascals, i.e. one bar (b). A very common unit of stress is the megapascal (MPa), which is equal to 10 bars, or 10^6 pascals.

In the Imperial System a pressure is measured in pounds per square inch (psi), with 1 psi $\simeq$ 0.0069 MPa, or 1 Kb = 100 MPa $\simeq$ 14 500 psi.

Isobars are domains for which the principal stresses are constant. Isopacs are domains with uniform spherical stress. Because of past conventions in photoelasticity, domains where the deviatoric stress is constant are known as isochromatics.

Stress trajectories, sometimes called isostatics, form a set of orthogonal curves that are tangent at every point to a principal stress direction.

3.6 Equilibrium conditions

A system of forces applied to the system under consideration results in a displacement field, and it is precisely the object of mechanics to study the relations between forces and displacements. The law of motion must be formulated for this purpose and we address this issue in chapter 6; we discuss displacement fields in the next chapter.

The stresses that exist today in geological formations have become a topic of interest *per se* (chapter 13) even without a consideration of the processes that have led to the present situation. Geological materials are at rest at the scale of the study, and it is necessary to verify that the stress measurements that have been obtained are consistent with these stability conditions.

We determine here the conditions that must be satisfied by the stress field in any small volume so that the system under study is at rest. We consider as the body forces only vertical gravity forces.

3.6.1 Cartesian coordinates

Let us consider the stress tensor $\tilde{\sigma}$ at a point $\boldsymbol{X}$ with coordinates x_1, x_2 and x_3 and a small volume $dx_1 dx_2 dx_3$ centered on $\boldsymbol{X}$ (fig. 3.10). The stress field varies in space and we consider its first-order variations.

For the small volume to be at rest, the resultant $\boldsymbol{R}$ of the surface and body forces must be zero. The faces of the small volume are numbered from 1 to 6, with 1 and 4 normal to the $\hat{\boldsymbol{I}}_1$ direction, 2 and 5 normal to the $\hat{\boldsymbol{I}}_2$ direction, 3 and 6 normal to the $\hat{\boldsymbol{I}}_3$ direction. Faces 1, 2 and 3 are respectively the faces with their normals in the same sense as $\hat{\boldsymbol{I}}_1$, $\hat{\boldsymbol{I}}_2$ and $\hat{\boldsymbol{I}}_3$(fig. 3.10). On face 1 the components of the surface force dt_1 are, using a Taylor expansion,

$$dt_{11} = \left(\sigma_{11} + \frac{\partial \sigma_{11}}{\partial x_1}\frac{dx_1}{2}\right) dx_2 dx_3, \tag{3.86}$$

$$dt_{12} = \left(\sigma_{12} + \frac{\partial \sigma_{12}}{\partial x_1}\frac{dx_1}{2}\right) dx_2 dx_3, \tag{3.87}$$

$$dt_{13} = \left(\sigma_{13} + \frac{\partial \sigma_{13}}{\partial x_1}\frac{dx_1}{2}\right) dx_2 dx_3, \tag{3.88}$$

while on face 4 the components of the surface force $\boldsymbol{dt}_4$ are

$$dt_{41} = \left(\sigma_{11} - \frac{\partial \sigma_{11}}{\partial x_1}\frac{dx_1}{2}\right) dx_2 dx_3, \tag{3.89}$$

$$dt_{42} = \left(\sigma_{12} - \frac{\partial \sigma_{12}}{\partial x_1}\frac{dx_1}{2}\right) dx_2 dx_3, \tag{3.90}$$

$$dt_{43} = \left(\sigma_{13} - \frac{\partial \sigma_{13}}{\partial x_1}\frac{dx_1}{2}\right) dx_2 dx_3, \tag{3.91}$$

with similar relations for the pair of faces 2 and 5 and for the pair 3 and 6.

Let us consider the component R_1 of $\boldsymbol{R}$ in the $\hat{\boldsymbol{I}}_1$ direction. It must be null, so that

$$\begin{aligned} R_1 = &-\left(\sigma_{11} - \frac{\partial \sigma_{11}}{\partial x_1}\frac{dx_1}{2}\right) dx_2 dx_3 + \left(\sigma_{11} + \frac{\partial \sigma_{11}}{\partial x_1}\frac{dx_1}{2}\right) dx_2 dx_3 \\ &-\left(\sigma_{21} - \frac{\partial \sigma_{21}}{\partial x_2}\frac{dx_2}{2}\right) dx_1 dx_3 + \left(\sigma_{21} + \frac{\partial \sigma_{21}}{\partial x_2}\frac{dx_2}{2}\right) dx_1 dx_3 \\ &-\left(\sigma_{31} - \frac{\partial \sigma_{31}}{\partial x_3}\frac{dx_3}{2}\right) dx_1 dx_2 + \left(\sigma_{31} + \frac{\partial \sigma_{31}}{\partial x_3}\frac{dx_3}{2}\right) dx_1 dx_2 \\ = &\ 0, \end{aligned} \tag{3.92}$$

which reduces to

$$R_1 = \frac{\partial \sigma_{11}}{\partial x_1} + \frac{\partial \sigma_{21}}{\partial x_2} + \frac{\partial \sigma_{31}}{\partial x_3} = 0 \tag{3.93}$$

We obtain a similar equation for the component R_2 in the $\hat{\boldsymbol{I}}_2$ direction, but for the $\hat{\boldsymbol{I}}_3$ direction we must include the component of the body force so that, given that the stress tensor is symmetrical, the equilibrium conditions may be written in vectorial form:

$$\nabla \cdot \tilde{\sigma} - \varrho \boldsymbol{b} = 0 \tag{3.94}$$

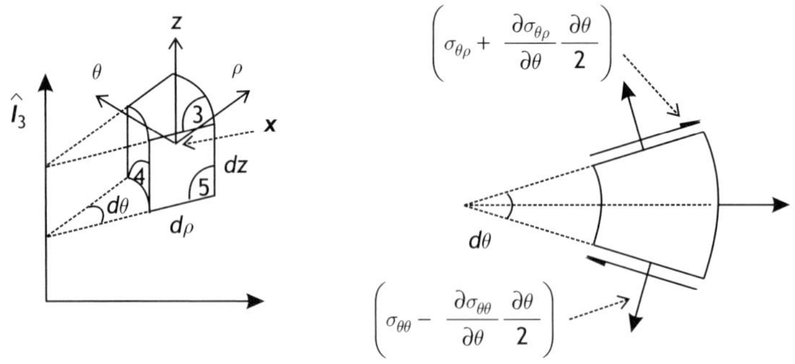

Fig. 3.11 Equilibrium conditions in cylindrical coordinates.

3.6.2 Cylindrical and spherical coordinates

When we consider a small elementary volume appropriate for writing down the equilibrium conditions in cylindrical coordinates, the areas of the different faces are not the same, hence slightly more lengthy equations in terms of the coordinates are encountered; however, in vectorial form the equation remains the same.

Let us consider the small elementary volume $dv = \rho d\rho d\theta dz$, centered on a point $\boldsymbol{X}$ with cylindrical coordinates (ρ, θ, z), as shown in fig. 3.11.

The faces of this elementary volume are numbered from 1 to 6. The area of face 1 is $(\rho + d\rho/2)\, d\theta dz$, while that of face 4 is $(\rho - d\rho/2)\, d\theta dz$. The area of faces 2 and 5 is $d\rho dz$, while that of faces 3 and 6 is $\rho\, d\theta d\rho$.

The components of the surface force $\boldsymbol{dt}_1$, on face 1, normal to the ρ axis, are

$$dt_{1\rho} = \left(\sigma_{\rho\rho} + \frac{\partial\sigma_{\rho\rho}}{\partial\rho}\frac{d\rho}{2}\right)\left(\rho + \frac{d\rho}{2}\right) d\theta dz \tag{3.95}$$

$$dt_{1\theta} = \left(\sigma_{\rho\theta} + \frac{\partial\sigma_{\rho\theta}}{\partial\rho}\frac{d\rho}{2}\right)\left(\rho + \frac{d\rho}{2}\right) d\theta dz \tag{3.96}$$

$$dt_{1z} = \left(\sigma_{\rho z} + \frac{\partial\sigma_{\rho z}}{\partial\rho}\frac{d\rho}{2}\right)\left(\rho + \frac{d\rho}{2}\right) d\theta dz \tag{3.97}$$

while for face 4, also normal to the ρ axis, the components of the surface force $\boldsymbol{dt}_4$ are

$$dt_{4\rho} = \left(\sigma_{\rho\rho} - \frac{\partial\sigma_{\rho\rho}}{\partial\rho}\frac{d\rho}{2}\right)\left(\rho - \frac{d\rho}{2}\right) d\theta dz \tag{3.98}$$

$$dt_{4\theta} = \left(\sigma_{\rho\theta} - \frac{\partial\sigma_{\rho\theta}}{\partial\rho}\frac{d\rho}{2}\right)\left(\rho - \frac{d\rho}{2}\right) d\theta dz \tag{3.99}$$

$$dt_{4z} = \left(\sigma_{\rho z} - \frac{\partial\sigma_{\rho z}}{\partial\rho}\frac{d\rho}{2}\right)\left(\rho - \frac{d\rho}{2}\right) d\theta dz \tag{3.100}$$

Similar relations are written for the other faces, except that the face areas must be set to the appropriate values.

Let us note that neither of the faces 2 and 5 is parallel to the ρ axis, so that the ρ component of the force $\boldsymbol{F}_s$ resulting from the surface forces acting on dv must include projections of the $dt_{i\rho}$ and $dt_{i\theta}$ components ($i = 2$ or 5) on to the ρ axis. Further, for dt

it may be observed that while the components $dt_{2\rho}$ and $dt_{5\rho}$ have different signs when projected onto the ρ axis, the components $dt_{2\theta}$ and $dt_{5\theta}$ are both negative when projected onto this axis.

Hence, the ρ component of the resultant surface force $\boldsymbol{F}_s$ is

$$\begin{aligned} F_{s\rho} = {} & dt_{1\rho} - dt_{4\rho} \\ & + dt_{3\rho} - dt_{6\rho} \\ & + (dt_{2\rho} - dt_{5\rho}) \cos \frac{d\theta}{2} \\ & + (-dt_{2\theta} - dt_{5\theta}) \sin \frac{d\theta}{2} \end{aligned} \tag{3.101}$$

However, $d\theta$ is small, so $\cos d\theta/2 \simeq 1$ and $\sin d\theta/2 \simeq d\theta/2$.

Hence, finally, the ρ component of the resultant is

$$F_{s\rho} = \frac{\partial \sigma_{\rho\rho}}{\partial \rho} + \frac{1}{\rho}\frac{\partial \sigma_{\rho\theta}}{\partial \theta} + \frac{\partial \sigma_{\rho z}}{\partial z} + \frac{\sigma_{\rho\rho} - \sigma_{\theta\theta}}{\rho} \tag{3.102}$$

We may proceed in exactly the same manner for the θ component and the z component of the resultant surface force. For equilibrium, we simply require that the resultant of the surface and body forces is zero.

For simplicity, it is common to change the notation for partial derivatives so that "$\partial/\partial x$" is simply written as the subscript "$,x$". Thus $\partial y/\partial x$ is written as $y_{,x}$. Adopting this new notation convention from now on, we will write down the equilibrium conditions in cylindrical coordinates. If b_ρ, b_θ, b_z are the components of the body force $\boldsymbol{b}$ in the corresponding frame of reference (the axis z may not be vertical) and ϱ is the material density, the equilibrium conditions are

$$\sigma_{\rho\rho,\rho} + \frac{1}{\rho}\sigma_{\theta\rho,\theta} + \sigma_{z\rho,z} + \frac{\sigma_{\rho\rho} - \sigma_{\theta\theta}}{\rho} - \varrho b_\rho = 0 \tag{3.103}$$

$$\sigma_{\rho\theta,\rho} + \frac{1}{\rho}\sigma_{\theta\theta,\theta} + \sigma_{z\theta,z} + \frac{2}{\rho}\sigma_{\rho\theta} - \varrho b_\theta = 0 \tag{3.104}$$

$$\sigma_{\rho z,\rho} + \frac{1}{\rho}\sigma_{\theta z,\theta} + \sigma_{zz,z} + \frac{1}{\rho}\sigma_{z\rho} - \varrho b_z = 0 \tag{3.105}$$

A similar approach may be followed when working in spherical coordinates (fig. 3.12). The equilibrium conditions become

$$\sigma_{\rho\rho,\rho} + \frac{1}{\rho}\sigma_{\rho\theta,\theta} + \frac{1}{\rho \sin\theta}\sigma_{\rho\phi,\phi} + \frac{1}{\rho}(2\sigma_{\rho\rho} - \sigma_{\theta\theta} - \sigma_{\phi\phi} + \sigma_{\rho\theta}\cot\theta) = 0 \tag{3.106}$$

$$\sigma_{\rho\theta,\rho} + \frac{1}{\rho}\sigma_{\theta\theta,\theta} + \frac{1}{\rho \sin\theta}\sigma_{\theta\phi,\phi} + \frac{1}{\rho}[3\sigma_{\rho\theta} + (\sigma_{\theta\theta} - \sigma_{\phi\phi})\cot\theta] = 0 \tag{3.107}$$

$$\sigma_{\rho\phi,\rho} + \frac{1}{\rho}\sigma_{\theta\phi,\theta} + \frac{1}{\rho \sin\theta}\sigma_{\phi\phi,\phi} + \frac{1}{\rho}(3\sigma_{\rho\phi} + 2\sigma_{\theta\phi}\cot\theta) = 0 \tag{3.108}$$

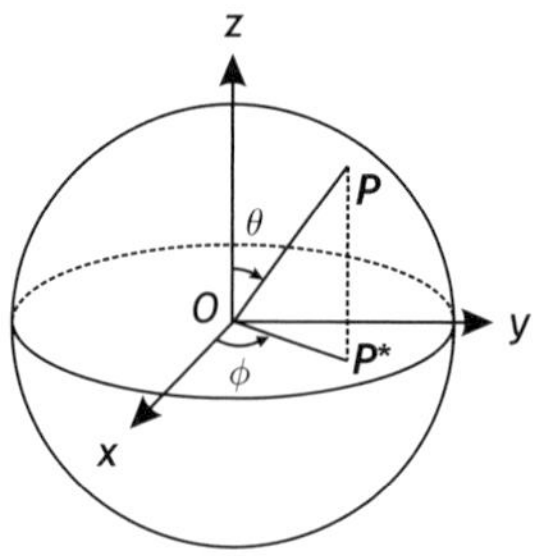

Fig. 3.12 Spherical coordinates.

3.7 Exercises

1. Practising index notation: write out explicitly the following expression:

$$t_i = T_{ij} n_j, \qquad i, j = 1, 2, 3$$

Show that $\delta_{ii} = 3$. Write out explicitly $\det \tilde{T} = \varepsilon_{ijk} T_{i1} T_{j2} T_{k3}$ and demonstrate that $\varepsilon_{rst} \det \tilde{T} = \varepsilon_{ijk} T_{ir} T_{js} T_{kt}$.

2. Consider the three unit vectors that make up a cartesian frame of reference. Write their components in a cylindrical frame of reference and in a spherical frame of reference. Find the coordinates of the stress tensor in a spherical frame of reference when the components T_{ij}; $i, j = 1, 2, 3$, in a cartesian frame of reference are known.

3. Consider an infinite half space with its horizontal surface free of stress and its vertical axis $\hat{\boldsymbol{I}}_3$ oriented positive downward. In this half space there exists a stress field characterized by a tensor that varies linearly with depth according to the equation

$$\tilde{\sigma} = \tilde{\sigma}^0 + (x_3 - x_3^0)\tilde{\alpha} \tag{3.109}$$

where $\tilde{\sigma}^0$ is the stress at depth x^0 whilst $\tilde{\alpha}$ is the vertical stress gradient. Compressions are reckoned as positive. The half space is filled with a material with density $\varrho = 2.5$ g/cm^3.

3.1. Given that this half space is under equilibrium, show that the vertical is a principal direction for both $\tilde{\alpha}$ and $\tilde{\sigma}$.

3.2. What is the vertical component α_{33}, what are the vertical components of the stress gradient and of $\tilde{\sigma}^0$, i.e. σ_{33}^0?

3.3. For the stress tensor $\tilde{\sigma}^0$ at a depth of 200 m the horizontal maximum and minimum principal components are respectively equal to 3 and 2 MPa and the orientation of the maximum horizontal principal stress direction is N30°E. The horizontal maximum and minimum principal components of the vertical stress gradient $\tilde{\alpha}$ are respectively equal to 0.022 and 0.020 MPa/m and the orientation of the eigenvector associated with the maximum horizontal principal stress gradient is N0°E.

Determine the components of the stress vector supported by a plane P oriented N45°E and dipping 45° to the south, for depths equal to 200 m and 5000 m.

3.4. Plot the Mohr circles for the stress tensors at 200 m and at 5000 m. On the same diagram plot the stress vectors supported by the plane P at both these depths.

3.5. We define an *effective stress* by the relation $\tilde{\sigma}' = \tilde{\sigma} - P\tilde{1}$, with $P = \varrho_f g x_3$ where $\varrho_f = 1$ g/cm^3 and $g = 9.81$ m/s^2. Plot the Mohr circles for the effective stress $\tilde{\sigma}^0$. Plot the *effective* resolved shear stress supported by the plane P at 200 m.

3.6. What is the difference in orientation of the resolved shear stress components supported by plane P at 200 m and at 5000 m?

4. Consider an empty cylindrical hole in a solid massif. What are the three components (in cylindrical coordinates) of the stress vector that exists in the solid at the wall of the cylinder when a uniform pressure P is applied to the walls of the cylinder?

4 Elements of kinematics

With the development of satellite geodesy, whether through global positioning systems (GPS) or short aperture radar interferometry (InSAR), the accuracy of displacement measurements at the scale of hundreds, if not thousands, of kilometers has reached the millimeter scale. This renders possible the direct measurement of geological deformations at time scales reaching merely a few years.

In some cases (like slope stability configurations, for example), such displacements may not be small. In fact, when the displacements become large, as in fluid mechanics problems, particle displacements are generally not considered any longer; our attention turns to particle velocities.

We introduce in this chapter some elements of kinematics in order to describe the motion of particles. Because geomaterials involve not only solids but also fluids we follow the classical approach of continuum mechanics, as presented for example by Malvern (1969), for discussing the deformation and strain of geomaterials and their variation with time. The reader is expected to be familiar with the elements of linear algebra as given in the appendix, for we shall apply the index notation defined there. This notation convention is readily adapted to straightforward numerical programming.

Our objective is to introduce general concepts without any hypothesis on the relative scale of displacements. These concepts are then applied to the conditions of small displacement gradients that hold in many solid mechanics problems. After that the definitions of the rate of deformation, the strain rate and the spin are given.

First, however, elementary two-dimensional examples of deformation are introduced in order to familiarize the reader with the classical small-strain concept.

4.1 Two-dimensional elementary definitions of strain

Let us consider a small rectangular surface element *ABCD* with sides of length ΔX_1 and ΔX_2 (fig. 4.1).

Let side *BC* be pulled to the right by a uniform displacement Δu_1 parallel to *AB* while side *AD* is kept fixed. Thus the surface element is stretched by a uniaxial extension process in direction x_1. This extension induces a uniaxial strain $\varepsilon_{11} = \Delta u_1/\Delta x_1$. A uniform displacement Δu_2 in the x_2 direction, while *AB* is kept fixed, yields a uniaxial strain $\varepsilon_{22} = \Delta u_2/\Delta x_2$.

However, if the displacement Δu_1 is applied at point *D* whilst the displacement Δu_2 is applied at point *B*, so that the rectangle *ABCD* becomes a quadrilateral $AB^1C^1D^1$, the

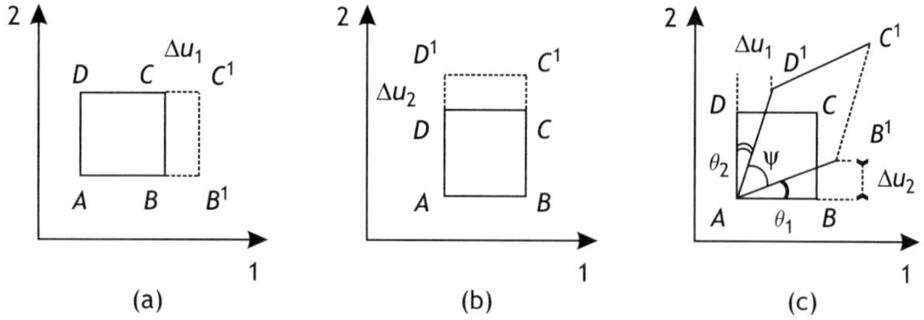

Fig. 4.1 (a), (b) Uniaxial extension in the 1 or the 2 direction, respectively; (c) shear strain.

displacements give rise to a rotation θ_1 of side AB and a rotation θ_2 of side AD, with $\theta_1 \simeq \Delta u_2/\Delta x_1$ and $\theta_2 \simeq \Delta u_1/\Delta x_2$ if Δu_1 and Δu_2 are small compared with Δx_1 and Δx_2. The shear strain is defined as

$$\gamma_{12} \equiv \frac{\pi}{2} - \psi = \theta_1 + \theta_2 \tag{4.1}$$

where ψ is the angle between AB^1 and AD^1 after the displacements have been applied. When $\theta_1 = \theta_2$ the strain is a pure shear (i.e. there is no rotation). When $\Delta u_2/\Delta x_1 = \theta_1 \neq 0$ and $\Delta u_1/\Delta x_1 = \Delta u_1/\Delta x_2 = 0$, the strain is said to be a simple shear.

We define

$$\varepsilon_{12} \equiv \frac{1}{2}\gamma_{12} = \frac{1}{2}\left(\frac{\Delta u_1}{\Delta x_2} + \frac{\Delta u_2}{\Delta x_1}\right) = \varepsilon_{21} \tag{4.2}$$

The array of coefficients $\begin{pmatrix} \varepsilon_{11} & \varepsilon_{12} \\ \varepsilon_{21} & \varepsilon_{22} \end{pmatrix}$ is the matrix of the two-dimensional small-strain tensor. Note that the coefficients be simply given as

$$\varepsilon_{ij} = \frac{1}{2}\left(\frac{\Delta u_i}{\Delta x_j} + \frac{\Delta u_j}{\Delta x_i}\right), \quad i, j = 1, 2 \tag{4.3}$$

It will be shown in section 4.3.4 that the three-dimensional version of the small-strain tensor is simply defined by

$$\varepsilon_{ij} = \frac{1}{2}\left(\frac{\partial u_i}{\partial x_j} + \frac{\partial u_j}{\partial x_i}\right), \qquad i, j = 1, 2, 3,$$

where the three coordinates of displacement $u_i; i = 1, 2, 3$, are functions of the coordinates x_i of the point where displacement occurs. This tensor $\tilde{\varepsilon}$ enables relative changes in length of vectors with any orientation to be determined, as will be shown in section 4.3.4.

More generally, we will separate the so-called rigid body motion and the strain, the superposition of which results in the deformation field as further discussed in the following sections.

Sign convention

It is customary in continuum mechanics to consider an extension as positive, i.e. a length increase or stretching is taken as positive. If, however, compressions are reckoned positive, as is the case in many rock mechanics problems, then shortening is reckoned as a positive strain. The positive direction of rotation is often chosen as being counterclockwise, i.e. the classical trigonometric convention applies, but that changes when compression is reckoned positive as discussed in the next chapter.

4.2 Lagrangian and Eulerian frames of reference; material time derivative

We define a particle as a point with a mass. In this analysis we do not consider the atomic structure of the material. The particle is the physical representation of the mathematical point. Any small volume dv of a given material is assimilated to a set of particles such that the mean properties that characterize the small volume dv, for example the material density, remain the same as the volume shrinks to zero, i.e. to a point. The critical volume size below which the theory fails to represent the deformation process properly depends on the material and is defined rigorously by the homogenization theory discussed at the start of chapter 1.

All the functions that describe the material in the volume dv are supposed to be continuous, except possibly at a finite number of interior surfaces. Their derivatives exist and are continuous. This defines a continuum material, referred to more simply as a continuum.

We discuss in chapter 7 how fractures correspond to discontinuities in the displacement field and the consequences for the local stress distributions.

A continuum is *homogeneous* when its properties are the same at all points. A material is *isotropic* with respect to a given property when at any given point the property is independent of direction.

Two different kinds of frame of reference are commonly used in continuum mechanics to map continuum materials: the Lagrangian frame of reference, also known as the *material description*, and the Eulerian frame of reference, also known as the *spatial description*.

Now consider a body B made up of particles p and a subset of these particles that makes up part P of B.

Let $\boldsymbol{B}_\kappa$ be the position occupied by the body B at time $t = 0$, $\boldsymbol{P}_\kappa$ the position of part P of B and $\boldsymbol{X}$ the position of particle p (fig. 4.2). Let $\boldsymbol{B}_\chi$ be the position occupied by the body B at time t, $\boldsymbol{P}_\chi$ the position of part P and $\boldsymbol{x}$ the position of particle p (fig. 4.2) also at time t.

We define a function χ that maps the position $\boldsymbol{x}$ of any particle p at time t as a function of its position at time $t = 0$; thus $\boldsymbol{x} = \chi(\boldsymbol{X}, t)$. This is the Lagrangian representation of the body B; the position of each particle at time t is described as a function of its position at time $t = 0$. In many cases $t = 0$ is chosen to correspond to a time when the loading process or the motion under consideration has not yet occurred. The Lagrangian representation is generally used in solid mechanics, where the paths of particles are of interest.

In the Eulerian (spatial) description the independent variable is the present position $\boldsymbol{x}$ of a particle p. This is the description most commonly used in fluid mechanics. As time passes, different particles occupy point $\boldsymbol{x}$, and the objective is to describe the motion of the

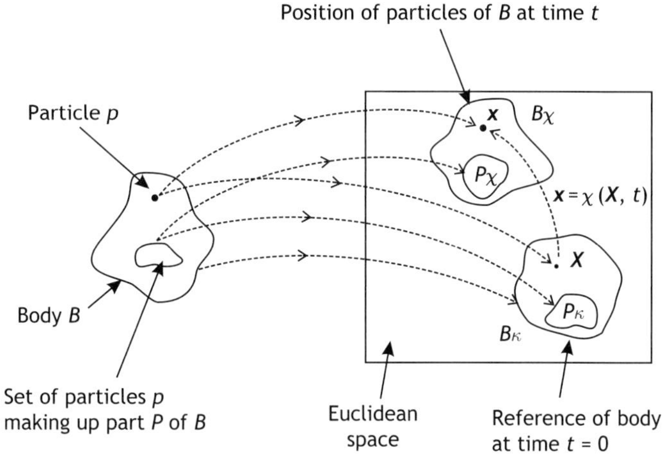

Fig. 4.2 Lagrangian and Eulerian descriptions.

particles at point $\boldsymbol{x}$ and in its neighborhood. The position at time $t = 0$ of the particle that occupies point $\boldsymbol{x}$ at time t is then $\boldsymbol{X} = \boldsymbol{\chi}^{-1}(\boldsymbol{x}, t)$.

The *material time derivative* refers to the time derivative with the material coordinates $\boldsymbol{X}$ held constant. For example, in the Lagrangian frame of reference the velocity of particle p at $\boldsymbol{x}$ is $\boldsymbol{v} = \dot{\boldsymbol{\chi}}(\boldsymbol{X}, t) = \boldsymbol{v}(\boldsymbol{X}, t) = \partial \boldsymbol{\chi}(\boldsymbol{X}, t)/\partial t$ and the acceleration is simply $\boldsymbol{a} = \dot{\boldsymbol{v}}(\boldsymbol{X}, t) = \ddot{\boldsymbol{\chi}}(\boldsymbol{X}, t) = \boldsymbol{a}(\boldsymbol{X}, t) = \partial^2 \boldsymbol{\chi}(\boldsymbol{X}, t)/\partial t^2$.

In spatial (Eulerian) coordinates the time derivative of the velocity at $\boldsymbol{x}$ is simply the local rate of change of velocity at $\boldsymbol{x}$ for the various particles that pass through $\boldsymbol{x}$ during time dt. Yet the particle acceleration is often of interest. It may be determined by applying the chain rule of differentiation if the gradient of the particle velocity at $\boldsymbol{x}$ is known:

$$\boldsymbol{v} \equiv \hat{\boldsymbol{v}}(\boldsymbol{x}, t) = \hat{\boldsymbol{v}}\left(\boldsymbol{\chi}(\boldsymbol{X}, t), t\right) = \bar{\boldsymbol{v}}(\boldsymbol{X}, t) \tag{4.4}$$

so that the particle acceleration (the material time derivative of the velocity) is

$$\boldsymbol{a}(\boldsymbol{X}, t) = \left.\frac{\partial \bar{\boldsymbol{v}}}{\partial t}\right|_{\boldsymbol{X}} = \left.\frac{\partial \hat{\boldsymbol{v}}}{\partial t}\right|_{\boldsymbol{x}} + \frac{\partial \hat{\boldsymbol{v}}}{\partial \boldsymbol{x}} \frac{\partial \boldsymbol{x}}{\partial t} \tag{4.5}$$

or, in terms of components,

$$\left.\frac{\partial \bar{v}_i}{\partial t}\right|_{\boldsymbol{X}} = \left.\frac{\partial \hat{v}_i}{\partial t}\right|_{\boldsymbol{x}} + \left.\frac{\partial \hat{v}_i}{\partial x_j} \frac{\partial x_j}{\partial t}\right|_{\boldsymbol{X}} \tag{4.6}$$

For simplicity, we write

$$\left.\frac{\partial \bar{v}_i}{\partial t}\right|_{\boldsymbol{X}} = \frac{dv_i}{dt} = \dot{v}_i$$

and

$$\left.\frac{\partial \hat{v}_i}{\partial t}\right|_{\boldsymbol{x}} = \frac{\partial v_i}{\partial t},$$

so that the components of acceleration of a particle at point $\boldsymbol{x}$ are

$$a_i = \frac{dv_i}{dt} = \dot{v}_i = \frac{\partial v_i}{\partial t} + \frac{\partial v_i}{\partial x_j} v_j \tag{4.7}$$

or, in vector form,

$$\boldsymbol{a} = \frac{d\boldsymbol{v}}{dt} \equiv \dot{\boldsymbol{v}} = \frac{\partial \boldsymbol{v}}{\partial t} + (\text{grad}\, \boldsymbol{v})\, \boldsymbol{v} \tag{4.8}$$

where the notation grad $\boldsymbol{v}$ will be taken to refer to the gradient of velocity with respect to $\boldsymbol{x}$, in the Eulerian frame of reference, while the notation $\nabla \boldsymbol{v}$ will be taken to refer to the gradient of velocity with respect to $\boldsymbol{X}$, in the Lagrangian frame of reference.

More generally, for any function, if $\hat{\boldsymbol{\psi}}$ is the expression for the function in the Eulerian frame of reference (i.e. as a function of $\boldsymbol{x}$ and t), and if $\bar{\psi}$ is its expression in the Lagrangian frame of reference (i.e. as a function of $\boldsymbol{X}$ and t) the following relation is satisfied:

$$\dot{\boldsymbol{\psi}} = \frac{\partial \hat{\boldsymbol{\psi}}}{\partial t} + (\text{grad}\, \hat{\boldsymbol{\psi}}) \boldsymbol{v} \tag{4.9}$$

As an example, let us consider the motion of a point for which the function $\boldsymbol{x} = \boldsymbol{\chi}(\boldsymbol{X}, t)$ that maps the configuration of body B at time t with respect to its position at time $t = 0$ is defined by $x_1 = X_1(1 + t)$, $x_2 = X_2(1 + t)^2, x_3 = X_3(1 + t^2)$; here t is dimensionless.

The components of the velocity in the Lagrangian frame of reference are $\bar{v}_1 = X_1$, $\bar{v}_2 = 2X_2(1 + t)$, $\bar{v}_3 = 2X_3 t$, while those of the acceleration are $\bar{a}_1 = 0$, $\bar{a}_2 = 2X_2$, $\bar{a}_3 = 2X_3$.

In the spatial (Eulerian) frame of reference, the coordinates of velocity are

$$\hat{v}_1 = \frac{x_1}{1+t}, \qquad \hat{v}_2 = \frac{2x_2}{1+t}, \qquad \hat{v}_3 = \frac{2x_3 t}{1+t^2}. \tag{4.10}$$

The gradient of the velocity in the Eulerian frame of reference is given by

$$\begin{pmatrix} 1/(1+t) & 0 & 0 \\ 0 & 2/(1+t) & 0 \\ 0 & 0 & 2t/(1+t^2) \end{pmatrix} \tag{4.11}$$

while the components of the time derivative of the velocity are

$$\frac{\partial \hat{v}_1}{\partial t} = -\frac{x_1}{(1+t)^2}, \qquad \frac{\partial \hat{v}_2}{\partial t} = -\frac{2x_2}{(1+t)^2}, \qquad \frac{\partial \hat{v}_3}{\partial t} = \frac{2x_3(1-t^2)}{(1+t^2)^2} \tag{4.12}$$

So, finally, the coordinates of the particle acceleration are obtained from equation (4.9): $\hat{a}_1 = 0$, $\hat{a}_2 = 2x_2/(1+t)^2$, $\hat{a}_3 = 2x_3/(1+t^2)$, which, once expressed in terms of X_1, X_2 and X_3, are precisely the same values as those already obtained in the Lagrangian frame of reference.

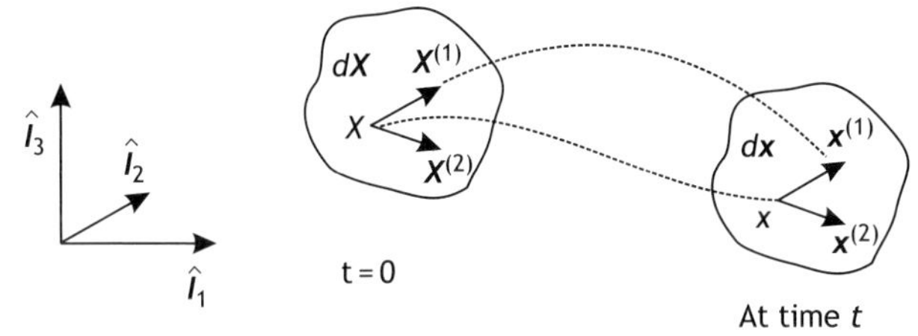

Fig. 4.3 Gradient of deformation.

4.3 Deformation and strain

4.3.1 Deformation gradient, displacement gradient

Let us consider a body in a Lagrangian frame of reference with reference orthogonal unit vectors $\hat{\boldsymbol{I}}_i$. At time $t = 0$ a particle p occupies point $\boldsymbol{X}$ and at time t it occupies point $\boldsymbol{x}$, such that $\boldsymbol{x} = \boldsymbol{\chi}(\boldsymbol{X}, t)$. Particle $p^{(1)}$ occupies $\boldsymbol{X}^{(1)}$ at time $t = 0$ and $\boldsymbol{x}^{(1)}$ at time t, so that $\boldsymbol{x}^{(1)} = \boldsymbol{\chi}(\boldsymbol{X}^{(1)}, t)$ (fig. 4.3). The displacement of particle p at time t is simply $\boldsymbol{u} = \boldsymbol{x} - \boldsymbol{X}$ while that of particle $p^{(1)}$ is $\boldsymbol{u}^{(1)} = \boldsymbol{x}^{(1)} - \boldsymbol{X}^{(1)}$. The vector $d\boldsymbol{X}^{(1)} = \boldsymbol{X}^{(1)} - \boldsymbol{X}$ that joins the two particles p and $p^{(1)}$ at time $t = 0$ becomes $d\boldsymbol{x}^{(1)} = \boldsymbol{x}^{(1)} - \boldsymbol{x}$ at time t. Similarly, we may define for a particle $p^{(2)}$ the vectors $d\boldsymbol{X}^{(2)}$ and $d\boldsymbol{x}^{(2)}$.

A motion is said to be *rigid* if the scalar product of any two displacement vectors remains constant at any time t:

$$d\boldsymbol{X}^{(1)} \cdot d\boldsymbol{X}^{(2)} = d\boldsymbol{x}^{(1)} \cdot d\boldsymbol{x}^{(2)}, \qquad \forall\, d\boldsymbol{X}^{(1)}, \forall\, d\boldsymbol{X}^{(2)}, \forall\, t \tag{4.13}$$

Rigid motion preserves distances and angles. It may be shown that rigid motion includes translations, rotations or both together.

In the remaining part of the chapter we will consider nonrigid motions, i.e. motions during which changes of distances and changes of angles do occur.

Let us assume now that particle $p^{(1)}$ is very close to particle p, so that the point $\boldsymbol{x}^{(1)}$ is very close to point $\boldsymbol{x}$ and its coordinates may be obtained from those of $\boldsymbol{x}$ by a first-order Taylor expansion:

$$\boldsymbol{x}^{(1)} = \boldsymbol{\chi}(\boldsymbol{X}, t) + \nabla\boldsymbol{\chi}(\boldsymbol{X}, t)\, d\boldsymbol{X} + \boldsymbol{\varepsilon} d\boldsymbol{X} \tag{4.14}$$

where $\boldsymbol{\varepsilon}(d\boldsymbol{X})$, contains terms in $d\boldsymbol{X}$, with exponents larger than 1, that are negligible compared with the other terms of the right-hand side expression in equation (4.14). Then we obtain a simple relation between $d\boldsymbol{x}$ and $d\boldsymbol{X}$:

$$d\boldsymbol{x} = \boldsymbol{x}^{(1)} - \boldsymbol{x} = \boldsymbol{\chi}(\boldsymbol{X}, t) + \nabla\boldsymbol{\chi}(\boldsymbol{X}, t)\, d\boldsymbol{X} + \boldsymbol{\varepsilon}(d\boldsymbol{X}) - \boldsymbol{\chi}(\boldsymbol{X}, t) \tag{4.15}$$

$$\simeq \nabla\boldsymbol{\chi}(\boldsymbol{X}, t)\, d\boldsymbol{X} \equiv \tilde{F}\, d\boldsymbol{X} \tag{4.16}$$

We define the tensor $\tilde{F} \equiv \nabla\boldsymbol{\chi}(\boldsymbol{X}, t)$ as the deformation gradient tensor. Its components are $F_{ij} = \chi_{i,j}$. It operates in the immediate vicinity of $\boldsymbol{X}$ (i.e. $d\boldsymbol{X}$ is small) but no assumption is made on the amplitude of the displacement $\boldsymbol{u} = \boldsymbol{x} - \boldsymbol{X}$.

Let us call $\tilde{H}$ the displacement gradient: $\tilde{H} = \nabla \boldsymbol{u}$. It is a second-order tensor with components $H_{ij} = u_{i,j}$. Since $\boldsymbol{u} = \boldsymbol{x} - \boldsymbol{X} = \boldsymbol{\chi}(\boldsymbol{X}, t) - \boldsymbol{X}$, the displacement gradient $\tilde{H}$ is related to the deformation gradient $\tilde{F}$:

$$u_{i,j} = \chi_{i,j} - \delta_{ij} \qquad \text{or} \qquad \tilde{H} = \tilde{F} - \tilde{1} \tag{4.17}$$

Let us consider three small noncoplanar vectors $d\boldsymbol{X}_1, d\boldsymbol{X}_2$ and $d\boldsymbol{X}_3$. They are transformed at time t into three vectors $d\boldsymbol{x}_1, d\boldsymbol{x}_2$ and $d\boldsymbol{x}_3$ such that

$$d\boldsymbol{x}_i = \tilde{F}\, d\boldsymbol{X}_i, \qquad i = 1, 2, 3. \tag{4.18}$$

We now consider the small volume $dV = d\boldsymbol{X}_1 \wedge d\boldsymbol{X}_2 \cdot d\boldsymbol{X}_3$ defined at time $t = 0$; dV becomes the volume $dv = d\boldsymbol{x}_1 \wedge d\boldsymbol{x}_2 \cdot d\boldsymbol{x}_3$ at time t. Following the geometrical definition of the determinant of a second-order tensor (see the appendix), we may write:

$$\det \tilde{F} = \frac{d\boldsymbol{x}_1 \wedge d\boldsymbol{x}_2 \cdot d\boldsymbol{x}_3}{d\boldsymbol{X}_1 \wedge d\boldsymbol{X}_2 \cdot d\boldsymbol{X}_3} \tag{4.19}$$

Let us observe that for rigid body motion $\det \tilde{F} = 1$. The fundamental hypothesis of continuum mechanics is that no volume disappears ($\det \tilde{F} \neq 0$) and that no volume becomes infinitely large ($\det \tilde{F} \neq \infty$). Hence $0 < \det \tilde{F} < \infty$ may be considered as a paradigm (a fundamental hypothesis) of continuum mechanics.

Various mechanisms such as dissolution or precipitation, for which this paradigm is not satisfied are mentioned in chapter 12 and in chapter 14 we will see how this can influence natural regional stress fields in the earth.

4.3.2 Local polar decomposition of the deformation gradient

The tensor $\tilde{C} = \tilde{F}^T \tilde{F}$ is symmetrical. Indeed, $\tilde{C}^T = (\tilde{F}^T \tilde{F})^T = \tilde{F}^T \tilde{F} = \tilde{C}$ since $(\tilde{F}^T)^T = \tilde{F}$. Further, the scalar product $\boldsymbol{a} \cdot (\tilde{C}\boldsymbol{a})$ is always positive for any vector $\boldsymbol{a} \neq 0$: $\boldsymbol{a} \cdot \tilde{F}^T \tilde{F} \boldsymbol{a} = \tilde{F}\boldsymbol{a} \cdot \tilde{F}\boldsymbol{a} > 0$.

The tensor $\tilde{C}$ is positive definite and for such a tensor we can define a square root tensor $\tilde{U} = \sqrt{\tilde{C}}$, whose eigenvalues are the square roots of the eigenvalues of $\tilde{C}$ and whose eigenvectors are the same as those of $\tilde{C}$. The tensor $\tilde{U}$ is symmetrical.

Let us define now a tensor $\tilde{R}$ by the relation $\tilde{R} = \tilde{F}\tilde{U}^{-1}$, where $\tilde{U}^{-1}$ is the inverse of $\tilde{U}$, i.e. it has the same eigenvectors as $\tilde{U}$ and its eigenvalues are the inverses of the eigenvalues of $\tilde{U}$, so that $\tilde{U}\tilde{U}^{-1} = \tilde{1}$.

One may note that $\tilde{R}^T = (\tilde{F}\tilde{U}^{-1})^T = (\tilde{U}^{-1})^T \tilde{F}^T = \tilde{U}^{-1}\tilde{F}^T$, since $(\tilde{U}^{-1})^T = \tilde{U}^{-1}$ as shown in the appendix.

As a consequence $\tilde{R}^T \tilde{R} = \tilde{U}^{-1}(\tilde{F}^T \tilde{F})\tilde{U}^{-1} = \tilde{U}^{-1}\tilde{U}\tilde{U}\tilde{U}^{-1} = \tilde{1}$. Hence $\tilde{R}$ is an orthogonal tensor.

Finally, one may define a tensor $\tilde{V}$ by the relation $\tilde{V} \equiv \tilde{R}\tilde{U}\tilde{R}^T$, so that

$$\tilde{V}\tilde{R} = \tilde{R}\tilde{U}\tilde{R}^T\tilde{R} = \tilde{R}\tilde{U} = \tilde{F} \tag{4.20}$$

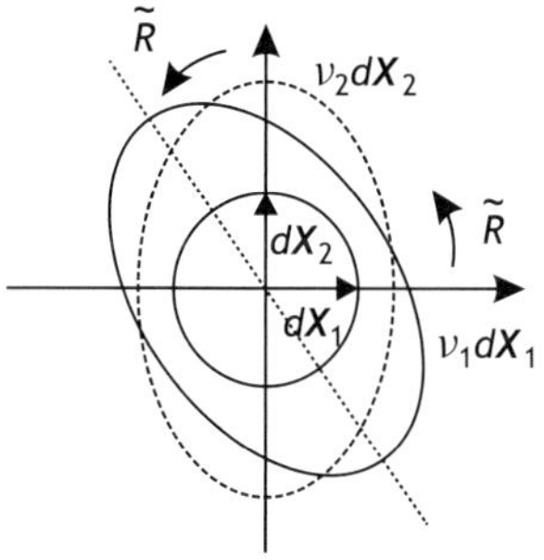

Fig. 4.4 Transformation by the right stretch tensor $\tilde{U}$ and the rotation tensor $\tilde{R}$.

Hence $\tilde{F}$ may be decomposed into the product of a symmetrical tensor and an orthogonal tensor: $\tilde{F} = \tilde{V}\tilde{R} = \tilde{R}\tilde{U}$. This is a general property of second-order tensors and now we explore its physical implication for the deformation gradient tensor.

Let us note that $\det \tilde{F} = \det \tilde{R} \det \tilde{U}$. However, we showed in the previous subsection that $\det \tilde{F} > 0$ and $\det \tilde{U}$ is always positive; hence $\det \tilde{R} > 0$. Because $\tilde{R}$ is orthogonal, $\tilde{R}\tilde{R}^T = \tilde{1}$ and $(\det \tilde{R})^2 = 1$. We conclude that $\det \tilde{R} = +1$ and that $\tilde{R}$ corresponds to a rotation in the direct sense. The tensor $\tilde{R}$ does not involve any deformation (change in length or in internal angle); it is called the *rotation tensor*.

Let $\nu; i = 1, 2, 3$, and $\boldsymbol{a}^{(i)}; i = 1, 2, 3$, be respectively the eigenvalues and the eigenvectors of $\tilde{U}$ and let $\boldsymbol{b}^{(i)}$ be the eigenvectors of $\tilde{V}$. We note that

$$\tilde{F}\boldsymbol{a}^{(i)} = \tilde{R}\tilde{U}\boldsymbol{a}^{(i)} = \tilde{R}\nu_i\boldsymbol{a}^{(i)} \qquad (i \text{ is not summed here}) \tag{4.21}$$

$$= \nu_i(\tilde{R}\boldsymbol{a}^{(i)}) = (\tilde{V}\tilde{R})\boldsymbol{a}^{(i)} = \tilde{V}(\tilde{R}\boldsymbol{a}^{(i)}) \tag{4.22}$$

Equation (4.22) defines the eigenvectors of the tensor $\tilde{V}$, so the eigenvalues and eigenvectors of $\tilde{V}$ must be respectively ν_i and $\boldsymbol{b}^{(i)} = \tilde{R}\boldsymbol{a}^{(i)}$; $i = 1, 2, 3$.

Consider the transformation $d\boldsymbol{x} = \tilde{F}\, d\boldsymbol{X} = \tilde{R}\tilde{U}\, d\boldsymbol{X}$. We may treat it as follows: first the vector $d\boldsymbol{X}$ is transformed by $\tilde{U}$ into the vector $\tilde{U}\, d\boldsymbol{X}$ and then the vector $\tilde{U}\, d\boldsymbol{X}$ is transformed by $\tilde{R}$ into the vector $\tilde{R}\tilde{U}\, d\boldsymbol{X} = d\boldsymbol{x}$.

We have already shown that the transformation effected by $\tilde{R}$ is a simple rotation. Now let us analyze the transformation caused by $\tilde{U}$ and choose $d\boldsymbol{X}^{(i)}$ parallel to $\boldsymbol{a}^{(i)}$. Then we have

$$d\boldsymbol{X}^{(i)} \rightarrow \tilde{U}\, d\boldsymbol{X}^{(i)} = \nu_i\, d\boldsymbol{X}^{(i)} \qquad (i \text{ is not summed here}) \tag{4.23}$$

If the original volume was a sphere, the deformed volume is an ellipsoid for which there exist three perpendicular directions that have not changed orientation (fig. 4.4). The symmetrical tensor $\tilde{U}$ is called the right stretch tensor.

Then an application of the tensor $\tilde{R}$ simply rotates the ellipsoid by a rigid motion.

The decomposition of $\tilde{F}$ into $\tilde{V}\tilde{R}$ implies that first the rigid body rotation is applied to the sphere and then the sphere is stretched into the final position. The final deformation is identical in both cases. The symmetrical tensor $\tilde{V}$ is called the left stretch tensor.

4.3.3 Finite deformation: the Cauchy–Green tensors

We consider a deformation in which a vector $d\boldsymbol{X}$ in the direction $\boldsymbol{N} = d\boldsymbol{X}/|d\boldsymbol{X}|$ at time $t = 0$ becomes $d\boldsymbol{x}$ in the direction $\boldsymbol{n} = d\boldsymbol{x}/|d\boldsymbol{x}|$ at time t such that $d\boldsymbol{x} = \tilde{F}\, d\boldsymbol{X}$ (equation (4.18)).

The stretch is defined by the ratio $|d\boldsymbol{x}|/|d\boldsymbol{X}|$. It may be described either as a function $\Lambda_{\boldsymbol{N}}$ of the orientation $\boldsymbol{N}$, in the Lagrange representation, or as a function $\lambda_{\boldsymbol{n}}$ of the orientation $\boldsymbol{n}$ in the Euler representation.

Let us consider first the stretch $\Lambda_{\boldsymbol{N}}$ in the Lagrange representation. We note from the definition of a transposed tensor (see the appendix) that

$$|d\boldsymbol{x}|^2 = d\boldsymbol{x} \cdot d\boldsymbol{x} = \tilde{F}\, d\boldsymbol{X} \cdot \tilde{F}\, d\boldsymbol{X} = (\tilde{F}^T \tilde{F}\, d\boldsymbol{X}) \cdot d\boldsymbol{X} \tag{4.24}$$

In section 4.3.2 the product $\tilde{F}^T\tilde{F}$ was denoted $\tilde{C}$. It is called the right Cauchy–Green tensor. We recall that

$$\tilde{C} = \tilde{F}^T\tilde{F} = \tilde{U}^T\tilde{R}^T\tilde{R}\tilde{U} = \tilde{U}^T\tilde{U} = \tilde{U}^2 \tag{4.25}$$

The tensor $\tilde{C}$ is positive definite and symmetrical and provides a measurement of the stretch in the Lagrange frame of reference:

$$\frac{|d\boldsymbol{x}|^2}{|d\boldsymbol{X}|^2} = \left(\tilde{C}\frac{d\boldsymbol{X}}{|d\boldsymbol{X}|}\right) \cdot \frac{d\boldsymbol{X}}{|d\boldsymbol{X}|} = \tilde{C}\boldsymbol{N} \cdot \boldsymbol{N} \tag{4.26}$$

$$\Lambda_{\boldsymbol{N}} = \frac{|d\boldsymbol{x}|}{|d\boldsymbol{X}|} = \sqrt{\tilde{C}\boldsymbol{N} \cdot \boldsymbol{N}} \tag{4.27}$$

Here $\boldsymbol{N}$ is a unit vector collinear with $d\boldsymbol{X}$.

In the Euler frame of reference we may write $d\boldsymbol{X} = \tilde{F}^{-1} d\boldsymbol{x}$, so that

$$d\boldsymbol{X} \cdot d\boldsymbol{X} = \tilde{F}^{-1} d\boldsymbol{x} \cdot \tilde{F}^{-1} d\boldsymbol{x} = [(\tilde{F}^{-1})^T \tilde{F}^{-1} d\boldsymbol{x}] \cdot d\boldsymbol{x} \tag{4.28}$$

$$= [(\tilde{F}^T)^{-1}\tilde{F}^{-1}]\, d\boldsymbol{x} \cdot d\boldsymbol{x} = [(\tilde{F}\tilde{F}^T)^{-1} d\boldsymbol{x}] \cdot d\boldsymbol{x} \tag{4.29}$$

Let us define the left Cauchy–Green tensor $\tilde{B} = \tilde{F}\tilde{F}^T$. It may be used to evaluate the stretch in the Euler frame of reference:

$$d\boldsymbol{X} \cdot d\boldsymbol{X} = (\tilde{B}^{-1} d\boldsymbol{x}) \cdot d\boldsymbol{x}, \tag{4.30}$$

so that

$$\frac{|d\boldsymbol{X}|^2}{|d\boldsymbol{x}|^2} = \tilde{B}^{-1}\boldsymbol{n} \cdot \boldsymbol{n} \qquad \text{and} \qquad \lambda_{\boldsymbol{n}} = \frac{|d\boldsymbol{x}|}{|d\boldsymbol{X}|} = \frac{1}{\sqrt{\tilde{B}^{-1}\boldsymbol{n} \cdot \boldsymbol{n}}} \tag{4.31}$$

Here $\boldsymbol{n}$ is a unit vector collinear with $d\boldsymbol{x}$.

We recall from equation (4.20) that $\tilde{B} = \tilde{F}\tilde{F}^T = \tilde{V}\tilde{R}\tilde{R}^T\tilde{V}^T = \tilde{V}^2$. Hence the eigenvectors of $\tilde{B}$ are those of $\tilde{V}$ (i.e. $\boldsymbol{b}_i = \tilde{R}\boldsymbol{a}_i$) and the eigenvalues are ν_i^2. If $\boldsymbol{n}$ is parallel to any of the eigenvectors $\boldsymbol{b}_i$ then the stretch is $\lambda_{\boldsymbol{b}_i} = 1/\nu_i$.

4.3.4 Finite-strain and small-strain tensors

Finite strain

An important concept is that of the relative change in length, also called the strain. Instead of considering the ratio of the length in the deformed and in the undeformed states ($\Lambda_N = |d\boldsymbol{x}|/|d\boldsymbol{X}|$) we consider now the relative change in length, characterized by

$$\frac{d\boldsymbol{x}^2 - d\boldsymbol{X}^2}{d\boldsymbol{X}^2} = (\Lambda_N)^2 - 1. \tag{4.32}$$

Let us observe that if $|d\boldsymbol{X}|$ is stretched to $|d\boldsymbol{X}| + \delta|d\boldsymbol{X}|$ then, if $\delta|d\boldsymbol{X}|$ is small,

$$\frac{(d\boldsymbol{x}^2 - d\boldsymbol{X}^2)}{d\boldsymbol{X}^2} = \frac{(|d\boldsymbol{X}| + \delta|d\boldsymbol{X}|)^2 - |d\boldsymbol{X}|^2}{|d\boldsymbol{X}|^2} \simeq \frac{2\delta|d\boldsymbol{X}|}{|d\boldsymbol{X}|} \tag{4.33}$$

So we note that the quantity $(d\boldsymbol{x}^2 - d\boldsymbol{X}^2)/d\boldsymbol{X}^2$ yields twice the relative change in length δ of the vector $d\boldsymbol{X}$ when the change in length is small.

We consider first the strain in a Lagrangian representation and observe that

$$|d\boldsymbol{x}|^2 - |d\boldsymbol{X}|^2 = \tilde{C}\, d\boldsymbol{X} \cdot d\boldsymbol{X} - |d\boldsymbol{X}|^2 = (\tilde{C} - \tilde{1})\, d\boldsymbol{X} \cdot d\boldsymbol{X} \tag{4.34}$$

The strain tensor $\tilde{E}$ is defined as $\tilde{E} \equiv \frac{1}{2}(\tilde{C} - \tilde{1}) = \tilde{E}^T$, from which we conclude that

$$\frac{|d\boldsymbol{x}|^2 - |d\boldsymbol{X}|^2}{|d\boldsymbol{X}|^2} = (\tilde{C} - \tilde{1})\boldsymbol{N} \cdot \boldsymbol{N} = 2\tilde{E}\boldsymbol{N} \cdot \boldsymbol{N} \tag{4.35}$$

Similarly, in the Eulerian frame of reference we may represent the strain with reference to the length of the vector $d\boldsymbol{x}$ at time t:

$$\frac{|d\boldsymbol{x}|^2 - |d\boldsymbol{X}|^2}{|d\boldsymbol{x}|^2} = (\tilde{1} - \tilde{B}^{-1})\boldsymbol{n} \cdot \boldsymbol{n} = 2\tilde{e}\boldsymbol{n} \cdot \boldsymbol{n} \tag{4.36}$$

with $\tilde{e} \equiv \frac{1}{2}(\tilde{1} - \tilde{B}^{-1}) = \tilde{e}^T$; for $\tilde{B}$ see the text after (4.29).

The strain tensors may be expressed as functions of the displacement gradient either in a Lagrangian frame of reference or in an Eulerian frame of reference:

$$\boldsymbol{u} = \boldsymbol{x} - \boldsymbol{X} = \begin{cases} \chi(\boldsymbol{X}, t) - \boldsymbol{X} = \boldsymbol{u}(\boldsymbol{X}, t) & \text{(Lagrange)} \\ \boldsymbol{x} - \chi^{-1}(\boldsymbol{x}, t) = \boldsymbol{u}(\boldsymbol{x}, t) & \text{(Euler)} \end{cases}$$

Let us note that

$$(\nabla \boldsymbol{u})_{ij} = \frac{\partial u(\boldsymbol{X}, t)_i}{\partial X_j} = u_{i,j} \tag{4.37}$$

$$= \frac{\partial u(\boldsymbol{x}, t)_i}{\partial x_k} \frac{\partial x_k}{\partial X_j} = (\text{grad}\, \boldsymbol{u})_{ik} F_{kj} \tag{4.38}$$

In vector form the displacement gradient tensor $\tilde{H}$ is given by

$$\tilde{H} \equiv \nabla \boldsymbol{u} = \tilde{F} - \tilde{1} = (\text{grad}\, \boldsymbol{u})\tilde{F} \tag{4.39}$$

and

$$\tilde{F}^{-1} = 1 - \operatorname{grad} \boldsymbol{u} \tag{4.40}$$

Equation (4.39) may be generalized to the functions $\bar{\psi}(\boldsymbol{X}, t)$ and $\hat{\psi}(\boldsymbol{x}, t)$ defined in section 4.2 (see equation (4.9)), so that $\nabla\bar{\psi}(\boldsymbol{X}, t) = [\operatorname{grad}\hat{\psi}(\boldsymbol{x}, t)]\tilde{F}$ or, in terms of coordinates,

$$\bar{\psi}_{,i} = \frac{\partial\hat{\psi}}{\partial x_j} F_{ji}$$

We see from equation (4.35) that

$$\frac{|d\boldsymbol{x}|^2 - |d\boldsymbol{X}|^2}{|d\boldsymbol{X}|^2} = 2\tilde{E}\boldsymbol{N}\cdot\boldsymbol{N}$$

so that, using (4.39), the strain $\tilde{E}$ in the Lagrange representation is

$$\begin{aligned}\tilde{E} &= \tfrac{1}{2}(\tilde{F}^T\tilde{F} - \tilde{1}) = \tfrac{1}{2}[(\tilde{1} + \tilde{H})^T(\tilde{1} + \tilde{H}) - \tilde{1}] \\ &= \tfrac{1}{2}(\tilde{H}^T\tilde{H} + \tilde{H}^T + \tilde{H})\end{aligned} \tag{4.41}$$

or, in terms of coordinates,

$$E_{ij} = \tfrac{1}{2}(H_{ki}H_{kj} + H_{ji} + H_{ij}) = \tfrac{1}{2}(u_{k,i}u_{k,j} + u_{j,i} + u_{i,j})$$

We saw in section 4.3.3 that the eigenvalues of $\tilde{C}$ are $\nu_i^2; i = 1, 2, 3$, while its eigenvectors are $\boldsymbol{a}_i; i = 1, 2, 3$, i.e. the same as those of the tensor $\tilde{U}$ defined in the polar decomposition of $\tilde{F}$. Consequently, the eigenvalues of $\tilde{E}$ are $\frac{1}{2}(\nu_i^2 - 1)$ whilst its eigenvectors are the same as those of $\tilde{C}$ and therefore of $\tilde{U}$.

If $I_{\tilde{C}} = \operatorname{tr}\tilde{C}$, $II_{\tilde{C}} = \frac{1}{2}[(\operatorname{tr}\tilde{C})^2 - \operatorname{tr}\tilde{C}^2]$ and $III_{\tilde{C}} = \det\tilde{C}$ are the three invariants of $\tilde{C}$, while $I_{\tilde{E}}$, $II_{\tilde{E}}$ and $III_{\tilde{E}}$ are those of $\tilde{E}$, then the following relations are easily obtained:

$$I_{\tilde{C}} = 3 + 2I_{\tilde{E}} \tag{4.42}$$

$$II_{\tilde{C}} = 3 + 4I_{\tilde{E}} + 4II_{\tilde{E}} \tag{4.43}$$

$$III_{\tilde{C}} = 1 + 2I_{\tilde{E}} + 4II_{\tilde{E}} + 8III_{\tilde{E}} \tag{4.44}$$

Following a very similar line of reasoning for an Eulerian frame of reference, we obtain (see (4.36))

$$\begin{aligned}\tilde{e} &= \tfrac{1}{2}(\tilde{1} - \tilde{B}^{-1}) = \tfrac{1}{2}[\tilde{1} - (\tilde{F}\tilde{F}^T)^{-1})] = \tfrac{1}{2}[\tilde{1} - (\tilde{F}^T)^{-1}\tilde{F}^{-1}] \\ &= \tfrac{1}{2}[\tilde{1} - (\tilde{1} - (\operatorname{grad}\boldsymbol{u}^T))(\tilde{1} - \operatorname{grad}\boldsymbol{u})] \\ &= \tfrac{1}{2}(\operatorname{grad}\boldsymbol{u} + (\operatorname{grad}\boldsymbol{u})^T - (\operatorname{grad}\boldsymbol{u})^T\operatorname{grad}\boldsymbol{u})\end{aligned} \tag{4.45}$$

or, in terms of components,

$$e_{ij} = \frac{1}{2}\left(\frac{\partial u_i}{\partial x_j} + \frac{\partial u_j}{\partial x_i} - \frac{\partial u_m}{\partial x_i}\frac{\partial u_m}{\partial x_j}\right) \tag{4.46}$$

with the following relations between the invariants of $\tilde{e}$ and $\tilde{B}$:

$$I_{\tilde{B}^{-1}} = 3 - 2I_{\tilde{e}} \tag{4.47}$$

$$II_{\tilde{B}^{-1}} = 3 - 4I_{\tilde{e}} + 4II_{\tilde{e}} \tag{4.48}$$

$$III_{\tilde{B}^{-1}} = 1 - 2I_{\tilde{e}} + 4II_{\tilde{e}} - 8III_{\tilde{e}} \tag{4.49}$$

Small strain: linearization

In this section we consider the case of small displacement gradients. The norm of $\tilde{H}$, the displacement gradient tensor, expressed in a Lagrangian frame of reference, is defined as follows:

$$|\tilde{H}| \equiv \sqrt{\mathrm{tr}(\tilde{H}\tilde{H}^T)} = \sqrt{u_{i,j}u_{j,i}} \tag{4.50}$$

and we define $\|\tilde{H}\| \equiv \max|\tilde{H}|, \forall \boldsymbol{X}$. We assume now that $\|\tilde{H}\| = \epsilon \ll 1, \forall \boldsymbol{x}$. Then $\tilde{F} = \tilde{1} + \tilde{H} \simeq \tilde{1}$. Consequently,

$$\tilde{H} = \nabla \boldsymbol{u}(\boldsymbol{X}, t) = [\mathrm{grad}\, \boldsymbol{u}(\boldsymbol{x}, t)]\tilde{F} = [\mathrm{grad}\, \boldsymbol{u}(\boldsymbol{x}, t)](\tilde{1} + \tilde{H}) \simeq \mathrm{grad}\, \boldsymbol{u}(\boldsymbol{x}, t) \tag{4.51}$$

Hence the strain tensors $\tilde{E}$ and $\tilde{e}$ defined respectively in the Lagrangian and Eulerian frames of reference become

$$\tilde{E} = \tfrac{1}{2}(\tilde{H} + \tilde{H}^T + \tilde{H}^T\tilde{H}) \simeq \tfrac{1}{2}(\tilde{H} + \tilde{H}^T) \tag{4.52}$$

and

$$\tilde{e} = \tfrac{1}{2}(\mathrm{grad}\, \boldsymbol{u} + \mathrm{grad}\, \boldsymbol{u}^T + \mathrm{grad}\, \boldsymbol{u}\; \mathrm{grad}\, \boldsymbol{u}^T) \simeq \tfrac{1}{2}(\tilde{H} + \tilde{H}^T) \simeq \tilde{E} \tag{4.53}$$

or, in terms of components,

$$E_{ij} \simeq \tfrac{1}{2}(u_{i,j} + u_{j,i}) \equiv \varepsilon_{ij} \simeq e_{ij} \tag{4.54}$$

The tensor $\tilde{\varepsilon}$ with components $\varepsilon_{ij} = \frac{1}{2}(u_{i,j} + u_{j,i})$ is known as the small-strain tensor.

As shown by equations (4.33) and (4.35), the small-strain tensor $\tilde{\varepsilon}$ provides a means to evaluate the relative change in length of a vector with any orientation $\boldsymbol{N}$:

$$\frac{d\boldsymbol{x}^2 - d\boldsymbol{X}^2}{d\boldsymbol{X}^2} \simeq \frac{2\delta|d\boldsymbol{X}|}{|d\boldsymbol{X}|} \simeq 2\tilde{\varepsilon}\boldsymbol{N} \cdot \boldsymbol{N} \tag{4.55}$$

This is precisely the result that was proposed in section 4.1. An important result is that for small displacement gradients the strain tensor components are the same whether they are expressed in a Lagrangian frame of reference or in an Eulerian frame of reference.

Let us recall that

$$\tilde{U} = \sqrt{\tilde{F}^T\tilde{F}} = \sqrt{(\tilde{1} + \tilde{H}^T)(\tilde{1} + \tilde{H})}$$

or, in terms of components,

$$U_{ij} \simeq \sqrt{\delta_{ij} + H_{ij} + H_{ji}} \tag{4.56}$$

However, H_{ij} is of the order of $\epsilon \ll 1$, so that $U_{ij} \simeq \delta_{ij} + \frac{1}{2}(H_{ij} + H_{ji})$ and therefore $(\tilde{U}^{-1})_{ij} \simeq \delta_{ij} - \frac{1}{2}(H_{ij} + H_{ji})$ or $\tilde{U}^{-1} \simeq \tilde{1} - \frac{1}{2}(\tilde{H} + \tilde{H}^T)$.

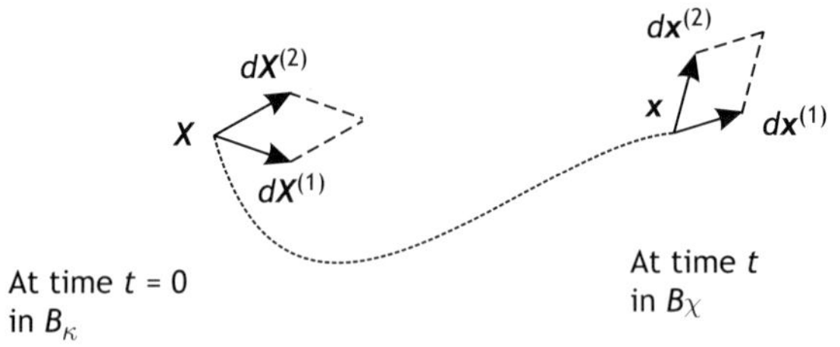

Fig. 4.5 Strain of a surface element.

From the polar decomposition of the deformation gradient tensor we have $\tilde{F} = \tilde{R}\tilde{U}$ or $\tilde{R} = \tilde{F}\tilde{U}^{-1}$. Replacing $\tilde{F}$ and $\tilde{U}$ by their expressions in terms of the displacement gradient $\tilde{H}$, we obtain an expression for the rotation tensor $\tilde{R}$ in the context of small displacement gradients: $\tilde{R} \simeq (\tilde{1} + \tilde{H})[\tilde{1} - \frac{1}{2}(\tilde{H} + \tilde{H}^T)]$. From this we obtain $\tilde{R} \simeq \tilde{1} + \frac{1}{2}(\tilde{H} - \tilde{H}^T)$.

The tensor $\tilde{\omega} \equiv \frac{1}{2}(\tilde{H} - \tilde{H}^T)$ is called the linear rotation tensor and we may write, for small displacement gradients,

$$\tilde{F} \simeq \tilde{1} + \tilde{\varepsilon} + \tilde{\omega} \qquad \text{or} \qquad d\boldsymbol{x} \simeq (\tilde{1} + \tilde{\varepsilon} + \tilde{\omega})d\boldsymbol{X} \tag{4.57}$$

4.3.5 Surface strain, volumetric strain

Surface strain

Let us consider at time $t = 0$ and at a point $\boldsymbol{X}$ two noncollinear vectors $d\boldsymbol{X}^{(1)}$ and $d\boldsymbol{X}^{(2)}$, which become $d\boldsymbol{x}^{(1)} = \tilde{F}\, d\boldsymbol{X}^{(1)}$ and $d\boldsymbol{x}^{(2)} = \tilde{F}\, d\boldsymbol{X}^{(2)}$ at time t (fig. 4.5).

The surface element $d\boldsymbol{S} = d\boldsymbol{X}^{(1)} \wedge \boldsymbol{X}^{(2)}$ becomes $d\boldsymbol{s} = d\boldsymbol{x}^{(1)} \wedge d\boldsymbol{x}^{(2)}$. In terms of components this is written as

$$ds_i = \varepsilon_{ijk} dx_j^{(1)} dx_k^{(2)}, \qquad i,j,k = 1,2,3 \tag{4.58}$$

$$= \varepsilon_{ijk} F_{jp} dX_p^{(1)} F_{km} dX_m^{(2)}, \qquad i,j,k,p,m = 1,2,3 \tag{4.59}$$

which, keeping in mind that $\delta_{si} = F_{st}F_{ti}^{-1}$ and that $\varepsilon_{sjk}F_{st}F_{jp}F_{km} = \varepsilon_{tpm}\det\tilde{F}$, as shown in the appendix, may be rewritten as

$$ds_i = \varepsilon_{sjk} F_{jp} F_{km} dX_p^{(1)} dX_m^{(2)} \delta_{si}, \qquad i,j,k,p,m,s = 1,2,3 \tag{4.60}$$

$$= (\det\tilde{F})\varepsilon_{tpm} F_{ti}^{-1} dX_p^{(1)} dX_m^{(2)} \tag{4.61}$$

$$= (\det\tilde{F}) F_{ti}^{-1} dS_t \tag{4.62}$$

In vector form this becomes

$$d\boldsymbol{s} = (\det\tilde{F})(\tilde{F}^{-1})^T d\boldsymbol{S} \tag{4.63}$$

or

$$d\boldsymbol{S} = \frac{1}{\det\tilde{F}}\tilde{F}^T d\boldsymbol{s} \tag{4.64}$$

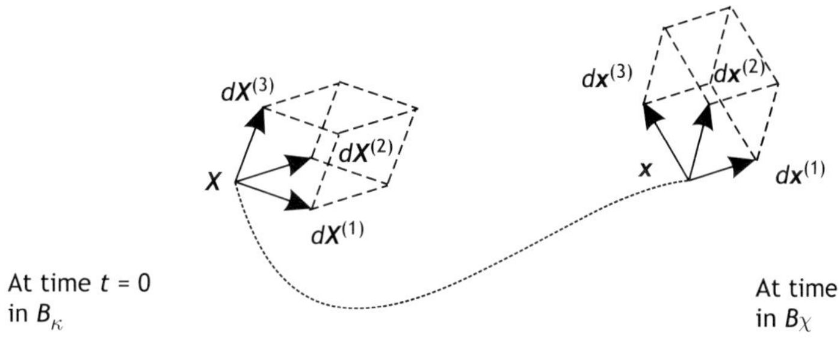

Fig. 4.6 Strain of a small volume.

We may compute the relative change in area by evaluating ds^2/dS^2 in a Lagrange frame of reference. We have

$$|ds|^2 = (\det \tilde{F})^2(\tilde{F}^{-1})^T d\boldsymbol{S} \cdot (\tilde{F}^{-1})^T d\boldsymbol{S} \tag{4.65}$$

$$= (\det \tilde{F})^2[\tilde{F}^{-1}(\tilde{F}^{-1})^T d\boldsymbol{S}] \cdot d\boldsymbol{S} \tag{4.66}$$

$$= (\det \tilde{C})\tilde{C}^{-1} d\boldsymbol{S} \cdot d\boldsymbol{S} \tag{4.67}$$

so that, finally,

$$\frac{|ds|^2}{|d\boldsymbol{S}|^2} = (\det \tilde{C})\tilde{C}^{-1}\boldsymbol{N} \cdot \boldsymbol{N} \tag{4.68}$$

Similar lines of reasoning may be followed to obtain an expression for the surface strain in an Eulerian frame of reference.

Volumetric strain

We consider now at time $t = 0$ and at a point $\boldsymbol{X}$, three noncoplanar and noncollinear vectors $d\boldsymbol{X}^{(1)}$, $d\boldsymbol{X}^{(2)}$ and $d\boldsymbol{X}^{(3)}$, which define a small volume $dV = d\boldsymbol{X}^{(1)} \wedge \boldsymbol{X}^{(2)} \cdot d\boldsymbol{X}^{(3)}$.

At time t the particle that was in $\boldsymbol{X}$ at $t = 0$ is now at $\boldsymbol{x}$ and the vectors $d\boldsymbol{X}^{(1)}$, $d\boldsymbol{X}^{(2)}$ and $d\boldsymbol{X}^{(3)}$ have become $d\boldsymbol{x}^{(1)} = \tilde{F}\, d\boldsymbol{X}^{(1)}$, $d\boldsymbol{x}^{(2)} = \tilde{F}\, d\boldsymbol{X}^{(2)}$ and $d\boldsymbol{x}^{(3)} = \tilde{F}\, d\boldsymbol{X}^{(3)}$; these define the small volume $dv = d\boldsymbol{x}^{(1)} \wedge d\boldsymbol{x}^{(2)} \cdot d\boldsymbol{x}^{(3)}$ (fig. 4.6).

The determinant of the tensor $\tilde{F}$ (see the appendix) is given by

$$\det \tilde{F} = \frac{d\boldsymbol{x}^{(1)} \wedge d\boldsymbol{x}^{(2)} \cdot d\boldsymbol{x}^{(3)}}{d\boldsymbol{X}^{(1)} \wedge d\boldsymbol{X}^{(2)} \cdot d\boldsymbol{X}^{(3)}} = \frac{dv}{dV} \tag{4.69}$$

However, since $\tilde{C} = \tilde{F}^T\tilde{F}$ and $\tilde{B} = \tilde{F}\tilde{F}^T$ we observe that $\det \tilde{C} = \det \tilde{F}^2 = \det \tilde{B}$.

In a Lagrangian frame of reference the change in volume dv/dV is $\det \tilde{F} = \sqrt{\det \tilde{C}}$. Recall from equation (4.44) that $\det \tilde{C} = 1 + 2I_{\tilde{E}} + 4II_{\tilde{E}} + 8III_{\tilde{E}}$, so that

$$\frac{dv}{dV} = \sqrt{III_{\tilde{C}}} = \sqrt{1 + 2I_{\tilde{E}} + 4II_{\tilde{E}} + 8III_{\tilde{E}}} \tag{4.70}$$

For small displacement gradients we saw (equation (4.54)) that $\tilde{E} \simeq \tilde{\epsilon}$. Further, since $I_{\tilde{\varepsilon}}$ is of order $\epsilon \ll 1$, $II_{\tilde{\varepsilon}}$ is of order ϵ^2 and $III_{\tilde{\varepsilon}}$ is of order ϵ^3, it follows that $II_{\tilde{\varepsilon}}$ and $III_{\tilde{\varepsilon}}$ are negligible in comparison with $I_{\tilde{\varepsilon}}$. Hence, in a Lagrangian frame of reference,

$$\frac{dv}{dV} = 1 + I_{\tilde{\varepsilon}} \tag{4.71}$$

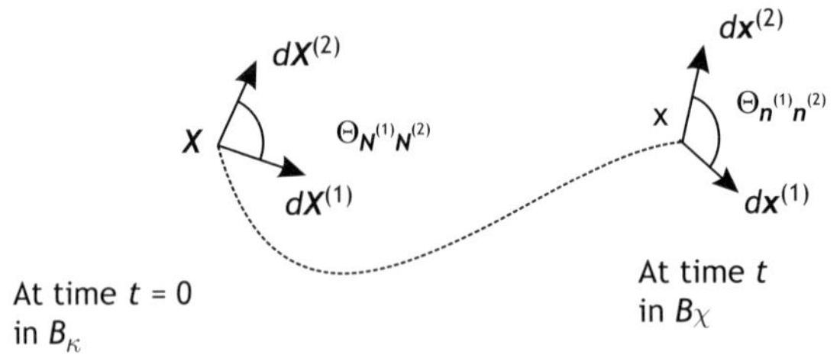

Fig. 4.7 Shear strain in terms of angular variations.

The relative change in volume is $(dv - dV)/dV$, which is simply the trace of the tensor $\tilde{\varepsilon}$ for small displacement gradients:

$$\frac{dv - dV}{dV} \simeq I_{\tilde{\varepsilon}} = \varepsilon_{ii} = u_{i,i} = \nabla \cdot \boldsymbol{u} \tag{4.72}$$

Thus, for small displacement gradients the relative volume variation has the same expression in the Lagrangian and Eulerian frames of reference ($\nabla \boldsymbol{u} \simeq \operatorname{grad} \boldsymbol{u}$).

4.3.6 Shear strain

At time $t = 0$ and a point $\boldsymbol{X}$ the two noncollinear vectors $d\boldsymbol{X}^{(1)}$ and $d\boldsymbol{X}^{(2)}$ define the angle $\Theta_{\boldsymbol{N}^{(1)}\boldsymbol{N}^{(2)}}$. At time t the two vectors $d\boldsymbol{x}^{(1)} = \tilde{F}\, d\boldsymbol{X}^{(1)}$ and $d\boldsymbol{x}^{(2)} = \tilde{F}\, d\boldsymbol{X}^{(2)}$ define the angle $\theta_{\boldsymbol{n}^{(1)}\boldsymbol{n}^{(2)}}$ (fig. 4.7).

The *shear strain* Γ_{12} is defined as the difference between the angles $\Theta_{\boldsymbol{N}^{(1)}\boldsymbol{N}^{(2)}}$ and $\theta_{\boldsymbol{n}^{(1)}\boldsymbol{n}^{(2)}}$:

$$\Gamma_{12} = \Theta_{\boldsymbol{N}^{(1)}\boldsymbol{N}^{(2)}} - \theta_{\boldsymbol{n}^{(1)}\boldsymbol{n}^{(2)}} \tag{4.73}$$

The cosines of $\Theta_{\boldsymbol{N}^{(1)}\boldsymbol{N}^{(2)}}$ and $\theta_{\boldsymbol{n}^{(1)}\boldsymbol{n}^{(2)}}$ are given by scalar products:

$$\cos\Theta_{\boldsymbol{N}^{(1)}\boldsymbol{N}^{(2)}} = \frac{d\boldsymbol{X}^{(1)} \cdot d\boldsymbol{X}^{(2)}}{|d\boldsymbol{X}^{(1)}||d\boldsymbol{X}^{(2)}|}, \qquad \text{and} \qquad \cos\theta_{\boldsymbol{n}^{(1)}\boldsymbol{n}^{(2)}} = \frac{d\boldsymbol{x}^{(1)} \cdot d\boldsymbol{x}^{(2)}}{|d\boldsymbol{x}^{(1)}||d\boldsymbol{x}^{(2)}|} \tag{4.74}$$

By using relation (4.27) we obtain

$$\cos\theta_{\boldsymbol{n}^{(1)}\boldsymbol{n}^{(2)}} = \frac{\tilde{F}d\boldsymbol{X}^{(1)} \cdot \tilde{F}d\boldsymbol{X}^{(2)}}{|d\boldsymbol{x}^{(1)}||d\boldsymbol{x}^{(2)}|} = \tilde{C}\boldsymbol{N}^{(1)} \cdot \boldsymbol{N}^{(2)} \Big/ \frac{|d\boldsymbol{x}^{(1)}|}{|d\boldsymbol{X}^{(1)}|}\frac{|d\boldsymbol{x}^{(2)}|}{|d\boldsymbol{X}^{(2)}|} \tag{4.75}$$

$$= \frac{\tilde{C}\boldsymbol{N}^{(1)} \cdot \boldsymbol{N}^{(2)}}{\Lambda_{\boldsymbol{N}^{(1)}}\Lambda_{\boldsymbol{N}^{(2)}}} \tag{4.76}$$

For example, if $\boldsymbol{N}^{(1)}$ is parallel to the unit vector $\hat{\boldsymbol{I}}^{(1)}$ and $\boldsymbol{N}^{(2)}$ to $\hat{\boldsymbol{I}}^{(2)}$, whilst $\boldsymbol{n}^{(1)}$ is parallel to the unit vector $\hat{\boldsymbol{I}}'^{(1)}$ and $\boldsymbol{n}^{(2)}$ to $\hat{\boldsymbol{I}}'^{(2)}$, equation (4.76) yields

$$\cos\theta_{\hat{\boldsymbol{I}}'^{(1)}\hat{\boldsymbol{I}}'^{(2)}} = \frac{C_{12}}{\sqrt{C_{11}}\sqrt{C_{22}}} \tag{4.77}$$

but

$$\Gamma_{12} = \pi/2 - \theta_{\hat{I}'^{(1)}\hat{I}'^{(2)}}, \tag{4.78}$$

so that

$$\sin\Gamma_{12} = \cos\theta_{\hat{I}'^{(1)}\hat{I}'^{(2)}} = \frac{C_{12}}{\sqrt{C_{11}}\sqrt{C_{22}}} \tag{4.79}$$

In the theory of small displacement gradients, $C_{ij} \simeq \delta_{ij} + 2\varepsilon_{ij}$ so that $C_{12} \simeq 2\varepsilon_{12}$, $C_{11} \simeq 1 + 2\varepsilon_{11}$, $C_{22} \simeq 1 + 2\varepsilon_{22}$ and

$$\frac{1}{\sqrt{C_{11}C_{22}}} \simeq \frac{1}{\sqrt{1 + 2\left(\varepsilon_{11} + \varepsilon_{22}\right)}} \simeq 1 - (\varepsilon_{11} + \varepsilon_{22}) \tag{4.80}$$

from which we conclude that

$$\sin\Gamma_{12} \simeq 2\varepsilon_{12}[1 - (\varepsilon_{11} + \varepsilon_{22})] \simeq 2\varepsilon_{12} \tag{4.81}$$

The shear strain for small-displacement gradients is

$$\sin\Gamma_{12} \simeq \Gamma_{12} \simeq 2\varepsilon_{12} \tag{4.82}$$

This is precisely the result already mentioned in section 4.1.

4.3.7 Compatibility conditions

When considering small-displacement gradients, we defined two symmetrical tensors, the small-strain tensor with six components and the rotation tensor with three components. These nine quantities involve partial derivatives of the three displacement components, which are the only independent variables.

This implies that six relations must exist between the components of the small-strain and small-rotation tensors. These six conditions are known as the strain compatibility equations.

For example, let us consider the strain component $\varepsilon_{12} = \frac{1}{2}\left(u_{1,2} + u_{2,1}\right)$. If we differentiate this relation with respect to x_1 and then with respect to x_2, we obtain

$$(\varepsilon_{12})_{,12} \equiv \varepsilon_{12,\,12} = \tfrac{1}{2}(u_{1,2} + u_{2,1})_{,12} = \tfrac{1}{2}(u_{1,212} + u_{2,112}) \tag{4.83}$$

so that

$$\varepsilon_{12,12} = \tfrac{1}{2}(u_{1,122} + u_{2,211}) = \tfrac{1}{2}(\varepsilon_{11,22} + \varepsilon_{22,11}) \tag{4.84}$$

Two other similar conditions are easily verified, namely $\varepsilon_{23,23} = \frac{1}{2}(\varepsilon_{22,33} + \varepsilon_{33,22})$ and $\varepsilon_{31,31} = \frac{1}{2}(\varepsilon_{11,33} + \varepsilon_{33,11})$.

In addition it may be verified that the three following conditions must also be satisfied (Malvern, 1969):

$$\varepsilon_{11,23} = (-\varepsilon_{23,1} + \varepsilon_{31,2} + \varepsilon_{12,3})_{,1} \tag{4.85}$$

$$\varepsilon_{22,31} = (\varepsilon_{23,1} - \varepsilon_{31,2} + \varepsilon_{12,3})_{,2} \tag{4.86}$$

$$\varepsilon_{33,12} = (\varepsilon_{23,1} + \varepsilon_{31,2} - \varepsilon_{12,3})_{,3} \tag{4.87}$$

4.3.8 Small strains in cylindrical and spherical coordinates

Small-strain results for cylindrical coordinates are obtained from the expression for a small strain in cartesian coordinates, $\varepsilon_{ij} = \frac{1}{2}(u_{i,j} + u_{j,i})$, with the following changes of variables (e.g. Jaeger and Cook, 1979):

$$x_1 = \rho\cos\theta, \quad x_2 = \rho\sin\theta, \quad x_3 = z \tag{4.88}$$

$$\rho = (x_1^2 + x_2^2)^{\frac{1}{2}}, \quad \theta = \arctan\frac{x_2}{x_1} \tag{4.89}$$

with corresponding results for the partial derivatives:

$$\rho_{,1} = (x_1^2 + x_2^2)^{1/2}_{,1} = \frac{x_1}{(x_1^2 + x_2^2)^{1/2}} = \frac{x_1}{\rho} = \cos\theta, \tag{4.90}$$

$$\rho_{,2} = (x_1^2 + x_2^2)^{1/2}_{,2} = \frac{x_2}{(x_1^2 + x_2^2)^{1/2}} = \frac{x_2}{\rho} = \sin\theta, \tag{4.91}$$

$$\theta_{,1} = \arctan\left(\frac{x_2}{x_1}\right)_{,1} = \frac{-x_2/x_1^2}{[1 + (x_2/x_1)^2]} = \frac{-x_2}{\rho^2} = \frac{-\sin\theta}{\rho} \tag{4.92}$$

$$\theta_{,2} = \arctan\left(\frac{x_2}{x_1}\right)_{,2} = \frac{1/x_1}{[1 + (x_2/x_1)^2]} = \frac{x_1}{\rho^2} = \frac{\cos\theta}{\rho} \tag{4.93}$$

$$\rho_{,3} = \theta_{,3} = 0 \tag{4.94}$$

The components of the displacement vector in the two frames of reference are related by

$$u_\rho = u_1\cos\theta + u_2\sin\theta, \quad u_\theta = -u_1\sin\theta + u_2\cos\theta, \quad u_z = u_3 \tag{4.95}$$

$$u_1 = u_\rho\cos\theta - u_\theta\sin\theta, \quad u_2 = u_\rho\sin\theta + u_\theta\cos\theta, \quad u_3 = u_z \tag{4.96}$$

Hence, when we apply the chain rule of derivation to the small-strain component ε_{11} we obtain

$$\varepsilon_{11} = u_{1,1} = (u_{1,\rho})\rho_{,1} + (u_{1,\theta})\theta_{,1} = (\cos\theta)u_{1,\rho} - \left(\frac{\sin\theta}{\rho}\right)u_{1,\theta} \tag{4.97}$$

However, from (4.96) we get

$$u_{1,\rho} = (\cos\theta)u_{\rho,\rho} - (\sin\theta)u_{\theta,\rho} \tag{4.98}$$

$$u_{1,\theta} = (\cos\theta)u_{\rho,\theta} - (\sin\theta)u_\rho - (\sin\theta)u_{\theta,\theta} - (\cos\theta)u_\theta \tag{4.99}$$

Substituting (4.98) and (4.99) into (4.97) we get

$$\varepsilon_{11} = u_{\rho,\rho}\cos^2\theta - \left(u_{\theta,\rho} + \frac{1}{\rho}u_{\rho,\theta} - \frac{u_\theta}{\rho}\right)\sin\theta\cos\theta + \left(\frac{1}{\rho}u_{\theta,\theta} + \frac{u_\rho}{\rho}\right)\sin^2\theta \tag{4.100}$$

The small-strain components in the cartesian frame of reference may also be obtained from the components in a cylindrical frame of reference using the expressions for a change of reference frame (see equation (3.67)), so that:

$$\begin{pmatrix} \cos\theta & -\sin\theta & 0 \\ \sin\theta & \cos\theta & 0 \\ 0 & 0 & 1 \end{pmatrix} \begin{pmatrix} \varepsilon_{\rho\rho} & \varepsilon_{\rho\theta} & \varepsilon_{\rho z} \\ \varepsilon_{\theta\rho} & \varepsilon_{\theta\theta} & \varepsilon_{\theta z} \\ \varepsilon_{z\rho} & \varepsilon_{z\theta} & \varepsilon_{zz} \end{pmatrix} \begin{pmatrix} \cos\theta & \sin\theta & 0 \\ -\sin\theta & \cos\theta & 0 \\ 0 & 0 & 1 \end{pmatrix} \tag{4.101}$$

with the following result for ε_{11}:

$$\varepsilon_{11} = \varepsilon_{\rho\rho}\cos^2\theta - 2\varepsilon_{\rho\theta}\sin\theta\cos\theta + \varepsilon_{\theta\theta}\sin^2\theta \tag{4.102}$$

Comparing (4.100) and (4.102) we conclude that

$$\varepsilon_{\rho\rho} = u_{\rho,\rho}, \quad \varepsilon_{\theta\theta} = \frac{1}{\rho}(u_{\theta,\theta} + u_\rho), \quad \varepsilon_{\rho\theta} = \tfrac{1}{2}(u_{\theta,\rho} + \frac{1}{\rho}u_{\rho,\theta} - \frac{u_\theta}{\rho}) \tag{4.103}$$

Expressions for ε_{zz}, $\varepsilon_{\theta z}, \varepsilon_{\rho z}$ may be obtained following the same procedure:

$$\varepsilon_{zz} = u_{z,z}, \quad \varepsilon_{\theta z} = \tfrac{1}{2}\left(\frac{1}{\rho}u_{z,\theta} + u_{\theta,z}\right), \quad \varepsilon_{\rho z} = \tfrac{1}{2}(u_{z,\rho} + u_{\rho,z}) \tag{4.104}$$

When the displacement gradient is small, the volumetric strain in cylindrical coordinates is

$$\varepsilon_v = \varepsilon_{\rho\rho} + \varepsilon_{\theta\theta} + \varepsilon_{zz} = u_{\rho,\rho} + \frac{u_\rho}{\rho} + \frac{1}{\rho}u_{\theta,\theta} + u_{z,z} \tag{4.105}$$

Similar reasoning may be applied to obtain the small-strain results in spherical coordinates. If ρ, φ and θ are spherical coordinates (fig. 3.12) such that

$$x_1 = \rho\sin\theta\cos\phi, \quad x_2 = \rho\sin\theta\sin\phi, \quad x_3 = \rho\cos\theta \tag{4.106}$$

then the small-strain components are (e.g. Fung, 1965):

$$\varepsilon_{\rho\rho} = u_{\rho,\rho}, \quad \varepsilon_{\theta\theta} = \frac{1}{\rho}(u_{\theta,\theta} + u_\rho), \quad \varepsilon_{\rho\theta} = \tfrac{1}{2}\left(\frac{1}{\rho}u_{\rho,\theta} + u_{\theta,\rho} - \frac{u_\theta}{\rho}\right); \tag{4.107}$$

$$\varepsilon_{\phi\phi} = \frac{1}{\rho}\left(\frac{u_{\phi,\phi}}{\sin\theta} + u_\rho + u_\theta\cot\theta\right); \quad \varepsilon_{\rho\phi} = \tfrac{1}{2}\left(\frac{u_{\rho,\phi}}{\rho\sin\theta} + u_{\phi,\rho} - \frac{u_\phi}{\rho}\right), \tag{4.108}$$

$$\varepsilon_{\theta\phi} = \tfrac{1}{2}\left(\frac{u_{\theta,\phi}}{\rho\sin\theta} + \frac{u_{\phi,\theta}}{\rho} - \frac{u_\phi\cot\theta}{\rho}\right) \tag{4.109}$$

4.4 Motion

4.4.1 Particle paths, streamlines, streaklines

As time passes, a given particle of a solid or fluid material may occupy different positions in space. The locus of all positions that have been occupied by the particle is called the particle path. At each position, the particle exhibits a given velocity. At a given time t, one may define a curve passing through different particles and such that, at each point, the curve is tangent to the local particle velocity at the corresponding time (fig. 4.8). Such a curve is called a streamline.

Hence, by definition, at every point a streamline is tangent to the instantaneous particle velocity at the corresponding point. A flow is defined as being steady when the streamlines

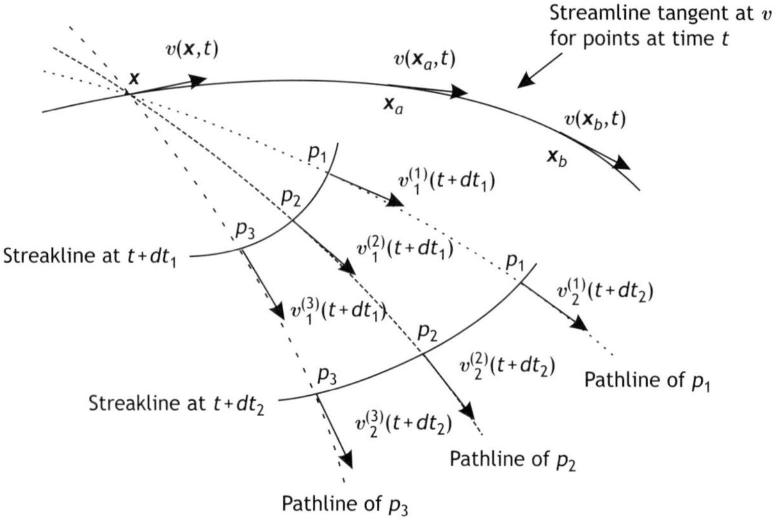

Fig. 4.8 Pathline, streamline and streakline.

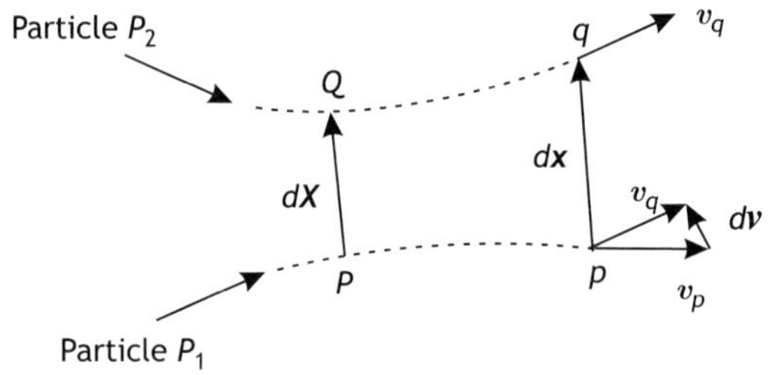

Fig. 4.9 Relative velocity of particle p_2 at time t with respect to that of particle p_1.

do not change with time. The locus, at time $t + dt$, of all particles that passed through a point $\boldsymbol{x}$ at any time before the time $t + dt$ is called a streakline (fig. 4.8).

The material derivative of the velocity along a streamline, which is the particle acceleration, takes the very simple form

$$\boldsymbol{a} = \frac{d\boldsymbol{v}}{dt} = \frac{\partial \boldsymbol{v}}{\partial t} + \frac{d\boldsymbol{v}}{ds}\frac{ds}{dt} = \frac{\partial \boldsymbol{v}}{\partial t} + v\frac{dv}{ds} \tag{4.110}$$

where ds is a short arc length along the streamline in the direction of the motion and v is the magnitude (modulus) of the velocity.

In steady conditions the particle path, the stream line and the streak line are superposed.

4.4.2 Rate of deformation (stretching) and spin

Consider two particles P_1 and P_2, occupying points p and q at time t, which are the origin and extremity of a displacement vector $d\boldsymbol{x}$ and which move respectively with velocity $\boldsymbol{v}_p$ and velocity $\boldsymbol{v}_q$ (fig. 4.9).

It was shown in section 4.3.1 that if particle P_1 is close to particle P_2 then

$$dx = \tilde{F} dX \tag{4.111}$$

where $\tilde{F}$ is the deformation gradient and dX is the vector that joins the points P and Q occupied at time $t = 0$ by the particles p_1 and p_2. The time derivative of equation (4.111) is

$$\dot{\overline{dx}} = \dot{\tilde{F}} dX \tag{4.112}$$

In a Lagrangian frame of reference

$$\dot{\tilde{F}} = \overline{\left(\frac{\partial \chi(X,t)}{\partial X}\right)}^{\bullet} = \frac{\partial\, \overline{\chi(X,t)}^{\bullet}}{\partial X} = \nabla v(X,t) \tag{4.113}$$

but in an Eulerian frame of reference

$$\dot{\tilde{F}} = \frac{\partial v(x,t)}{\partial x} \frac{\partial x}{\partial X} = (\text{grad } v)\, \tilde{F} \tag{4.114}$$

so that

$$\dot{\overline{dx}} = (\text{grad } v)\, \tilde{F}\, dX = (\text{grad } v)\, dx \tag{4.115}$$

Let $\tilde{L}$ be the velocity gradient tensor in an Eulerian frame of reference ($\tilde{L} \equiv$ grad v). It may be decomposed into the sum of a symmetrical tensor $\tilde{D}$ and a skew tensor $\tilde{W}$, with

$$\tilde{D} = \tilde{D}^T = \tfrac{1}{2}(\tilde{L} + \tilde{L}^T) \qquad \text{and} \qquad \tilde{W} = -\tilde{W}^T = \tfrac{1}{2}(\tilde{L} - \tilde{L}^T)$$

Then

$$\dot{\overline{dx}} = \tilde{D}\, dx + \tilde{W}\, dx \tag{4.116}$$

where $\tilde{D}$ is the stretch tensor and $\tilde{W}$ is the spin tensor. They are defined in an Eulerian frame of reference.

In terms of components we may write

$$L_{mn} = v_{m,n}, \quad D_{mn} = \tfrac{1}{2}(v_{m,n} + v_{n,m}), \quad W_{mn} = \tfrac{1}{2}(v_{m,n} - v_{n,m}) \tag{4.117}$$

The physical significance of the spin tensor may be seen by considering a rigid body motion, i.e. a motion in which all the components D_{mn} are null at all points of the body. Then the components of the velocity of particle P_2 with respect to that of P_1 are given by the components of the vector $\tilde{W} dx$, which are the same as the components of the rigid body rotation defined by $w \wedge x$, where

$$w_k = -\tfrac{1}{2}\epsilon_{ijk} W_{jk} = \epsilon_{ijk} v_{j,k} \qquad \text{or} \qquad w = \tfrac{1}{2}\nabla \wedge v \tag{4.118}$$

Here v_i is the ith component of the particle velocity (see Malvern, 1969).

A motion is said to be irrotational when all the components W_{mn}, or w_i, are zero for all points in the volume under consideration.

4.4.3 Rate of deformation and strain rate

As was shown in section 4.3.3, the square of the modulus of the vector $\boldsymbol{dx}$ at time t may be evaluated as a function of its value at time $t = 0$:

$$\boldsymbol{dx}^2 = \tilde{C}\, \boldsymbol{dX} \cdot \boldsymbol{dX} \qquad \text{with} \qquad \tilde{C} = \tilde{F}^T\tilde{F} \tag{4.119}$$

Its variation with time is

$$\overline{\dot{\boldsymbol{dx}^2}} = \dot{\tilde{C}}\, \boldsymbol{dX} \cdot \boldsymbol{dX} \tag{4.120}$$

Furthermore,

$$\dot{\tilde{C}} = \overline{\dot{F}}^{\,T}\tilde{F} + \tilde{F}^T\dot{\tilde{F}} \qquad \text{with} \qquad \dot{\tilde{F}} = \tilde{L}\tilde{F} \tag{4.121}$$

Equation (4.121) may be rewritten as

$$\dot{\tilde{C}} = \tilde{F}^T\tilde{L}^T\tilde{F} + \tilde{F}^T\tilde{L}\tilde{F} = 2\tilde{F}^T\tilde{D}\tilde{F} \tag{4.122}$$

so that if we remember that the finite strain $\tilde{E}$ is $\frac{1}{2}(\tilde{C} - \tilde{1})$, we obtain the relationship between the finite-strain rate and the deformation rate:

$$\dot{\tilde{E}} = \tilde{F}^T\tilde{D}\tilde{F} \tag{4.123}$$

Both quantities, the finite-strain rate (defined in a Lagrangian frame of reference) and the deformation rate (defined in an Eulerian frame of reference) may be significantly different when the deformation gradient $\tilde{F}$ differs from unity. However, for very small displacements $\tilde{F}$ is indeed very close to unity, as shown in section 4.3.4, so that both quantities are nearly equal (see Malvern, 1969).

4.5 Exercises

1. Give the expression for the components of the small-strain tensor as a function of the partial derivatives of the displacement components. What are the physical interpretations of the diagonal terms and of the nondiagonal terms ?
2. If $\tilde{\epsilon}$ is the small-strain tensor, what is the change in length δl of a segment with length l oriented parallel to the unit vector $\boldsymbol{n}$?
3. A 4000 m deep borehole provides the opportunity to collect an oriented rock core at this depth. Immediately after the core has been collected a small cubic sample is prepared with two of its faces perpendicular to the core axis. On each of three faces, three strain gauges are glued as shown in fig. 4.10; a strain gauge is a device that provides a means to measure very accurately changes in length. The unperturbed length of the gauges is 2 cm. After 48 hours the following readings have been collected from the various strain gauges:
   for strain gauge 1, 82.0 μm; for strain gauge 2, 98.0 μm; for strain gauge 3, 27.0 μm; for strain gauge 4, 82.5 μm; for strain gauge 5, 41.0 μm; for strain gauge 6, 4.0 μm; for strain gauge 7, 3.0 μm; for strain gauge 8, −10.0 μm; for strain gauge 9, 26.5. μm;

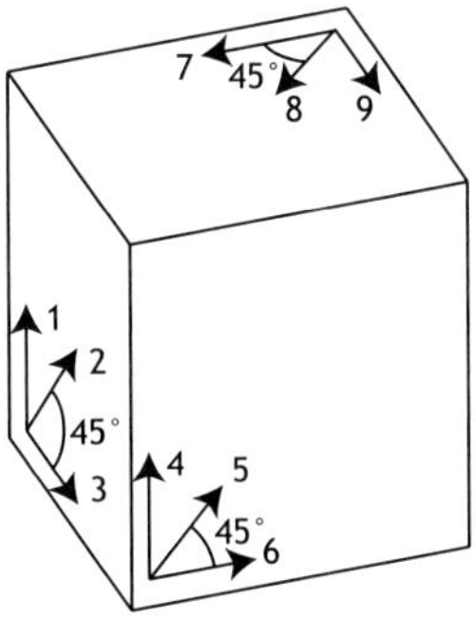

Fig. 4.10 Strain gauges on a cube of the core. The core axis is parallel to strain gauges 1 and 4.

3.1. What can you say of the rheological behavior of this material ?
3.2. Determine all the components of the small-strain tensor after 48 h.
3.3. Strain gauges 1 and 4 are parallel to the core axis whilst strain gauges 3 and 9 have been prepared perpendicular to the N0°E direction. What are the orientations of the three principal strain directions?
3.4. What are the principal components of the mean strain-rate tensor?
3.5. Is the rock sample still cubic after 48 h? What are the changes?
3.6. What is the total change in volume?

5 Elements of linear elasticity

Linear elasticity theory describes stress and strain fields in solid materials in which the particle displacements remain very small, so that the strain is related to the displacement gradient by equation (4.54), $\varepsilon_{ij} = \frac{1}{2}(u_{i,j} + u_{j,i})$, where ε_{ij} are the small-strain tensor components and $u_{i,j}; i,j = 1,2,3$, are the displacement gradient components. Consequently, all equations can be written in a Lagrangian frame of reference, i.e. particle displacements are expressed with respect to the particle's initial position at time $t = 0$.

Solids are characterized by the fact that they can sustain shear stresses indefinitely. In addition, the property of elasticity assumes perfect synchronization between stress variation and strain variation: in an elastic material, the deformation is instantaneous.

Because many geomaterials exhibit a solid-type viscoelastic behavior (section 2.10) and because in many problems only the long-term deformation is of interest, this long-term deformation of linear solid-type viscoelastic materials may be conveniently approximated by the solution for truly elastic solids, provided that the elastic constants are properly adjusted and the small-displacement-gradient hypothesis remains valid.

Further, when the stress–strain relationship for elastic solids is linear, the solution to boundary value problems can take advantage of the superposition principle, according to which solutions to simpler problems are superposed in such a way as to reconstruct the boundary conditions of the problem of interest.

In this chapter we address the basic concepts of linear elasticity and consider solutions for some problems discussed in later chapters.

5.1 Hooke's law for isotropic materials

Let us consider a cubic sample of homogeneous and isotropic rock with side a, to which we apply a uniaxial tensile stress $\sigma_a = F_a/a^2$ in the $\hat{\boldsymbol{I}}_3$ direction, with F_a the force applied at both ends of the sample (fig. 5.1).

The cube deforms under this load in such a way that it gets slightly longer by $\delta_3 a$ in the $\hat{\boldsymbol{I}}_3$ direction but shrinks in both the $\hat{\boldsymbol{I}}_1$ and $\hat{\boldsymbol{I}}_2$ directions by the same small amount $\delta_1 a = \delta_2 a$, while all faces of the cube remain perpendicular to each other. Hence the three nonzero components of the small-strain tensor $\tilde{\varepsilon}$ induced by the uniaxial stress σ_a are

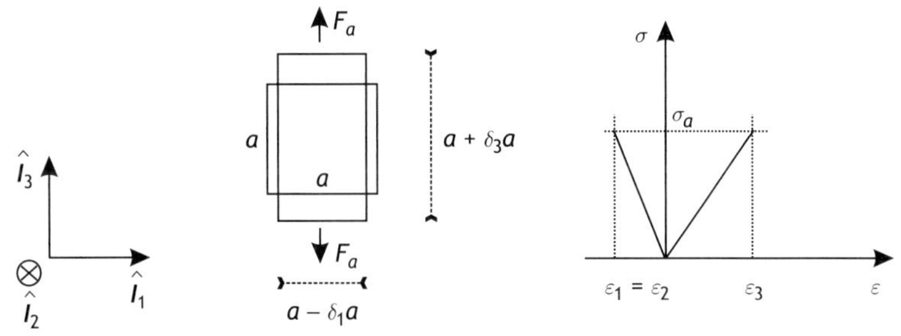

Fig. 5.1 Elastic deformation of a cube under a uniaxial stress.

$\varepsilon_3 = \delta_3 a/a, \varepsilon_1 = \varepsilon_2 = -\delta_1 a/a = -\delta_2 a/a$. We can express this by saying that the stress tensor gives rise to a strain tensor:

$$\begin{pmatrix} 0 & 0 & 0 \\ 0 & 0 & 0 \\ 0 & 0 & \sigma_a \end{pmatrix} \rightarrow \begin{pmatrix} \varepsilon_1 & 0 & 0 \\ 0 & \varepsilon_2 & 0 \\ 0 & 0 & \varepsilon_3 \end{pmatrix} \tag{5.1}$$

Experiences show that the relative changes in length $\varepsilon_1, \varepsilon_2, \epsilon_3$ are all proportional to σ_a:

$$\varepsilon_3 = \frac{\sigma_a}{E}, \qquad \varepsilon_1 = \varepsilon_2 = -\nu\varepsilon_3 = -\frac{\nu}{E}\sigma_a \tag{5.2}$$

where E is the Young's modulus of the material while ν is its Poisson's ratio.

For rocks, Young's modulus is in the range $10^4 - 10^5$ MPa and it will be shown that Poisson's ratio varies between -1 and 0.5. Consequently, the strains induced by stresses in the range of tens or hundreds of MPa are always very small, and this is even more true for the stresses associated with teleseismic waves, which are in the range $10^{-3} - 10^{-2}$ MPa . Hence, only the small-strain tensor $\varepsilon_{ij} = \frac{1}{2}(u_{i,j} + u_{j,i})$ needs to be considered.

Let us observe that the principal directions of strain and stress are the same. This is a property of isotropic materials. Anisotropic materials are considered in section 5.4.

Had we applied the uniaxial load F_a in the $\hat{\boldsymbol{I}}_1$ direction, the resulting relation between stress and strain would have been as follows:

$$\begin{pmatrix} \sigma_a & 0 & 0 \\ 0 & 0 & 0 \\ 0 & 0 & 0 \end{pmatrix} \rightarrow \begin{pmatrix} \varepsilon'_1 & 0 & 0 \\ 0 & \varepsilon'_2 & 0 \\ 0 & 0 & \varepsilon'_3 \end{pmatrix} \tag{5.3}$$

with

$$\varepsilon'_1 = \frac{\sigma_a}{E}, \qquad \varepsilon'_2 = \varepsilon'_3 = -\frac{\nu}{E}\sigma_a \tag{5.4}$$

For a uniaxial stress in the $\hat{\boldsymbol{I}}_2$ direction the resulting strain is

$$\varepsilon''_2 = \frac{\sigma_a}{E}, \qquad \varepsilon''_1 = \varepsilon''_3 = -\frac{\nu}{E}\sigma_a \tag{5.5}$$

and applying loads simultaneously in all three directions results in a strain that is equal to the sum of all three strains. We have

$$\begin{pmatrix} \sigma_1 & 0 & 0 \\ 0 & \sigma_2 & 0 \\ 0 & 0 & \sigma_3 \end{pmatrix} \rightarrow \begin{pmatrix} \varepsilon_1 & 0 & 0 \\ 0 & \varepsilon_2 & 0 \\ 0 & 0 & \varepsilon_3 \end{pmatrix} \tag{5.6}$$

with

$$\varepsilon_1 = \frac{\sigma_1}{E} - \frac{\nu}{E}(\sigma_2 + \sigma_3) \tag{5.7}$$

$$\varepsilon_2 = \frac{\sigma_2}{E} - \frac{\nu}{E}(\sigma_1 + \sigma_3) \tag{5.8}$$

$$\varepsilon_3 = \frac{\sigma_3}{E} - \frac{\nu}{E}(\sigma_1 + \sigma_2) \tag{5.9}$$

The linear relationship between small strain and stress is known as Hooke's law. Equations (5.7)–(5.9) may be rewritten as

$$\begin{pmatrix} \varepsilon_1 \\ \varepsilon_2 \\ \varepsilon_3 \end{pmatrix} = \begin{pmatrix} 1/E & -\nu/E & -\nu/E \\ -\nu/E & 1/E & -\nu/E \\ -\nu/E & -\nu/E & 1/E \end{pmatrix} \begin{pmatrix} \sigma_1 \\ \sigma_2 \\ \sigma_3 \end{pmatrix} \tag{5.10}$$

or $\tilde{\varepsilon} = [D]\tilde{\sigma}$, where $[D]$ is known as the compliance matrix of the material. It involves only two independent coefficients for isotropic materials, here E and ν.

Similarly, we may write $\tilde{\sigma} = [D]^{-1}\tilde{\varepsilon} = [C]\tilde{\varepsilon}$, where $[C]$ is known as the material stiffness matrix. It is generally defined with respect to Lamé's coefficients λ and G:

$$\sigma_1 = (\lambda + 2G)\varepsilon_1 + \lambda\varepsilon_2 + \lambda\varepsilon_3 = \lambda\varepsilon_v + 2G\varepsilon_1 \tag{5.11}$$

$$\sigma_2 = \lambda\varepsilon_1 + (\lambda + 2G)\varepsilon_2 + \lambda\varepsilon_3 = \lambda\varepsilon_v + 2G\varepsilon_2 \tag{5.12}$$

$$\sigma_3 = \lambda\varepsilon_1 + \lambda\varepsilon_2 + (\lambda + 2G)\varepsilon_3 = \lambda\varepsilon_v + 2G\varepsilon_3 \tag{5.13}$$

where $\varepsilon_v = \varepsilon_1 + \varepsilon_2 + \varepsilon_3$ is the volumetric strain. The constant G is also called the shear modulus or Coulomb modulus (sometimes referred to as μ).

For a uniaxial stress σ_1 ($\sigma_2 = \sigma_3 = 0$), equations (5.12) and (5.13) yield

$$\lambda\varepsilon_v + 2G\varepsilon_2 = \lambda\varepsilon_v + 2G\varepsilon_3 = 0 \tag{5.14}$$

and

$$\varepsilon_2 = \varepsilon_3 = \frac{-\lambda}{2(\lambda + G)}\varepsilon_1 \tag{5.15}$$

so that $\nu = -\varepsilon_2/\varepsilon_1 = \lambda/[2(\lambda + G)]$.

Adding equations (5.11)–(5.13) provides a relationship between the spherical stress component σ_m and the volumetric strain:

$$\varepsilon_v = \frac{\sigma_1 + \sigma_2 + \sigma_3}{3\lambda + 2G} = \frac{\operatorname{tr}\tilde{\sigma}}{3K} = \frac{\sigma_m}{K} \tag{5.16}$$

Here K is the modulus of incompressibility or *bulk modulus*: $K = (3\lambda + 2G)/3$.

Equation (5.10) for Hooke's law is expressed in the principal stress frame of reference. An expression for any other frame of reference is obtained by applying the rule for a change of frame of reference (see equation (3.61), discussed in section 3.4 and the appendix). Let $Q_{ij}; j = 1, 2, 3$, be the components of eigenvector $\boldsymbol{e}_i$ in the frame of reference $\hat{\boldsymbol{I}}_j; j = 1, 2, 3$. If σ_1, σ_2 and σ_3 are the principal stress components, the stress tensor components σ_{ij} in the $\hat{\boldsymbol{I}}_j; j = 1, 2, 3$, frame of reference are given by

$$\begin{pmatrix} \sigma_{11} & \sigma_{12} & \sigma_{13} \\ \sigma_{21} & \sigma_{22} & \sigma_{23} \\ \sigma_{31} & \sigma_{32} & \sigma_{33} \end{pmatrix} = \begin{pmatrix} Q_{11} & Q_{21} & Q_{31} \\ Q_{12} & Q_{22} & Q_{32} \\ Q_{13} & Q_{23} & Q_{33} \end{pmatrix} \begin{pmatrix} \sigma_1 & 0 & 0 \\ 0 & \sigma_2 & 0 \\ 0 & 0 & \sigma_3 \end{pmatrix} \begin{pmatrix} Q_{11} & Q_{12} & Q_{13} \\ Q_{21} & Q_{22} & Q_{23} \\ Q_{31} & Q_{32} & Q_{33} \end{pmatrix} \tag{5.17}$$

Let us consider for example the components σ_{11} and σ_{12}:

$$\sigma_{11} = Q_{11}^2\sigma_1 + Q_{21}^2\sigma_2 + Q_{31}^2\sigma_3 \tag{5.18}$$

$$\sigma_{12} = Q_{11}Q_{12}\sigma_1 + Q_{21}Q_{22}\sigma_2 + Q_{31}Q_{32}\sigma_3 \tag{5.19}$$

In (5.18) and (5.19) we may replace the values of σ_1, σ_2 and σ_3 by their expressions in terms of the principal strain component, (5.11)–(5.13):

$$\sigma_{11} = Q_{11}^2(\lambda\varepsilon_v + 2G\varepsilon_1) + Q_{21}^2(\lambda\varepsilon_v + 2G\varepsilon_2) + Q_{31}^2(\lambda\varepsilon_v + 2G\varepsilon_3) \tag{5.20}$$

$$\begin{aligned}\sigma_{12} = {} & Q_{11}Q_{12}(\lambda\varepsilon_v + 2G\varepsilon_1) + Q_{21}Q_{22}(\lambda\varepsilon_v + 2G\varepsilon_2) \\ & + Q_{31}Q_{32}(\lambda\varepsilon_v + 2G\varepsilon_3)\end{aligned} \tag{5.21}$$

Regrouping terms we may write

$$\sigma_{11} = (Q_{11}^2 + Q_{21}^2 + Q_{31}^2)\lambda\varepsilon_v + 2G(Q_{11}^2\varepsilon_1 + Q_{21}^2\varepsilon_2 + Q_{31}^2\varepsilon_3) \tag{5.22}$$

$$\begin{aligned}\sigma_{12} = {} & (Q_{11}Q_{12} + Q_{21}Q_{22} + Q_{31}Q_{32})\lambda\varepsilon_v + 2G(Q_{11}Q_{12}\varepsilon_1 \\ & + Q_{21}Q_{22}\varepsilon_2 + Q_{31}Q_{32}\varepsilon_3)\end{aligned} \tag{5.23}$$

We note that

$$Q_{11}^2 + Q_{21}^2 + Q_{31}^2 = \hat{\boldsymbol{I}}_1^2 = 1 \qquad \text{and} \qquad Q_{11}^2\varepsilon_1 + Q_{21}^2\varepsilon_2 + Q_{31}^2\varepsilon_3 = \varepsilon_{11}$$

Further,

$$Q_{11}Q_{12} + Q_{21}Q_{22} + Q_{31}Q_{32} = \hat{\boldsymbol{I}}_1 \cdot \hat{\boldsymbol{I}}_2 = 0$$

and

$$Q_{11}Q_{12}\varepsilon_1 + Q_{21}Q_{22}\varepsilon_2 + Q_{31}Q_{32}\varepsilon_3 = \varepsilon_{12}$$

Hence equations (5.18) and (5.19) may be rewritten as

$$\sigma_{11} = \lambda\varepsilon_v + 2G\varepsilon_{11}, \quad \sigma_{12} = 2G\varepsilon_{12} \tag{5.24}$$

More generally, we conclude that Hooke's law is expressed in an arbitrary frame of reference by

$$\sigma_{ij} = \lambda\varepsilon_v\delta_{ij} + 2G\varepsilon_{ij} = \lambda\varepsilon_{kk}\delta_{ij} + 2G\varepsilon_{ij} \tag{5.25}$$

We may derive a very similar expression for the inverse relationship, in which strain is expressed as a function of stress:

$$\varepsilon_1 = \frac{\sigma_1}{E} - \frac{\nu}{E}(\sigma_2 + \sigma_3) = \frac{1+\nu}{E}\sigma_1 - \frac{\nu}{E}\,\mathrm{tr}\,\tilde{\sigma} \tag{5.26}$$

When $\tilde{\varepsilon}$ is expressed in an arbitrary frame of reference, following the same procedure as for stress we obtain

$$\varepsilon_{ij} = \frac{1+\nu}{E}\sigma_{ij} - \frac{\nu}{E}\sigma_{kk}\delta_{ij} \tag{5.27}$$

We can express the relationship between the spherical component of stress and the volumetric strain by summing all equations, using the summation convention for repeated indices:

$$\varepsilon_{ii} = \varepsilon_v = \frac{1+\nu}{E}\sigma_{kk} - \frac{3\nu}{E}\sigma_{ll} = \frac{1-2\nu}{E}\sigma_{ll} = 3\sigma_m\left(\frac{1-2\nu}{E}\right) \tag{5.28}$$

where $\varepsilon_{ii} = \varepsilon_{11} + \varepsilon_{22} + \varepsilon_{33}$ etc. and from (5.16) we conclude that

$$\frac{\Delta V}{V} = \frac{\sigma_m}{K} = \frac{1-2\nu}{E}\sigma_{ll} \qquad \text{so that} \qquad K = \frac{E}{3(1-2\nu)} \tag{5.29}$$

Further, we have

$$\varepsilon_{12} = \frac{1+\nu}{E}\sigma_{12} = \frac{\sigma_{12}}{2G} \qquad \text{so that} \qquad G = \frac{E}{2(1+\nu)} \tag{5.30}$$

However, K and G are always positive (the strain has the same sign as the stress causing it), so we can deduce that $-1 < \nu < 0.5$. It may be observed that as ν gets closer to 0.5, the value of K gets larger and larger. When $\nu = 0.5$ the material is incompressible. For such materials G is zero, so $\nu = \lambda/[2(\lambda + G)] = 0.5$.

Let us observe that for a linearly elastic material there is no change in volume associated with a shear stress. It will be shown that rocks may in fact experience a volume change with shear (an effect that has very important consequences for the stability of rock deformation); this may be modeled with nonlinear elasticity (see chapter 8).

For many rocks we have $0.2 \le \nu \le 0.35$, so that sometimes we may simply take ν as equal to 0.25. In such instances, $G = \lambda$, $K = 5G/3$ and $E = 5G/2$.

In summary, we have introduced five different constants, E, ν, K, G, λ, to describe the linearly elastic behavior of an isotropic material but only two of these constants are independently defined and the following relationships exist between the various constants:

$$\begin{gathered} E = \frac{(3\lambda + 2G)G}{\lambda + G}, \qquad \nu = \frac{\lambda}{2(\lambda + G)} \\ K = \frac{E}{3(1-2\nu)} = \frac{3\lambda + 2G}{3} \\ G = \frac{E}{2(1+\nu)}, \qquad \lambda = \frac{E\nu}{(1+\nu)(1-2\nu)} \end{gathered} \tag{5.31}$$

Finally, because we are considering isotropic materials, the expression for Hooke's law does not depend on the frame of reference. For example, in cylindrical coordinates, Hooke's law is simply written in matrix notation

$$\begin{pmatrix} \sigma_{\rho\rho} \\ \sigma_{\theta\theta} \\ \sigma_{zz} \\ \sigma_{\rho\theta} \\ \sigma_{\theta z} \\ \sigma_{z\rho} \end{pmatrix} = \begin{pmatrix} \lambda + 2G & \lambda & \lambda & 0 & 0 & 0 \\ \lambda & \lambda + 2G & \lambda & 0 & 0 & 0 \\ \lambda & \lambda & \lambda + 2G & 0 & 0 & 0 \\ 0 & 0 & 0 & 2G & 0 & 0 \\ 0 & 0 & 0 & 0 & 2G & 0 \\ 0 & 0 & 0 & 0 & 0 & 2G \end{pmatrix} \begin{pmatrix} \varepsilon_{\rho\rho} \\ \varepsilon_{\theta\theta} \\ \varepsilon_{zz} \\ \varepsilon_{\rho\theta} \\ \varepsilon_{\theta z} \\ \varepsilon_{z\rho} \end{pmatrix} \tag{5.32}$$

5.2 Navier's equation

Problems in elasticity generally involve mixed boundary value problems, i.e. some components of stress and/or some components of displacement are known on parts of the boundary of a body B. The objective is to determine the missing components of stress and displacement on the boundary and to characterize the stress field and the displacement field in the complete volume of the body.

In chapter 3 we saw that the equilibrium conditions for the body to be at rest are described by

$$\tilde{\sigma} = \tilde{\sigma}^T \tag{5.33}$$

$$\nabla \cdot \tilde{\sigma} + \varrho \boldsymbol{b} = 0 \tag{5.34}$$

for all points $\boldsymbol{x}$ in the body B. In section 6.1.2 we will demonstrate that when the body is in motion, the law of motion must satisfy, for all $\boldsymbol{x}$ in B,

$$\nabla \cdot \tilde{\sigma} + \varrho \boldsymbol{b} = \varrho \ddot{\boldsymbol{u}} \tag{5.35}$$

or, in terms of components,

$$\sigma_{ji,i} + \varrho b_j = \varrho \ddot{u}_j, \quad i,j = 1,2,3 \tag{5.36}$$

where $\ddot{\boldsymbol{u}}$ is the acceleration at point $\boldsymbol{x}$.

We consider here the general form of the equation of motion (5.35), rather than its expression when the body is at rest, (5.33).

From Hooke's law (equations (5.25)) we can express the stress components in terms of the strain and from equation (4.54) we may express the strain components as functions of the partial derivatives of the displacement, when displacements are very small:

$$\sigma_{ij} = 2G\varepsilon_{ij} + \lambda \varepsilon_v \delta_{ij} \tag{5.37}$$

$$\varepsilon_{ij} = \tfrac{1}{2}(u_{i,j} + u_{j,i}) \tag{5.38}$$

Inserting (5.38) into (5.37) we can express Hooke's law as a function of the partial derivatives of the displacement:

$$\sigma_{ij} = \lambda u_{k,k} \delta_{ij} + G(u_{i,j} + u_{j,i}) \tag{5.39}$$

If we differentiate equation (5.39) with respect to x_l we obtain

$$\sigma_{ij,l} = \lambda u_{k,kl}\delta_{ij} + G(u_{i,jl} + u_{j,il}) \tag{5.40}$$

Then, summing the equations for $l = 1, 2$ and 3 gives

$$\sigma_{ij,i} = \lambda u_{k,kj} + G(u_{i,ji} + u_{j,ii}) \tag{5.41}$$

Thus we may express the law of motion in terms of the partial derivatives of the displacement after observing that $u_{i,ji} = u_{i,ij} = u_{k,kj}$:

$$(\lambda + G)u_{i,ij} + Gu_{j,ii} + \varrho b_j = \varrho \ddot{u}_j, \quad i, j = 1, 2, 3 \tag{5.42}$$

or, in vector form,

$$(\lambda + G)\nabla(\nabla \cdot \boldsymbol{u}) + G\nabla^2\boldsymbol{u} + \varrho\boldsymbol{b} = \varrho\ddot{\boldsymbol{u}} \tag{5.43}$$

where

$$\nabla^2 u_j = \frac{\partial^2 u_j}{\partial x_1^2} + \frac{\partial^2 u_j}{\partial x_2^2} + \frac{\partial^2 u_j}{\partial x_3^2} = u_{j,ii}$$

is the Laplacian of u_j.

For a body at rest under surface forces only (i.e. with negligible body forces), equation (5.43) becomes

$$(\lambda + G)\nabla(\nabla \cdot \boldsymbol{u}) + G\,\nabla^2\boldsymbol{u} = 0 \tag{5.44}$$

If one differentiates equation (5.44) with respect to x_l and then sums the component equations for $i, j = 1, 2, 3$ one obtains

$$\begin{aligned}(\lambda + G)u_{i,ijj} + Gu_{j,iij} &= (\lambda + 2G)u_{i,ijj} = 0 \\ \Rightarrow \qquad \nabla^2\varepsilon_v &= 0\end{aligned} \tag{5.45}$$

which shows that, in the absence of body forces, the volumetric strain in an elastic body satisfies the Laplace equation once the body has reached equilibrium.

In cylindrical coordinates, the equilibrium conditions (3.103)–(3.105), together with the constitutive equation for elasticity (equation (5.32)) and the definition of a small strain (equation (4.103)) yield Navier's equation. In components, this becomes

$$(\lambda + G)\Delta_{,\rho} + G\left(u_{\rho,\rho\rho} + \frac{1}{\rho}u_{\rho,\rho} - \frac{u_\rho}{\rho^2} + \frac{1}{\rho^2}u_{\rho,\theta\theta} - \frac{2}{\rho^2}u_{\theta,\theta} + u_{\rho,zz}\right) + \varrho f b_\rho = 0 \tag{5.46}$$

$$\frac{(\lambda + G)}{\rho}\Delta_{,\theta} + G\left(u_{\theta,\rho\rho} + \frac{1}{\rho}u_{\theta,\rho} - \frac{u_\theta}{\rho^2} + \frac{1}{\rho^2}u_{\theta,\theta\theta} + \frac{2}{\rho^2}u_{\rho,\theta} + u_{\theta,zz}\right) + \varrho f b_\theta = 0 \tag{5.47}$$

$$(\lambda + G)\Delta_{,z} + G\left(u_{z,\rho\rho} + \frac{1}{\rho}u_{z,\rho} + \frac{1}{\rho^2}u_{z,\theta\theta} + u_{z,zz}\right) + \varrho f b_z = 0 \tag{5.48}$$

where

$$\Delta = u_{\rho,\rho} + \frac{u_\rho}{\rho} + \frac{1}{\rho}u_{\theta,\theta} + u_{z,z}$$

is the volumetric strain.

5.3 Simple problems in elasticity

In truly three-dimensional elasticity problems one has to solve partial differential equations with three or four variables (x_i; $i = 1, 2, 3; t$) depending on whether the time t is to be considered. For many practical problems of elasticity encountered in static geomechanics, however, simple considerations of symmetry or boundary conditions provide a means to decrease the number of variables to two or even one. We examine in what follows a few simple problems of practical use in geomechanics.

5.3.1 Uniaxial stress

A simple case is that of uniaxial stress ($\sigma_2 = \sigma_3 = 0$). It corresponds to conditions encountered in tests designed for measuring in the laboratory Young's modulus and Poisson's ratio for rocks, as already mentioned:

$$\varepsilon_1 = \frac{\sigma_1}{E}, \qquad \varepsilon_2 = \varepsilon_3 = -\frac{\nu}{E}\sigma_1 \tag{5.49}$$

This is also considered to be the condition encountered near the center of pillars left in place by the room-and-pillar mining technique. It is the condition encountered in beam theory as well.

The bending and buckling of beams

A beam is a structural element with one dimension (in the x_1 direction) much larger than the other two (fig. 5.2). Here we consider beams with cross sections that are symmetrical with respect to both the x_2 and x_3 directions.

Beam-shaped rock samples are frequently tested in the laboratory in investigations of tensile fracturing processes (chapter 7). Furthermore, understanding the bending and buckling of beams is a good introduction to understanding stresses in folds or the deformation process in thin sedimentary layers loaded parallel to the sedimentary planar interfaces.

We adopt the Euler–Bernoulli hypothesis, i.e. that planar cross sections of a beam remain planar when the beam is being bent. Each longitudinal fiber of the beam behaves as if in uniaxial compression or tension, independently of the other fibers.

Compressions are taken as positive so that *shortening* is a positive strain.

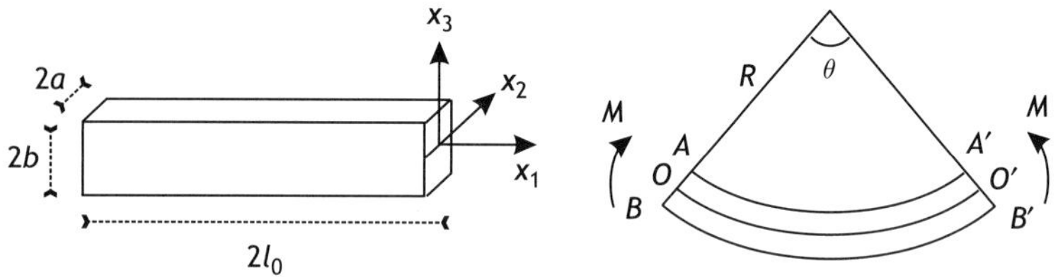

Fig. 5.2 A rectangular beam loaded with a moment M at both ends of the beam. The right-hand diagram gives a plan view of the beam.

Bending

Let $2l_0$ be the original length of the beam and $2a$ and $2b$ its width and thickness (fig. 5.2). We consider a moment M applied at both ends of the beam in the (x_1, x_3) plane such that the angle between the extreme sections AOB and $A'O'B'$ is equal to θ.

The region $x_3 > 0$ is under compression while the region $x_3 < 0$ is in extension. For $x_3 = 0$ (the OO' axis) no change of length occurs. It is called the neutral fiber.

Let R be the radius of curvature of the beam for the neutral fiber OO': $2l_0 = R\theta$. For any fiber $x_3 > 0$ with cross section dx_2dx_3, the length after bending is $2l_0 - dl = (R - x_3)\theta$, so that the strain component in the fiber is $\varepsilon_1 = dl/(2l_0) = x_3\theta/(R\theta) = \sigma_1/E$. The sign "plus" for ε_1 results from our convention that compressions and therefore shortenings are positive.

Hence

$$\sigma_1 = \frac{Ex_3}{R}, \qquad \text{or} \qquad \frac{\sigma_1}{x_3} = \frac{E}{R} \tag{5.50}$$

The further away from the neutral fiber, the larger the stress, which is compressive or tensile depending on the sign of x_3.

The moment M that generates such a stress field is

$$M = \int_S \sigma_1 \, dx_2 \, dx_3 \, x_3 = \int_{x_2} \int_{x_3} \frac{E}{R} x_3^2 \, dx_2 \, dx_3 \tag{5.51}$$

where S is the beam's cross section. The quantity $\int_{x_2} \int_{x_3} x_3^2 \, dx_2dx_3$ is defined as the moment of inertia I of the beam's cross section with respect to the x_2 axis. It depends only on the geometry of the beam.

If $u_3 = u_3(x_1)$ describes the shape of the beam once it is bent, the radius of curvature R is approximately equal to $(d^2u_3/dx_1^2)^{-1}$. Hence the relationship between the moment M, the beam radius of curvature R and the moment of inertia I of the beam's cross section is

$$M = \frac{EI}{R} = EI\frac{d^2u_3}{dx_1^2}. \tag{5.52}$$

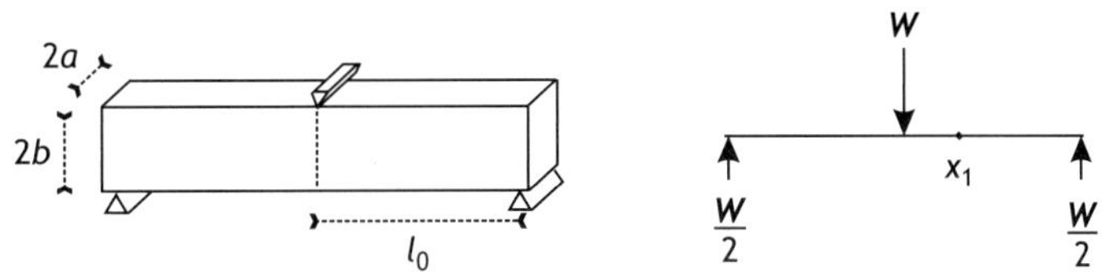

Fig. 5.3 The three-point bending test.

Using (5.52) in (5.50), we obtain the stress as a function of the moment, $\sigma_1 = x_3 M/I$ and find that it reaches its maximum values for $x_3 = \pm b$, so that

$$\sigma_1 = \pm \frac{bM}{I} \tag{5.53}$$

An example of bending is the so-called *three-point* bending test used to investigate the tensile strength of rock in the laboratory (fig. 5.3).

A beam of length $2l_0$ and cross-sectional area $S = 4ab$ is freely supported at both ends. It is submitted to a uniformly distributed load at the center of its upper surface, and the situation may be modeled by the one dimensional diagram shown in fig. 5.3: W is the load applied at the center of the beam and x_1 is the location where the moment is evaluated, $x_1 \in [-l_0, l_0]$.

The beam is at rest, so the moment M of the forces to the left of x_1 is equal in magnitude, but with the opposite sense of rotation, to that exerted by the forces to the right of point x_1. Because of symmetry we consider only the domain $x_1 \in [0, l_0]$. The moment

$$M = (l_0 - x_1)\frac{W}{2} \tag{5.54}$$

is null for $x_1 = l_0$ and reaches its maximum value for $x_1 = 0$, where it is equal to $Wl_0/2$. The moment of inertia for the cross section of the beam is

$$I = \int_{-b}^{+b} \int_{-a}^{+a} x_3^2 \, dx_3 \, dx_2 = \frac{4ab^3}{3} \tag{5.55}$$

Hence the maximum tensile stress in the beam is reached for $x_1 = 0$, $x_3 = -b$ and is equal to

$$\sigma_1 = -\frac{3Wl_0}{8ab^2} \tag{5.56}$$

The differential equation that describes the deflection $u_3(x_1)$ of the beam caused by the moments M is obtained by combining equations (5.54) and (5.52):

$$EI\frac{d^2u_3}{dx_1^2} = (l_0 - x_1)\frac{W}{2} \tag{5.57}$$

The deflection of the beam is obtained by integrating equation (5.57) twice so as to satisfy the boundary condition $u_3 = 0$, $du_3/dx_1 = 0$ at the center of the beam (i.e. for $x_1 = 0$):

$$EIu_3 = \frac{W}{2}\left(\tfrac{1}{2}l_0 x_1^2 - \tfrac{1}{6}x_1^3\right) \tag{5.58}$$

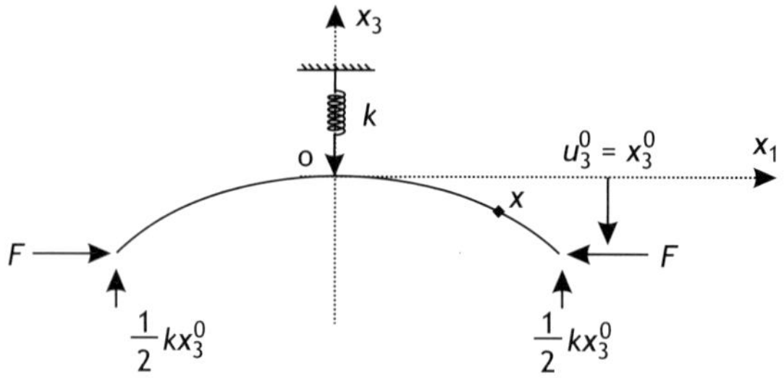

Fig. 5.4 Buckling of a beam.

The value obtained for $x_1 = l_0$ gives the deflection at the center of the beam as $u_3 = wl_0^3/(6EI)$. Measuring the deflection of the beam as the load W is increased thus provides a direct measurement of the Young's modulus of the material.

Buckling

Suppose now that the beam is freely hinged at both its extremities, where it is loaded by equal and opposite uniaxial loads F in the direction of the beam (fig. 5.4).

The undeflected solution $u_3(x_1) = 0$ is always a possible solution. However, for high enough values of F this solution becomes unstable and the beam is said to buckle, i.e. a deflection appears.

This is best modeled by assuming that, in addition to the axial load F, a vertical load $W = ku_3$ is exerted by a spring of stiffness k in the direction perpendicular to the beam and acting at its center O. Assume that the point O has been displaced by $u_3^0 = x_3^0$. In order to maintain equilibrium, the force $W = kx_3^0$ must be balanced by forces $-\frac{1}{2}kx_3^0$ applied at both extremities of the beam.

From equation (5.52), taking the origin at O, the moment acting at point x_1 is

$$EI\frac{d^2u_3}{dx_1^2} = F(x_3^0 - u_3) - \frac{1}{2}kx_3^0(l_0 - x_1) \tag{5.59}$$

or

$$\frac{d^2u_3}{dx_1^2} + \omega^2 u_3 = \omega^2 x_3^0 - \frac{k}{2F}\omega^2 x_3^0(l_0 - x_1) \tag{5.60}$$

where

$$\omega^2 = \frac{F}{EI} \tag{5.61}$$

The general solution of equation (5.60) is

$$u_3 = A\sin\omega x_1 + B\cos\omega x_1 + x_3^0 - \frac{k}{2F}x_3^0(l_0 - x_1) \tag{5.62}$$

where A and B are constants that are determined by the boundary conditions,

$$u_3 = 0, \qquad \frac{du_3}{dx_1} = 0 \quad \text{for} \quad x_1 = 0 \tag{5.63}$$

$$u_3 = x_3^0 \quad \text{for} \quad x_1 = l_0 \tag{5.64}$$

Hence the constants A and B are such that

$$B + x_3^0 - \frac{kl_0 x_3^0}{2F} = 0 \qquad \text{and} \qquad \omega A + \frac{kx_3^0}{2F} = 0 \tag{5.65}$$

and

$$A \sin \omega l_0 + B \cos \omega l_0 = 0 \tag{5.66}$$

The solution to the problem is then given by

$$x_3^0 \left[\sin \omega l_0 - \omega l_0 \left(1 - \frac{2EI\omega^2}{kl_0} \right) \cos \omega l_0 \right] = 0 \tag{5.67}$$

The only solution of (5.67) is $x_3^0 = 0$ unless ωl_0 is a solution of

$$\tan \omega l_0 = \omega l_0 \left(1 - \frac{2EI\omega^2}{kl_0} \right) \tag{5.68}$$

If $k = 0$, conditions (5.65) and (5.66) reduce to $B \cos \omega l_0 = 0$ or $\omega l_0 = \pi/2$, so that from (5.61) we get

$$F = \frac{\pi^2 EI}{4l_0^2} \tag{5.69}$$

The critical load given by equation (5.69) is called the Euler load in the case of a freely hinged column. When the axial load F is smaller than the Euler load, the only solution is the undeflected solution. When the load becomes equal to or larger than the Euler load the undeflected solution becomes unstable and buckling occurs.

This has consequences for defining the critical length of rock samples tested under uniaxial compression in an investigation of their so-called unconfined compressive strength (see chapter 8). Generally the length of cylindrical samples tested in uniaxial stress conditions is equal to 2 or 3 times the sample diameter.

Similarly, when a thin plate is compression loaded parallel to its main dimension, the plate buckles beyond a critical load. For horizontal sedimentary formation, loads parallel to the bedding plane result in folding.

Once buckling or folding starts, the displacement gradients are no longer infinitesimal and the small-strain approximation does not hold. The analysis of geological folding requires a finite-strain approach, for which a linear elasticity formulation is not valid.

5.3.2 Uniaxial strain

For conditions of uniaxial strain, $\varepsilon_1 = \varepsilon_2 = 0$, the Hooke equation yields

$$\sigma_1 = \lambda\varepsilon_3 = \sigma_2 \tag{5.70}$$

and from equation (5.27) we conclude

$$\varepsilon_1 = \frac{\sigma_1}{E} - \frac{\nu}{E}\sigma_1 - \frac{\nu}{E}\sigma_3 = 0$$

so that

$$\sigma_1 = \sigma_2 = \frac{\nu}{1-\nu}\sigma_3 \tag{5.71}$$

Such conditions are encountered in the problem of a semi-infinite space bounded by the horizontal plane $x_3 = 0$ and occupied by material with density ϱ loaded under gravity. The solution cannot depend on the x_1 and x_2 coordinates, and therefore all derivatives with respect to x_1 and x_2 must vanish ($\varepsilon_1 = u_{1,1} = \varepsilon_2 = u_{2,2} = 0,\ \sigma_{3,1} = \sigma_{3,2} = 0$). From the equilibrium conditions (equations (5.33)) we note that

$$\sigma_{3,3} - \varrho g = 0 \tag{5.72}$$

from which we conclude that

$$\sigma_3 = \varrho g x_3 \qquad \text{and} \qquad \sigma_1 = \sigma_2 = \frac{\nu}{1-\nu}\varrho g x_3 \tag{5.73}$$

In addition, if it is assumed that $\nu = 0.25$ then $\sigma_1 = \sigma_2 = \varrho g x_3/3$.

This solution has often been used to estimate the stress field at depth in homogeneous rock masses where the topography is not significant. However, this situation is never observed in reality, the main reason being that rocks are not formed without gravity and then loaded under gravity. Rocks always result from a complex diagenesis process that develops continuously under gravity. This is discussed in more detail in chapter 13.

5.3.3 Biaxial or plane stress conditions

In biaxial or plane stress conditions one stress component, say σ_3, is zero, i.e. the $\hat{\boldsymbol{I}}_3$ direction of the $\hat{\boldsymbol{I}}_i;\ i = 1, 2, 3$, frame of reference is a principal stress direction and therefore also a principal strain direction for isotropic materials.

This is the condition that prevails at surfaces that are free of load such as the ground surface. Hence in mountainous areas where the ground surface is inclined to the horizontal, no principal stress direction is vertical.

In biaxial stress conditions Hooke's equation yields

$$E\varepsilon_1 = \sigma_1 - \nu\sigma_2, \quad E\varepsilon_2 = \sigma_2 - \nu\sigma_1 \tag{5.74}$$

$$E\varepsilon_3 = -\nu(\sigma_1 + \sigma_2) \tag{5.75}$$

or, alternatively,

$$(\lambda + 2G)\sigma_1 = 4G(\lambda + G)\varepsilon_1 + 2\lambda G\varepsilon_2 \tag{5.76}$$

$$(\lambda + 2G)\sigma_2 = 2\lambda G\varepsilon_1 + 4G(\lambda + G)\varepsilon_2 \tag{5.77}$$

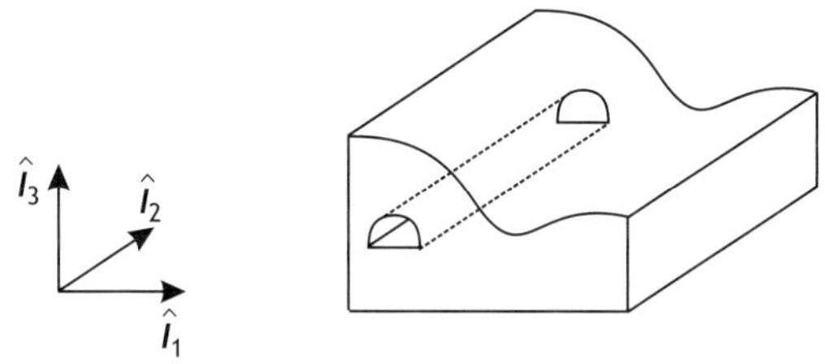

Fig. 5.5 Example of a geometry that leads to plane strain conditions.

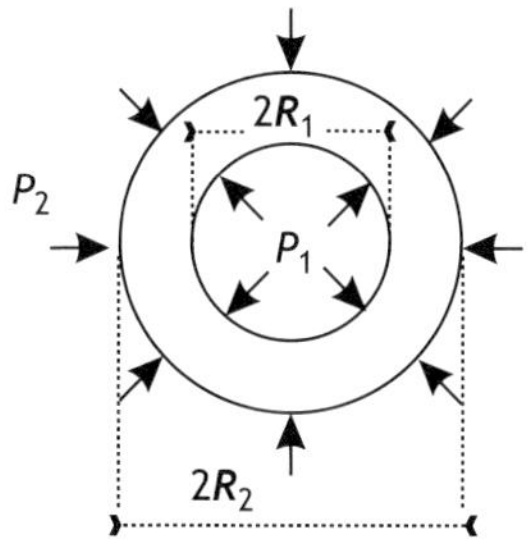

Fig. 5.6 Internally and externally pressurized thick cylinder.

5.3.4 Biaxial or plane strain conditions

In biaxial or plane strain conditions, one strain component is null, say $\varepsilon_2 = 0$, so that

$$\varepsilon_2 = \frac{\sigma_2}{E} - \frac{\nu}{E}(\sigma_1 + \sigma_3) = 0 \tag{5.78}$$

or

$$\sigma_2 = \nu(\sigma_1 + \sigma_3) \tag{5.79}$$

The $\hat{\boldsymbol{I}}_2$ direction of the $\hat{\boldsymbol{I}}_i$; $i = 1, 2, 3$, frame of reference is a principal strain and principal stress direction. This is typically the case for problems in which the geometry does not vary along one direction, such as a long valley, mountain range, tunnel or borehole (see fig. 5.5).

An example of plane strain conditions occurs when an infinite thick cylinder is loaded by both internal and external pressures (fig. 5.6). Let P_1 be the pressure exerted on the inner surface, of radius R_1, and P_2 the pressure exerted on the external surface, of radius R_2.

Because the problem is axisymmetrical, it is solved in cylindrical coordinates (ρ, θ, z) and the solution must not depend on either θ or z. Hence the equilibrium conditions given by equation (4.96) reduce to

$$\sigma_{\rho\rho,\rho} + \frac{\sigma_{\rho\rho} - \sigma_{\theta\theta}}{\rho} = 0 \tag{5.80}$$

$$\sigma_{\rho\theta,\rho} + \frac{2}{\rho}\sigma_{\rho\theta} = 0 \tag{5.81}$$

$$\sigma_{\rho z,\rho} + \frac{\sigma_{z\rho}}{\rho} = 0 \tag{5.82}$$

From the boundary conditions we know that

$$\sigma_{\rho\theta}|_{\rho=R1} = \sigma_{\rho\theta}|_{\rho=R2} = 0 \tag{5.83}$$

so that, from equation (5.81) we conclude that $\sigma_{\rho\theta} = 0\ \forall\rho$. Similarly, from the boundary conditions and from equation (5.82) we conclude $\sigma_{\rho z} = 0\ \forall\rho$ and, from Hooke's law, $\varepsilon_{z\rho} = \sigma_{z\rho}/(2G) = \varepsilon_{\rho\theta} = \sigma_{\rho\theta}/(2G) = 0$. Furthermore, because the solution does not depend on z or θ we note that $\varepsilon_{zz} = \varepsilon_{\theta z} = 0$.

By using equation (4.103) for a small strain in terms of the partial derivatives of displacement, as expressed in cylindrical coordinates, the problem reduces to

$$\sigma_{\rho\rho,\rho} + \frac{\sigma_{\rho\rho} - \sigma_{\theta\theta}}{\rho} = 0 \tag{5.84}$$

$$\sigma_{\rho\rho} = \lambda\left(u_{\rho,\rho} + \frac{u_\rho}{\rho}\right) + 2Gu_{\rho,\rho} \tag{5.85}$$

$$\sigma_{\theta\theta} = \lambda\left(u_{\rho,\rho} + \frac{u_\rho}{\rho}\right) + 2G\frac{u_\rho}{\rho} \tag{5.86}$$

Using equations (5.85) and (5.86) in (12.68) yields

$$u_{\rho,\rho\rho} + \frac{1}{\rho}u_{\rho,\rho} - \frac{u_\rho}{\rho^2} = 0 \tag{5.87}$$

which may be rewritten as

$$\left(u_{\rho,\rho} + \frac{u_\rho}{\rho}\right)_{,\rho} = 0$$

Its general solution is

$$u_\rho = A\rho + \frac{B}{\rho} \tag{5.88}$$

where A and B are constants that are defined so as to satisfy the boundary conditions (compressions are taken positive)

$$\sigma_{\rho\rho}|_{\rho=R1} = P_1, \qquad \sigma_{\rho\rho}|_{\rho=R2} = P_2 \tag{5.89}$$

From (5.85) we obtain

$$2A(\lambda + G) - \frac{2GB}{R_1^2} = P_1, \qquad 2A(\lambda + G) - \frac{2GB}{R_2^2} = P_2 \tag{5.90}$$

which yields

$$A = \frac{P_2R_2^2 - P_1R_1^2}{2(\lambda + G)(R_2^2 - R_1^2)}, \qquad B = \frac{(P_2 - P_1)R_2^2R_1^2}{2G(R_2^2 - R_1^2)} \tag{5.91}$$

The solution to the problem is

$$u_\rho = \frac{P_2R_2^2 - P_1R_1^2}{2(\lambda + G)(R_2^2 - R_1^2)}\rho + \frac{(P_2 - P_1)R_2^2R_1^2}{2G\rho(R_2^2 - R_1^2)} \tag{5.92}$$

$$\sigma_{\rho\rho} = \sigma_\rho = \frac{P_2R_2^2 - P_1R_1^2}{R_2^2 - R_1^2} - \frac{(P_2 - P_1)R_2^2R_1^2}{\rho^2(R_2^2 - R_1^2)} \tag{5.93}$$

$$\sigma_{\theta\theta} = \sigma_\theta = \frac{P_2R_2^2 - P_1R_1^2}{R_2^2 - R_1^2} + \frac{(P_2 - P_1)R_2^2R_1^2}{\rho^2(R_2^2 - R_1^2)} \tag{5.94}$$

A case of particular interest occurs when the external radius R_2 becomes infinite and the pressure P_2 is null. This corresponds to the problem of a long pressurized borehole in an infinite space. The radial displacement at the borehole wall caused by the pressure is

$$u_\rho|_{\rho=R_1} = -\frac{P_1R_1}{2G} \tag{5.95}$$

This method is used to measure the shear modulus G in geotechnical reconnaissance boreholes.

The nonzero stress components are

$$\sigma_{\rho\rho}|_{\rho=R_1} = P_1, \quad \sigma_{\theta\theta}|_{\rho=R_1} = -P_1 \tag{5.96}$$

When a borehole is pressurized, the tangential stress at the borehole wall corresponds to a tensile stress with the same magnitude as the applied pressure. This is used in the oil industry to stimulate unproductive wells, a process known as hydraulic fracturing. It is discussed in chapter 12.

5.3.5 Plane elastic waves

From equation (5.42), in the absence of body forces the Navier equation becomes

$$(\lambda + G)u_{j,ji} + Gu_{i,jj} = \varrho\ddot{u}_i \tag{5.97}$$

where ϱ is the rock density.

Let us consider the motion of a particle described by

$$u_2 = u_3 = 0, \quad u_1 = A\sin\left[\frac{2\pi}{l}(x_1 - Ct)\right] \tag{5.98}$$

where A, C and l are constants. The partial derivatives of u_1 are

$$u_{1,1} = \frac{2\pi A}{l}\cos\left[\frac{2\pi}{l}(x_1 - Ct)\right], \quad u_{1,11} = -\frac{4\pi^2A}{l^2}\sin\left[\frac{2\pi}{l}(x_1 - Ct)\right] \tag{5.99}$$

$$\dot{u}_1 = -CA\frac{2\pi}{l}\cos\left[\frac{2\pi}{l}(x_1 - Ct)\right], \quad \ddot{u}_1 = -AC^2\frac{4\pi^2}{l^2}\sin\left[\frac{2\pi}{l}(x_1 - Ct)\right] \tag{5.100}$$

The displacement satisfies the Navier equation if $C = C_L = \sqrt{(\lambda + 2G)/\varrho}$

The quantity C_L is the phase velocity of longitudinal waves with wavelength l. All particles for which $x_1 - Ct$ is constant sustain the same displacement u_1. The particle motion is coaxial to the direction of wave front propagation. Given the relations between the various elastic constants (see (5.31)), we may write

$$C_L = \sqrt{\frac{\lambda + 2G}{\varrho}} = \sqrt{\frac{E(1-\nu)}{(1+\nu)(1-2\nu)\varrho}} \tag{5.101}$$

Let us consider now a motion described by

$$u_1 = u_3 = 0, \quad u_2 = A \sin\left[\frac{2\pi}{l}(x_1 - Ct)\right] \tag{5.102}$$

Here $u_{j,ji} = 0$,

$$u_{2,11} = -4\pi^2 \frac{A}{l^2} \sin\left[\frac{2\pi}{l}(x_1 - Ct)\right] \tag{5.103}$$

and

$$\ddot{u}_2 = -AC^2 \frac{4\pi^2}{l^2} \sin\left[\frac{2\pi}{l}(x_1 - Ct)\right] \tag{5.104}$$

so that the Navier equation is satisfied if

$$C = C_T = \sqrt{\frac{G}{\varrho}} \tag{5.105}$$

Here C_T is the phase velocity for transverse, or shear, waves. The motion of particles is perpendicular to the direction of wave propagation.

Let us observe that the motion described by

$$u_1 = u_2 = 0, \qquad u_3 = A \sin\left[\frac{2\pi}{l}(x_1 \pm Ct)\right] \tag{5.106}$$

results in exactly the same type of wave except that the particle moves in a direction perpendicular to that of the previous example.

The plane parallel to which particles move is called the plane of polarization. Shear waves are polarized in a plane perpendicular to the direction of wave propagation.

Longitudinal (L) waves are generally called *primary* (P) waves, for they propagate fastest and arrive first, whilst shear or transverse (T) waves are called *secondary* (S) waves, for they arrive after the P waves.

The ratio between the P and S wave velocities depends on the elastic parameters of the material:

$$C_S = C_P \sqrt{\frac{1-2\nu}{2(1-\nu)}}$$

When $\nu = 0.25$, $C_P = \sqrt{3} C_S$.

Both P and S waves are body waves. When body waves reach an interface they may be reflected, refracted or both. This is discussed in chapter 11.

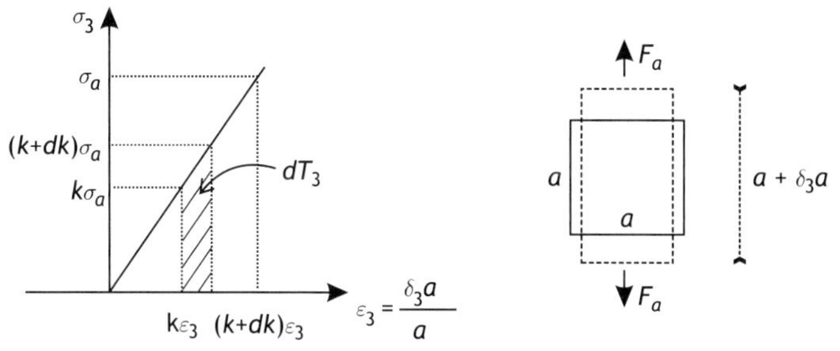

Fig. 5.7 Elastic strain energy.

5.4 Elastic strain energy and a new definition of elasticity

5.4.1 Elastic strain energy

Let us return to the discussion in the first section of this chapter, where a cube of rock is loaded progressively along the three principal stress directions. We will assume that the load is very slowly incremented so that no inertial force appears (quasistatic conditions; see fig. 5.7).

We wish to evaluate the amount of work done on the sample during the loading process. Let us consider first the work done by the $\sigma_3 = \sigma_a$ principal stress component.

The stress is incremented very slowly from 0 to σ_a. Let $k\sigma_a$, with $0 < k < 1$, be the magnitude of the stress at a given time during this loading process. The corresponding load is $k\sigma_a a^2$ and the corresponding deformation is $k\varepsilon_3 = k(\delta a)/a$; a is the original dimension of the sample in the $\boldsymbol{e}_3$ direction before loading and $k\delta a$ is the change in length caused by $k\sigma_a$.

When the stress component is incremented from $k\sigma_a$ to $(k + dk)\sigma_a$, the deformation is incremented to $(k + dk)\delta a/a$ and the incremental work dT_3 done on the specimen is

$$dT_3 = \tfrac{1}{2}\sigma_a a^2(k + k + dk)(dk\varepsilon_3 a) \tag{5.107}$$

$$\simeq k\sigma_a a^3 dk\varepsilon_3 \tag{5.108}$$

So, the total work done in loading the specimen from 0 to σ_a is

$$T_3 = \int_0^1 \sigma_a \varepsilon_3 a^3 k dk = \tfrac{1}{2} a^3 \sigma_a \varepsilon_3 \tag{5.109}$$

A similar result is obtained for the work associated with the loading by σ_1 and for that associated with σ_2. Let T be the total work exerted on the sample when all loads are applied simultaneously ($T = T_1 + T_2 + T_3$).

The elastic strain energy density w in the sample is defined by the relation

$$w = \frac{T}{dv} = \tfrac{1}{2}(\sigma_1\varepsilon_1 + \sigma_2\varepsilon_2 + \sigma_3\varepsilon_3) = \tfrac{1}{2}\sigma_i\varepsilon_i \tag{5.110}$$

where $dv = a^3$.

For a linearly elastic body of volume V in which a stress field $\sigma(\boldsymbol{X})$ generates a deformation field $\varepsilon(\boldsymbol{X})$ at all points $\boldsymbol{X}$ of the body, the elastic strain energy stored in the body under load is

$$W_e = \tfrac{1}{2}\int_V (\sigma_1\varepsilon_1 + \sigma_2\varepsilon_2 + \sigma_3\varepsilon_3)\, dv \tag{5.111}$$

While equation (5.110) is expressed as a function of both the stress and strain components, it could be expressed only in terms of the stress components or only in terms of the strain components by taking advantage of Hooke's law. For example, since $\sigma_i = \lambda\varepsilon_v + 2G\varepsilon_i$, equation (5.110) may be rewritten as

$$w = \tfrac{1}{2}[(\lambda\varepsilon_v + 2G\varepsilon_1)\varepsilon_1 + (\lambda\varepsilon_v + 2G\varepsilon_2)\varepsilon_2 + (\lambda\varepsilon_v + 2G\varepsilon_3)\varepsilon_3] \tag{5.112}$$

$$= \tfrac{1}{2}[(\lambda + 2G)(\varepsilon_1^2 + \varepsilon_2^2 + \varepsilon_3^2) + 2\lambda(\varepsilon_1\varepsilon_2 + \varepsilon_2\varepsilon_3 + \varepsilon_3\varepsilon_1)] \tag{5.113}$$

Similarly, since

$$\varepsilon_i = \frac{1+\nu}{E}\sigma_i - \frac{\nu}{E}\operatorname{tr}\tilde{\sigma}$$

we may write

$$w = \frac{1}{2E}[\sigma_1^2 + \sigma_2^2 + \sigma_3^2 - 2\nu(\sigma_1\sigma_2 + \sigma_2\sigma_3 + \sigma_3\sigma_1)] \tag{5.114}$$

However, we recall from chapter 3 or the appendix that the first and second invariants of the stress are

$$I = \operatorname{tr}\tilde{\sigma} = \sigma_1 + \sigma_2 + \sigma_3 = \sigma_{11} + \sigma_{22} + \sigma_{33} \tag{5.115}$$

$$II = \sigma_1\sigma_2 + \sigma_2\sigma_3 + \sigma_3\sigma_1 = \tfrac{1}{2}[(\operatorname{tr}\tilde{\sigma})^2 - \operatorname{tr}(\tilde{\sigma}^2)] \tag{5.116}$$

$$= \sigma_{11}\sigma_{22} + \sigma_{22}\sigma_{33} + \sigma_{33}\sigma_{11} - (\sigma_{12}^2 + \sigma_{23}^2 + \sigma_{31}^2) \tag{5.117}$$

Since

$$\sigma_1^2 + \sigma_2^2 + \sigma_3^2 = (\sigma_1 + \sigma_2 + \sigma_3)^2 - 2(\sigma_1\sigma_2 + \sigma_2\sigma_3 + \sigma_3\sigma_1) = I^2 - 2II \tag{5.118}$$

we conclude that

$$w = \frac{1}{2E}[I^2 - 2(1+\nu)II] \tag{5.119}$$

Equation (5.119) may be used to write down an expression for the elastic strain energy density in any frame of reference $\hat{\boldsymbol{I}}_i$; $i = 1, 2, 3$:

$$w = \frac{1}{2E}\{\sigma_{11}^2 + \sigma_{22}^2 + \sigma_{33}^2 + 2(\sigma_{11}\sigma_{22} + \sigma_{22}\sigma_{33} + \sigma_{33}\sigma_{11})$$
$$- 2(1+\nu)[\sigma_{11}\sigma_{22} + \sigma_{22}\sigma_{33} + \sigma_{33}\sigma_{11} - (\sigma_{12}^2 + \sigma_{23}^2 + \sigma_{31}^2)]\} \tag{5.120}$$

$$= \frac{1}{2E}[\sigma_{11}^2 + \sigma_{22}^2 + \sigma_{33}^2 - 2\nu(\sigma_{11}\sigma_{22} + \sigma_{22}\sigma_{33} + \sigma_{33}\sigma_{11})$$
$$+ 2(1+\nu)(\sigma_{12}^2 + \sigma_{23}^2 + \sigma_{31}^2)] \tag{5.121}$$

Equation (5.121) may be rewritten so as to highlight the expressions for the strain components given by Hooke's law (equation (5.27)):

$$w = \frac{1}{2}\left\{\sigma_{11}\left[\frac{\sigma_{11}}{E} - \frac{\nu}{E}(\sigma_{22}+\sigma_{33})\right] + \sigma_{22}\left[\frac{\sigma_{22}}{E} - \frac{\nu}{E}(\sigma_{33}+\sigma_{11})\right]\right.$$
$$+ \sigma_{33}\left[\frac{\sigma_{33}}{E} - \frac{\nu}{E}(\sigma_{11}+\sigma_{22})\right] + 2\sigma_{12}\frac{(1+\nu)\sigma_{12}}{E}$$
$$\left. + 2\sigma_{23}\frac{(1+\nu)\sigma_{23}}{E} + 2\sigma_{31}\frac{(1+\nu)\sigma_{31}}{E}\right\} \tag{5.122}$$

which may be further rewritten as

$$w = \frac{1}{2}(\sigma_{11}\varepsilon_{11} + \sigma_{22}\varepsilon_{22} + \sigma_{33}\varepsilon_{33} + 2\sigma_{12}\varepsilon_{12} + 2\sigma_{23}\varepsilon_{23} + 2\sigma_{31}\varepsilon_{31}) \tag{5.123}$$

or

$$w = \frac{1}{2}\sigma_{ij}\varepsilon_{ij} \tag{5.124}$$

5.4.2 A general definition of elasticity

When the expression (5.121), which relates the elastic strain energy density w to the stress, is differentiated with respect to σ_{11} or σ_{12}, it yields

$$\frac{\partial w}{\partial \sigma_{11}} = \frac{1}{2E}[2\sigma_{11} - 2\nu(\sigma_{22}+\sigma_{33})] = \varepsilon_{11} \tag{5.125}$$

$$\frac{\partial w}{\partial \sigma_{12}} = \frac{1}{2E}4(1+\nu)\sigma_{12} = 2\varepsilon_{12} \tag{5.126}$$

When considering the expression (5.121) that relates the elastic strain energy density to the strain components, it may also be shown that $\partial w/\partial\varepsilon_{11} = \sigma_{11}$ and $\partial w/\partial\varepsilon_{12} = 2\sigma_{12}$.

Let us define quantities S_i and E_i; $i = 1, 2, \ldots, 6$, by

$$S_1 = \sigma_{11}, \quad S_2 = \sigma_{22}, \quad S_3 = \sigma_{33}, \quad S_4 = \sigma_{12}, \quad S_5 = \sigma_{23}, \quad S_6 = \sigma_{31} \tag{5.127}$$

and, somewhat similarly, for the strain components,

$$E_1 = \varepsilon_{11}, \quad E_2 = \varepsilon_{22}, \quad E_3 = \varepsilon_{33}$$
$$E_4 = 2\varepsilon_{12}, \quad E_5 = 2\varepsilon_{23}, \quad E_6 = 2\varepsilon_{31} \tag{5.128}$$

We may define now an elastic body as a body for which the elastic strain energy density $w = \frac{1}{2}S_iE_i$; $i = 1, 2, \ldots, 6$, satisfies the conditions

$$S_i = \frac{\partial w}{\partial E_i}, \quad E_j = \frac{\partial w}{\partial S_j}, \quad i,j = 1, 2, \ldots, 6 \tag{5.129}$$

For small strains in linearly elastic solids, the strain energy can be assumed to be a quadratic function of the strain components:

$$w = C_0 + C_iE_i + \frac{1}{2}\bar{C}_{jk}E_jE_k \tag{5.130}$$

Regrouping terms such as $\frac{1}{2}(\bar{C}_{12}E_1E_2 + \bar{C}_{21}E_2E_1)$ into a single term $C_{12}E_1E_2$, with $C_{12} \equiv \frac{1}{2}(\bar{C}_{12} + \bar{C}_{21})$, the strain energy function becomes

$$\begin{aligned} w = {} & C_0 + C_1E_1 + C_2E_2 + C_3E_3 + C_4E_4 + C_5E_5 + C_6E_6 \\ & + \tfrac{1}{2}C_{11}E_1^2 + C_{12}E_1E_2 + C_{13}E_1E_3 + C_{14}E_1E_4 + C_{15}E_1E_5 + C_{16}E_1E_6 \\ & + \tfrac{1}{2}C_{22}E_2^2 + C_{23}E_2E_3 + C_{24}E_2E_4 + C_{25}E_2E_5 + C_{26}E_2E_6 \\ & + \tfrac{1}{2}C_{33}E_3^2 + C_{34}E_3E_4 + C_{35}E_3E_5 + C_{36}E_3E_6 \\ & + \tfrac{1}{2}C_{44}E_4^2 + C_{45}E_4E_5 + C_{46}E_4E_6 \\ & + \tfrac{1}{2}C_{55}E_5^2 + C_{56}E_5E_6 \\ & + \tfrac{1}{2}C_{66}E_6^2 \end{aligned} \tag{5.131}$$

The strain energy is generally taken to be zero in the unstrained state (i.e. when $E_i = 0$), so that $C_0 = 0$. The stress component S_i is given by

$$S_i = \frac{\partial w}{\partial E_i} = C_i + C_{1i}E_1 + C_{2i}E_2 + C_{3i}E_3 + C_{4i}E_4 + C_{5i}E_5 + C_{6i}E_6 \tag{5.132}$$

If the stress is also taken to be zero in the unstrained state ($E_i = 0\ \forall i$) then $C_i = 0; i = 1, 2, \ldots, 6$, and the stress–strain equation becomes

$$\begin{pmatrix} S_1 \\ S_2 \\ S_3 \\ S_4 \\ S_5 \\ S_6 \end{pmatrix} = \begin{pmatrix} C_{11} & C_{12} & C_{13} & C_{14} & C_{15} & C_{16} \\ C_{21} & C_{22} & C_{23} & C_{24} & C_{25} & C_{26} \\ C_{31} & C_{32} & C_{33} & C_{34} & C_{35} & C_{36} \\ C_{41} & C_{42} & C_{43} & C_{44} & C_{45} & C_{46} \\ C_{51} & C_{52} & C_{53} & C_{54} & C_{55} & C_{56} \\ C_{61} & C_{62} & C_{63} & C_{64} & C_{65} & C_{66} \end{pmatrix} \begin{pmatrix} E_1 \\ E_2 \\ E_3 \\ E_4 \\ E_5 \\ E_6 \end{pmatrix} \tag{5.133}$$

or, in index notation, $S_i = C_{ij}E_j$ and in shorthand matrix notation $S = [C]E$, where $[C]$ is the stiffness matrix of the material. But we know that

$$w = \tfrac{1}{2}S_iE_i = \tfrac{1}{2}C_{ij}E_jE_i \tag{5.134}$$

If we differentiate w twice with respect to E_i and E_j using either (5.131) or (5.134) then we obtain

$$\frac{\partial^2 w}{\partial E_i \partial E_j} = C_{ij} \tag{5.135}$$

$$= \frac{\partial^2(\frac{1}{2}S_kE_k)}{\partial E_i \partial E_j} = \frac{1}{2}(C_{ij} + C_{ji}) \tag{5.136}$$

which implies that $C_{ij} = C_{ji}$, i.e. the matrix $[C]$ is symmetrical.

In tensor notation, using (5.127) and (5.128) equation (5.133) becomes

$$\sigma_{ij} = C_{ijkl}\varepsilon_{kl}, \qquad i, j, k, l = 1, 2, 3 \tag{5.137}$$

with the condition $C_{ijkl} = C_{jikl} = C_{ijlk} = C_{klij}$, since both the stress tensor and the strain tensor are symmetrical.

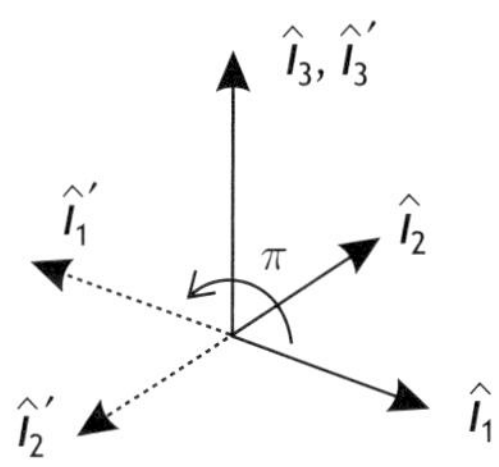

Fig. 5.8 Symmetry in orthorombic materials.

For the most anisotropic case the stiffness matrix involves 21 coefficients. Such a material is called a triclinic material.

For isotropic materials we have already seen that the stiffness matrix involves only two coefficients. It is given in equation 5.32. We consider now two examples of anisotropy frequently encountered in geomechanics.

5.4.3 Examples of anisotropic elasticity

It was shown in section 5.3.5 that the body-wave velocity depends on the elastic coefficients of the material. When the material is anisotropic, the wave velocities vary with the direction of the wave propagation (the ray orientation). Hence, for shear waves, polarized perpendicularly to the ray direction, the velocity will vary with the direction of polarization within the plane normal to the ray and different shear-wave arrivals are observed. This is called *shear-wave splitting* and has important applications in geomechanics; it will be discussed further in chapters 11 and 13.

Orthorombic symmetry, or orthotropy

Orthotropy or orthorombic symmetry refers to materials that possess three planes of symmetry (fig. 5.8).

This is typically the situation in rocks submitted to triaxial stress fields. Rocks and rock masses are always the site of microfractures and when these microdefaults are randomly oriented the material is apparently isotropic. When a truly triaxial stress field is applied to this material, the closure of microfractures depends on their relative orientation with respect to the principal stress directions. This, in turn, influences the elastic properties of the equivalent material. Hence the three planes perpendicular to the principal stress directions are planes of symmetry for the equivalent homogeneous material. This is discussed further in chapter 9.

We choose as the frame of reference $\hat{\boldsymbol{I}}_i$; $i = 1, 2, 3$, the three axes perpendicular to the planes of symmetry. From the symmetry properties, the stress–strain relation expressed in a frame of reference $\hat{\boldsymbol{I}}'_i$; $i = 1, 2, 3$, that has been rotated 180° around the $\hat{\boldsymbol{I}}_3$ axis does not change.

We adopt a stress–strain relation expressed using the notation (5.127) and (5.128) and use primes to refer to the components obtained in the frame of reference $\hat{\boldsymbol{I}}'_i$; $i = 1, 2, 3$.

Let us consider for example the component $S_4 = \sigma_{12}$:

$$S_4 = C_{41}E_1 + C_{42}E_2 + C_{43}E_3 + C_{44}E_4 + C_{45}E_5 + C_{46}E_6 \tag{5.138}$$

$$S'_4 = C_{41}E'_1 + C_{42}E'_2 + C_{43}E'_3 + C_{44}E'_4 + C_{45}E'_5 + C_{46}E'_6 \tag{5.139}$$

With the sign convention discussed in chapter 3, we note that

$$\begin{gathered} S_1 = S'_1, \quad S_2 = S'_2, \quad S_3 = S'_3, \quad S_4 = S'_4 \\ S_5 = -S'_5, \quad S_6 = -S'_6 \end{gathered} \tag{5.140}$$

with similar relationships for the strain components.

Replacing all the *primed* values by their *unprimed* values, given by (5.140), in equation (5.139) and subtracting equations (5.138) and (5.139) yields

$$2C_{45}E_5 + 2C_{46}E_6 = 0 \tag{5.141}$$

which must hold true whatever the values of E_5 and E_6. Hence $C_{45} = C_{46} = 0$. The same demonstration may be conducted for S_1, S_2 and S_3, so we deduce that $C_{15} = C_{16} = C_{25} = C_{26} = C_{35} = C_{36} = 0$.

Proceeding along precisely the same lines but for a rotation around the $\hat{\boldsymbol{I}}_2$ axis, we find that $C_{14} = C_{24} = C_{34} = C_{56} = 0$.

For such orthotropic materials, the stress–strain equation expressed in the frame of reference associated with the symmetry axes of the material is therefore

$$\begin{pmatrix} S_1 \\ S_2 \\ S_3 \\ S_4 \\ S_5 \\ S_6 \end{pmatrix} = \begin{pmatrix} C_{11} & C_{12} & C_{13} & 0 & 0 & 0 \\ C_{12} & C_{22} & C_{23} & 0 & 0 & 0 \\ C_{13} & C_{23} & C_{33} & 0 & 0 & 0 \\ 0 & 0 & 0 & C_{44} & 0 & 0 \\ 0 & 0 & 0 & 0 & C_{55} & 0 \\ 0 & 0 & 0 & 0 & 0 & C_{66} \end{pmatrix} \begin{pmatrix} E_1 \\ E_2 \\ E_3 \\ E_4 \\ E_5 \\ E_6 \end{pmatrix} \tag{5.142}$$

which involves nine independent elastic constants.

The compliance matrix $[D]$ is obtained by inverting equation (5.142). The compliance coefficients are

$$\begin{gathered} D_{11} = (C_{22}C_{33} - C_{23}^2)/\det, \qquad D_{12} = (C_{13}C_{23} - C_{12}C_{33})/\det \\ D_{13} = (C_{12}C_{23} - C_{22}C_{13})/\det \end{gathered} \tag{5.143}$$

$$\begin{gathered} D_{22} = (C_{11}C_{33} - C_{13}^2)/\det, \qquad D_{23} = (C_{12}C_{13} - C_{11}C_{23})/\det \\ D_{33} = (C_{11}C_{22} - C_{12}^2)/\det \end{gathered} \tag{5.144}$$

$$D_{44} = 1/C_{44}, \quad D_{55} = 1/C_{55}, \quad D_{66} = 1/C_{66} \tag{5.145}$$

where $\det = \begin{vmatrix} C_{11} & C_{12} & C_{13} \\ C_{12} & C_{22} & C_{23} \\ C_{13} & C_{23} & C_{33} \end{vmatrix}$.

For example, the ratio of E_1 and S_1 is D_{11}, which is the inverse of the Young's modulus for this direction. Hence the material exhibits different Young's moduli when compressed along the axes of symmetry. Similarly, the values of Poisson's ratio, defined as E_2/E_1 or E_3/E_1, differ between directions 2 and 3.

Planar isotropy

In sedimentary rocks, the elastic properties do not vary for directions parallel to the bedding planes but exhibit different properties in the direction perpendicular to the bedding planes. This corresponds to the case of planar isotropy. The stiffness matrix simplifies further, since the elastic properties are the same for all orientations parallel to bedding plane. The expression for the stiffness matrix may be obtained by using the fact that any rotation θ around the direction perpendicular to the bedding planes does not change the stress–strain equation. This yields the following stiffness matrix:

$$\begin{pmatrix} C_{11} & C_{11}-2C_{66} & C_{13} & 0 & 0 & 0 \\ C_{11}-2C_{66} & C_{11} & C_{13} & 0 & 0 & 0 \\ C_{13} & C_{13} & C_{33} & 0 & 0 & 0 \\ 0 & 0 & 0 & C_{44} & 0 & 0 \\ 0 & 0 & 0 & 0 & C_{44} & 0 \\ 0 & 0 & 0 & 0 & 0 & C_{66} \end{pmatrix} \tag{5.146}$$

Thus, such a material exhibits five different elastic coefficients.

5.4.4 Change of frame of reference

While for isotropic materials the stress–strain law is the same in all frames of reference, this is no longer the case for anisotropic materials.

For example, let us determine the expression for Hooke's law for an orthorombic material, where the frame of reference does not correspond to the axis of symmetry of the material. In the frame of reference of the axes of symmetry, $\boldsymbol{e}_i$; $i = 1, 2, 3$, Hooke's law is $\sigma_{ij} = C_{ijkl}\varepsilon_{kl}$, while in the frame of reference $\hat{\boldsymbol{I}}_i$; $i = 1, 2, 3$, with $\boldsymbol{e}_i = Q_{ij}\hat{\boldsymbol{I}}_j$, it becomes $\sigma'_{ij} = C'_{ijkl}\varepsilon'_{kl}$. As was shown in section 3.4, the relations between the stress and strain components in the two frames of reference are

$$\sigma'_{ij} = Q^T_{ir}\sigma_{rs}Q_{sj} \qquad \text{and} \qquad \varepsilon'_{kl} = Q^T_{km}\varepsilon_{mn}Q_{nl} \tag{5.147}$$

so that

$$Q^T_{ir}\sigma_{rs}Q_{sj} = C'_{ijkl}Q^T_{km}\varepsilon_{mn}Q_{nl} \tag{5.148}$$

or

$$\sigma_{pq} = Q_{pi}Q^T_{jq}C'_{ijkl}Q^T_{km}Q_{nl}\varepsilon_{mn} = Q_{pi}Q_{qj}Q_{mk}Q_{nl}C'_{ijkl}\varepsilon_{mn} \tag{5.149}$$

$$= C_{pqmn}\varepsilon_{mn} \tag{5.150}$$

so that finally we have

$$C'_{rstu} = Q_{pr}Q_{qs}Q_{mt}Q_{nu}C_{pqmn} \tag{5.151}$$

In equation (5.151) none of the components C'_{rstu} is zero and the corresponding 6×6 matrix is full.

Hence, when the principal stress directions are not collinear with the axis of symmetry of the material, the principal strain directions are not collinear with the principal stress directions.

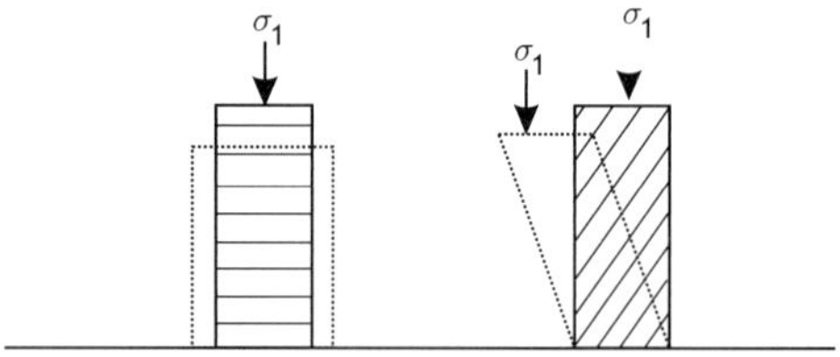

Fig. 5.9 In an anisotropic elastic material with schist bands (on the right), the principal directions of stress and strain do not coincide when the bands are inclined with respect to the symmetry axes of the material.

This is illustrated in fig. 5.9, where a rock sample has been prepared in such a way that its axis makes an angle with the *schistosity* direction (the direction of bands of schists in a material). When the sample is loaded with a uniaxial stress, the axis of the sample becomes inclined to the uniaxial stress direction. This implies that special attention must be given to the manner in which the axial load is exerted experimentally. In order to avoid generating horizontal forces at the top of the specimen during deformation, the top of the sample must be free to move horizontally during loading.

5.5 Theorems of elastostatics

A linear relationship between the stress and strain components implies simple relationships between the work done by external forces and the increase in internal energy of the body (the elastic strain energy) caused by the corresponding loading. Further, it leads to unique solutions when the displacement gradients remain small (the small-strain tensor was defined in (4.54)), a theorem not valid for buckling situations.

5.5.1 Clapeyron's theorem

Let Q be the work done during a deformation caused by body and surface forces applied to a body B of volume V and external surface S. If $\boldsymbol{u}$ is the displacement supported by point $\boldsymbol{X}$ when the body is at rest after loading, we may write

$$Q = \int_V (\varrho \boldsymbol{b} \cdot \boldsymbol{u})\, dv + \int_S [(\tilde{\sigma} \boldsymbol{n}) \cdot \boldsymbol{u}]\, da \tag{5.152}$$

$$= \int_V (\varrho b_i u_i)\, dv + \int_S (\sigma_{ij} n_j u_i)\, da \tag{5.153}$$

However, from Green's theorem,

$$\int_S (\sigma_{ij} n_j u_i)\, da = \int_V (\sigma_{ij} u_i)_{,j}\, dv = \int_V (\sigma_{ij,j} u_i + \sigma_{ij} u_{i,j})\, dv \tag{5.154}$$

so that

$$Q = \int_V (\varrho b_i + \sigma_{ij,j}) u_i\, dv + \int_V (\sigma_{ij} u_{i,j})\, dv = \int_V (\sigma_{ij} u_{i,j})\, dv \tag{5.155}$$

$$= \int_V \sigma_{ij}\varepsilon_{ij}\, dv = 2W_e \quad \text{since } \varrho b_i + \sigma_{ij,j} = 0 \tag{5.156}$$

where W_e is the elastic strain energy of the body.

The elastic strain energy of the body is equal to half the work done by the external forces, which is precisely the result that we obtained in section 5.4.1 for the loading of an isotropic sample.

5.5.2 Betti's reciprocal work theorem

Let us consider two loading systems applied on a body B of volume V and external surface S. For the first the stress field is $\tilde{\sigma}^{(1)}$, with which is associated the displacement field $\boldsymbol{u}^{(1)}$ corresponding to in a small strain $\tilde{\varepsilon}^{(1)}$, while for the second the stress field is $\tilde{\sigma}^{(2)}$, with which is associated the displacement field $\boldsymbol{u}^{(2)}$ corresponding to small strain $\tilde{\varepsilon}^{(2)}$.

Let us compute the work $Q_{(1,2)}$ done by the forces of loading system 1 through the displacement caused by loading system 2:

$$Q_{(1,2)} = \int_V (\varrho \boldsymbol{b} \cdot \boldsymbol{u}^{(2)})\, dv + \int_S (\tilde{\sigma}^{(1)}\boldsymbol{n} \cdot \boldsymbol{u}^{(2)})\, da \tag{5.157}$$

$$= \int_V (\varrho b_i + \sigma_{ij,j}^{(1)}) u_i^{(2)} dv + \int_V (\sigma_{ij}^{(1)} u_{i,j}^{(2)})\, dv = \int_V (C_{ijkl}\varepsilon_{kl}^{(1)}\varepsilon_{ij}^{(2)})\, dv \tag{5.158}$$

We may compute similarly the work $Q_{(2,1)}$ done by the forces of loading system 2 through the displacement caused by loading system 1:

$$Q_{(2,1)} = \int_V (\varrho \boldsymbol{b} \cdot \boldsymbol{u}^{(1)})\, dv + \int_S (\tilde{\sigma}^{(2)}\boldsymbol{n} \cdot \boldsymbol{u}^{(1)})\, da \tag{5.159}$$

$$= \int_V (C_{ijkl}\varepsilon_{kl}^{(2)}\varepsilon_{ij}^{(1)})\, dv \tag{5.160}$$

Comparing (5.158) and (5.160) we conclude that $Q_{(1,2)} = Q_{(2,1)}$, i.e. the work done by loading system 1 through the displacement associated with loading system 2 is equal to the work done by loading system 2 through the displacement associated with loading system 1.

5.5.3 Uniqueness of solution for elastostatic problems

We now consider an elastic body with strain energy density function w such that, from equations (5.129), $\sigma_{ij} = \partial w/\partial \varepsilon_{ij}$ and $w = \frac{1}{2}C_{ijkl}\varepsilon_{kl}\varepsilon_{ij}$. The equilibrium equations $\sigma_{ij,j} + \varrho b_i = 0$ may be rewritten as $(\partial w/\partial \varepsilon_{ij})_{,j} + \varrho b_i = 0$.

We call $\int_{\boldsymbol{u}}$ that part of surface S of the body for which displacement components are specified and $\int_{\boldsymbol{t}}$ that part of S for which the components of the stress vector are specified ($t_i = (\partial w/\partial \varepsilon_{ij})n_j$).

Let us assume that two solutions for the displacement field may exist; we call them $u_i^{(1)}$ and $u_i^{(2)}$. The difference between these two solutions, $u_i = u_i^{(1)} - u_i^{(2)}$, satisfies the condition $w = w^{(1)} - w^{(2)}$, where $w^{(1)}$ and $w^{(2)}$ are the strain energy density functions associated respectively with solutions $u_i^{(1)}$ and $u_i^{(2)}$ while w is the strain energy density function associated with solution u_i. It satisfies the equilibrium condition $(\partial w/\partial \varepsilon_{ij})_{,j} = 0$, with boundary conditions $u_i = 0$ on $\int_{\boldsymbol{u}}$ and $(\partial w/\partial \varepsilon_{ij})n_j = 0$ on $\int_{\boldsymbol{t}}$.

However, we have

$$\int_V u_i \left(\frac{\partial w}{\partial \varepsilon_{ij}} \right)_{,j} dv = 0 \tag{5.161}$$

The left-hand side may be integrated by parts:

$$\int_V u_i \left(\frac{\partial w}{\partial \varepsilon_{ij}} \right)_{,j} dv = \int_S u_i \frac{\partial w}{\partial \varepsilon_{ij}} n_j \, da - \int_V \frac{\partial w}{\partial \varepsilon_{ij}} u_{i,j} \, dv = 0 \tag{5.162}$$

with

$$\int_S u_i \frac{\partial w}{\partial \varepsilon_{ij}} n_j \, da = \int_{\boldsymbol{u}} u_i \frac{\partial w}{\partial \varepsilon_{ij}} n_j \, da + \int_{\boldsymbol{t}} u_i \frac{\partial w}{\partial \varepsilon_{ij}} n_j \, da = 0 \tag{5.163}$$

given the boundary conditions of the solution u_i. But

$$\int_V \frac{\partial w}{\partial \varepsilon_{ij}} u_{i,j} \, dv = \int_V \sigma_{ij} \varepsilon_{ij} \, dv = \int_V 2w \, dv = 0 \tag{5.164}$$

Since the elastic strain energy density function is always positive, we conclude that the only possible solution for equation (5.164) is $w = 0$, which implies that $\varepsilon_{ij} = 0$. This solution is the unstrained state, that is $u_i^{(1)} - u_i^{(2)} = 0$, which proves that the solution is unique.

5.6 Solutions for two-dimensional problems of elastostatics

In section 5.3.3 we saw that for plane stress problems we have

$$\sigma_3 = 0 \qquad \text{and} \qquad \varepsilon_3 = -\frac{\nu}{E}(\sigma_1 + \sigma_2),$$

so that

$$(\lambda + 2G)\sigma_1 = 4G(\lambda + G)\varepsilon_1 + 2\lambda G \varepsilon_2 \tag{5.165}$$

$$(\lambda + 2G)\sigma_2 = 2\lambda G \varepsilon_1 + 4G(\lambda + G)\varepsilon_2 \tag{5.166}$$

For plane strain problems ($\varepsilon_3 = 0$, see section 5.3.4) the Hooke's law equations are

$$\sigma_1 = (\lambda + 2G)\varepsilon_1 + \lambda \varepsilon_2 \tag{5.167}$$

$$\sigma_2 = (\lambda + 2G)\varepsilon_2 + \lambda \varepsilon_1 \tag{5.168}$$

$$\sigma_3 = \nu(\sigma_1 + \sigma_2) \tag{5.169}$$

These two sets of equations may be simply written as

$$8G\varepsilon_1 = (\kappa + 1)\sigma_1 + (\kappa - 3)\sigma_2 \tag{5.170}$$

$$8G\varepsilon_2 = (\kappa - 3)\sigma_1 + (\kappa + 1)\sigma_2 \tag{5.171}$$

with $\kappa = 3 - 4\nu$ for plane strain conditions and $\kappa = (3 - \nu)/(1 + \nu)$ for plane stress conditions.

For any cartesian frame of reference, $\hat{\boldsymbol{I}}_i$, $i = 1, 2, 3$, these equations become

$$8G\varepsilon_{11} = (\kappa + 1)\sigma_{11} + (\kappa - 3)\sigma_{22} \tag{5.172}$$

$$8G\varepsilon_{22} = (\kappa - 3)\sigma_{11} + (\kappa + 1)\sigma_{22} \tag{5.173}$$

$$\varepsilon_{12} = \frac{\sigma_{12}}{2G} \tag{5.174}$$

5.6.1 Beltrami–Michell compatibility conditions and Airy stress function

In section 4.3.7 it was shown that the small-strain components are not independent quantities and that some relations must exist between their partial derivatives. In particular it was shown that $2\varepsilon_{12,12} = \varepsilon_{11,22} + \varepsilon_{22,11}$. Taking advantage of equations (5.172)–(5.174), this strain compatibility condition becomes a stress compatibility condition,

$$\frac{1}{G}\sigma_{12,12} = \frac{1}{8G}[(\kappa + 1)\sigma_{11,22} + (\kappa - 3)\sigma_{22,22} + (\kappa - 3)\sigma_{11,11} + (\kappa + 1)\sigma_{22,11}] \tag{5.175}$$

or

$$8\sigma_{12,12} = (\kappa + 1)(\sigma_{11,22} + \sigma_{22,11}) + (\kappa - 3)(\sigma_{11,11} + \sigma_{22,22}) \tag{5.176}$$

In two dimensions, the equilibrium conditions are

$$\sigma_{11,1} + \sigma_{12,2} = -\varrho b_1 \quad \text{and} \quad \sigma_{21,1} + \sigma_{22,2} = -\varrho b_2 \tag{5.177}$$

so that

$$\sigma_{21,21} = -\sigma_{11,11} - \varrho b_{1,1} \quad \text{and} \quad \sigma_{12,12} = -\sigma_{22,22} - \varrho b_{2,2} \tag{5.178}$$

Using (5.178) in (5.176) yields

$$(\kappa + 1)\nabla^2(\sigma_{11} + \sigma_{22}) = 8\sigma_{12,12} + 4\sigma_{11,11} + 4\sigma_{22,22} \tag{5.179}$$

$$= 4(\sigma_{12,12} + \sigma_{11,11}) + 4(\sigma_{12,12} + \sigma_{22,22}) \tag{5.180}$$

$$= -4\varrho b_{1,1} - 4\varrho b_{2,2} \tag{5.181}$$

so that combining the linear elasticity conditions with the equilibrium and strain compatibility conditions yields the Beltrami–Michell equation

$$(\kappa + 1)\nabla^2(\sigma_{11} + \sigma_{22}) + 4\varrho(b_{1,1} + b_{2,2}) = 0 \tag{5.182}$$

When body forces may be neglected ($b_1 = b_2 = 0$), the Beltrami–Michell equation (5.182) reduces to

$$\nabla^2(\sigma_{11} + \sigma_{22}) = 0 \tag{5.183}$$

A similar result is obtained if the body force $\boldsymbol{b}$ is derived from a potential V, so that $\nabla^2 V = 0$ with $b_1 = -V_{,1}$ and $b_2 = -V_{,2}$.

Suppose that a function U exists such that

$$\sigma_{11} = U_{,22} + \varrho V, \qquad \sigma_{22} = U_{,11} + \varrho V, \qquad \sigma_{12} = -U_{,12} \tag{5.184}$$

It satisfies the equilibrium conditions (5.177). In the absence of body forces, when equations (5.184) are substituted into (5.182) we get

$$\nabla^2(\nabla^2 U) = 0 \tag{5.185}$$

$$\text{or} \quad U_{,1111} + 2U_{,1122} + U_{,2222} = 0 \tag{5.186}$$

Equation (5.185) is the biharmonic equation, and we have just shown that any solution of it automatically satisfies the equilibrium conditions and the strain compatibility conditions.

The function U is called the *Airy stress function*.

In polar coordinates equations (5.184) become

$$\sigma_{\rho\rho} = \frac{1}{\rho} U_{,\rho} + \frac{1}{\rho^2} U_{,\theta\theta}, \qquad \sigma_{\theta\theta} = U_{,\rho\rho}, \qquad \sigma_{\rho\theta} = -\left(\frac{1}{\rho} U_{,\theta}\right)_{,\rho} \tag{5.187}$$

and the biharmonic equation (5.185) becomes

$$\left(\frac{\partial^2}{\partial \rho^2} + \frac{1}{\rho}\frac{\partial}{\partial \rho} + \frac{1}{\rho^2}\frac{\partial^2}{\partial \theta^2}\right)\left(U_{,\rho\rho} + \frac{1}{\rho} U_{,\rho} + \frac{1}{\rho^2} U_{,\theta\theta}\right) = 0 \tag{5.188}$$

5.6.2 Elements of analytic functions

The complex plane

Let us consider the complex variable z defined as follows:

$$z = x_1 + ix_2 = \rho(\cos\theta + i\sin\theta) = \rho e^{i\theta} \tag{5.189}$$

Its complex conjugate is $\bar{z} = x_1 - ix_2 = \rho e^{-i\theta}$.

The new stress tensor components in two dimensions following a change in the frame of reference are given by

$$\begin{pmatrix} \sigma'_{11} & \sigma'_{12} \\ \sigma'_{21} & \sigma'_{22} \end{pmatrix} = \begin{pmatrix} \cos\theta & \sin\theta \\ -\sin\theta & \cos\theta \end{pmatrix} \begin{pmatrix} \sigma_{11} & \sigma_{12} \\ \sigma_{12} & \sigma_{22} \end{pmatrix} \begin{pmatrix} \cos\theta & -\sin\theta \\ \sin\theta & \cos\theta \end{pmatrix} \tag{5.190}$$

which yields

$$\sigma'_{11} = \sigma_{11}\cos^2\theta + 2\sigma_{12}\sin\theta\cos\theta + \sigma_{22}\sin^2\theta \tag{5.191}$$

$$\sigma'_{22} = \sigma_{11}\sin^2\theta - 2\sigma_{12}\sin\theta\cos\theta + \sigma_{22}\cos^2\theta \tag{5.192}$$

$$\sigma'_{12} = (\sigma_{22} - \sigma_{11})\sin\theta\cos\theta + \sigma_{12}(\cos^2\theta - \sin^2\theta) \tag{5.193}$$

so that

$$\sigma'_{22} - \sigma'_{11} + 2i\sigma'_{12} = (\sigma_{22} - \sigma_{11} + 2i\sigma_{12})e^{2i\theta} \tag{5.194}$$

In addition we recall that

$$\sigma'_{11} + \sigma'_{22} = \sigma_{11} + \sigma_{22} \tag{5.195}$$

Subtracting (5.194) from (5.195) yields

$$2(\sigma'_{11} - i\sigma'_{12}) = (\sigma_{11} + \sigma_{22}) - (\sigma_{22} - \sigma_{11} + 2i\sigma_{12})e^{2i\theta} \tag{5.196}$$

For points located on the boundary of a body, if we choose $\hat{\boldsymbol{I}}'_1$ to be parallel to the normal to the boundary and $\hat{\boldsymbol{I}}'_2$ to be tangent to it, then σ'_{11} and σ'_{12} are simply the components N and T of the stress vector acting on the boundary at this point (see also section 3.4.1). Hence equation (5.196) is a convenient way to relate the stress vector boundary conditions to the local stress tensor components.

For the new displacement components following a rotation of the frame of reference through an angle θ around the origin, we get

$$u'_1 + iu'_2 = (u_1 + iu_2)e^{-i\theta} \tag{5.197}$$

These expressions are also convenient for change from cartesian to polar coordinates:

$$\sigma_{\rho\rho} + \sigma_{\theta\theta} = \sigma_{11} + \sigma_{22} \tag{5.198}$$

$$\sigma_{\theta\theta} - \sigma_{\rho\rho} + 2i\sigma_{\rho\theta} = (\sigma_{22} - \sigma_{11} + 2i\sigma_{12})e^{2i\theta} \tag{5.199}$$

Our objective is to determine two analytic functions of the complex variable $z = x_1 + ix_2 = \rho e^{i\theta}$ which will be used to characterize the Airy stress function and therefore the components of the stress tensor and of the displacement at all points of the body.

Cauchy–Riemann conditions for analytic functions

A function $\zeta(z)$ of the complex variable z is said to be analytic in a domain D if it is finite, single valued and differentiable at all points z of the domain.

Let $\chi(x_1,x_2)$ and $\psi(x_1,x_2)$ be two real functions of the real variables x_1 and x_2 such that

$$\zeta(z) = \chi(x_1,x_2) + i\psi(x_1,x_2) \tag{5.200}$$

$$\bar{\zeta}(z) = \chi(x_1,x_2) - i\psi(x_1,x_2) \tag{5.201}$$

where $\bar{\zeta}(z)$ denotes the complex conjugate of $\zeta(z)$.

The derivative of $\zeta(z)$ at z_0 is defined by

$$\zeta'(z_0) = \lim_{\Delta z \to 0} \frac{\zeta(z_0 + \Delta z) - \zeta(z_0)}{\Delta z} \tag{5.202}$$

with $z_0 = x_{10} + ix_{20}$, $\Delta z = \Delta x_1 + i\Delta x_2$ and

$$\frac{\zeta(z_0 + \Delta z) - \zeta(z_0)}{\Delta z} = \frac{\chi(x_{10} + \Delta x_1, x_{20} + \Delta x_2) - \chi(x_{10}, x_{20}) + i[\psi(x_{10} + \Delta x_1, x_{20} + \Delta x_2) - \psi(x_{10}, x_{20})]}{\Delta x_1 + i\Delta x_2} \tag{5.203}$$

It follows that

$$\operatorname{Re}\zeta'(z_0) = \lim_{\Delta x_1, \Delta x_2 \to 0} \operatorname{Re} \frac{\zeta(z_0 + \Delta z) - \zeta(z_0)}{\Delta z} \tag{5.204}$$

$$\operatorname{Im}\zeta'(z_0) = \lim_{\Delta x_1, \Delta x_2 \to 0} \operatorname{Im} \frac{\zeta(z_0 + \Delta z) - \zeta(z_0)}{\Delta z} \tag{5.205}$$

Let us consider the particular case $\Delta z = \Delta x_1 + 0i$, so that $z_0 + \Delta z = x_{10} + \Delta x_1 + ix_{20}$. Then

$$\operatorname{Re}\zeta'(z_0) = \lim_{\Delta x_1 \to 0} \frac{\chi(x_{10} + \Delta x_1\, , x_{20}) - \chi(x_{10}\, , x_{20})}{\Delta x_1} = \chi_{,1} \tag{5.206}$$

$$\operatorname{Im}\zeta'(z_0) = \lim_{\Delta x_1 \to 0} \frac{\psi(x_{10} + \Delta x_1\, , x_{20}) - \psi(x_{10}\, , x_{20})}{\Delta x_1} = \psi_{,1} \tag{5.207}$$

Thus we may write

$$\zeta'(z_0) = \chi_{,1} + i\psi_{,1} \tag{5.208}$$

However, if one considers the case $\Delta z = 0 + i\Delta x_2$, so that $z_0 + \Delta z = x_{10} + i(x_{20} + \Delta x_2)$, then

$$\zeta'(z_0) = \psi_{,2} - i\chi_{,2} \tag{5.209}$$

In order that equations (5.208) and (5.209) be satisfied simultaneously, the following conditions must be met:

$$\chi_{,1} = \psi_{,2} \qquad \text{and} \qquad \psi_{,1} = -\chi_{,2} \tag{5.210}$$

The relations (5.210) correspond to the Cauchy–Riemann conditions required for a function to be analytic.

In polar coordinates, we have $x_1 = \rho\cos\theta$, $x_2 = \rho\sin\theta$, $z = \rho e^{i\theta}$ and $\zeta(z) = \chi(\rho,\theta) + i\psi(\rho,\theta)$. From the chain rule for differentiation we obtain

$$\chi_{,\rho} = \chi_{,1}\cos\theta + \chi_{,2}\sin\theta, \qquad \chi_{,\theta} = -\rho\sin\theta\,\chi_{,1} + \rho\cos\theta\,\chi_{,2} \tag{5.211}$$

since $x_{1,\rho} = \cos\theta$, $x_{1,\theta} = -\rho\sin\theta$, $x_{2,\rho} = \sin\theta$, $x_{2,\theta} = \rho\cos\theta$.

Hence we have

$$\chi_{,1} = \chi_{,\rho}\cos\theta - \frac{\chi_{,\theta}}{\rho}\sin\theta \tag{5.212}$$

$$\chi_{,2} = \chi_{,\rho}\sin\theta + \frac{\chi_{,\theta}}{\rho}\cos\theta \tag{5.213}$$

with similar results for the partial derivatives of ψ. Using equations (5.212) and (5.213) in (5.210) we obtain the Cauchy–Riemann conditions in polar coordinates:

$$\chi_{,\rho} = \frac{1}{\rho}\psi_{,\theta} \qquad \text{and} \qquad \frac{1}{\rho}\chi_{,\theta} = -\psi_{,\rho} \tag{5.214}$$

From equations (5.208) and (5.209) together with (5.214) we conclude that

$$\zeta'(z_0) = e^{-i\theta_0}(\chi_{,\rho} + i\psi_{,\rho}) \tag{5.215}$$

$$= \frac{1}{\rho_0}e^{-i\theta_0}(\psi_{,\theta} - i\chi_{,\theta}) \tag{5.216}$$

A consequence is that if a function of a complex variable is analytic then its real and imaginary parts are harmonic, i.e. $\nabla^2\chi = \nabla^2\psi = 0$. Indeed, since $\chi_{,1} = \psi_{,2}$ and $\chi_{,2} = -\psi_{,1}$ it follows that $\chi_{,11} = \psi_{,21}$ and $\chi_{,22} = -\psi_{,12}$ so that $\nabla^2\chi = \chi_{,11} + \chi_{,22} = 0$, with similar results for ψ. The functions χ and ψ are said to be conjugate harmonic.

5.6.3 General form of Airy stress function

We concluded in section 5.6.1 that solutions to elastostatic problems may be assimilated to finding a so-called Airy stress function U, which satisfies the biharmonic equation $\nabla^2(\nabla^2 U) = 0$. In the absence of body forces the function is such that

$$\sigma_{11} = U_{,22}, \qquad \sigma_{22} = U_{,11}, \qquad \sigma_{12} = -U_{,12} \tag{5.217}$$

Let $P = \sigma_{11} + \sigma_{22}$ be a harmonic function ($\nabla^2 P = 0$) and assume that $P = \nabla^2 U$. Let Q be the harmonic conjugate of P. Then the function $f(z) = P + iQ$ is an analytic function, as is its integral $\phi(z)$, defined as

$$\phi(z) = \tfrac{1}{4}\int f(z)dz = p + iq \tag{5.218}$$

where p and q are the real and imaginary parts of $\phi(z)$, which satisfy the Cauchy–Riemann conditions. Hence

$$\phi'(z) = p_{,1} + iq_{,1} = \tfrac{1}{4}f(z) = \frac{1}{4}(P + iQ) \tag{5.219}$$

with

$$p_{,1} = q_{,2} = \tfrac{1}{4}P, \quad q_{,1} = -p_{,2} = \tfrac{1}{4}Q \tag{5.220}$$

Let us show that the function $U - px_1 - qx_2$ is harmonic. Indeed,

$$\nabla^2(U - px_1 - qx_2) = \nabla^2 U - x_1\nabla^2 p - 2p_{,1} - x_2\nabla^2 q - 2q_{,2} = 0 \tag{5.221}$$

since

$$(px_1)_{,11} = (p + x_1 p_{,1})_{,1} = (p + x_1 q_{,2})_{,1} \tag{5.222}$$

and $2(p_{,1} + q_{,2}) = P$.

We may choose $U - px_1 - qx_2$ as the real part of an analytic complex function $\chi(z)$. We note further that $px_1 + qx_2$ is the real part of $\bar{z}\phi(z)$ and we conclude that

$$U = \mathrm{Re}[\bar{z}\phi(z) + \chi(z)] \tag{5.223}$$
$$= \tfrac{1}{2}[\bar{z}\phi(z) + z\overline{\phi(z)} + \chi(z) + \overline{\chi(z)}] \tag{5.224}$$

is the general solution of the biharmonic equation (5.185). It involves two analytic functions, $\phi(z)$ and $\chi(z)$.

Now we derive the various partial derivatives required for computing the stress and displacement components:

$$U_{,1} = \tfrac{1}{2}[\phi(z) + \bar{z}\phi'(z) + \overline{\phi(z)} + z\overline{\phi'(z)} + \chi'(z) + \overline{\chi'(z)}] \tag{5.225}$$
$$U_{,2} = \tfrac{1}{2}[-i\phi(z) + i\bar{z}\phi'(z) + i\overline{\phi(z)} - iz\overline{\phi'(z)} + i\chi'(z) - i\overline{\chi'(z)}] \tag{5.226}$$
$$U_{,11} = \tfrac{1}{2}[2\phi'(z) + \bar{z}\phi''(z) + 2\overline{\phi'(z)} + z\overline{\phi''(z)} + \chi''(z) + \overline{\chi''(z)}] \tag{5.227}$$
$$U_{,22} = \tfrac{1}{2}[2\phi'(z) - \bar{z}\phi''(z) + 2\overline{\phi'(z)} - z\overline{\phi''(z)} - \chi''(z) - \overline{\chi''(z)}] \tag{5.228}$$
$$U_{,12} = \tfrac{1}{2}[i\bar{z}\phi''(z) - iz\overline{\phi''(z)} + i\chi''(z) - i\overline{\chi''(z)}] \tag{5.229}$$

The following useful result is obtained from these relationships:

$$U_{,1} + iU_{,2} = \phi(z) + z\overline{\phi'(z)} + \overline{\chi'(z)} \tag{5.230}$$

Further, we note that

$$\sigma_{22} - \sigma_{11} + 2i\sigma_{12} = U_{,11} - U_{,22} - 2iU_{,12} = 2[\bar{z}\phi''(z) + \chi''(z)] \tag{5.231}$$

$$\sigma_{11} + \sigma_{22} = U_{,11} + U_{,22} = 4\,\mathrm{Re}[\phi'(z)] \tag{5.232}$$

The components of displacement are obtained by introducing Hooke's law:

$$8G\varepsilon_{11} = 8Gu_{1,1} = (\kappa + 1)\sigma_{11} + (\kappa - 3)\sigma_{22} \tag{5.233}$$

$$8G\varepsilon_{22} = 8Gu_{2,2} = (\kappa - 3)\sigma_{11} + (\kappa + 1)\sigma_{22} \tag{5.234}$$

$$\varepsilon_{12} = \tfrac{1}{2}(u_{1,2} + u_{2,1}) = \frac{\sigma_{12}}{2G} \tag{5.235}$$

Equations (5.233) and (5.234) may be rewritten as

$$2Gu_{1,1} = -\sigma_{22} + \tfrac{1}{4}(\kappa + 1)(\sigma_{11} + \sigma_{22}) \tag{5.236}$$

$$2Gu_{2,2} = -\sigma_{11} + \tfrac{1}{4}(\kappa + 1)(\sigma_{11} + \sigma_{22}) \tag{5.237}$$

with $\sigma_{11} + \sigma_{22} = P = 4p_{,1} = 4q_{,2}$. Hence

$$2Gu_{1,1} = -U_{,11} + (\kappa + 1)p_{,1} \tag{5.238}$$

$$2Gu_{2,2} = -U_{,22} + (\kappa + 1)q_{,2} \tag{5.239}$$

These two equations may be integrated to give

$$2Gu_1 = -U_{,1} + (\kappa + 1)p + f_1(x_2) \tag{5.240}$$

$$2Gu_2 = -U_{,2} + (\kappa + 1)q + f_2(x_1) \tag{5.241}$$

But $2G\varepsilon_{12} = G(u_{1,2} + u_{2,1}) = \sigma_{12} = -U_{,12}$ so that, because of equation (5.220), $f_1'(x_2) + f_2'(x_1) = 0$, which implies that

$$f_1(x_2) = \omega x_2 + a \tag{5.242}$$

$$f_2(x_1) = -\omega x_1 + b \tag{5.243}$$

The solution $u_1 = \omega x_2 + a$, $u_2 = \omega x_1 + a$ corresponds to a rigid body motion that does not generate strain or, therefore, stress.

We conclude that

$$2G(u_1 + iu_2) = -(U_{,1} + iU_{,2}) + (\kappa + 1)(p + iq) \tag{5.244}$$

or

$$2G(u_1 + iu_2) = \kappa\phi(z) - z\overline{\phi'(z)} - \overline{\chi'(z)} \tag{5.245}$$

in terms of the complex conjugates of ϕ' and χ'; see (5.223) and (5.224). The function $\chi(z)$ does not appear in any of the basic equations; we define $\psi(z) = \chi'(z)$ so that the general solution of the two-dimensional problems of elastostatics may be summarized as

$$\sigma_{11} + \sigma_{22} = 2[\phi'(z) + \overline{\phi'(z)}] = 4\,\mathrm{Re}[\phi'(z)] \tag{5.246}$$

$$\sigma_{22} - \sigma_{11} + 2i\sigma_{12} = 2[\bar{z}\phi''(z) + \psi'(z)] \tag{5.247}$$

$$2G(u_1 + iu_2) = \kappa\phi(z) - z\overline{\phi'(z)} - \overline{\psi(z)} \tag{5.248}$$

where $\phi(z)$ and $\psi(z)$ are analytic functions.

5.6.4 Practical applications for geomechanics

Many practical elastostatic problems of geomechanics may be solved by superposing well-known simple solutions. We describe hereafter some useful solutions obtained by taking the functions $\phi(z)$ and $\psi(z)$ to be polynomial or power series in z or $1/z$.

Simple examples revisited

Homogeneous stresses

We consider first the case $\phi(z) = \alpha z,\ \psi(z) = dz$ where α and d are constants that may be complex. From equations (5.246) and (5.247) we note that

$$\sigma_{11} + \sigma_{22} = 2[\phi'(z) + \overline{\phi'(z)}] = 2(\alpha + \bar{\alpha}) \tag{5.249}$$

$$\sigma_{22} - \sigma_{11} + 2i\sigma_{12} = 2[\bar{z}\phi''(z) + \psi'(z)] = 2d \tag{5.250}$$

But $\alpha + \bar{\alpha}$ is always real and we write $\alpha = 2c$ with c real, so that $\sigma_{11} + \sigma_{22} = 4c$.

If σ_1 and σ_2 are the principal stress components then

$$\sigma_{22} + \sigma_{11} = 4c = \sigma_1 + \sigma_2 \tag{5.251}$$

$$[\sigma_{22} - \sigma_{11} + 2i\sigma_{12}]e^{2i\beta} = 2de^{2i\beta} = \sigma_2 - \sigma_1 \tag{5.252}$$

where β is the angle between the σ_1 direction and the x_1 axis direction. For all points of the space

$$\frac{\sigma_1 + \sigma_2}{4} = c, \qquad \left(\frac{\sigma_2 - \sigma_1}{2}\right)e^{-2i\beta} = d \tag{5.253}$$

which is a homogeneous stress field.

If the stress field is hydrostatic ($\sigma_1 = \sigma_2 = p$) then $d = 0$ and $c = p/2$. If it is uniaxial ($\sigma_2 = 0$) then $c = \sigma_1/4$ and $d = -\sigma_1/2$. If it corresponds to a pure shear ($\sigma_1 = -\sigma_2$) with maximum shear stress in the x_1 axis direction (i.e. $\beta = \pi/4$) then choose $c = 0$ and $d = i\sigma_1$.

The displacements are given by equation (5.248), so that

$$2G(u_1 + iu_2) = (\kappa - 1)cz - \bar{d}\bar{z} \tag{5.254}$$

that is

$$2Gu_1 = (\kappa - 1)cx_1 - \mathrm{Re}(\bar{d}\bar{z}), \qquad 2Gu_2 = (\kappa - 1)cx_2 - \mathrm{Im}(\bar{d}\bar{z}) \tag{5.255}$$

When the x_1 and x_2 axes are the principal directions then $c = \frac{1}{4}(\sigma_1 + \sigma_2)$ and $d = \frac{1}{2}(\sigma_2 - \sigma_1)$, so that

$$8Gu_1 = [(\kappa + 1)\sigma_1 + (\kappa - 3)\sigma_2]x_1 \tag{5.256}$$

$$8Gu_2 = [(\kappa - 3)\sigma_1 + (\kappa + 1)\sigma_2]x_2 \tag{5.257}$$

from which we obtain

$$8G\frac{u_1}{x_1} = 8G\varepsilon_1 = (\kappa + 1)\sigma_1 + (\kappa - 3)\sigma_2 \tag{5.258}$$

$$8G\frac{u_2}{x_2} = 8G\varepsilon_2 = (\kappa - 3)\sigma_1 + (\kappa + 1)\sigma_2 \tag{5.259}$$

which are precisely equations (5.170) and (5.171), proposed for Hooke's law at the beginning of this section.

A pressurized infinite cylindrical hole

We consider here the case $\phi(z) = 0$, $\psi(z) = d/z$ where d is real and z is expressed in polar coordinates, $z = \rho e^{i\theta}$.

Then $\psi(z) = (d/\rho)e^{-i\theta}$ and $\overline{\psi(z)} = (d/\rho)e^{i\theta}$. The change from cartesian to polar coordinates is given by

$$\sigma_{\rho\rho} + \sigma_{\theta\theta} = \sigma_{11} + \sigma_{22} = 2[\phi'(z) + \overline{\phi'(z)}] = 0 \tag{5.260}$$

$$(\sigma_{\theta\theta} - \sigma_{\rho\rho} + 2i\sigma_{\rho\theta}) = (\sigma_{22} - \sigma_{11} + 2i\sigma_{12})e^{2i\theta} \tag{5.261}$$

However, from equation (5.247) we have $\sigma_{22} - \sigma_{11} + 2i\sigma_{12} = 2[\bar{z}\phi''(z) + \psi'(z)]$, so that

$$\sigma_{\theta\theta} - \sigma_{\rho\rho} + 2i\sigma_{\rho\theta} = 2\psi'(z)e^{2i\theta} \tag{5.262}$$

We saw in equation (5.215) that if $w = u(\rho,\theta) + iv(\rho,\theta)$ then the derivative is $w' = e^{-i\theta}(u_{,\rho} + iv_{,\rho})$; thus if $\psi(z) = (d/\rho)e^{-i\theta}$ then $\psi'(z) = (-d)/\rho^2 e^{-2i\theta}$ and, accordingly,

$$\sigma_{\theta\theta} - \sigma_{\rho\rho} + 2i\sigma_{\rho\theta} = -\frac{2d}{\rho^2} \tag{5.263}$$

so that

$$\sigma_{\rho\theta} = 0, \quad \sigma_{\theta\theta} - \sigma_{\rho\rho} = -\frac{2d}{\rho^2}, \quad \sigma_{\theta\theta} + \sigma_{\rho\rho} = 0 \tag{5.264}$$

which yields $\sigma_{\rho\rho} = d/\rho^2$ and $\sigma_{\theta\theta} = -d/\rho^2$.

In order to satisfy the boundary condition $\sigma_{\rho\rho} = p$ for $\rho = R$, where R is the radius of a hollow cylinder in an infinite body, we choose $d = pR^2$. The displacements are obtained from equation (5.248):

$$2G(u_\rho + iu_\theta) = 2G(u_1 + iu_2)e^{-i\theta} \tag{5.265}$$

with

$$2G(u_1 + iu_2) = \kappa\phi(z) - z\overline{\phi'(z)} - \overline{\psi(z)} = -\overline{d/z} \tag{5.266}$$

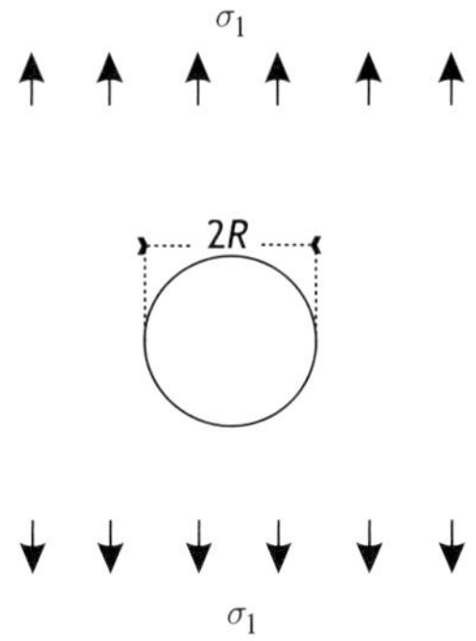

Fig. 5.10 Stress concentrations near an infinite cylindrical borehole with a uniaxial stress field in a direction perpendicular to the borehole axis.

Hence

$$2G(u_\rho + iu_\theta) = \frac{-d}{\rho} \tag{5.267}$$

With $d = pR^2$, the solution $\sigma_{\rho\rho} = pR^2/\rho^2$, $\sigma_{\theta\theta} = -pR^2/\rho^2, u_\rho = -pR^2/(2G\rho)$ is precisely that derived in section 5.3.4 for a pressurized infinite cylinder. It may be noticed that we chose d to be positive, taking compressions as positive, i.e. *shortening* deformations are positive; hence the negative sign for u_ρ.

It may be verified following similar reasoning that the analytic functions $\phi(z) = cz$, $\psi(z) = d/z$ yield the solution for an internally and externally pressurized thick cylinder (equations (5.92)–(5.94)).

Infinite empty circular hole with a given stress field at infinity

We address now the case of an infinite hollow cylinder of circular cross section and radius R, in an infinite body for which there exists a given stress field at infinity. This may be considered as an approximation for a tunnel or for a deep borehole and constitutes one of the most useful solutions in rock mechanics (fig. 5.10).

We consider first a uniaxial stress field σ_1 at infinity in a direction perpendicular to the cylinder axis. We saw when discussing equation (5.253) that the solution $\phi(z) = (\sigma_1/4)z$, $\psi(z) = -\sigma_1/2$ yields a homogeneous uniaxial stress field in the x_1 direction.

This solution will now be perturbed so as to introduce a circular geometry. We assume that

$$\phi(z) = \frac{\sigma_1}{4}\left(z + \frac{A}{z}\right), \qquad \psi(z) = -\frac{\sigma_1}{2}\left(z + \frac{B}{z} + \frac{C}{z^3}\right) \tag{5.268}$$

so that, differentiating, we get

$$\phi'(z) = \frac{\sigma_1}{4}\left(1 - \frac{A}{z^2}\right), \qquad \phi''(z) = \frac{\sigma_1 A}{2z^3} \tag{5.269}$$

$$\psi'(z) = -\frac{\sigma_1}{2}\left(1 - \frac{B}{z^2} - \frac{3C}{z^4}\right) \tag{5.270}$$

A change to polar coordinates $z = \rho e^{i\theta}$ yields

$$\sigma_{\rho\rho} + \sigma_{\theta\theta} = \sigma_{11} + \sigma_{22} = 2[\phi'(z) + \overline{\phi'(z)}] \tag{5.271}$$

$$= \frac{\sigma_1}{2}\left(2 - \frac{A}{\rho^2}e^{-2i\theta} - \frac{A}{\rho^2}e^{2i\theta}\right) \tag{5.272}$$

$$= \sigma_1\left(1 - \frac{A}{\rho^2}\cos 2\theta\right) \tag{5.273}$$

Further, since $\sigma_{\theta\theta} - \sigma_{\rho\rho} + 2i\sigma_{\rho\theta} = (\sigma_{22} - \sigma_{11} + 2i\sigma_{12})e^{2i\theta}$, with

$$\sigma_{22} - \sigma_{11} + 2i\sigma_{12} = 2[\bar{z}\phi''(z) + \psi'(z)] \tag{5.274}$$

$$= 2\left[\rho e^{-i\theta}\frac{A\sigma_1}{2\rho^3}e^{-3i\theta} - \frac{\sigma_1}{2}\left(1 - \frac{B}{\rho^2}e^{-2i\theta} - \frac{3C}{\rho^4}e^{-4i\theta}\right)\right] \tag{5.275}$$

we have

$$\sigma_{\theta\theta} - \sigma_{\rho\rho} + 2i\sigma_{\rho\theta} = \sigma_1\left[\frac{B}{\rho^2} - e^{2i\theta} + \left(\frac{A}{\rho^2} + \frac{3C}{\rho^4}\right)e^{-2i\theta}\right]. \tag{5.276}$$

Separating the real and imaginary parts we get

$$\sigma_{\theta\theta} - \sigma_{\rho\rho} = \sigma_1\left[\frac{B}{\rho^2} - \left(1 - \frac{A}{\rho^2} - \frac{3C}{\rho^4}\right)\cos 2\theta\right] \tag{5.277}$$

$$\sigma_{\rho\theta} = -\frac{\sigma_1}{2}\left[1 + \frac{A}{\rho^2} + \frac{3C}{\rho^4}\right]\sin 2\theta \tag{5.278}$$

Subtracting the expression for $\sigma_{\theta\theta} - \sigma_{\rho\rho}$ from that for $\sigma_{\theta\theta} + \sigma_{\rho\rho}$ we obtain

$$2\sigma_{\rho\rho} = \sigma_1\left[1 - \frac{B}{\rho^2} + \left(1 - \frac{2A}{\rho^2} - \frac{3C}{\rho^4}\right)\cos 2\theta\right] \tag{5.279}$$

The constants A, B and C must satisfy the boundary conditions $\sigma_{\rho\rho} = \sigma_{\rho\theta} = 0$ for $\rho = R$, independently of the value of θ, so that

$$1 + \frac{A}{R^2} + \frac{3C}{R^4} = 0, \quad 1 - \frac{B}{R^2} = 0, \quad 1 - \frac{2A}{R^2} - \frac{3C}{R^4} = 0 \tag{5.280}$$

the solution of which is $A = 2R^2, \quad B = R^2, \quad C = -R^4$. Thus the solution for the stress components is

$$\sigma_{\rho\rho} = \frac{\sigma_1}{2}\left(1 - \frac{R^2}{\rho^2}\right) + \frac{\sigma_1}{2}\left(1 - \frac{4R^2}{\rho^2} + 3\frac{R^4}{\rho^4}\right)\cos 2\theta \tag{5.281}$$

$$\sigma_{\theta\theta} = \frac{\sigma_1}{2}\left(1 + \frac{R^2}{\rho^2}\right) - \frac{\sigma_1}{2}\left(1 + 3\frac{R^4}{\rho^4}\right)\cos 2\theta \tag{5.282}$$

$$\sigma_{\rho\theta} = -\frac{\sigma_1}{2}\left(1 + 2\frac{R^2}{\rho^2} - 3\frac{R^4}{\rho^4}\right)\sin 2\theta \tag{5.283}$$

The displacements are obtained from (5.248) and (5.197):

$$2G(u_\rho + iu_\theta) = 2G(u_1 + iu_2)e^{-i\theta} = [\kappa\phi(z) - z\overline{\phi'(z)} - \overline{\psi(z)}]e^{-i\theta} \tag{5.284}$$

$$= \frac{\sigma_1}{4}\left[\kappa\left(\rho e^{i\theta} + \frac{A}{\rho}e^{-i\theta}\right) - \rho e^{i\theta}\left(1 - \frac{A}{\rho^2}e^{2i\theta}\right)\right.$$
$$\left. + 2\left(\rho e^{-i\theta} + \frac{B}{\rho}e^{i\theta} + \frac{C}{\rho^3}e^{3i\theta}\right)\right]e^{-i\theta} \tag{5.285}$$

With the values for A, B and C already computed we obtain

$$\frac{8G}{R\sigma_1}(u_\rho + iu_\theta) = \kappa\left(\frac{\rho}{R} + \frac{2R}{\rho}e^{-2i\theta}\right) - \frac{\rho}{R}\left(1 - \frac{2R^2}{\rho^2}e^{2i\theta}\right) + 2\left(\frac{R}{\rho} + \frac{\rho}{R}e^{-2i\theta} - \frac{R^3}{\rho^3}e^{2i\theta}\right) \tag{5.286}$$

from which we can determine the solution,

$$\frac{8G}{R\sigma_1}u_\rho = (\kappa - 1 + 2\cos 2\theta)\frac{\rho}{R} + \frac{2R}{\rho}\left[1 + \left(\kappa + 1 - \frac{R^2}{\rho^2}\right)\cos 2\theta\right] \tag{5.287}$$

$$\frac{8G}{R\sigma_1}u_\theta = \left[-\frac{2\rho}{R} + \frac{2R}{\rho}\left(1 - \kappa - \frac{R^2}{\rho^2}\right)\right]\sin 2\theta \tag{5.288}$$

For an infinite cylinder that yields plane strain conditions ($\kappa = 3 - 4\nu$) the displacements at the borehole wall ($\rho = R$) are

$$\frac{u_\rho}{R} = \frac{1 - \nu^2}{E}(\sigma_1 + 2\sigma_1\cos 2\theta) \tag{5.289}$$

$$\frac{u_\theta}{R} = -2\frac{1 - \nu^2}{E}\sigma_1 \sin 2\theta \tag{5.290}$$

If a component σ_2 is applied in the direction perpendicular to that of σ_1 then the corresponding solution is obtained by replacing θ by $\theta + \pi/2$ in the above solutions. For the case in which both σ_1 and σ_2 are applied simultaneously, in two perpendicular directions, we simply add the two solutions and obtain, for the stress components,

$$\sigma_{\rho\rho} = \frac{\sigma_1 + \sigma_2}{2}\left(1 - \frac{R^2}{\rho^2}\right) + \frac{\sigma_1 - \sigma_2}{2}\left(1 - \frac{4R^2}{\rho^2} + 3\frac{R^4}{\rho^4}\right)\cos 2\theta \tag{5.291}$$

$$\sigma_{\theta\theta} = \frac{\sigma_1 + \sigma_2}{2}\left(1 + \frac{R^2}{\rho^2}\right) - \frac{\sigma_1 - \sigma_2}{2}\left(1 + 3\frac{R^4}{\rho^4}\right)\cos 2\theta \tag{5.292}$$

$$\sigma_{\rho\theta} = -\frac{\sigma_1 - \sigma_2}{2}\left(1 + 2\frac{R^2}{\rho^2} - 3\frac{R^4}{\rho^4}\right)\sin 2\theta \tag{5.293}$$

while for the displacements at the borehole wall the solution is

$$\frac{u_\rho}{R} = \frac{1 - \nu^2}{E}[(\sigma_1 + \sigma_2) + 2(\sigma_1 - \sigma_2)\cos 2\theta] \tag{5.294}$$

$$\frac{u_\theta}{R} = -2\frac{1 - \nu^2}{E}(\sigma_1 - \sigma_2)\sin 2\theta \tag{5.295}$$

It is worth noting that the stress components vary as $1/\rho^2$ and thus the stress perturbation caused by the borehole becomes negligible at distances larger than five to six times the borehole radius.

This solution may be extended to the case of an infinite cylinder with its axis parallel to the z direction, for which the displacement components u_ρ and u_θ are independent of z and the strain component $\varepsilon_z = u_{z,z}$ is also independent of z ($\varepsilon_z = \varepsilon$). In the absence of body forces equation (5.48) yields

$$u_{z,\rho\rho} + \frac{1}{\rho}u_{z,\rho} + \frac{1}{\rho^2}u_{z,\theta\theta} = 0 \tag{5.296}$$

while equations (5.46) and (5.47) are independent of u_z. Then the solutions given by equations (5.291)–(5.295) hold and we must add to them the solution of (5.296), which yields the shear stress components

$$\sigma_{\rho z} = 2\,G\varepsilon_{\rho z} = G\,u_{z,\rho}, \qquad \sigma_{\theta z} = 2G\,\varepsilon_{\theta z} = \frac{G}{\rho}u_{z,\theta} \tag{5.297}$$

A general solution of (5.296) is

$$u_z = \left(A\rho + \frac{B}{\rho}\right)\cos\theta + \left(D\rho + \frac{E}{\rho}\right)\sin\theta + \varepsilon z \tag{5.298}$$

where A, B, D and E are constants that are determined by the boundary conditions. The shear stress component $\sigma_{\rho z}$ is given by

$$\sigma_{\rho z} = G\left(A - \frac{B}{\rho^2}\right)\cos\theta + G\left(D - \frac{E}{\rho^2}\right)\sin\theta \tag{5.299}$$

However, $\sigma_{\rho z} = 0$ at the free surface of the cylinder ($\rho = R$), so that $B = AR^2$ and $E = DR^2$. The general solution of (5.298) is

$$u_z = A\left(\rho + \frac{R^2}{\rho}\right)\cos\theta + D\left(\rho + \frac{R^2}{\rho}\right)\sin\theta + \varepsilon z \tag{5.300}$$

$$\sigma_{\rho z} = GA\left(1 - \frac{R^2}{\rho^2}\right)\cos\theta + GD\left(1 - \frac{R^2}{\rho^2}\right)\sin\theta \tag{5.301}$$

$$\sigma_{\theta z} = -GA\left(1 + \frac{R^2}{\rho^2}\right)\sin\theta + GD\left(1 + \frac{R^2}{\rho^2}\right)\cos\theta \tag{5.302}$$

The constants A and D are determined by the values of the stress components $\sigma_{z\rho}$ and $\sigma_{z\theta}$ in the far field ($\rho \to \infty$).

Hence the stress perturbation caused by an infinite empty circular cylinder inclined to the principal stress directions is defined in the cylindrical frame of reference by the six equations (5.303)–(5.308):

$$\begin{aligned}\sigma_{\rho\rho} = {} & \frac{\sigma_{11}+\sigma_{22}}{2}\left(1 - \frac{R^2}{\rho^2}\right) \\ & + \left(\frac{\sigma_{11}-\sigma_{22}}{2}\cos 2\theta + \sigma_{12}\sin 2\theta\right)\left(1 - \frac{4R^2}{\rho^2} + 3\frac{R^4}{\rho^4}\right)\end{aligned} \tag{5.303}$$

$$\sigma_{\theta\theta} = \frac{\sigma_{11}+\sigma_{22}}{2}\left(1 + \frac{R^2}{\rho^2}\right) - \left(\frac{\sigma_{11}-\sigma_{22}}{2}\cos 2\theta + \sigma_{12}\sin 2\theta\right)\left(1 + 3\frac{R^4}{\rho^4}\right) \tag{5.304}$$

$$\sigma_{zz} = \sigma_{33} - 4\nu\frac{R^2}{\rho^2}\left(\frac{\sigma_{11}-\sigma_{22}}{2}\cos 2\theta + \sigma_{12}\sin 2\theta\right) \tag{5.305}$$

$$\sigma_{\rho\theta} = -\left(\frac{\sigma_{11}-\sigma_{22}}{2}\sin 2\theta - \sigma_{12}\cos 2\theta\right)\left(1 + 2\frac{R^2}{\rho^2} - 3\frac{R^4}{\rho^4}\right) \tag{5.306}$$

$$\sigma_{\theta z} = \left(1 + \frac{R^2}{\rho^2}\right)(\sigma_{23}\cos\theta - \sigma_{31}\sin\theta) \tag{5.307}$$

$$\sigma_{z\rho} = \left(1 - \frac{R^2}{\rho^2}\right)(\sigma_{31}\cos\theta + \sigma_{23}\sin\theta) \tag{5.308}$$

where $\sigma_{ij}, i,j = 1,2,3$, are the far-field stress components in a cartesian coordinate system with axis $\hat{\boldsymbol{I}}_3$ parallel to the cylinder axis and the $\hat{\boldsymbol{I}}_1$ axis oriented parallel to the $\theta = 0$ direction. Equations (5.303)–(5.308) correspond to the solution published by Hiramatsu and Oka (1968). This solution is used in chapter 12 in our discussion of hydraulic fracturing.

The displacement field associated with these boundary conditions (an empty cylindrical hole in an infinite body with a uniform far-field stress state defined by its six components $\sigma_{ij}, i,j = 1,2,3$) is

$$\begin{aligned} u_\rho = {} & \frac{1+\nu}{E}\left(\frac{1-\nu}{1+\nu}\rho + \frac{R^2}{\rho}\right)\frac{\sigma_{11}+\sigma_{22}}{2} - \frac{\nu}{E}\rho\sigma_{33} \\ & + \frac{1+\nu}{E}\left[\rho + \frac{4(1-\nu)R^2}{\rho} - \frac{R^4}{\rho^3}\right]\left(\frac{\sigma_{11}-\sigma_{22}}{2}\cos 2\theta + \sigma_{12}\sin 2\theta\right) \end{aligned} \tag{5.309}$$

$$u_\theta = \frac{1+\nu}{E}\left[\rho + \frac{2(1-2\nu)R^2}{\rho} + \frac{R^4}{\rho^3}\right]\left(\frac{\sigma_{11}-\sigma_{22}}{2}\sin 2\theta + \sigma_{12}\cos 2\theta\right) \tag{5.310}$$

$$u_z = \frac{2(1+\nu)}{E}\left(\rho + \frac{R^2}{\rho}\right)(\sigma_{31}\cos\theta + \sigma_{23}\sin\theta) - \frac{2\nu}{E}z\frac{\sigma_{11}+\sigma_{22}}{2} + \frac{z}{E}\sigma_{33} \tag{5.311}$$

This solution is applied in the overcoring method for *in situ* stress measurements, as will be discussed in chapter 13.

Infinite empty square hole with a given stress field at infinity

The solution for the infinite empty circular hole with infinite stress field at infinity, as derived above, may be applied to cavities with various shapes thanks to *conformal mapping*. Indeed, conformal mapping can be used to transform the $z = x_1 + ix_2$ plane for which a circular-opening solution has been derived into the $\zeta = \xi + i\eta$ plane in which a cavity with the desired shape is defined (Jaeger *et al.*, 2007, chapter 8).

For example, the circular shape $z = \exp(i\alpha)$ with radius $r = 1$ may be transformed into an approximately square shape with sides parallel to axis ξ and η by the mapping function

$$\xi = a(\cos\alpha - \tfrac{1}{6}\cos 3\alpha); \quad \eta = -a(\sin\alpha + \tfrac{1}{6}\sin 3\alpha) \tag{5.312}$$

With this transformation, the lengths of the sides of the equivalent square opening are equal to $5a/3$ whilst the radius of curvature at the corners is approximately $a/10$ (fig. 5.11).

The values of the tangential stress $\sigma_{\theta\theta}$ for $\beta = 0$ (a uniaxial stress at infinity parallel to the ξ axis) and for $0 < \theta < 45°$ in fig. 5.11 illustrate how the stress component varies for points along the face normal to the ξ axis, as well as for points along the face normal to the η axis ($45° < \theta < 90°$). The following analytical expression may be used for any orientation β:

$$\sigma_{\theta\theta} = \sigma_1(105 - 360\cos 2\beta\cos 2\theta + 168\sin 2\beta\sin 2\theta)/[35(5 + 4\cos 4\theta)] \tag{5.313}$$

An application of this solution occurs when one is discussing the influence of the loading geometry in laboratory investigations on rock behavior under compression (chapter 8).

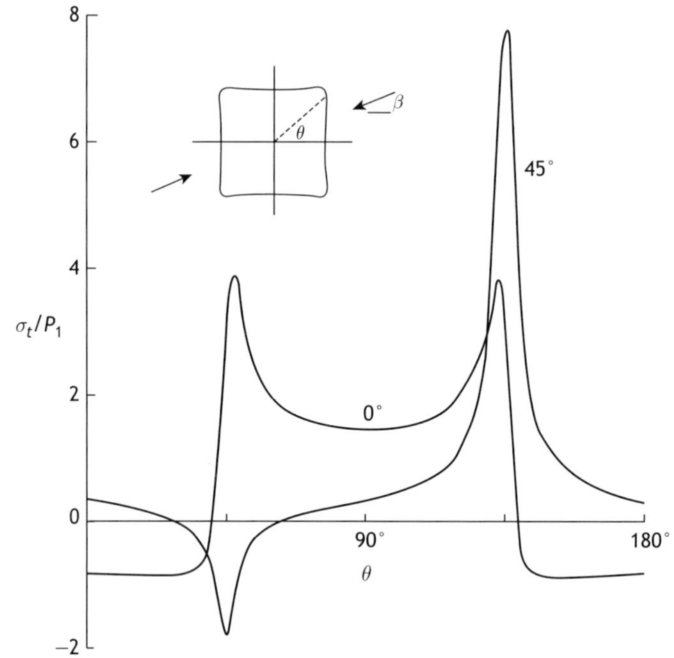

Fig. 5.11 Stress concentration in the vicinity of an infinite square opening loaded by a uniaxial stress at infinity. The orientation of the uniaxial stress is β. The plots give the ratio $\sigma_{\theta\theta}/\sigma_1$ for $\beta = 0°$ and $\beta = 45°$ (Reproduced from Jaeger, 1969, fig. 73, with permission from Wiley).

Conformal mapping has been used also for evaluating the influence of fractures (very thin elliptical cavities) on local stresses, as will be discussed in chapter 7. Another interesting application of conformal mapping and Airy stress functions to geomechanics was proposed by Savage and Swolfs (1986) for evaluating stress fields in long symmetric ridges loaded by gravity and tectonics.

Finally, we may add that all these analytic solutions are helpful for testing the accuracy of numerical codes developed for more complex three-dimensional geometries.

5.7 Exercises

1. How many independent elastic constants are needed to characterize an isotropic linearly elastic material? How many are needed for a material that exhibits a planar isotropy?
2. What are the units for Lamé coefficients, for Young's modulus and for the Poisson ratio?
3. What are the Young's modulus, the Poisson ratio and the shear modulus for a granite with density 2.7 g/cm^3, in which P waves propagate with a velocity of 5.7 km/s and S waves propagate with a velocity of 3300 m/s? Repeat the question for a limestone with density 2.5 g/cm^3 for which the P and S wave velocities are respectively 3800 m/s and 2240 m/s.

4. Consider a long cylindrical rock core of circular cross section with radius r_c and axis X_1. The core is placed horizontally on two knife supports separated by a distance $2L$ from each other. The core is loaded by two vertical point forces, downward oriented and applied respectively at a distance $2L/3$ from the two knife supports.
   4.1. Draw a simplified diagram of the loading system (inspired from that shown on the right of fig. 5.3) and compute the moment M supported by the vertical cross sections of the core as a function of the distance x_1, from the center of the core (which is a distance L from either knife support). Discuss the variation of M as a function of x_1 as x_1 varies from 0 to +L.
   4.2. Taking advantage of the fact that the radius of curvature R of the deformed core is equal to the inverse of d^2u_3/dx_1^2, compute the moment M as a function of R and of the moment of inertia I of the core (see equation (5.52)). This moment of inertia is computed with respect to the horizontal axis X_2 perpendicular to X_1. Compute I as a function of the radius of the core r_c.
   4.3. Determine the extreme values of the stress component σ_{11} in the core.
   4.4. Draw the Mohr circles that describe the stress tensor at the points $x_1 = x_2 = 0,\ x_3 = \pm r_c$ as well as for the point $x_1 = x_2 = x_3 = 0$.
   4.5. Show why measurements of the deflection of the beam at its center provide a means to measure the Young's modulus of the rock.
5. Consider a vertical cylindrical rock specimen with radius r_c and length $l = 6r_c$. Recall that the small-strain components in cylindrical coordinates $\rho,\ \theta$ and z, are given by equations (4.103) and (4.104). This core is loaded in such a way that the displacement field inside the specimen is characterized by

$$u_\rho = 0, \quad u_\theta = \Omega\rho z/l, \quad u_z = 0$$

   5.1. Compute the stress tensor components in the specimen as a function of the cylindrical coordinates.
   5.2. Compute the moment of the forces acting on the specimen's extremity, located at $z = l$, and discuss how this loading process may be applied to measure the shear modulus of the rock.
   5.3. In addition to the above loading, the specimen is loaded by a uniaxial stress σ_a, applied to both extremities of the specimen, as well as by a uniform pressure P_c applied around the specimen. Give the new stress components inside the specimen. Determine the principal stress direction orientations for an arbitrary point located on the cylindrical surface of the specimen (with coordinate, say, $\rho = r_c$).
   5.4. Consider a plane P inclined at 60° to the specimen's axis. Define $\theta = 0$ as the angular coordinate of the upper point of the ellipse defined by the intersection of plane P with the sample. Compute the resolved shear stress in the plane, for $\theta = 0$ and for $\theta = \pi/2$.
6. Consider a horizontal tunnel of circular cross section through a massif. For short enough horizontal distances within the tunnel the far-field stress can be considered uniform; it is defined by its components $(\sigma_{ij}^{\infty}; i, j = 1, 2, 3)$, where the x_1 axis is parallel to the tunnel axis.

6.1. Given that the stress field at the wall of the tunnel is defined by equations (5.303)–(5.308), compute the normal stress supported by a plane normal to the tunnel axis, at a point located at the top of the tunnel (with angular coordinate $\theta = \pi/2$).

6.2. Repeat exercise 6.1 for a plane inclined at 45° to the tunnel axis.

6.3. We will discuss in section 13.1.1 how flatjacks may be used to measure the normal stress supported by a plane. Discuss how such flatjacks may be used to determine the components of the far-field stress $\tilde{\sigma}^{\infty}$.

6.4. A vertical plane P intersects the tunnel at an angle of 60°. Compute the normal stress supported by the plane for $\theta = 0$, $\theta = \pi/2$ and $\theta = \pi$.

6.5. Assume now that the tunnel is filled with water under pressure. How would you identify the maximum pressure that may be applied to the water before it leaks into the rock massif?

6 From continuum mechanics to fluid mechanics

Fluids are characterized by the fundamental property that they cannot support any shear stress when they are at rest. Hence, the stress tensor at any point in a fluid at rest is simply $p\tilde{1}$, where p is the fluid's static pressure at the point under consideration. We may also conclude from this property that the most viscous fluid-type material keeps deforming as long as the three principal stress components of the stress tensor at any point in the body are not respectively equal to each other. We saw in chapter 1 that crustal materials are usually multiphasic, i.e. they involve both solids and fluids, whether the fluid is liquid (with some viscosity but negligible compressibility) or gas (with negligible viscosity but highly compressible). In chapter 12 we discuss some aspects of the interactions between solids and fluids in geomechanics. Before discussing these interactions, however, some elementary considerations on fluid mechanics are necessary.

In fluid mechanics, the displacements of the particles are generally not known and attention turns to the velocity of the particles at a given time. The frame of reference chosen for analyzing fluid mechanics problems is not a *Lagrangian* frame of reference, as in elasticity, but an *Eulerian* frame of reference, in which all equations are written with respect to the spatial coordinates at a given time.

The objectives of this chapter are, first, to introduce basic conservation principles that constitute a paradigm for continuum mechanics; then considerations for deriving the constitutive equations for fluids are very briefly stated, with particular attention to incompressible fluids, and finally we present solutions useful in geomechanics, namely for laminar flow through parallel plates and flow through cylindrical conduits.

Some material presented in this chapter was inspired by L. E. Malvern's book on continuum mechanics (Malvern, 1969, chapters 5–7) and the book by Guyon *et al.* on fluid mechanics (Guyon *et al.*, 1991, chapters 4 and 5).

6.1 Paradigm of continuum mechanics: the conservation principles

The paradigm of continuum mechanics includes the conservation of mass, the conservation of momentum and the conservation of energy. The corresponding basic equations are written down for each geomaterial that makes up the system under concern. However, in some instances mass may be transferred from the solid to the fluid and conversely. This is the case, for example, when atoms of one constituent migrate from the solid phase to the fluid phase (as in chemical dissolution or partial melting) or conversely when they migrate from the fluid phase to the solid phase (as in scaling or magma partial crystallization). In such instances some additional differential equations are needed to describe these mass transfers. This would take us to the field of thermodynamics, which is not addressed in this book.

6.1.1 The conservation of mass and the continuity equation

The divergence (or Green or Gauss) theorem and Stokes' theorem

The divergence theorem, also known as the *Green* or *Gauss* theorem, transforms the integral of a function defined over a closed surface S into an integral over the volume V enclosed by the surface S. If the scalar function $f(x_1, x_2, x_3)$, defined in cartesian coordinates, has continuous derivatives in volume V then

$$\int_S fn_i \, da = \int_V f_{,i} \, dv \tag{6.1}$$

where n_i is the direction cosine of the normal to the surface with respect to the unit vector $\hat{\boldsymbol{I}}_i$ of the cartesian frame of reference; da and dv are respectively elements of surface area and volume. A derivation of this result may be found in books of advanced calculus. If the scalar function f is the component v_i of a vector function $\boldsymbol{v}$, and if we sum over the $i = 1, 2, 3$, equation (6.1) becomes

$$\int_S v_i n_i \, da = \int_V v_{i,i} \, dv \tag{6.2}$$

or, in vector form,

$$\int_S \boldsymbol{v} \cdot \boldsymbol{n} \, da = \int_V \nabla \cdot \boldsymbol{v} \, dv \tag{6.3}$$

where $\boldsymbol{v}$ is any vector.

Note that in this chapter, we keep to the convention discussed in chapter 4 according to which grad, div and rot involve partial derivatives with respect to coordinates x_i in an Eulerian frame of reference whilst the notation ∇, $\nabla \cdot$ and $\nabla\wedge$ refers to operators involving partial derivatives with respect to coordinates X_i in a Lagrangian frame of reference.

If the vector $\boldsymbol{v}$ is a linear function of $\boldsymbol{n}$ ($v_i = T_{ij}n_j$) then equation (6.1) becomes

$$\int_S T_{ij} n_j \, da = \int_V T_{ij,i} \, dv \tag{6.4}$$

Another important vector integral transformation is described by Stokes' theorem, which relates an integral around a closed contour C with that over an oriented surface S bounded by the curve C (fig. 6.1).

If A is a continuous field function (scalar, vector or tensor) then, if ds is an element of the curve C,

$$\int_S (\boldsymbol{n} \wedge \nabla) * A \, da = \int_C \boldsymbol{t} * A \, ds \tag{6.5}$$

where $*$ indicates the appropriate form of multiplication. The normal vector $\boldsymbol{n}$ defines the positive side of the surface S and $\boldsymbol{t}$ is a unit vector tangent to the curve C. The positive

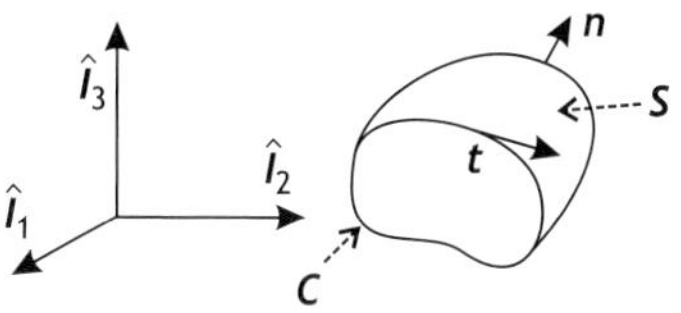

Fig. 6.1 Closed contour C and oriented surface S.

sense of C is implied by the right-hand screw rule applied to a screw lying along $\boldsymbol{n}$. For example, for a vector $\boldsymbol{v}$ equation (6.5) becomes

$$\int_S \boldsymbol{n} \cdot (\nabla \wedge \boldsymbol{v})\, da \;=\; \int_C \boldsymbol{t} \cdot \boldsymbol{v}\, ds \tag{6.6}$$

This theorem also applies to so-called multiply connected surfaces (e.g. Malvern, 1969).

If the vector $\boldsymbol{v}$ in equation (6.6) is the particle velocity at a point in a deforming fluid and the contour C is a closed contour in the fluid, the quantity

$$\oint_C \boldsymbol{v} \cdot d\boldsymbol{s} = \oint_C \boldsymbol{v} \cdot \boldsymbol{t}\, ds \tag{6.7}$$

is called the circulation of the velocity $\boldsymbol{v}$ around C.

When the spin tensor $\tilde{W}$ (see section 4.4.2) is null at all points of C, i.e. when

$$\nabla \wedge \boldsymbol{v} \cdot d\boldsymbol{s} = 0$$

the flow is said to be circulation preserving, and the quantity $\boldsymbol{v} \cdot d\boldsymbol{s}$ is an exact differential of a scalar function. For a circulation-preserving fluid motion, the acceleration and velocity derive from a scalar potential.

Isochoric motion

Isochoric motion is flow at constant volume, i.e. the flow of incompressible materials. The isochoric condition takes different forms, depending on whether it is expressed in a Lagrangian or an Eulerian frame of reference.

Lagrangian frame of reference

Let us consider a body B and any smaller part of it, referred to as P. At time $t = 0$, a particle of B occupies point $\boldsymbol{X}$ so that the volume of part P is vol $P_\kappa = \int_{P_\kappa} dV(\boldsymbol{X}, 0)$. At time t, particle p occupies point $\boldsymbol{x} = \boldsymbol{\chi}(\boldsymbol{X}, t)$ and, according to equation (4.19), the volume of part P is

$$\text{vol}\, P_\chi = \int_{P_\chi} dv(\boldsymbol{x}, t) \tag{6.8}$$

$$= \int_{P_\kappa} \det \tilde{F}(\boldsymbol{X}, t)\, dV(\boldsymbol{X}, 0) \tag{6.9}$$

with $F_{ij}(\boldsymbol{X},t) = \chi_{i,j}(\boldsymbol{X},t)$. Hence, if the flow is isochoric,

$$\int_{P_\kappa} dV - \int_{P_\kappa} \det \tilde{F}\, dV = \int_{P_\kappa} (1 - \det \tilde{F})\, dV = 0 \tag{6.10}$$

Equation (6.10) must hold whatever the volume of P; hence $1 - \det \tilde{F} = 0$ and therefore $\det \tilde{F}(\boldsymbol{X},t) = 1$.

Eulerian frame of reference

In an *Eulerian* frame of reference we compute the variation with time of the fluid properties in a small volume surrounding the point $\boldsymbol{x}$ and express the fact that this variation is zero.

At time t_1, the volume of part P, surrounding $\boldsymbol{x}$, is

$$V_{t_1}(P_\chi) = \int_{P_\chi} dv(\boldsymbol{x},t_1) = \int_{P_\kappa} \det \tilde{F}(\boldsymbol{X},t_1)\, dV(\boldsymbol{X},0) \tag{6.11}$$

At time t_2 the volume of part P is

$$V_{t_2}(P_\chi) = \int_{P_\chi} dv(\boldsymbol{x},t_2) = \int_{P_\kappa} \det \tilde{F}(\boldsymbol{X},t_2)\, dV(\boldsymbol{X},0) \tag{6.12}$$

The difference in volume between times t_2 and t_1 is

$$V_{t_2}(P_\chi) - V_{t_1}(P_\chi) = \int_{P_\kappa} [\det \tilde{F}(\boldsymbol{X},t_2) - \det \tilde{F}(\boldsymbol{X},t_1)] dV(\boldsymbol{X},0) \tag{6.13}$$

If $dt = t_2 - t_1$, we must evaluate

$$\lim_{dt\to 0} \frac{V_{t_2}(P_\chi) - V_{t_1}(P_\chi)}{dt} = \overline{\int_{P_\chi} dv(\boldsymbol{x},t)}^{\,\bullet} = \int_{P_\kappa} \overline{\det \tilde{F}}^{\,\bullet}\, dV(\boldsymbol{X},0) \tag{6.14}$$

The time derivative of $\det \tilde{F}$ is

$$\overline{\det \tilde{F}}^{\,\bullet} = \lim_{dt\to 0} \frac{\det \tilde{F}(t+dt) - \det \tilde{F}(t)}{dt} \tag{6.15}$$

A Taylor expansion of $\tilde{F}(t+dt)$ gives

$$\tilde{F}(t+dt) = \tilde{F}(t) + \dot{\tilde{F}}(t)\, dt + 0(dt^2) \tag{6.16}$$

$$\text{with } \lim_{dt\to 0} \left[\frac{0(dt^2)}{dt}\right] = 0$$

Equation (6.16) may be rewritten as

$$\tilde{F}(t+dt) = dt\, \tilde{F}(t) \left[\tilde{F}^{-1}\dot{\tilde{F}} + \tilde{F}^{-1}\frac{0(dt^2)}{dt} + \frac{1}{dt}\tilde{1}\right] \tag{6.17}$$

Let us refer to the tensor $\{\tilde{F}^{-1}\dot{\tilde{F}} + \tilde{F}^{-1}[0(dt^2)/dt]\}$ as $\tilde{A}$. The determinant of $\tilde{F}(t+dt)$ is

$$\det \tilde{F}(t+dt) = (dt)^3 \det \tilde{F}(t) \det \left(\tilde{A} + \frac{1}{dt}\tilde{1}\right) \tag{6.18}$$

If we remember from the determination of the eigenvalues of tensors that (see the appendix)

$$\det(\tilde{T} - \lambda\tilde{1}) = -\lambda^3 + \lambda^2 I_{\tilde{T}} - \lambda II_{\tilde{T}} + III_{\tilde{T}} \tag{6.19}$$

where $I_{\tilde{T}}$, $II_{\tilde{T}}$ and $III_{\tilde{T}}$ are the first, second and third invariants of $\tilde{T}$, then we may write:

$$\det\left(\tilde{A} + \frac{1}{dt}\tilde{1}\right) = \left(\frac{1}{dt}\right)^3 + \left(\frac{1}{dt}\right)^2 I_{\tilde{A}} + \frac{1}{dt} II_{\tilde{A}} + III_{\tilde{A}} \tag{6.20}$$

so that

$$\det\tilde{F}(t+dt) = (dt)^3 \det\tilde{F}\left[\left(\frac{1}{dt}\right)^3 + \left(\frac{1}{dt}\right)^2 I_{\tilde{A}} + \frac{1}{dt} II_{\tilde{A}} + III_{\tilde{A}}\right] \tag{6.21}$$

and

$$\frac{\det\tilde{F}(t+dt) - \det\tilde{F}(dt)}{dt} = \det\tilde{F}(I_{\tilde{A}} + II_{\tilde{A}}\,dt + III_{\tilde{A}}\,dt^2) \tag{6.22}$$

Hence, when $dt \to 0$ we obtain

$$\dot{\overline{\det\tilde{F}}} = \det\tilde{F}\,\mathrm{tr}\left(\tilde{F}^{-1}\dot{\tilde{F}} + \tilde{F}^{-1}\frac{0(dt^2)}{dt}\right) \tag{6.23}$$

$$= \det\tilde{F}\,\mathrm{tr}(\tilde{F}^{-1}\dot{\tilde{F}}) \tag{6.24}$$

since $0(dt^2)/dt \to 0$ when $dt \to 0$.

The volume variation of part P during time dt is therefore

$$\int_{P_\kappa} \dot{\overline{\det\tilde{F}}}\,dV = \int_{P_\kappa} \det\tilde{F}\,\mathrm{tr}(\tilde{F}^{-1}\dot{\tilde{F}})\,dV(\boldsymbol{X},0) = \int_{P_\chi} \mathrm{tr}(\tilde{F}^{-1}\dot{\tilde{F}})\,dv(\boldsymbol{x},t) \tag{6.25}$$

But we saw in section 4.4.2 that $\dot{\tilde{F}} = \nabla\boldsymbol{v}(\boldsymbol{X},t) = (\mathrm{grad}\,\boldsymbol{v})\,\tilde{F}$, so that

$$\mathrm{tr}(\tilde{F}^{-1}\dot{\tilde{F}}) = \mathrm{tr}[\tilde{F}^{-1}(\mathrm{grad}\,\boldsymbol{v})\tilde{F}] \tag{6.26}$$

or, in terms of components,

$$\mathrm{tr}(\tilde{F}^{-1}\dot{\tilde{F}}) = F^{-1}_{im}(\mathrm{grad}\,\boldsymbol{v})_{mn}F_{ni} = \delta_{mn}(\mathrm{grad}\,\boldsymbol{v})_{mn} \tag{6.27}$$

$$= \mathrm{div}\,\boldsymbol{v}(\boldsymbol{x},t) \tag{6.28}$$

For the motion of part P of B to be isochoric we require no volume variation during flow, for arbitrary P, and therefore

$$\mathrm{div}\,\boldsymbol{v} = 0 \tag{6.29}$$

In conclusion, the condition for a motion to be isochoric is that $\det\tilde{F} = 1$ if the motion is described in a Lagrangian frame of reference or $\mathrm{div}\,\boldsymbol{v} = 0$ if the motion is described in an Eulerian frame of reference.

Conservation of mass: the continuity equation

A paradigm of continuum mechanics is that mass is neither destroyed nor is it created during motion. In other words there is conservation of mass during a deformation process. As in the previous section, the conservation of mass principle may be expressed either in a Lagrangian (material) frame of reference or in an Eulerian (spatial) frame of reference.

Lagrangian frame of reference

The mass M of part P considered in the previous section may be evaluated at time $t = 0$ and then at any other time t:

$$M(P) = \int_{P_\kappa} \varrho_\kappa \, dV(\boldsymbol{X}, 0) = \int_{P_\chi} \varrho_\kappa \frac{dv(\boldsymbol{x}, t)}{\det \tilde{F}} \tag{6.30}$$

$$= \int_{P_\chi} \varrho \, dv(\boldsymbol{x}, t) \tag{6.31}$$

where ϱ_κ and ϱ are respectively the material density at time $t = 0$ and at any other time t. Hence in a Lagrangian frame of reference, the conservation of mass is simply expressed by $\varrho = \varrho_\kappa / \det \tilde{F}$.

Eulerian frame of reference

In an Eulerian frame of reference we simply express the fact that the mass does not vary with time. For any part P of the body under consideration, we may write:

$$\overline{M(P)}^{\boldsymbol{\cdot}} = \overline{\int_{P_\chi} \varrho \, dv}^{\boldsymbol{\cdot}} = 0 \qquad \forall P_\chi \tag{6.32}$$

$$= \int_{P_\kappa} \overline{\varrho \det \tilde{F}}^{\boldsymbol{\cdot}} \, dV(\boldsymbol{X}, 0) = \int_{P_\kappa} (\dot{\varrho} \det \tilde{F} + \varrho \, \overline{\det \tilde{F}}^{\boldsymbol{\cdot}}) \, dV \tag{6.33}$$

Taking advantage of the result in equation (6.24), we conclude that

$$\overline{M(P)}^{\boldsymbol{\cdot}} = \int_{P_\kappa} \left[\dot{\varrho} + \varrho \, \mathrm{tr}(\tilde{F}^{-1} \dot{\tilde{F}}) \right] \det \tilde{F} \, dV(\boldsymbol{X}, 0) \tag{6.34}$$

$$= \int_{P_\chi} (\dot{\varrho} + \varrho \, \mathrm{div} \, \boldsymbol{v}) \, dv(\boldsymbol{x}, t) = 0 \qquad \forall P_\chi \tag{6.35}$$

Hence, the conservation of mass in an Eulerian frame of reference, also known as the *continuity equation*, is

$$\dot{\varrho} + \varrho \, \mathrm{div} \, \boldsymbol{v} = 0 \tag{6.36}$$

However, from equation (4.9) we observe that $\dot{\varrho} = \partial \varrho / \partial t + (\mathrm{grad} \, \varrho) \boldsymbol{v}$, so that

$$\dot{\varrho} + \varrho \, \mathrm{div} \, \boldsymbol{v} = \partial \varrho / \partial t + (\mathrm{grad} \, \varrho) \boldsymbol{v} + \varrho \, \mathrm{div} \, \boldsymbol{v} \tag{6.37}$$

$$= \partial \varrho / \partial t + \mathrm{div}(\varrho \boldsymbol{v}) = 0 \tag{6.38}$$

For steady flow conditions, so that there is no dependence on time, the continuity equation implies that $\mathrm{div}(\varrho \boldsymbol{v}) = 0$.

We may also derive the continuity equation in the Eulerian frame of reference by direct application of the divergence theorem. Let us first determine the flux of volume through a small surface $d\boldsymbol{S} = \boldsymbol{n} \, da$, where $\boldsymbol{n}$ is a unit vector normal to $d\boldsymbol{S}$, which is of area da. The

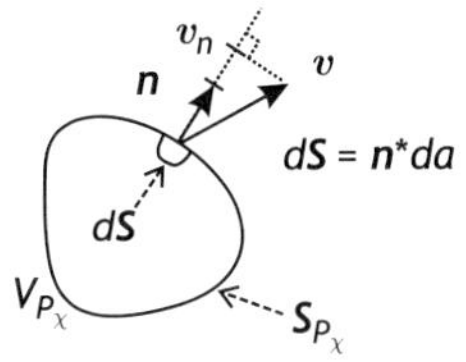

Fig. 6.2 Mass of part P of a body filled with material of density ϱ.

volume of fluid moving with velocity $\boldsymbol{v}$ through the small surface element dS during a time dt is $dv = \boldsymbol{v}\, dt \cdot \boldsymbol{n}\, da$.

For the total surface S, the volume flux, i.e. the volume per unit time that flows through S, is given by

$$\text{volume flux} = \int_{S_{P_X}} \boldsymbol{v} \cdot \boldsymbol{n}\, da = \int_{S_{P_X}} v_i n_i\, da \tag{6.39}$$

and the mass flux is given by

$$\text{mass flux} = \int_{S_{P_X}} \varrho \boldsymbol{v} \cdot \boldsymbol{n}\, da = \int_{S_{P_X}} \varrho v_i n_i\, da \tag{6.40}$$

Note that, associated with the mass flux, we may define a flux of momentum (which is a vector) and a flux of kinetic energy (which is a scalar):

$$\text{flux of momentum} = \int_{S_{P_X}} \varrho\, \boldsymbol{v}\, (\boldsymbol{v} \cdot \boldsymbol{n})\, da = \hat{\boldsymbol{I}}_k \int_{S_{P_X}} \varrho v_k v_i n_i\, da \tag{6.41}$$

$$\text{flux of kinetic energy} = \int_{S_{P_X}} \tfrac{1}{2} \varrho v^2 (\boldsymbol{v} \cdot \boldsymbol{n})\, da = \int_{S_{P_X}} \tfrac{1}{2} \varrho v_i v_i v_j n_j\, da \tag{6.42}$$

Let us now consider the volume V of any part P of the body, at time t (fig. 6.2). The mass of the medium that fills up volume V is

$$M = \int_{V_{P_X}} \varrho\, dv \tag{6.43}$$

The rate of increase of the total mass in volume V is

$$\frac{\partial M}{\partial t} = \int_{V_{P_X}} \frac{\partial \varrho}{\partial t}\, dv \tag{6.44}$$

Since it is assumed that mass is neither created nor destroyed, this change in mass must be equal to the rate of mass flux:

$$\frac{\partial M}{\partial t} = \int_{V_{P_X}} \frac{\partial \varrho}{\partial t}\, dv = -\int_{S_{P_X}} \varrho\, \boldsymbol{v} \cdot \boldsymbol{n}\, da \tag{6.45}$$

$$= -\int_{V_{P_X}} \operatorname{div}(\varrho\, \boldsymbol{v})\, dv \tag{6.46}$$

Since equation (6.46) must hold for any part P, we obtain the continuity equation (6.38) already derived

$$\frac{\partial \varrho}{\partial t} + \mathrm{div}(\varrho \boldsymbol{v}) = 0$$

6.1.2 Momentum principles and the equation of motion

In chapter 3 we introduced the concept of body forces and surface forces, which led us to introduce the concept of the stress tensor and stress vector (the surface force per unit area). In particular we showed that the stress tensor allows us to determine the stress vector acting on a surface with unit area and unit normal $\boldsymbol{n}$ according to the relation $\boldsymbol{t} = \tilde{\sigma}^T \boldsymbol{n}$. Then we derived an equilibrium equation, i.e. we wrote down conditions that must be satisfied by the body forces and surface forces for a body B with volume V to be at rest at all points ($\ddot{\boldsymbol{u}}(\boldsymbol{x}) = 0$ for all $\boldsymbol{x}$ lying in V). Similar equilibrium conditions had been considered in section 3.2.2 for demonstrating the symmetry of the stress tensor.

Now we will consider a body under motion. We derive the relations that must exist between forces, masses and acceleration and discuss the symmetry of the stress tensor.

Reynolds' theorem or the material time derivative of a volume integral

Here we want to define the material time derivative of a volume integral in order to measure the rate of change of the total amount of a quantity carried by a given mass system in space. Let us consider any volume V_{P_X}, bounded by a controlled surface S_{P_X}, fixed in space (the Eulerian frame of reference) and let A be a quantity tied to the mass (A could be a scalar, vector or tensor). The rate of increase of the total amount of A inside the control surface S_{P_X} is equal to the rate of increase of the amount of A possessed by the material instantaneously inside the control surface minus the net rate of outward flux of A carried by mass transport through the control surface S_{P_X}:

$$\int_{P_X} \frac{\partial}{\partial t}(A\varrho)\, dv = \frac{d}{dt}\left(\int_{P_X} A\varrho\, dv\right) - \int_{S_{P_X}} A\varrho\, \boldsymbol{v} \cdot \boldsymbol{n}\, da \tag{6.47}$$

The first term on the right-hand side of equation (6.47) is precisely the time derivative of the volume integral for which we are looking:

$$\frac{d}{dt}\int_{P_X} A\varrho\, dv = \int_{P_X} \frac{\partial}{\partial t}(A\varrho)\, dv + \int_{S_{P_X}} A\varrho\, v_j n_j\, da \tag{6.48}$$

The surface integral may be transformed into a volume integral using the divergence theorem:

$$\frac{d}{dt}\int_{P_X} A\varrho\, dv = \int_{P_X}\left[\frac{\partial}{\partial t}(A\varrho) + \frac{\partial}{\partial x_j}(A\varrho\, v_j)\right] dv \tag{6.49}$$

The right-hand term may be expanded to obtain

$$\int_{P_X} \varrho\left(\frac{\partial A}{\partial t} + v_j \frac{\partial A}{\partial x_j}\right) + A\left[\frac{\partial \varrho}{\partial t} + \frac{\partial(\varrho v_j)}{\partial x_j}\right] dv \tag{6.50}$$

However, because of the principle of mass conservation, the bracketed term in equation (6.50) vanishes, and we obtain the classical form of Reynolds' theorem:

$$\overline{\int_{P_\chi} \varrho A \, dv}^{\,\cdot} = \int_{P_\chi} \varrho \dot{A} \, dv \tag{6.51}$$

The material derivative of a contour integral is also useful. We note that

$$\int_C f \, d\boldsymbol{x} = \int_C f \, dx_k \hat{\boldsymbol{I}}_k \tag{6.52}$$

where f is a function defined at a point $\boldsymbol{x}$ on the contour C at time t and $\hat{\boldsymbol{I}}_k$ are unit vectors of a cartesian base.

In an Eulerian frame of reference we saw (equation (4.115)) that $d\boldsymbol{x} = \tilde{F} \, d\boldsymbol{X}$, so that

$$\dot{\overline{d\boldsymbol{x}}} = (\text{grad } \boldsymbol{v})\tilde{F} \, d\boldsymbol{X} = (\text{grad } \boldsymbol{v}) \, d\boldsymbol{x}$$

or, in terms of components,

$$\dot{\overline{dx_k}} = \frac{\partial v_k}{\partial x_m} dx_m = L_{km} \, dx_m$$

if $\tilde{L} \equiv \text{grad } \boldsymbol{v}$. The time derivative of the contour integral becomes

$$\overline{\int_C f \, d\boldsymbol{x}}^{\,\cdot} = \int_C \dot{f} \, d\boldsymbol{x} + \int_C f \, \dot{\overline{d\boldsymbol{x}}} \tag{6.53}$$

$$= \int_C \dot{f} \, d\boldsymbol{x} + \int_C f \tilde{L} \, d\boldsymbol{x} \tag{6.54}$$

The momentum principles

Let us consider a subpart P of body B in an Eulerian frame of reference. It is subject to both body forces $\boldsymbol{f}_b$ and surface forces $\boldsymbol{f}_s$ (see section 3.1.1):

$$\boldsymbol{f}_b = \int_{P_\chi} \varrho \boldsymbol{b} \, dv, \qquad \boldsymbol{f}_s = \int_{SP_\chi} d\boldsymbol{t} = \int_{SP_\chi} \boldsymbol{t} \, da \tag{6.55}$$

where $\boldsymbol{b}$ is the body force per unit volume and $\boldsymbol{t}$ is the stress vector per unit area. If $P_\chi \to 0$ then $\boldsymbol{f}_b \to 0$. But, as $da \to 0$, we define the stress tensor $\tilde{\sigma}$ with component σ_{ij}, where $\sigma_{ij} \equiv \lim_{da \to 0} dt_{ij}/da, i, j = 1, 2, 3$; here dt_{ij} is the jth component of the surface force acting on the surface element by $\boldsymbol{ds}$ with normal parallel to the unit vector $\hat{\boldsymbol{I}}_i$ of the reference base. The resultant force acting on part P_χ of the body $\boldsymbol{B}$ is

$$\boldsymbol{R}(P_\chi) = \boldsymbol{f}_b + \boldsymbol{f}_s = \int_{P_\chi} \varrho \boldsymbol{b} \, dv + \int_{SP_\chi} \boldsymbol{t} \, da \tag{6.56}$$

The resulting moment with respect to the point $\boldsymbol{x}_0$ is

$$\boldsymbol{m}(P_\chi, \boldsymbol{x}_0) = \int_{P_\chi} [(\boldsymbol{x} - \boldsymbol{x}_0) \wedge \varrho \boldsymbol{b}] \, dv + \int_{SP_\chi} [(\boldsymbol{x} - \boldsymbol{x}_0) \wedge \boldsymbol{t}] \, da \tag{6.57}$$

The momentum of P_χ is

$$\boldsymbol{l}(P_\chi) = \int_{P_\chi} \varrho \boldsymbol{v}\, dv \tag{6.58}$$

whilst the moment of momentum (angular momentum) for P_χ is

$$\boldsymbol{h}(P_\chi, \boldsymbol{x}_0) = \int_{P_\chi} [(\boldsymbol{x} - \boldsymbol{x}_0) \wedge \varrho \boldsymbol{v}]\, dv \tag{6.59}$$

From Euler's laws we know that $\boldsymbol{R}(P_\chi) = \dot{\boldsymbol{l}}(P_\chi)$ and $\boldsymbol{m}(P_\chi, \boldsymbol{x}_0) = \dot{\boldsymbol{h}}(P_\chi)$.

We wish to show now that

$$\boldsymbol{R}(P_\chi) = \int_{P_\chi} \frac{\partial}{\partial t}(\varrho \boldsymbol{v})\, dv + \int_{SP_\chi} \boldsymbol{v}(\varrho\, \boldsymbol{v} \cdot \boldsymbol{n})\, da \tag{6.60}$$

$$\boldsymbol{m}(P_\chi, \boldsymbol{x}_0) = \int_{P_\chi} (\boldsymbol{x} - \boldsymbol{x}_0) \wedge \frac{\partial}{\partial t}(\varrho \boldsymbol{v})\, dv + \int_{SP_\chi} (\boldsymbol{x} - \boldsymbol{x}_0) \wedge \boldsymbol{v}(\varrho\, \boldsymbol{v} \cdot \boldsymbol{n})\, da \tag{6.61}$$

Considering the time derivative of momentum, we may write in a Lagrangian frame of reference

$$\boldsymbol{R}(P_\chi) = \dot{\boldsymbol{l}}(P_\chi) = \overline{\left(\int_{P_\chi} \varrho \boldsymbol{v}\, dv\right)}^{\,\cdot} = \overline{\int_{P_\kappa} \varrho \boldsymbol{v} \det\, \tilde{F}\, dV}^{\,\cdot} = \int_{P_\kappa} \overline{\varrho \boldsymbol{v} \det\, \tilde{F}}^{\,\cdot}\, dV \tag{6.62}$$

However, we have

$$\overline{\varrho \boldsymbol{v} \det \tilde{F}}^{\,\cdot} = \dot{\varrho} \boldsymbol{v} \det \tilde{F} + \varrho \dot{\boldsymbol{v}} \det \tilde{F} + \varrho \boldsymbol{v}\, \overline{\det \tilde{F}}^{\,\cdot} \tag{6.63}$$

and the right-hand side may be expanded into

$$\left[\frac{\partial \varrho}{\partial t} + (\operatorname{grad} \varrho)\boldsymbol{v}\right] \boldsymbol{v} \det \tilde{F} + \varrho \left[\frac{\partial \boldsymbol{v}}{\partial t} + (\operatorname{grad} \boldsymbol{v})\boldsymbol{v}\right] \det \tilde{F} + \varrho \boldsymbol{v} \det \tilde{F} \operatorname{div} \boldsymbol{v} \tag{6.64}$$

so that, by grouping the terms differently and using an Eulerian frame of reference, we can write

$$\boldsymbol{R}(P_\chi) = \int_{P_\chi} \frac{\partial(\varrho \boldsymbol{v})}{\partial t}\, dv \tag{6.65}$$

$$+ \int_{P_\chi} [(\operatorname{grad} \varrho \boldsymbol{v})\boldsymbol{v} + \varrho\, (\operatorname{grad} \boldsymbol{v})\boldsymbol{v} + \varrho \boldsymbol{v} \operatorname{div} \boldsymbol{v}]\, dv \tag{6.66}$$

But

$$(\operatorname{grad} \varrho \boldsymbol{v})\, \boldsymbol{v} + \varrho\, \boldsymbol{v} \operatorname{div} \boldsymbol{v} = \operatorname{div}(\varrho \boldsymbol{v})\, \boldsymbol{v} \tag{6.67}$$

and

$$\operatorname{div}(\varrho \boldsymbol{v})\, \boldsymbol{v} + \varrho\, (\operatorname{grad} \boldsymbol{v})\, \boldsymbol{v} = (\varrho v_i)_{,i} v_j \hat{\boldsymbol{I}}_j + \varrho\, v_{j,k} v_k \hat{\boldsymbol{I}}_j \tag{6.68}$$

so that

$$\int_{P_\chi} [(\varrho v_i)_{,i} v_j + \varrho v_k v_{j,k}] \hat{\boldsymbol{I}}_j\, dv = \int_{P\chi} (\varrho v_i v_j)_{,i} \hat{\boldsymbol{I}}_j\, dv \tag{6.69}$$

$$= \int_{SP_\chi} \boldsymbol{v}(\varrho\, \boldsymbol{v} \cdot \boldsymbol{n})\, da \tag{6.70}$$

This completes the derivation of equation (6.60). We follow the same line of reasoning to derive equation (6.61) and observe that

$$\boldsymbol{m}(P_\chi, \boldsymbol{x}_0) = \dot{\boldsymbol{h}}(P_\chi, \boldsymbol{x}_0) = \overline{\int_{P_\chi} (\boldsymbol{x} - \boldsymbol{x}_0) \wedge \varrho \boldsymbol{v} \, dv}^{\,\boldsymbol{\cdot}} \tag{6.71}$$

$$= \int_{P_\kappa} \overline{[(\boldsymbol{x} - \boldsymbol{x}_0) \wedge \varrho \boldsymbol{v} \det \tilde{F}]}^{\,\boldsymbol{\cdot}} \, dV \tag{6.72}$$

$$= \int_{P_\kappa} \varepsilon_{ijk} \overline{(x - x_0)_i \varrho v_j \det F)}^{\,\boldsymbol{\cdot}} \hat{\boldsymbol{I}}_k \, dV \tag{6.73}$$

Now we write

$$\overline{(\boldsymbol{x} - \boldsymbol{x}_0)_i \varrho v_j \det \tilde{F}}^{\,\boldsymbol{\cdot}} = \overline{(\boldsymbol{x} - \boldsymbol{x}_0)_i}^{\,\boldsymbol{\cdot}} (\varrho v_j \det \tilde{F}) + (\boldsymbol{x} - \boldsymbol{x}_0)_i \overline{(\varrho v_j \det \tilde{F})}^{\,\boldsymbol{\cdot}} \tag{6.74}$$

and observe that $\overline{(\boldsymbol{x} - \boldsymbol{x}_0)_i}^{\,\boldsymbol{\cdot}} = v_i$, so that since $\boldsymbol{v} \wedge \varrho \boldsymbol{v} = 0$ we get

$$\int_{P_\kappa} \varepsilon_{ijk} v_i \varrho v_j \det \tilde{F} \hat{\boldsymbol{I}}_k \, dV = 0 \tag{6.75}$$

from which we conclude that

$$\boldsymbol{M}(P, \boldsymbol{x}_0) = \int_{P_\kappa} \varepsilon_{ijk} (\boldsymbol{x} - \boldsymbol{x}_0)_i \overline{\varrho v_j \det \tilde{F}}^{\,\boldsymbol{\cdot}} \hat{\boldsymbol{I}}_k \, dV \tag{6.76}$$

$$= \int_{P_\kappa} \varepsilon_{ijk} (\boldsymbol{x} - \boldsymbol{x}_0)_i \, (\dot{\varrho} v_j \det \tilde{F} + \varrho \dot{v}_j \det \tilde{F} + \varrho v_j \det \tilde{F} \operatorname{div} \boldsymbol{v}) \, dV \tag{6.77}$$

Returning to the Eulerian frame of reference we obtain the desired result (equation (6.61)):

$$\boldsymbol{m}(P_\chi, \boldsymbol{x}_0) = \int_{P_\chi} (\boldsymbol{x} - \boldsymbol{x}_0) \wedge \frac{\partial}{\partial t} (\varrho \boldsymbol{v}) \, dv + \int_{SP_\chi} (\boldsymbol{x} - \boldsymbol{x}_0) \wedge \boldsymbol{v} (\varrho \, \boldsymbol{v} \cdot \boldsymbol{n}) \, da$$

For steady flow conditions, $\partial(\varrho \boldsymbol{v})/\partial t = 0$, equations (6.60) and (6.61) become

$$\boldsymbol{R}(P_\chi) = \int_{SP_\chi} \boldsymbol{v} (\varrho \, \boldsymbol{v} \cdot \boldsymbol{n}) \, da \tag{6.78}$$

$$\boldsymbol{m}(P_\chi, \boldsymbol{x}_0) = \int_{SP_\chi} (\boldsymbol{x} - \boldsymbol{x}_0) \wedge \boldsymbol{v} (\varrho \, \boldsymbol{v} \cdot \boldsymbol{n}) \, da \tag{6.79}$$

In such steady flow conditions, the resultant force acting on any fixed volume defined in an Eulerian frame of reference and through which flow occurs is equal to the flux of momentum through the surface that limits this volume.

As an example, let us consider a flexible hose in which flow occurs under steady conditions (fig. 6.3). Since $\boldsymbol{v}$ is normal to $\boldsymbol{n}$ along part 2 of the surface S_{P_χ}, the only forces acting on the hose are those at the ends. For a flexible gardening hose connected to a tap at one end and resting on the ground at the other end, the flow of water results in some pressure being applied by the hose to the water flowing through the free end of the hose. The action–reaction principle implies that a force is therefore exerted on the free end of the hose which makes it move in a somewhat random manner, as we all have noticed.

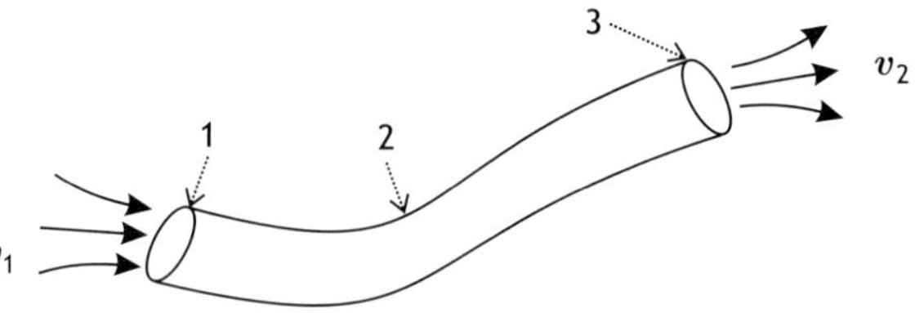

Fig. 6.3 Forces and moments acting on a hose in which the flow is steady.

Cauchy equation of motion

Consider the resultant force acting on any fixed part P (bounded by a surface S) of a body B defined in the Eulerian frame of reference and filled by a deforming fluid at time t. From Euler's law and Reynolds' theorem, we saw in chapter 3 that

$$\boldsymbol{R}(P_\chi) = \int_{P_\chi} \varrho \boldsymbol{b}\, dv + \int_{SP_\chi} \boldsymbol{t}\, da \; = \dot{\boldsymbol{l}}(P_\chi) \tag{6.80}$$

$$= \overline{\int_{SP_\chi} \varrho \boldsymbol{v}\, dv}^{\,\cdot} = \int_{SP_\chi} \varrho \dot{\boldsymbol{v}}\, dv \tag{6.81}$$

where $\boldsymbol{t} = \tilde{\sigma}^T \boldsymbol{n}$. Applying the divergence theorem, we may write

$$\int_{SP_\chi} \boldsymbol{t}\, da \; = \int_{SP_\chi} \tilde{\sigma}^T \boldsymbol{n}\, da \tag{6.82}$$

$$= \int_{SP_\chi} (\sigma_{ji} n_j) \hat{\boldsymbol{I}}_i\, da \; = \int_{P_\chi} \sigma_{ji,j} \hat{\boldsymbol{I}}_i\, dv \tag{6.83}$$

so that, finally,

$$\int_{P_\chi} (\varrho b_i + \sigma_{ji,j})\, dv \; = \int_{P_\chi} \varrho \dot{v}_i\, dv \tag{6.84}$$

Given that equation (6.84) must hold for any part P, we conclude that

$$\operatorname{div} \tilde{\sigma}^T \; + \varrho \boldsymbol{b} \; = \varrho \dot{\boldsymbol{v}} \; = \varrho \ddot{\boldsymbol{u}} \tag{6.85}$$

Symmetry of the stress tensor

Let us now turn to the symmetry of the stress tensor $\tilde{\sigma}$. When we consider the second Euler law (equation (3.9)), the moment applied to any part P_χ of a given body is

$$\boldsymbol{m}(P_\chi, \boldsymbol{x}_0) = \dot{\boldsymbol{h}}(P_\chi, \boldsymbol{x}_0) \tag{6.86}$$

$$= \int_{SP_\chi} [(\boldsymbol{x} - \boldsymbol{x}_0) \wedge \boldsymbol{t}]\, da + \int_{P_\chi} [(\boldsymbol{x} - \boldsymbol{x}_0) \wedge \varrho \boldsymbol{b}]\, dv \tag{6.87}$$

$$= \overline{\int_{P_\chi} [(\boldsymbol{x} - \boldsymbol{x}_0) \wedge \varrho \boldsymbol{v}]\, dv}^{\,\cdot} \tag{6.88}$$

If we apply Reynolds' theorem to the right-hand side of equation (6.88) we obtain

$$\overline{\int_{P_\chi} [(\boldsymbol{x}-\boldsymbol{x}_0)\wedge \varrho \boldsymbol{v}]\,dv}^{\,\bullet} = \int_{P_\chi} \overline{[(\boldsymbol{x}-\boldsymbol{x}_0)\wedge \boldsymbol{v}}^{\,\bullet}]\varrho\,dv \tag{6.89}$$

But $\boldsymbol{x}_0$ is fixed and $\overline{(\boldsymbol{x}-\boldsymbol{x}_0)}^{\,\bullet} = \boldsymbol{v}$; $\boldsymbol{v}\wedge\boldsymbol{v} = 0$ and so

$$\overline{\int_{P_\chi} [(\boldsymbol{x}-\boldsymbol{x}_0)\wedge \varrho \boldsymbol{v}]\,dv}^{\,\bullet} = \int_{P_\chi} (\boldsymbol{x}-\boldsymbol{x}_0)\wedge \varrho\dot{\boldsymbol{v}}\,dv \tag{6.90}$$

Applying the divergence theorem, the first term on the right-hand side of equation (6.87) may be rewritten as

$$\int_{SP_\chi} (\boldsymbol{x}-\boldsymbol{x}_0)\wedge \boldsymbol{t}\,da = \int_{SP_\chi} (\boldsymbol{x}-\boldsymbol{x}_0)\wedge(\tilde{\sigma}^T\boldsymbol{n})\,da \tag{6.91}$$

$$= \int_{SP_\chi} \varepsilon_{ijk}(\boldsymbol{x}-\boldsymbol{x}_0)_i\sigma_{mj}n_m\hat{\boldsymbol{I}}_k\,da \tag{6.92}$$

$$= \int_{P_\chi} [\varepsilon_{ijk}(\boldsymbol{x}-\boldsymbol{x}_0)_i\sigma_{mj}]_{,m}\hat{\boldsymbol{I}}_k\,dv \tag{6.93}$$

$$= \int_{P_\chi} \varepsilon_{ijk}[\delta_{im}\sigma_{mj} + (\boldsymbol{x}-\boldsymbol{x}_0)_i\sigma_{mj,m}]\hat{\boldsymbol{I}}_k\,dv \tag{6.94}$$

$$= \int_{P_\chi} \varepsilon_{ijk}\sigma_{ij}\hat{\boldsymbol{I}}_k\,dv + \int_{P_\chi} (\boldsymbol{x}-\boldsymbol{x}_0)\wedge \operatorname{div}\tilde{\sigma}^T\,dv \tag{6.95}$$

Hence, substituting equations (6.90) and (6.95) into equations (6.87) and (6.88), we obtain

$$\int_{P_\chi} \varepsilon_{ijk}\sigma_{ij}\hat{\boldsymbol{I}}_k\,dv + \int_{P_\chi} (\boldsymbol{x}-\boldsymbol{x}_0)\wedge(\operatorname{div}\tilde{\sigma}^T + \varrho\boldsymbol{b} - \varrho\dot{\boldsymbol{v}})\,dv = 0 \tag{6.96}$$

The second term on the left-hand side of equation (6.96), which corresponds to the law of motion (see equation (6.85)), is zero and we conclude from the evaluation of the moment of momentum (angular momentum) that, for all parts P,

$$\int_{P_\chi} \varepsilon_{ijk}\sigma_{ij}\hat{\boldsymbol{I}}_k\,dv = 0 \qquad \forall P_\chi \tag{6.97}$$

Therefore $\varepsilon_{ijk}\sigma_{ij}\hat{\boldsymbol{I}}_k = 0$; $i,j,k = 1,2,3$, which implies the symmetry of the stress tensor ($\sigma_{ij} = \sigma_{ji}$ or $\tilde{\sigma} = \tilde{\sigma}^T$). Hence the equation of motion may be written as

$$\operatorname{div}\tilde{\sigma} + \varrho\boldsymbol{b} = \varrho\dot{\boldsymbol{v}} = \varrho\ddot{\boldsymbol{u}} \tag{6.98}$$

It is also known as *Cauchy's law of motion.*

In fluid mechanics it is customary to separate the spherical part of the tensor, which corresponds to a pressure, from the deviatoric part (see section 3.5). With the convention that pressures are negative, the stress tensor may be written

$$\sigma_{ij} = \sigma'_{ij} - p\delta_{ij}$$

Accordingly,

$$\operatorname{div}\tilde{\sigma} = \operatorname{div}\tilde{\sigma'} - \operatorname{grad} p \tag{6.99}$$

Now, if we recall that $\dot{\boldsymbol{v}} = \partial \boldsymbol{v}/\partial t + (\operatorname{grad} \boldsymbol{v})\boldsymbol{v}$, Cauchy's equation of motion may be rewritten as

$$\varrho \frac{\partial \boldsymbol{v}}{\partial t} + \varrho(\operatorname{grad} \boldsymbol{v})\boldsymbol{v} = \varrho \boldsymbol{b} - \operatorname{grad} p + \operatorname{div} \sigma' \tag{6.100}$$

The first term on the left-hand side of equation (6.100) represents the influence of the acceleration of a fluid particle. It is zero in steady flow conditions.

The second term on the left-hand side of equation (6.100) corresponds to the change in velocity when a fluid particle moves through the fixed volume element considered in the Eulerian frame of reference. It exists even in steady flow conditions.

The first term on the right-hand side of equation (6.100) corresponds to the body force associated with the mass that flows through the volume of interest.

The second term on the right-hand side of equation (6.100) corresponds to the forces associated with the pressure that exist even when the fluid is at rest. In this last case ($\boldsymbol{v} = 0$) the motion law is simply the equilibrium equation $\varrho \boldsymbol{b} = \operatorname{grad} p$.

The third term on the right-hand side of (6.100) corresponds to the forces associated with the viscosity of the fluid. For Newtonian fluids they are proportional to the fluid velocity gradient. For example, on active andesitic volcanoes (with very viscous magma) the flow of magma during its vertical ascent through the volcanic structure results in a measurable displacement field. Its analysis may help determine the andesitic magma viscosity for this volcano (Beauducel and Cornet, 1999).

Conservation of momentum

The momentum per unit volume is $\varrho \boldsymbol{v}$, with time derivative

$$\frac{\partial(\varrho \boldsymbol{v})}{\partial t} = \boldsymbol{v}\frac{\partial \varrho}{\partial t} + \varrho \frac{\partial \boldsymbol{v}}{\partial t} \tag{6.101}$$

Combining equation (6.101) with the continuity equation (equation (6.38)) in Cauchy's law of motion (equation (6.100)) we obtain

$$\frac{\partial \varrho \boldsymbol{v}}{\partial t} = -(\operatorname{div} \varrho \boldsymbol{v})\boldsymbol{v} - (\operatorname{div} \boldsymbol{v})\varrho \boldsymbol{v} - \operatorname{grad} p + \operatorname{div} \tilde{\sigma}' + \varrho \boldsymbol{b} \tag{6.102}$$

or, in terms of components,

$$\frac{\partial(\varrho v_i)}{\partial t} = -(\varrho v_i v_j + p\delta_{ij} - \tilde{\sigma}'_{ij})_{,j} + \varrho b_i \tag{6.103}$$

Equation (6.103) holds for any part P, with volume V_P and surface S_P, so that

$$\int_{V_P} \frac{\partial(\varrho v_i)}{\partial t} dv = -\int_{V_P} (\varrho v_i v_j + p\delta_{ij} - \tilde{\sigma}'_{ij})_{,j}\, dv + \int_{V_P} \varrho b_i\, dv \tag{6.104}$$

Applying the divergence theorem to the first term on the right-hand side of equation (6.104) and observing that the volume of integration does not depend on time, we may write

$$\overline{\left(\int_{V_P} \varrho \boldsymbol{v}\, dv\right)}^{\boldsymbol{\cdot}} = -\int_{S_P} [(\varrho \boldsymbol{v} \cdot \boldsymbol{v})\boldsymbol{n} + p\boldsymbol{n} - \tilde{\sigma}'\boldsymbol{n})]\, da + \int_{V_P} \varrho \boldsymbol{b}\, dv \tag{6.105}$$

where $\boldsymbol{n}$ is a unit normal ($\boldsymbol{ds} = \boldsymbol{n}\, da$).

The surface integral on the right-hand side of equation (6.105) represents the contribution to the change in momentum because of the momentum flux through the bounding surface S_P. Momentum is a vector, so its flux is a tensor called the *flux of momentum tensor* $\tilde{\Pi}$, which is defined by

$$\Pi_{ij} \equiv \varrho v_i v_j + p\delta_{ij} - \sigma'_{ij} \tag{6.106}$$

The volume integral on the right-hand side of equation (6.105) represents the production of momentum per unit time in the corresponding volume V_P due to body forces.

Equation of motion in a Lagrangian frame of reference; the Piola–Kirchhoff stress tensor

In the previous section, we have expressed the law of motion in the deformed state, at time t, i.e. in an Eulerian frame of reference. However it may be written in the undeformed state at time $t = 0$, i.e. in a Lagrangian frame of reference. According to Euler's first law:

$$\boldsymbol{R}(P_\chi) = \int_{P_\chi} \varrho \boldsymbol{b}\, dv + \int_{S_{P_\chi}} \boldsymbol{t}\, da = \int_{S_{P_\chi}} \varrho \dot{\boldsymbol{v}}\, dv \tag{6.107}$$

However, we have already seen that $\int_{P_\chi} \varrho \boldsymbol{b}\, dv = \int_{P_\kappa} \varrho_\kappa \boldsymbol{b}\, dV$. Also, $\int_{S_{P_\chi}} \varrho \dot{\boldsymbol{v}}\, dv = \int_{S_{P_\kappa}} \varrho_\kappa \dot{\boldsymbol{v}}\, dV$, but $\boldsymbol{t}\, da = \tilde{\sigma}\boldsymbol{n}\, da = \tilde{\sigma}\, \boldsymbol{da}$. In chapter 4 we showed that the relationship between a deformed surface and an undeformed surface is $\boldsymbol{da} = (\det \tilde{F})(\tilde{F}^{-1})^T \boldsymbol{dA}$, so that

$$\boldsymbol{t}\, da = \tilde{\sigma}(\tilde{F}^{-1})^T \det \tilde{F}\, \boldsymbol{dA} \tag{6.108}$$

The Piola–Kirchhoff stress tensor $\tilde{\sigma}_\kappa$ is defined in such a way that

$$\boldsymbol{t}\, da = \tilde{\sigma}_\kappa \boldsymbol{dA} = \tilde{\sigma}_\kappa \boldsymbol{N} dA$$

$$\text{with} \qquad \tilde{\sigma}_\kappa \equiv \tilde{\sigma}^T (\tilde{F}^{-1})^T \det \tilde{F} \tag{6.109}$$

When applied to an undeformed surface element $\boldsymbol{dA}$, the Piola–Kirchhoff stress tensor $\tilde{\sigma}_\kappa$ generates the stress vector $\boldsymbol{t}$ that acts on the same surface element after it has been deformed.

Note that $\tilde{\sigma}_\kappa$ is not symmetrical. For very small displacement gradients ($\tilde{H} \simeq \tilde{0}$), the deformation gradient $\tilde{F}$ is nearly equal to $\tilde{1}$ and the stress tensor is the same in the initial and in the deformed state. The Piola–Kirchhoff stress tensor is of interest when one is considering large displacements.

6.1.3 Conservation of energy and the first law of thermodynamics

Power theorem and the conservation of kinetic energy

The power input $W(P)$ is defined as the rate at which the external surface forces per unit area $\boldsymbol{t}$ and the body forces per unit mass $\boldsymbol{b}$ are doing work on the mass of the particles that make up a part P of body B that instantaneously occupies volume V bounded by surface S at time t:

$$W(P) \equiv \int_{S_{P_\chi}} \boldsymbol{t} \cdot \boldsymbol{v} \, da + \int_{P_\chi} \varrho \, \boldsymbol{b} \cdot \boldsymbol{v} \, dv \tag{6.110}$$

while the kinetic energy of the system is

$$K(P) \equiv \tfrac{1}{2} \int_{P_\chi} \varrho \, \boldsymbol{v} \cdot \boldsymbol{v} \, dv \tag{6.111}$$

Let us show that

$$W(P) = \dot{K}(P) + \int_{P_\chi} \operatorname{tr}(\tilde{\sigma} \tilde{L}) \, dv \tag{6.112}$$

where the second term on the right-hand side of equation (6.112) is the rate at which external forces do work on P.

From the law of motion (6.98) we know that

$$\operatorname{div} \tilde{\sigma}^T + \varrho \, \boldsymbol{b} = \varrho \dot{\boldsymbol{v}}$$

so that we may write, after scalar-multiplying both sides of the equality by $\boldsymbol{v}$:

$$(\operatorname{div} \tilde{\sigma}^T) \cdot \boldsymbol{v} + \varrho \boldsymbol{b} \cdot \boldsymbol{v} - \varrho \dot{\boldsymbol{v}} \cdot \boldsymbol{v} = 0 \tag{6.113}$$

Equation (6.113) applies to any part P of the body under consideration. Hence

$$\int_{P_\chi} [(\operatorname{div} \tilde{\sigma}^T) \cdot \boldsymbol{v} + \varrho \boldsymbol{b} \cdot \boldsymbol{v} - \varrho \dot{\boldsymbol{v}} \cdot \boldsymbol{v}] \, dv = 0 \tag{6.114}$$

which may be rewritten as

$$\int_{P_\chi} [(\operatorname{div} \tilde{\sigma}^T) \cdot \boldsymbol{v} + \varrho \boldsymbol{b} \cdot \boldsymbol{v}] \, dv = \int_{P_\chi} \varrho \dot{\boldsymbol{v}} \cdot \boldsymbol{v} \, dv \tag{6.115}$$

However, the right-hand side of equation (6.115) is simply the rate of change of the kinetic energy, since

$$\int_{P_\chi} \varrho \dot{\boldsymbol{v}} \cdot \boldsymbol{v} \, dv = \int_{P_\kappa} \varrho_\kappa \dot{\boldsymbol{v}} \cdot \boldsymbol{v} \, dV \tag{6.116}$$

$$= \frac{1}{2} \overline{\int_{P_\kappa} \varrho_\kappa \boldsymbol{v}^2 \, dV}^{\,\cdot} = \frac{1}{2} \overline{\int_{P_\chi} \varrho \boldsymbol{v}^2 \, dv}^{\,\cdot} = \dot{K}(P) \tag{6.117}$$

Now, in the Eulerian frame of reference,

$$\int_{P_\chi} (\operatorname{div} \tilde{\sigma}^T \cdot \boldsymbol{v}) \, dv = \int_{P_\chi} \sigma_{ij,i} v_j \, dv \tag{6.118}$$

$$= \int_{P_\chi} [(\sigma_{ij} v_j]_{,i} - \sigma_{ij} v_{j,i}] \, dv \tag{6.119}$$

$$= \int_{S_{P_\chi}} \sigma_{ij}^T n_i v_j \, da - \int_{P_\chi} \sigma_{ij} v_{j,i} \, da \tag{6.120}$$

$$= \int_{S_{P_\chi}} t_j v_j \, da - \int_{P_\chi} \sigma_{ij} v_{j,i} \, dv \tag{6.121}$$

Hence equation (6.115) may be rewritten

$$\int_{S_{P_X}} t_j v_j \, da + \int_{P_X} \varrho \, b_j v_j \, dv = \frac{1}{2} \overline{\int_{P_X} \varrho v^2 \, dv}^{\boldsymbol{\cdot}} + \int_{P_X} \sigma_{ij} v_{j,i} \, dv \qquad (6.122)$$

The left-hand side of equation (6.122) is simply the power input, so we have proved the power theorem:

$$W(P) = \dot{K}(P) + \int_{P_X} \mathrm{tr}(\tilde{\sigma} \tilde{L}) \, dv = \dot{K}(P) + \int_{P_X} \mathrm{tr}(\tilde{\sigma} \tilde{D}) \, dv \qquad (6.123)$$

where $\tilde{D}$ is defined in Section 4.4.2. The work T done on part P between time t_1 and time t_2 is the integral of the power computed between t_1 and t_2:

$$T_P(t_1, t_2) = \int_{t_1}^{t_2} W(P) \, dt = \Delta K(P) + \int_{t_1}^{t_2} \int_{P_X} \mathrm{tr}(\tilde{\sigma} \tilde{D}) \, dv \, dt \qquad (6.124)$$

Returning to equation (6.116) we may write

$$\dot{K}(P) = \int_{P_X} \varrho \dot{\boldsymbol{v}} \cdot \boldsymbol{v} \, dv = \int_{P_X} \varrho \boldsymbol{v} \cdot \left[\frac{\partial \boldsymbol{v}}{\partial t} + (\mathrm{grad}\, \boldsymbol{v}) \, \boldsymbol{v} \right] dv \qquad (6.125)$$

Thus, for incompressible fluids (div $\boldsymbol{v} = 0$), equation (6.122) becomes

$$\begin{aligned} \frac{\partial}{\partial t} \left(\int_{P_X} \frac{\varrho}{2} \boldsymbol{v}^2 \, dv \right) = & - \int_{S_X} \varrho \frac{\boldsymbol{v}^2}{2} \boldsymbol{v} \cdot \boldsymbol{n} \, da - \int_{S_X} p \boldsymbol{v} \cdot \boldsymbol{n} \, da \\ & + \int_{S_X} \tilde{\sigma}' \boldsymbol{n} \cdot \boldsymbol{v} \, da + \int_{P_X} \varrho \boldsymbol{b} \cdot \boldsymbol{v} \, dv - \int_{P_X} \sigma'_{ij} v_{i,j} \, dv \end{aligned} \qquad (6.126)$$

if we remember that $\mathrm{div}(\alpha \boldsymbol{A}) = \boldsymbol{A} \, \mathrm{grad}\, \alpha + \alpha \, \mathrm{div} \boldsymbol{A}$.

- The first term on the right-hand side of equation (6.126) is the contribution of the flux of kinetic energy associated with the fluid flow.
- The second and third terms correspond respectively to the work done by the surface forces (pressure and shear) applied at the surface of part P.
- The fourth term corresponds to the work of the body forces.
- The fifth term corresponds to the transformation of kinetic energy into internal energy as heat, as will be discussed hereafter. Had the material been a solid, this term would have described the recoverable elastic strain energy.

Since equation (6.126) holds for any part P of the body, we may write the conservation of kinetic energy as

$$\frac{\partial}{\partial t} \left(\frac{\varrho \, \boldsymbol{v}^2}{2} \right) = -\mathrm{div} \left[\boldsymbol{v} \left(\frac{\varrho \boldsymbol{v}^2}{2} + p \right) - \tilde{\sigma}' \boldsymbol{v} \right] + \varrho \boldsymbol{v} \cdot \boldsymbol{b} - \sigma'_{ij} v_{i,j} \qquad (6.127)$$

Closed mechanical cycle

Suppose that a body B is subjected to a closed mechanical cycle during the period $[t_0, t_1]$. If each particle of B occupies the same position at time t_0 and at time t_1 and if its velocity is identical at time t_0 and at time t_1 then, for all $\boldsymbol{X}$ lying in B,

$$\chi(\boldsymbol{X}, t_1) = \chi(\boldsymbol{X}, t_0) \tag{6.128}$$

$$\boldsymbol{v}(\boldsymbol{X}, t_1) = \boldsymbol{v}(\boldsymbol{X}, t_0) \tag{6.129}$$

During a closed mechanical cycle, the work done on any part P of the body B is

$$T_P(t_0, t_1) = \int_{t_0}^{t_1} \oint_{P_\chi} \operatorname{tr}(\tilde{\sigma}\tilde{D})\, dv\, dt \tag{6.130}$$

Heat input

The heat input rate Q_{in} results from a heat flux $\boldsymbol{q}$ through the surface S of the body (see section 12.2.1 for the definition of heat flux) and from distributed internal heat sources of strength r per unit mass:

$$Q_{in} = -\int_{SP_\chi} \boldsymbol{q} \cdot \boldsymbol{n}\, da + \int_{P_\chi} \varrho r\, dv \tag{6.131}$$

where the negative sign is needed because the heat flux is positive when outward. It is generally observed that, during a closed mechanical cycle, there is an exchange between the rate at which work is done on the body and the heat input rate, if the total energy of the system is constant:

$$\int_{t_0}^{t_1} \oint_{P_\chi} [W(P) + Q_{in}]\, dt = 0 \tag{6.132}$$

In general, however, the rate of change $\dot{E}$ of the total energy of the system is defined as

$$\dot{E} \equiv W + Q_{in} \tag{6.133}$$

Thus the change in the total energy of a system between times t_1 and t_2 is $\Delta E = E(t_1) - E(t_2) = \int_{t_1}^{t_2} (W + Q_{in})\, dt$. The change in the total energy includes a change in the kinetic energy K and a change in the internal energy U.

Let u be the internal energy per unit mass. Then ϱu corresponds to the internal energy per unit volume, and we may write

$$\overset{\cdot}{\overline{\int_{P_\chi} [\tfrac{1}{2}\varrho \boldsymbol{v} \cdot \boldsymbol{v} + \varrho u]\, dv}} = \overset{\cdot}{\overline{\int_{P_\chi} \tfrac{1}{2}\varrho \boldsymbol{v} \cdot \boldsymbol{v}\, dv}} + \int_{P_\chi} \operatorname{tr}(\tilde{\sigma}\tilde{D})\, dv + \int_{P_\chi} \varrho r\, dv - \int_{SP_\chi} \boldsymbol{q} \cdot \boldsymbol{n}\, da \tag{6.134}$$

Applying the divergence theorem to equation (6.134), we get

$$\int_{P_\chi} [\varrho \dot{u} + \operatorname{div} \boldsymbol{q} - \operatorname{tr}(\tilde{\sigma}\tilde{D}) - \varrho r]\, dv = 0 \quad \forall P_\chi \tag{6.135}$$

and we finally obtain the conservation of energy:

$$\varrho \dot{u} = \mathrm{tr}(\tilde{\sigma}\tilde{D}) + \varrho r - \operatorname{div} \boldsymbol{q} \tag{6.136}$$

or, in terms of components,

$$\varrho \frac{du}{dt} = \sigma_{ij} D_{ij} + \varrho r - q_{i,i} \tag{6.137}$$

6.2 Constitutive equations for fluids

6.2.1 Constitutive equations

In chapter 2 we considered unidirectional models and introduced two ideal materials: the linearly elastic solid, in which the force required to impose a given displacement is proportional to that displacement, and the Newtonian viscous fluid, in which the force required to deform the body is proportional to the velocity of the deformation process. For both types of material we wrote down relationships respectively between forces and displacements or between forces and velocities; these were the constitutive equations for the corresponding material. Then we introduced more complex materials by combining these elementary ideal materials. Finally, we considered even more complex materials by introducing additional nonlinear elements. For all these materials the corresponding constitutive equations established relations between forces, displacements and velocities.

For three-dimensional situations, we introduced the concepts of stress, strain, rate of deformation and strain rate. In chapter 5 we discussed elasticity and showed that the constitutive equation for linearly elastic bodies relates the stress components $\sigma_{ij}; i,j = 1,2,3$, linearly to the small-strain components $\epsilon_{ij}; i,j = 1,2,3$:

$$\sigma_{ij} = C_{ijkl}\,\epsilon_{kl}\,, \qquad \epsilon_{kl} = \tfrac{1}{2}(u_{k,l} + u_{l,k}) \tag{6.138}$$

Now we will consider the relationships between stresses, strains and deformation rates or strain rates that describe the macroscopic responses of more complex bodies to loading processes.

Systems for which temperature changes are not considered

For systems in which the temperature is under consideration, the continuity equation

$$\frac{\partial \varrho}{\partial t} + \operatorname{div}(\varrho \boldsymbol{v}) = 0$$

introduces four independent variables (the density and the three components of velocity). The law of motion

$$\operatorname{div} \tilde{\sigma} + \varrho \boldsymbol{b} = \varrho \dot{\boldsymbol{v}}$$

introduces nine more unknowns (the stress components and the components of the body force per unit mass). The constitutive equations amount to six equations between the stress components and either the displacement gradient components or the velocity gradient components (three more unknowns). Hence we have ten equations for 16 unknowns. The compatibility equations discussed in section 4.3.7 provide an additional six equations, so now we have 16 equations for 16 unknowns and the problem is to solve this set of partial differential equations so as to satisfy the boundary conditions under consideration.

Systems with heat exchanges

For systems that involve heat exchanges, in addition to the continuity equation (one equation, four unknowns) and the law of motion (three equations, nine unknowns), we need to introduce the conservation of energy principle [$\mathrm{tr}(\tilde{\sigma}\tilde{D}) + \varrho r = \mathrm{div}\,\boldsymbol{q} + \varrho\dot{u}$]; this involves eight variables (the three components of $\boldsymbol{q}$, the temperature θ, the heat production per unit mass r and three more variables associated with the six components of the velocity gradient). Fourier's law ($\boldsymbol{q} = -K\,\mathrm{grad}\,\theta$) provides the relation between the temperature gradient and the heat flux (no new unknown but three more equations; see chapter 12). Two additional scalar equations are provided by thermodynamic considerations, namely the kinetic equation of state, which relates the pressure, density and temperature, $f(p, \varrho, \theta) = 0$, and the caloric equation of state, which expresses the internal energy as a function of density and temperature, $u = u(\varrho, \theta)$. Hence six constitutive equations are necessary so that, together with the six compatibility conditions, a set of 22 partial differential equations is available for solving the 22 unknowns.

6.2.2 Constitutive equations for Newtonian fluids

As stated in the introduction to this chapter, fluids are characterized by the fact that they cannot support any shear stress when at rest; thus the stress tensor at any point in a fluid at rest is isotropic:

$$\sigma_{ij} = -\bar{p}\delta_{ij} \qquad \text{or} \qquad \tilde{\sigma} = -\bar{p}\tilde{1} \tag{6.139}$$

where the minus sign corresponds to the convention that compressions are negative; it is important to differentiate the so-called static pressure $\bar{p} = -\frac{1}{3}\sigma_{kk}$, from the thermodynamic pressure in a gas as defined by the kinetic equation of state $f(p, \varrho, \theta) = 0$. For example, for a perfect gas $p = \varrho R\theta$, where R is the gas constant for the particular gas.

For incompressible fluids (i.e. liquids), the density is constant ($\varrho = \text{const}$). For perfect fluids the viscosity is always zero, so that $\tilde{\sigma} = -p\tilde{1}$ where p is the thermodynamic pressure.

For viscous fluids, the constitutive equation may be written

$$\tilde{\sigma} = -p\tilde{1} + F(\tilde{D}) \tag{6.140}$$

where F is a fourth-order tensor function and $\tilde{D} = \frac{1}{2}(\tilde{L} + \tilde{L}^T)$ with $\tilde{L} = \mathrm{grad}\,\boldsymbol{v}$ (see section 4.4.2). When F is linear, the fluid is called a Newtonian fluid. Hence, for Newtonian fluids

$$\sigma_{ij} = -p\delta_{ij} + C'_{ijkl}D_{kl} \tag{6.141}$$

with the conditions $C'_{ijkl} = C'_{jikl} = C'_{ijlk}$, since both $\tilde{\sigma}$ and $\tilde{D}$ are symmetrical tensors. Most fluids are isotropic, and thus equation (6.141) must be independent of the frame of reference.

It may be shown (Malvern, 1969) that the most general form for an isotropic fourth-order tensor is

$$C'_{ijkl} = \lambda\delta_{ij}\delta_{kl} + \mu_d(\delta_{ik}\delta_{jl} + \delta_{il}\delta_{jk}) \tag{6.142}$$

Then the constitutive equation for Newtonian fluids is

$$\sigma_{ij} = -p\delta_{ij} + \lambda D_{kk}\delta_{ij} + 2\mu_d D_{ij} \tag{6.143}$$

or

$$\tilde{\sigma} = -p\tilde{I} + \lambda \operatorname{tr} \tilde{D}\tilde{I} + 2\mu_d\tilde{D} \tag{6.144}$$

where p is the thermodynamic pressure and λ and μ_d are two independent constants that characterize the fluid viscosity. Let us consider only the deviatoric part of the tensors,

$$\sigma'_{ij} = \sigma_{ij} + \bar{p}\delta_{ij} \qquad (\sigma'_{kk} = 0 \text{ since } \bar{p} = -\sigma_{kk}/3) \tag{6.145}$$

$$D'_{ij} = D_{ij} - \frac{D_{kk}}{3}\delta_{ij} \qquad (D'_{jj} = 0) \tag{6.146}$$

Then the constitutive equation becomes

$$\sigma'_{ij} = (\bar{p} - p)\delta_{ij} + \left(\lambda + \frac{2\mu_d}{3}D_{kk}\right)\delta_{ij} + 2\mu_d D'_{ij} \tag{6.147}$$

from which we may compute the trace of the deviatoric stress tensor,

$$\sigma'_{jj} = 3\left[(\bar{p} - p) + \left(\lambda + \frac{2\mu_d}{3}D_{kk}\right)\right] = 0 \tag{6.148}$$

Accordingly, the constitutive equation of a Newtonian fluid may be written

$$\sigma'_{ij} = 2\mu_d D'_{ij} \qquad \text{or, in tensor form,} \qquad \tilde{\sigma}' = 2\mu_d\tilde{D}' \tag{6.149}$$

$$\bar{p} = p - \kappa D_{kk} = p + \kappa\frac{1}{\varrho}\dot{\varrho} \tag{6.150}$$

where $\kappa = \lambda + \frac{2}{3}\mu_d$ is called the bulk viscosity. The right-hand side part of equation (6.150) results from the continuity equation, $D_{kk} = \operatorname{div} \boldsymbol{v} = -(1/\varrho)\dot{\varrho}$.

We observe that the static pressure is equal to the thermodynamic pressure only for incompressible fluids (i.e. $\bar{p} = p$ for liquids and $\dot{\varrho} = 0$) or if $\kappa = 0$, namely if $\lambda = -\frac{2}{3}\mu_d$, which is known as the Stokes condition.

The constant μ_d is called the *dynamic viscosity*. Its dimension is

$$[\mu_d] = \left[\frac{F}{L^2}\right][T] = \left[\frac{M}{LT}\right] \tag{6.151}$$

On many occasions the quantity $\eta = \mu_d/\varrho$ is used rather than the dynamic viscosity; η is called the kinematic viscosity and its dimension is $[L^2/T]$.

Laminar and turbulent flow

The hypothesis that shear stress components are proportional to the velocity gradient is well verified for parallel-flow situations (flow in which the streamlines are parallel). Hence, for example, the dynamic viscosity of a liquid may be measured in a viscosimeter, in which flow occurs between concentric cylinders. Equation (6.149) applies in a satisfactory manner to flow situations in which very little turbulent mixing occurs between layers moving in the same direction, i.e. to laminar flow conditions.

Turbulent flow describes situations where the shear stress and velocity at a point fluctuate with time in a random manner. The transition from laminar to turbulent flow depends on the flow velocity but also on the thermodynamic conditions. It is associated with a set of characteristic numbers. For example, the Reynolds number characterizes for liquids the ratio of the forces associated with inertia and the forces associated with viscosity. Various tabulations have been established for various flow geometries. From dimensional analysis, other characteristic numbers have been introduced to take into account the influence of temperature, heat conduction and heat capacity on the onset of turbulent flow conditions (the Prandtl number and the Peclet number). These are discussed in classical fluid mechanics or continuum mechanics books.

6.2.3 Navier–Stokes equation for Newtonian fluids

All fluids must satisfy both the continuity equation and the law of motion:

$$\dot{\varrho} + \varrho \operatorname{div} \boldsymbol{v} = 0 \qquad \text{or} \qquad \frac{d\varrho}{dt} + \varrho v_{k,k} = 0$$

$$\operatorname{div} \tilde{\sigma} + \varrho \boldsymbol{b} = \varrho \dot{\boldsymbol{v}} \qquad \text{or} \qquad \sigma_{ij,i} + \varrho b_j = \varrho \dot{v}_j$$

In addition Newtonian fluids satisfy, for laminar flow, the condition that the shear stress is proportional to the velocity gradient at time t (there is no memory of the loading or deformation history):

$$\tilde{\sigma} = -p\tilde{1} + 2\lambda \operatorname{tr} \tilde{D} + 2\mu_d \tilde{D}$$

or, in terms of components,

$$\sigma_{ij} = -p\delta_{ij} + \lambda D_{kk}\delta_{ij} + 2\mu_d D_{ij}$$

However, we may observe that

$$2D_{ij,i} = (v_{i,j} + v_{j,i})_{,i} = (v_{i,i})_{,j} + (v_j)_{,ii} \tag{6.152}$$

$$= (\operatorname{div} \boldsymbol{v})_{,j} + \nabla^2 v_j \tag{6.153}$$

Accordingly, the constitutive equation may be written

$$\sigma_{ij,i} = -p_{,i}\delta_{ij} + \lambda(\operatorname{div} \boldsymbol{v})_{,i}\delta_{ij} + \mu_d[(\operatorname{div} \boldsymbol{v})_{,j} + \nabla^2 v_j] \tag{6.154}$$

$$= -p_{,j} + (\lambda + \mu_d)(\operatorname{div} \boldsymbol{v})_{,j} + \mu_d \nabla^2 v_j \tag{6.155}$$

so that the law of motion becomes

$$-\operatorname{grad} p + (\lambda + \mu_d) \operatorname{grad}(\operatorname{div} \boldsymbol{v}) + \mu_d \nabla^2 \boldsymbol{v} + \varrho \boldsymbol{b} = \varrho \dot{\boldsymbol{v}} \tag{6.156}$$

For incompressible fluids ($\operatorname{div} \boldsymbol{v} = 0$) the *Navier–Stokes equation* is

$$-\frac{1}{\varrho}\operatorname{grad}p + \eta\nabla^2 \boldsymbol{v} + \boldsymbol{b} = \dot{\boldsymbol{v}} \tag{6.157}$$

$$= \frac{\partial \boldsymbol{v}}{\partial t} + (\operatorname{grad}\boldsymbol{v})\,\boldsymbol{v} \tag{6.158}$$

and for compressible fluids with no bulk viscosity($\lambda = -\frac{2}{3}\mu_d$) we get

$$-\frac{1}{\varrho}\operatorname{grad}p + \frac{\eta}{3}\operatorname{grad}(\operatorname{div}\boldsymbol{v}) + \eta\nabla^2 \boldsymbol{v} + \boldsymbol{b} = \dot{\boldsymbol{v}} \tag{6.159}$$

with $\eta = \mu_d/\varrho$. If we recall (see section 4.4.3) that the angular velocity $\boldsymbol{w}$ is such that $\boldsymbol{w} = \frac{1}{2}\operatorname{rot}\boldsymbol{v}$ and if we note that

$$\nabla^2 \boldsymbol{v} \equiv \operatorname{grad}(\operatorname{div}\boldsymbol{v}) - \operatorname{rot}(\operatorname{rot}\boldsymbol{v}) \tag{6.160}$$

then equation (6.157) for incompressible fluids becomes

$$-\frac{1}{\varrho}\operatorname{grad}p - 2\eta\operatorname{rot}\boldsymbol{w} + \boldsymbol{b} = \dot{\boldsymbol{v}} \tag{6.161}$$

It simplifies for irrotational flow ($\boldsymbol{w} = 0$) to the same form of the Navier–Stokes equation as for perfect fluids ($\eta = 0$):

$$-\frac{1}{\varrho}\operatorname{grad}p + \boldsymbol{b} = \dot{\boldsymbol{v}} = \frac{\partial \boldsymbol{v}}{\partial t} + (\operatorname{grad}\boldsymbol{v})\,\boldsymbol{v} \tag{6.162}$$

which is known as the *Euler equation.*

6.2.4 The conservation of kinetic energy for incompressible perfect fluids: the Bernouilli equation

The Bernouilli equation corresponds to the integration along a streamline S of the Euler equation. Hence it applies to the steady flow ($\partial\boldsymbol{v}/\partial t = 0$) of perfect ($\eta = 0$) incompressible ($\operatorname{div}\boldsymbol{v} = 0$) fluids. If $\boldsymbol{ds}$ is a small element along the streamline we may evaluate the conservation of kinetic energy during motion $\boldsymbol{ds} = ds_i\hat{\boldsymbol{I}}_i$:

$$-\frac{1}{\varrho}(\operatorname{grad}p)\cdot\boldsymbol{ds} + \boldsymbol{b}\cdot\boldsymbol{ds} = (\operatorname{grad}\boldsymbol{v})\boldsymbol{v}\cdot\boldsymbol{ds} \tag{6.163}$$

But if $\boldsymbol{b} = -g\delta_{i3}\hat{\boldsymbol{I}}_3$ (thus the only body force is gravity) then

$$-\frac{1}{\varrho}(\operatorname{grad}p)\cdot\boldsymbol{ds} + \boldsymbol{b}\cdot\boldsymbol{ds} = -\frac{1}{\varrho}(p_{,i}\hat{\boldsymbol{I}}_i)(ds_i\hat{\boldsymbol{I}}_i) - g\delta_{i3}\hat{\boldsymbol{I}}_i\,ds = -\frac{1}{\varrho}dp - g\,ds_3 \qquad \text{along } S \tag{6.164}$$

and

$$\begin{aligned}(\operatorname{grad}\boldsymbol{v})\boldsymbol{v}\cdot\boldsymbol{ds} &= [\tfrac{1}{2}\operatorname{grad}(\boldsymbol{v}\cdot\boldsymbol{v}) - \boldsymbol{v}\wedge(\operatorname{rot}\boldsymbol{v})]\cdot\boldsymbol{ds} = \tfrac{1}{2}(\operatorname{grad}\boldsymbol{v}^2)\cdot\boldsymbol{ds}\\ &= \tfrac{1}{2}d(\boldsymbol{v}^2) \qquad \text{along } S\end{aligned} \tag{6.165}$$

since $\boldsymbol{v}$ is parallel to $d\boldsymbol{s}$. Hence

$$\frac{dp}{\varrho} + \frac{d(\boldsymbol{v}^2)}{2} + g\,ds_3 = 0 \qquad \text{along } S \tag{6.166}$$

Integrating along S, assuming that ϱ does not vary along S, we obtain the Bernouilli equation. This states that *along a streamline*, for the steady flow of an incompressible fluid under gravity,

$$\frac{p}{\varrho} + \frac{v^2}{2} + gx_3 = \text{const} \tag{6.167}$$

Generally, the constant on the right-hand side of equation (6.167) will vary from one streamline to the next, but when in addition the flow is irrotational (rot $\boldsymbol{v} = 0$) the same constant applies throughout the flow field. When gravity is the only force on the fluid, equation (6.167) is often written as

$$\frac{p}{\varrho g} + \frac{v^2}{2g} + h = \text{const} \qquad \text{or} \qquad h_p + h_v + h = \text{const} \tag{6.168}$$

where h is the vertical distance between the point of interest and a horizontal plane of reference; h_p and h_v are respectively the *pressure head* and the *velocity head*.

A direct application of Bernouilli's equation is the decrease in pressure observed when the velocity along a streamline increases. It may cause cavitation, i.e. the formation of vapor bubbles, when the increase in velocity is large enough.

For real, viscous, fluids, some energy is lost during flow because of friction, and Bernouilli's equation is then generalized by introducing the head loss Δh that occurs between two points A and B along a streamline:

$$\frac{p_A}{\varrho g} + \frac{v_A^2}{2g} + h_A = \frac{p_B}{\varrho g} + \frac{v_B^2}{2g} + h_B + \Delta h \tag{6.169}$$

The flow results in a decrease in pressure in the direction of flow. This observation will be recalled when we are discussing the seismicity induced by forced fluid flow in rock masses, in chapter 11.

6.3 Simple solutions for incompressible Newtonian fluids

6.3.1 Steady laminar flow between parallel plates

Let us consider the steady laminar flow of an incompressible liquid between two infinite parallel plates (fig. 6.4). The planes are perpendicular to the $\hat{\boldsymbol{I}}_3$ direction and extend infinitely in the $\hat{\boldsymbol{I}}_1$ and $\hat{\boldsymbol{I}}_2$ directions, while flow occurs in the $\hat{\boldsymbol{I}}_1$ direction ($v_3 = 0$). The only body force is gravity ($\boldsymbol{b} = -g\delta_{i3}\hat{\boldsymbol{I}}_3; i = 1, 2, 3$). The solution does not depend on x_2;

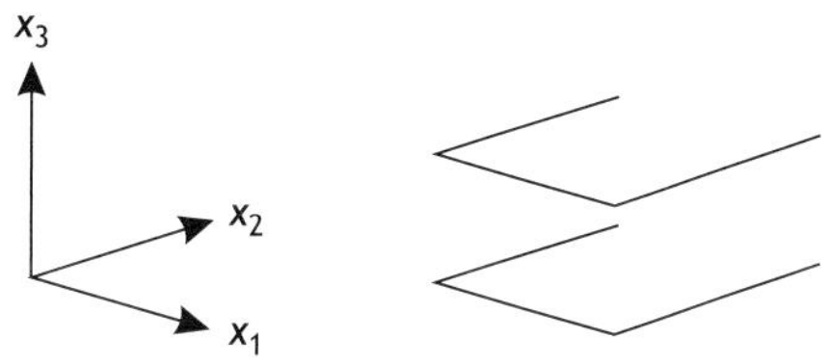

Fig. 6.4 Steady flow between parallel plates .

hence all partial derivatives with respect to x_2 are null and the continuity equation (6.36) and the Navier–Stokes equation (6.157) become

$$v_{1,1} + v_{3,3} = 0 \qquad \text{since} \qquad \frac{\partial \varrho}{\partial t} = 0 \tag{6.170}$$

and

$$-\frac{1}{\varrho}p_{,1} + \eta(v_{1,11} + v_{1,33}) = v_1 v_{1,1} + v_{1,3} v_3 \tag{6.171}$$

$$-\frac{1}{\varrho}p_{,3} + \eta(v_{3,11} + v_{3,33}) = v_1 v_{3,1} + v_{3,3} v_3 \tag{6.172}$$

Further, since $v_3 = 0$ everywhere, the continuity equation yields $v_{1,1} = 0$ so that v_1 does not depend on x_1 and equations (6.171) and (6.172) become

$$-\frac{1}{\varrho}p_{,1} + \eta v_{1,33} = 0 \tag{6.173}$$

$$-\frac{1}{\varrho}p_{,3} = 0 \tag{6.174}$$

From equation (6.174) we conclude that

$$p = \varrho g x_3 + f_1(x_1) \tag{6.175}$$

This result, inserted into equation (6.173), yields $\mu_d\, v_{1,33} = f_1'(x_1)$, the integral of which is

$$\mu_d\, v_1 = \frac{x_3^2}{2} f_1'(x_1) + x_3 f_2(x_1) + f_3(x_1) \tag{6.176}$$

Let us assume now that the lower plate is fixed, so that $v_1 = 0$ for $x_3 = 0$. Then $f_3(x_1) = 0$. However, from the continuity equation we know that $v_{1,1} = 0$, thus

$$\mu_d\, v_{1,1} = \frac{x_3^2}{2} f_1''(x_1) + x_3 f_2'(x_1) = 0 \tag{6.177}$$

from which we conclude that

$$x_3 f_1''(x_1) = -2 f_2'(x_1) \qquad \forall x_1\ \forall x_3 \tag{6.178}$$

This implies that

$$f_1''(x_1) = 0 \qquad \text{so that} \qquad f_1(x_1) = C_1 x_1 + C_3 \tag{6.179}$$

$$f_2'(x_1) = 0 \qquad \text{so that} \qquad f_2(x_1) = C_2 \tag{6.180}$$

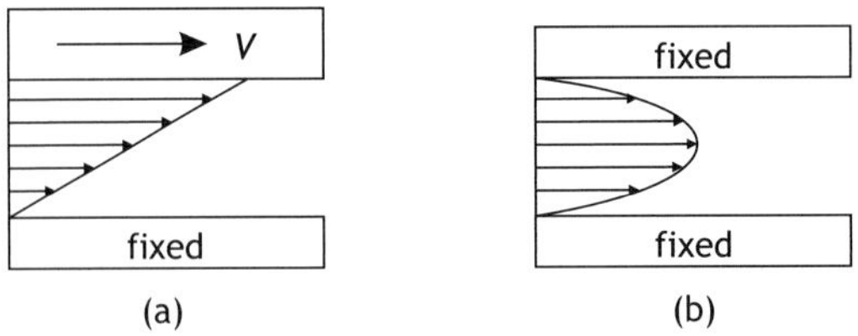

Fig. 6.5 (a) Couette flow; (b) Poiseuille flow.

Accordingly, equation (6.176) may be rewritten as

$$\mu_d v_1 = \frac{x_3^2}{2} C_1 + C_2 x_3 = x_3 \left(\frac{x_3}{2} C_1 + C_2 \right) \tag{6.181}$$

where C_1, C_2 and C_3 are constants which are determined from the boundary conditions.

However, from equation (6.175) we see that $p = \varrho g x_3 + f_1(x_1)$, and therefore $p_{,1} = C_1$: thus the pressure gradient in the direction of the flow is constant.

Further, we note from equation (6.181) that $C_2 = \mu_d\, v_{1,3}$ for $x_3 = 0$, so that the shear stress on the plate is also independent of x_1.

When the upper plate moves in the $\hat{\boldsymbol{I}}_1$ direction at a constant velocity V then, for a point in the fluid in contact with the upper plate $x_3 = h$, the velocity is $v_1 = V$ and, according to equation (6.181),

$$h C_2 = \mu_d\, V - C_1 \frac{h^2}{2} \tag{6.182}$$

so that $C_2 = (1/h)(\mu_d\, V - \frac{1}{2} C_1 h^2)$. The constant C_1 remains undetermined. It may be obtained experimentally by measuring the pressure at two points on either one of the plates and distant from each other by Δx_1, so that $C_1 = \Delta p / \Delta x_1$.

Then the solution for the flow is

$$\mu_d\, v_1 = \frac{x_3^2}{2} C_1 + \frac{x_3}{h} \left(\mu_d\, V - \frac{h^2}{2} C_1 \right) \tag{6.183}$$

Two cases are of particular interest for geomechanics. Couette flow conditions (see fig. 6.5), in which $p_{,1} = 0$, form a good approximation for the flow between two coaxial cylinders, the radii of which are large with respect to the gap between the cylinders. These are the conditions in a viscosimeter, and the solution for the velocity is $v_1 = V x_3 / h$.

The other case concerns Poiseuille flow, in which the velocity of the upper plate is null ($V = 0$ for $x_3 = h$). Then equation (6.183) becomes $\mu_d\, v_1 = \frac{1}{2} C_1 (x_3^2 - h x_3)$, which may be rewritten as

$$\mu_d\, v_1 = -\frac{1}{2} p_{,1} \left[\left(\frac{h}{2} \right)^2 - \left(x_3 - \frac{h}{2} \right)^2 \right] \tag{6.184}$$

The flow velocity shows a parabolic profile within the interval between the plates. If we wish to obtain the flux per unit length in the x_2 direction between the two fixed plates, we integrate with respect to x_3 and obtain the classical result

$$Q = \int_0^h v_1\, dx_3 = -\frac{p_{,1}}{2\mu_d} \left[\frac{x_3^3}{3} - \frac{h x_3^2}{2} \right]_0^h \tag{6.185}$$

$$Q = -\frac{h^3}{12\mu_d} p_{,1} \tag{6.186}$$

This solution is frequently used to describe the flow in natural fractures (see chapter 10). It implies that for steady laminar flow conditions the pressure loss along a fracture varies as the inverse of the cube of the fracture's hydraulic aperture for a constant flux.

6.3.2 Steady laminar flow through a circular pipe

Let us consider now the flow of an incompressible fluid through a circular pipe of radius R. Given that the problem is axisymmetrical with respect to the tube axis z, we may use polar coordinates, so that the Navier–Stokes equation becomes

$$p_{,z} = \mu_d v_{z,\rho\rho} + \frac{1}{\rho} u_{z,\rho} \tag{6.187}$$

Following the same procedure as for flow through parallel plates, integration of equation (6.190) shows that the velocity through the pipe's cross section varies as the square of the pipe radius. Integration of the velocity over the complete section of pipe shows that the flow rate through the pipe depends linearly on the pressure gradient along the pipe axis and varies as the fourth power of the pipe radius R:

$$q = \frac{\pi R^4}{8\mu_d} P_{,z} \tag{6.188}$$

For flow through circular pipes, the Reynolds number, which characterizes the onset of turbulent flow, is

$$Re = \frac{Vd}{\eta} \tag{6.189}$$

where V is the mean flow velocity, d is the diameter of the pipe and η is the kinematic viscosity ($\eta = \mu_d/\varrho$, where μ_d is the dynamic viscosity and ϱ is the fluid density).

6.4 Exercises

1. In polar coordinates the Navier–Stokes equation for incompressible Newtonian fluids is

$$p_{,z} = \mu_d v_{z,\rho\rho} + \frac{1}{\rho} u_{z,\rho}$$

   Following the same type of derivation as in section 6.3, demonstrate equation (6.188).
2. Let us approximate the pore space of a porous rock as a set of parallel tubes with radius r, so that there are n tubes normal to the surface with total area A. Determine the

equivalent porosity of the material and evaluate the flow rate through an area A of this porous rock caused by a pressure gradient $p_{,z}$. Show that it may be expressed as

$$Q = A \frac{k}{\mu_d} p_{,z} \tag{6.190}$$

and evaluate k when the tube diameter is equal to 0.1 mm and there are ten tubes per square centimeter. We call k the hydraulic conductivity of the material per unit area.

3. We consider now a granite with many sets of fractures. Each fracture is assimilated to a layer of independent parallel tubes.
   3.1. Assume that the fractures have a rectangular shape, so that the tubes are parallel to the long dimension of the rectangle, with N parallel tubes distributed per unit width of fracture. What is the hydraulic conductivity of the fracture?
   3.2. A normal load is applied to the fracture. The change in radius of the tubes may be evaluated from equation (5.309) assuming that a uniaxial far-field stress is applied to the fracture surface. Evaluate the change in hydraulic conductivity of the fracture if the normal stress is equal to 10 MPa for a rock with Young's modulus equal to 1000 MPa and Poisson's ratio equal to 0.25.

7 Elements of linear fracture mechanics

As pointed out in chapter 2, an elastic material may be modeled by a spring, and it is quite intuitive that a spring cannot be extended indefinitely: some maximum force, and correlatively some maximum spring extension, must exist that defines an elastic limit. When the force, or the spring extension, becomes larger than this maximum value the spring either breaks or extends further but with a nonreversible component, so that some residual deformation exists when the loading is relaxed. In a crude way, this defines two possible post-elastic behaviors: either the spring breaks, in which case the material used to make the spring is said to be brittle, or the spring deforms in a nonelastic manner, so that a permanent deformation is observed when the load is relaxed, in which case the material is said to be ductile.

While a spring may help illustrate elastic phenomena in one dimension, we saw in chapters 3–5 that for three-dimensional problems of elasticity we must introduce three-dimensional stresses and strains. These quantities are second-order tensors, and elasticity implies a linear relationship between the stress and small-strain components. By comparison with the spring's behavior, we may anticipate that when the principal stress (or principal strain) magnitudes, or some function of them, become larger than some critical values, either the material breaks into pieces or it deforms according to a nonreversible process. As for the case of a spring, we call the post-elastic behavior *brittle* when the material fails because of the extension of fractures. We call the post-elastic behavior *ductile* when the material deforms continuously with occurrence of a nonreversible component. The object of the present chapter is to discuss brittle failure; ductility is discussed in chapters 8 and 9.

Different types of laboratory experiments may be conducted on rock samples to determine their strength, or more precisely to investigate the phenomena that take place after the elastic limit has been reached (e.g. Jaeger and Cook, 1979, chapter 6). For example, a cylindrical rock specimen may be tested under a uniaxial stress field, either negative (traction) or positive (compression), if one adopts here the classical rock mechanics convention that compressions are reckoned positive (fig. 7.1). In both cases a maximum axial stress is measured and the rock specimen breaks into pieces, i.e. it fails in a brittle manner. The maximum uniaxial stress is called the tensile strength σ^T when negative and the uniaxial compressive strength σ^c when positive. For rocks, the tensile strength is an order of magnitude lower than the uniaxial compressive strength. Furthermore, different testing configurations lead to different values for the tensile strength of rock. For example, three-point bending tests (see section 5.3.1) may yield

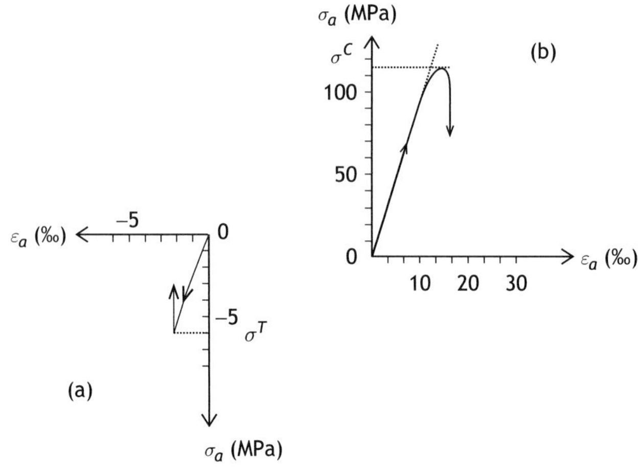

Fig. 7.1 Uniaxial strength of a rock: (a) the tensile strength σ^T; (b) the compressive strength σ^c. The tensile strength σ^T is often an order of magnitude smaller than σ^c, for rocks.

values of σ^T two to three times larger than those measured with a direct pull test (Jaeger, 1967). Also, different values are obtained depending on the volume of the specimens (the so-called size effect). This outlines the necessity to better understand the physics of brittle failure, i.e. the growth of fractures that lead to the failure of rocks.

At the beginning of the twentieth century, Griffith (1920) demonstrated the effect of scratches on the strength of steel; he showed that such scratches generate local stress concentrations that control the tensile strength of the specimens under test. He further developed the idea that all materials contain flaws that concentrate stresses locally, so that the brittle failure of materials is controlled by the propagation of these structural defaults rather than by the mean stress evaluated for an equivalent homogeneous continuum.

Griffith's observation thus raises doubts on the validity of defining equivalent continuum materials for analyzing the development of failure in rocks. Indeed, rocks are well known to be heterogeneous materials and are always affected by multiple defaults, which generate local stress concentrations when the rock is loaded. Accordingly the resistance of rock materials to failure, i.e. loading conditions for which microcracks extend, depends on these local stress concentrations.

It will be shown at the end of this chapter that the propagation of fractures subjected to a compressive stress field requires much higher stress levels than does the propagation of fractures subjected to tensile stresses. This explains why the uniaxial compressive strength for rocks is always much larger than their uniaxial tensile strength. So, investigating fracture propagation in rocks helps one to evaluate better the so-called rock elastic limit. It provides a means to analyze the fracturing processes that occur when loading reaches beyond this elastic limit, whether the stress components are tensile or compressive.

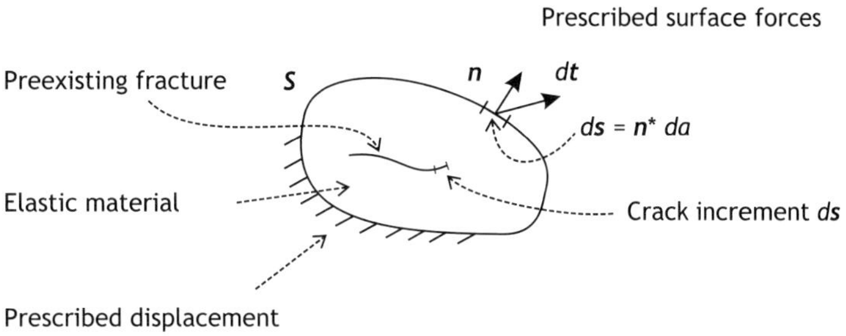

Fig. 7.2 Investigating fracture propagation in an elastic rocks: the brittle fracturing process.

In this chapter we discuss the propagation of fractures in homogeneous linearly elastic materials and try to answer three questions:

1. What is the maximum load that may be applied to a given rock system before a fracture propagates?
2. When a fracture propagates, what is the geometry of the fracture surface increment?
3. Once the fracture starts propagating, is the fracturing process unstable, i.e. does the velocity of the fracture extension increase with time, or is it stable, i.e. must the load be increased to propagate the fracture further after its initial extension phase?

From a mechanical point of view, a fracture is simply a discontinuity in the displacement field within an otherwise elastic body. The problem is schematized in fig. 7.2, where a body that involves a preexisting crack of length l is loaded by surface tension ($dt = \tilde{\sigma} \boldsymbol{n}\, da$) on some part of its surface and by imposed displacements $\boldsymbol{u}$ on other parts of its surface.

The objective is to define the loading conditions for which the fracture propagates by a crack increment ds and to characterize this propagation process. Defining the loading conditions for which a fracture propagates implies defining a fracture criterion; various fracture criteria have been proposed, involving maximum stress or maximum strain components. However, because it relates to considerations of energy conservation, Griffith's fracture criterion provides probably the most efficient general criterion for characterizing brittle failure processes, as we now discuss.

7.1 Fracture criteria

7.1.1 Griffith's energy fracture criterion

As stated above, Griffith (1920) understood that the computation of stress fields in equivalent homogeneous continua was not valid for analyzing fracture propagation. He realized that attention has to be given to the population of flaws and defects that exist in all materials, because of their efficiency at generating very high stress concentrations. Then, rather than considering the critical values of stresses or strains in an equivalent continuum, he proposed to consider the energy associated with the formation of fresh surfaces. This led him to postulate the following proposition (Griffith, 1924):

Just as in liquid, so in a solid the bounding surfaces possess a surface tension which implies the existence of a corresponding amount of potential energy. If owing to the action of a stress a crack is formed, or a pre-existing crack is caused to extend, therefore, a quantity of energy proportional to the area of the new surface must be added, and the condition that this shall be possible is that such addition of energy shall take place without any increase in the total potential energy of the system. This means that the increase of potential energy due to the surface tension of the crack must be balanced by a decrease in the potential of the strain energy and the applied forces.

The quantity of energy ΔD that Griffith reckoned must be added for propagating a pre-existing fracture $\boldsymbol{S}$ by an incremental surface element $\boldsymbol{ds} = \boldsymbol{n}\, da$, with unit normal $\boldsymbol{n}$ and area da, is referred to as a variation in the surface energy of the solid under consideration. This change in surface energy associated with fracture extension $\boldsymbol{ds}$ is a function of both the orientation and the area of the fracture increment: $\Delta D = f(\boldsymbol{ds})$.

During fracture extension the principle of energy conservation applies, so that

$$\Delta W(\boldsymbol{ds}) + \Delta Q_{in}(\boldsymbol{ds}) = \Delta U(\boldsymbol{ds}) + \Delta Q_1(\boldsymbol{ds}) + \Delta T(\boldsymbol{ds}) + \Delta D(\boldsymbol{ds}) \tag{7.1}$$

where, for a crack increment $\boldsymbol{ds}$,

$\Delta W(\boldsymbol{ds})$ is the external work done on the system,
$\Delta Q_{in}(\boldsymbol{ds})$ is the quantity of heat provided to the system,
$\Delta U(\boldsymbol{ds})$ is the change in the elastic energy;
$\Delta Q_1(\boldsymbol{ds})$ is the change in the internal heat,
$\Delta T(\boldsymbol{ds})$ is the variation of the kinetic energy,
$\Delta D(\boldsymbol{ds})$ is the variation of the surface energy.

Then, according to the theorem of the minimum potential energy (which states that a position of equilibrium corresponds to a local minimum of the potential energy), for a given loading process applied to a system a preexisting fracture extends when this extension results in an overall decrease of the system's potential energy. This is illustrated in fig. 7.3 for isothermal and adiabatic fracturing processes in which respectively the temperature remains constant throughout the system and there is no exchange of heat with the surroundings.

Let us consider a rock specimen with a preexisting crack $\boldsymbol{S}$ submitted to a uniaxial stress field σ_a and consider the loading condition for which the magnitude of the uniaxial stress component σ_a applied to the specimen has reached, in a quasistatic manner, the value σ_0, with corresponding axial strain ε_0 (the point A on fig. 7.3). The elastic strain energy in the rock specimen is represented by the triangle OAB area (see section 5.4.1). We consider now the uniaxial stress increment $\delta\sigma_a$ required to produce a change in axial strain $\delta\varepsilon_a$.

If the axial stress required to produce an elastic reversible deformation of the specimen is larger than that required to produce a crack increment $\boldsymbol{ds}$ then the crack increment occurs

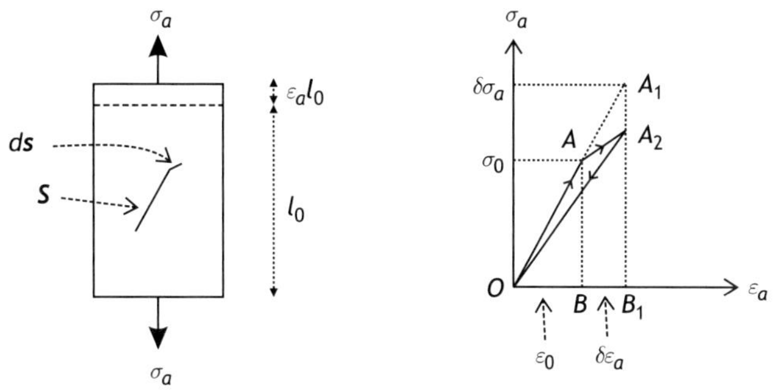

Fig. 7.3 Energy conservation during a purely brittle fracturing process; the only possibility for dissipating energy is through the formation of a fresh surface increment.

and results in a decrease in potential energy of the system; the pre-existing fracture ***S*** propagates by the increment ds.

Let us call A_1 the point that corresponds to the elastic response of the rock sample and A_2 the point that corresponds to the crack increment ds such that all points in the rock specimen are at rest after the crack increment has taken place (a quasistatic crack increment). We call B_1 the point on the horizontal axis that corresponds to the uniaxial strain $\varepsilon_0 + \delta\varepsilon_a$. We observe that, because the new equilibrium position corresponds to one for which the potential energy is a minimum, fracture propagates when $\widehat{OA_1B_1} > \widehat{OAA_2B_1}$, where $\widehat{OA_1B_1}$ is the area of triangle OA_1B_1 and $\widehat{OAA_2B_1}$ is the area of the quadrilateral element OAA_2B_1.

We notice that the quadrilateral area $\widehat{OAA_2B_1}$ may be computed in two different ways:

$$\widehat{OAA_2B_1} = \widehat{OAB} + \widehat{BAA_2B_1} = \widehat{OAA_2} + \widehat{OA_2B_1} \tag{7.2}$$

The area $\widehat{OAB}$ gives the elastic strain energy in the sample at point A, whilst the area $\widehat{BAA_2B_1}$ corresponds to the work done by the external forces while the crack grows by an increment ds. The area $\widehat{OA_2B_1}$ corresponds to the elastic strain energy in the specimen after the crack has extended by ds, so that the area $\widehat{OAA_2}$ gives the surface energy required to create the surface increment ds:

$$\widehat{OAA_2} = \widehat{BAA_2B_1} - (\widehat{OA_2B_1} - \widehat{OAB}) \tag{7.3}$$

or, in terms of energy,

$$\Delta D(ds) = \Delta W(ds) - \Delta U(ds) \tag{7.4}$$

If not all the points in the rock specimen are at rest after the creation of the crack increment ds, an amount of kinetic energy $\Delta T(ds)$ associated with the corresponding velocity field exists, so that, because of the conservation of energy,

$$\Delta D(ds) = \Delta W(ds) - \Delta U(ds) + \Delta T(ds) \tag{7.5}$$

when no heat is generated by the fracture extension process. When considering fracture propagation that could be either quasistatic or nonquasistatic, we may state that

$$\Delta W(ds) - \Delta U(ds) \geq \Delta D(ds) \tag{7.6}$$

As the crack increment area becomes infinitely small we define the limit of the term on the left-hand side of (7.6) as the strain energy release rate, $SERR(\boldsymbol{n})$, associated with an infinitesimally small crack extension in a direction perpendicular to its normal $\boldsymbol{n}$:

$$SERR(\boldsymbol{n}) \equiv \lim_{da \to 0} \frac{\Delta W(ds) - \Delta U(ds)}{da} \tag{7.7}$$

Since $SERR(\boldsymbol{n})$ depends on the crack orientation, a crack extension orientation exists for which $SERR(\boldsymbol{n})$ is a maximum. This largest value of $SERR(\boldsymbol{n})$ is called the strain energy release rate G. The fracture propagates when G reaches a critical value G_c that depends on the material characteristics.

Indeed, Griffith introduced the concept of the surface energy per unit area γ, defined in such a way that the crack area increment da is associated with a surface energy increment $2\gamma\ da$ since the crack increment has an upper and a lower face, each with area da. Hence, for the fracture to propagate by an increment $\boldsymbol{ds}$ it is necessary that

$$G = G_c = 2\gamma \tag{7.8}$$

Determining the value of the strain energy release rate G is a problem in elasticity, the solution of which depends on the crack geometry and on the boundary conditions applied to the body under consideration. However, the surface energy per unit area γ, or equivalently the critical strain energy release rate G_c, is a material property, which must be measured experimentally.

While in equation (7.8) the surface energy is considered to be independent of orientation, cases where some anisotropy or some heterogeneity affects the surface energy per unit area are equally well approached with the critical strain energy release rate criterion, as follows.

Let us call $\gamma(\boldsymbol{n})$ the surface energy per unit area for an orientation perpendicular to $\boldsymbol{n}$, and let us consider again a virtual fracture extension ds with area da and orientation perpendicular to $\boldsymbol{n}$. A virtual fracture extension is an extension for which the geometry is physically admissible but which may not occur in reality because it does not satisfy the minimum potential energy criterion. All virtual fracture extensions with area da must satisfy the condition

$$SERR(\boldsymbol{n}) - 2\gamma(\boldsymbol{n}) \leq 0 \tag{7.9}$$

The left-hand term of equation (7.9) is null only for a crack extension configuration that corresponds to the real fracture propagation path. It is negative for all other configurations. Indeed, for all virtual orientations that do not correspond to the real fracture orientation, the change in potential energy $dW(ds) - dU(ds)$ is smaller than the surface energy required to create a new surface. It is, though, exactly equal to the variation in surface energy associated with increment $\boldsymbol{ds}$ for the real fracture extension configuration, since no kinetic energy has yet been generated when the fracture extension is just beginning.

Equation (7.9) is particularly useful for investigating the effect of the preexisting fractures on fresh fracture extensions: for preexisting fractures, $\gamma(\boldsymbol{n}) \simeq 0$ when $\boldsymbol{n}$ is normal to the preexisting fracture plane.

We have just shown how the concept of the critical strain energy release rate helps to determine the loading conditions for which fracture propagates. Because the value of G depends on the crack increment's orientation, it provides a means to determine, through an incremental process, the geometry of fracture extension by computing after each new crack increment the orientation of the next increment.

In addition the strain energy release rate also helps one to determine whether the fracturing process is stable or unstable.

If $\partial G/\partial a < 0$, as the preexisting crack propagates by a small increment da the strain energy release rate decreases and therefore becomes smaller than the critical value G_c, so that crack extension stops. Further extension requires a further increase in the load. The fracture process is said to be stable.

If $\partial G/\partial a = 0$, as the preexisting crack propagates by a small increment the strain energy release rate remains equal to the critical value G_c, and therefore the fracture extends in a quasistatic manner, i.e. there is no acceleration of the fracturing process but the fracturing process does not stop.

If $\partial G/\partial a > 0$, as the preexisting crack propagates by a small increment the strain energy release rate becomes larger than the change in the surface energy associated with the crack increment, so that an amount of kinetic energy is generated consistent with the excess of internal energy in the system. The fracture process keeps accelerating and the fracture process is said to be unstable.

7.1.2 Irwin's basic modes of fracture and the stress intensity factor

Griffith's energy fracture criterion results from the observation that preexisting flaws exist in all materials and these flaws induce stress concentrations, and thus one difficulty in fracture mechanics is to evaluate these stress concentrations.

Irwin (1957) investigated the problem of fracture extension in thin plates and applied Griffith's energy fracture criterion. He observed that fractures correspond to discontinuities in the displacement field and analyzed the stress concentration associated with very thin fractures in order to evaluate the strain energy release rate for such plates. In the same way as for many problems in elasticity, he proposed to decompose the problem under consideration into simpler problems for which he could derive analytical solutions. Hence he introduced the so-called basic modes of fracture.

Consider the general problem of a preexisting planar fracture in an elastic body and define a local frame of reference associated with the fracture tip. Axis 1 is within the fracture plane and is normal to the fracture front; axis 2 is within the fracture plane and is tangent to the fracture front and axis 3 is normal to the two previous axes in such a way that the frame of reference is direct (fig. 7.4).

In this local frame of references, the planar crack increment corresponds to a discontinuity of the displacement field $\hat{\boldsymbol{U}}$, with components $\hat{U}_1$ with respect to axis 1, $\hat{U}_2$ with respect

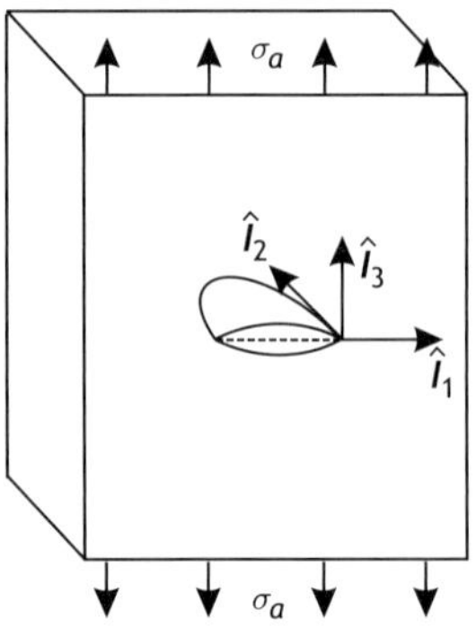

Fig. 7.4 Frame of reference associated with the fracture tip.

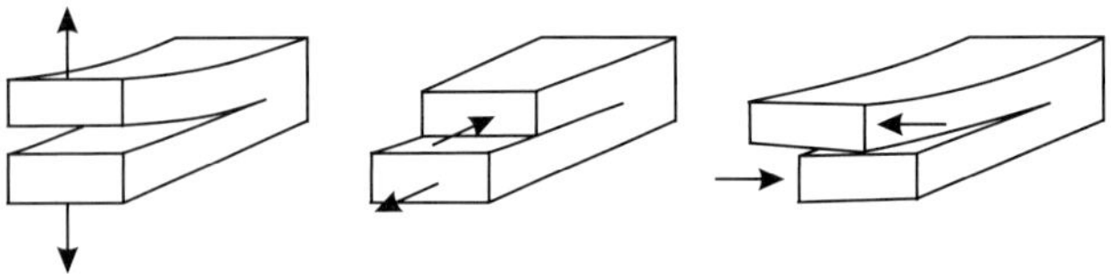

Fig. 7.5 Irwin's basic modes of fracture. (a) Mode I fracture or tensile mode; (b) mode II fracture or planar shear mode; (c) mode III fracture or antiplane shear fracture.

to axis 2 and $\hat{U}_3$ with respect to axis 3. Irwin observed that any crack increment $\hat{\boldsymbol{U}}$ may be decomposed into the superposition of three simple fracture modes (fig. 7.5)

Mode I fracture is such that $\hat{U}_1 = \hat{U}_2 = 0$, $\hat{U}_3 \neq 0$.
Mode II fracture is such that $\hat{U}_2 = \hat{U}_3 = 0$, $\hat{U}_1 \neq 0$.
Mode III fracture is such that $\hat{U}_1 = \hat{U}_3 = 0$, $\hat{U}_2 \neq 0$.

The objective is to investigate the stress singularities associated with each fracture mode.

The stress intensity factor for a mode I fracture

Inglis (1913) was possibly the first to recognize the significance of fractures and flaws in materials as sources of stress concentration. Making use of the Airy stress functions (section 5.6), he analyzed these stress concentrations by considering a so-called "penny-shaped crack", i.e. an elliptical cavity with one axis much smaller than the other, in an infinite body subjected to a uniaxial tensile stress σ_a at infinity (fig. 7.6).

If $2a$ is the length of the horizontal (largest) axis and $2b$ that of the vertical (smallest) axis then the maximum tensile stress component σ_a in the body is located at the tip of the crack, is oriented parallel to the uniaxial tensile stress component σ_a at infinity and is equal to

$$\sigma_{a\ \max} = \sigma_a \left(1 + \frac{2a}{b}\right) \tag{7.10}$$

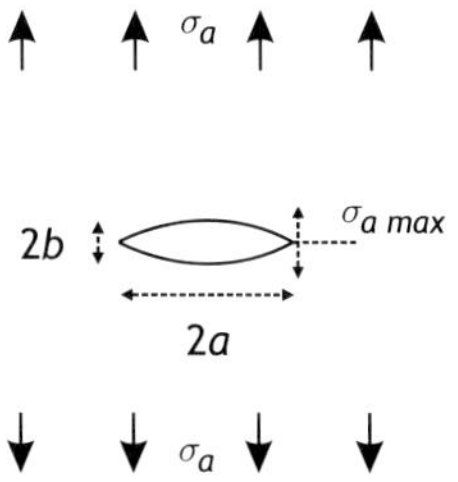

Fig. 7.6 Stresses at the edge of a penny-shaped crack.

Equation (7.10) shows that when the small-axis length $2b$ becomes infinitesimally small (corresponding to a very thin penny-shaped crack) the tensile stress becomes very large. Inglis further showed that at the apex of the major axis of the ellipse, where the radius of curvature $\rho_c = b^2/a$ is a minimum, the maximum tensile stress component may be expressed as a function of the radius of curvature:

$$\sigma_{a\ \max} = 2\sigma_a \left(\frac{a}{\rho_c}\right)^{1/2} \tag{7.11}$$

Interestingly, equation (7.11) implies that if the radius of curvature ρ_c is infinitely small then the tensile stress at the crack tip is infinitely large. Hence fractures should always propagate independently of the magnitude of the applied uniaxial stress at infinity, if the concept of a maximum stress is to be applied.

In order to resolve this paradox, Irwin applied Griffith's energy criterion of fracture and evaluated the strain energy release rate G for the mode I fracture configuration. For a plane stress problem (a preexisting fracture in a thin plate, which is a two-dimensional version of an axisymmetrical penny-shaped crack), he showed that the stress near the fracture tip may be expressed in polar coordinates as

$$\sigma_{ij} = \left(\frac{K_I}{\sqrt{\rho}}\right) f_{ij}(\theta) + \sum_{m=0}^{\infty} A_m \rho^{m/2} g_{ij}^{(m)}(\theta) \tag{7.12}$$

where σ_{ij} are the stress tensor components expressed in terms of polar coordinates ρ and θ referenced with respect to the fracture tip (fig. 7.7); K_I is a constant called the stress intensity factor and f_{ij} is a dimensionless finite function of θ. The higher-order terms depend on the geometry and become negligible in comparison with the first term as the distance ρ to the fracture tip gets smaller and smaller. For the specific geometry of fig. 7.7, the stress intensity factor K_I is

$$K_I = \sigma_a(\pi a)^{1/2} \tag{7.13}$$

The fundamental result obtained by Irwin was that, although the stress may become infinite very close to the fracture tip, the strain energy release rate G remains finite and for planar stress conditions is equal to

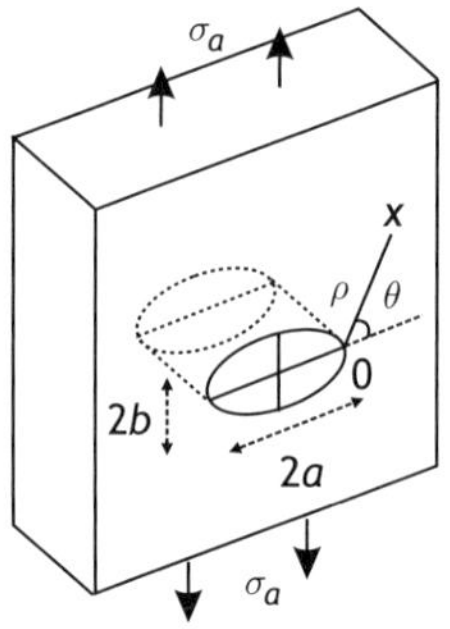

Fig. 7.7 Polar coordinates at the edge of a crack in an infinitely long plate (plane strain solution).

$$G = \frac{\pi \sigma_a^2 a}{E} = \frac{K_I^2}{E} \tag{7.14}$$

where E is the Young's modulus of the material.

For plane strain conditions, if ν is Poisson's ratio for the material, the relationship between the strain energy release rate and the stress intensity factor is (e.g. Rice, 1968)

$$G = \frac{\pi \sigma_a^2 a(1-\nu^2)}{E} = \frac{(1-\nu^2)K_I^2}{E} \tag{7.15}$$

Anderson (1995) proposed the adoption of the simple relationship $G = K_I^2/E^*$ for both plane stress and plane strain conditions, with the condition $E^* = E/(1-\nu^2)$ for plane strain conditions and $E^* = E$ for plane stress conditions. Let us note that this definition applies also to "penny-shaped" cracks if $E^* = \pi^2 E/[4(1-\nu^2)]$. Hence the stress singularity associated with a crack is only slightly dependent on its geometry.

The critical load required to extend a fracture of length $2a$ subject to a uniaxial stress field at infinity is determined by equating the strain energy release rate to the surface energy absorbed by the fracture extension process, namely 2γ, so that

$$G = G_c = 2\gamma = \frac{K_I^2}{E^*} \tag{7.16}$$

In a manner somewhat similar to the critical strain energy release rate criterion, the fracture criterion for mode I may be expressed in terms of a critical stress intensity factor $K_{Ic} = (2E^*\gamma)^{1/2}$. Whilst the determination of the stress intensity factor K_I is a problem in elasticity, the critical stress intensity factor value K_{Ic} for which fracture extends may be considered as a material property. It is often referred to as the "fracture toughness" of the material.

Stress intensity factors for mode II and mode III

Stress intensity factors may be defined for modes II and III, following the same approach as for the mode I case. According to equation (7.12), for both modes the singularity near the fracture tip is characterized by a term in $\rho^{1/2}$, if ρ is the local distance to the fracture tip:

$$\sigma_{12} = (k_{II}/\sqrt{\rho}) f_{12}(\theta) + O(1), \qquad \sigma_{23} = (k_{III}/\sqrt{\rho}) f_{23}(\theta) + O(1) \tag{7.17}$$

where f_{12} and f_{23} are nonsingular functions of the azimuthal coordinate θ and 0(1) are nonsingular terms in ρ and θ.

The demonstration of the equivalence between the surface energy and the fracture toughness (equation (7.16)) assumes that the fracture extends in its own plane; however, only for a pure mode I fracturing process does the fracture extend in its own plane and therefore only for mode I has the equivalence between the critical strain energy release rate and the critical stress intensity factor been demonstrated formally. For both mode II and mode III, and even more so for combined modes of fracture, the crack extension is not coplanar to the fracture plane, and a formal proof of the equivalence between the critical strain energy release rate and the critical stress intensity factor has not yet been provided. Also, for fractures that propagate in laminated materials (for example, sedimentary rocks), Comninou and Dundurs (1979) showed that the stress singularity in the vicinity of material interfaces is not a simple function of $\rho^{1/2}$, so that the local stress concentration may not be representable by a stress intensity factor formulation.

No such limitations have been encountered with the computation of the strain energy release rate G. Further, with present-day computers it is often fairly simple to produce numerical solutions for G (e.g. Anderson, 1995, chapter 11). Hence, today the so-called G_c energy fracture criterion is quite widely used in the earth sciences community when amended to take into account the effects of nonlinearities, as will be explained below.

7.1.3 Limits of linearly elastic fracture mechanics and the concept of the process zone

The surface energy concept as defined by Griffith (1924) concerns ideal atomically sharp cracks. The formulation is well suited for fractures in single crystals, but for polycrystalline materials, such as rocks, a more complex behavior is observed.

In laboratory measurements it is often convenient to initiate a fracture by first creating a sharp notch at the appropriate place in the material to be tested. This is particularly true for the three-point bending test often used for measuring the so-called "tensile strength" of rocks, i.e. the maximum tensile stress that can be applied to a rock before it breaks (see section 5.3.1 for the elastic analysis of stresses in the three-point bending test). However, when a load is being applied, some microcracking can be detected in the vicinity of the notch before a macroscopically visible fracture develops (fig. 7.8). Thus, a zone of intense microcracking is unavoidably generated simultaneously to the macroscopic fracture extension (e.g. Labuz *et al.*, 1985). This zone of intense microcracking is called the process zone, and for fracture propagation in polycrystalline rocks the surface energy concept must be replaced by that of the energy of the process zone. In practice, the macroscopic fracture must have extended some critical length before the measured value for the energy of the process zone stabilizes.

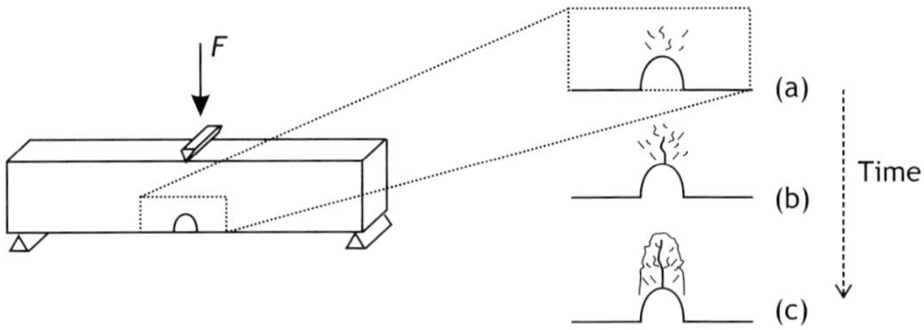

Fig. 7.8 Process zone associated with macroscopic fracture growth during a three-point bending test.

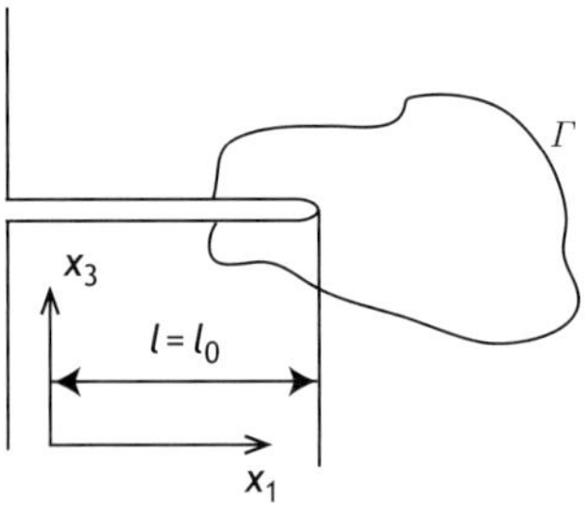

Fig. 7.9 The contour-independent J integral for the evaluation of energy variations associated with crack advance.

In addition, with the development of microcracking, the equivalent material at the fracture tip exhibits a nonlinear stress–strain relationship, and the effect of this local nonlinearity must be evaluated.

Rice (1968) introduced the so-called J integral for investigating the energy balance in the vicinity of a crack tip. In two dimensions, the J integral is defined by

$$J = \int_{\Gamma} \left(W dx_3 - \boldsymbol{t} \frac{\partial \boldsymbol{u}}{\partial x_1} ds \right) \tag{7.18}$$

where the axes x_1 and x_3 are respectively parallel and perpendicular to a pre-existing notch (the crack tip), as shown in fig. 7.9. In equation (7.18), Γ is a closed contour surrounding the crack tip which also includes the crack tip boudary. $\boldsymbol{t}$ is the stress vector acting on Γ and is assumed to be null along the crack tip; $\boldsymbol{u}$ is the displacement for points located on the contour Γ and W is the strain energy density. Rice showed that J is independent of the shape of Γ and provides an evaluation of the energy variation associated with a crack advance ds .

The contour-independent J integral has been shown to be very efficient for dealing with various sources of localized nonlinear effects, whether caused by microcracking or by plasticity. It is also very efficient for solving problems that involve time-dependent material behavior (e.g. Anderson, 1995, chapter 3).

7.2 On the dynamics of fracture propagation

During brittle fracture propagation, when the amount of energy absorbed by the formation of a new surface is smaller than the balance of energy between the applied external

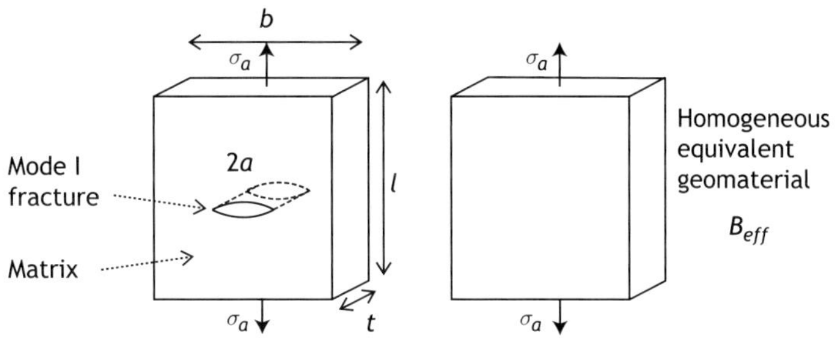

Fig. 7.10 Fractured material and equivalent continuum.

work and the change in internal elastic energy (equation (7.5)), the excess energy is transformed into kinetic energy. This implies that the fracture extension process generates an acceleration field.

The question arises whether this acceleration keeps increasing and leads to a complete disintegration of the body under consideration (like the complete rupture of a structural element) or whether a new stable position of equilibrium is reached, as is the case after an earthquake, whatever its size.

As an example, we discuss in the next section the case of a mode I fracturing process as observed in a rock specimen.

7.2.1 Griffith's locus

Let us consider a small sample with length l, width b and thickness t subject to a uniaxial stress σ_a. Within this volume, suppose that there exists a preexisting crack of length $2a$ loaded in mode I (fig. 7.10). The material filling the volume containing the crack is called the matrix, M, and B_{eff} is the equivalent continuum made up of the matrix together with the crack. There is no crack in the equivalent continuum B_{eff}, which fills completely the volume of length l, width b and thickness t.

According to equations (7.13) and (7.14), for plane stress conditions the critical load for which the crack starts propagating is

$$\sigma_{ac} = \left(\frac{2E\gamma}{\pi a} \right)^{1/2} \tag{7.19}$$

However, the presence of the crack in the volume under testing makes an equivalent continuum material filling up the same volume softer. We show in section 9.1.1 that the Young's modulus of the equivalent homogeneous continuum is

$$E_{eff} = E/(1 + 2\pi L a^2) \qquad \text{where} \qquad L = 1/(lb) \tag{7.20}$$

When crack propagation starts for a uniaxial stress σ_{ac}, the corresponding strain in the equivalent continuum is $\varepsilon_{ac} = \sigma_{ac}/E_{eff}$. Combining equations (7.19) and (7.20) to eliminate a, we get the stress–strain relation for the equivalent continuum when the crack of length $2a$ just starts propagating:

$$\varepsilon_{ac} = \frac{\sigma_{ac}}{E_{eff}} + \frac{8LE\gamma^2}{\pi \sigma_{ac}^3} \tag{7.21}$$

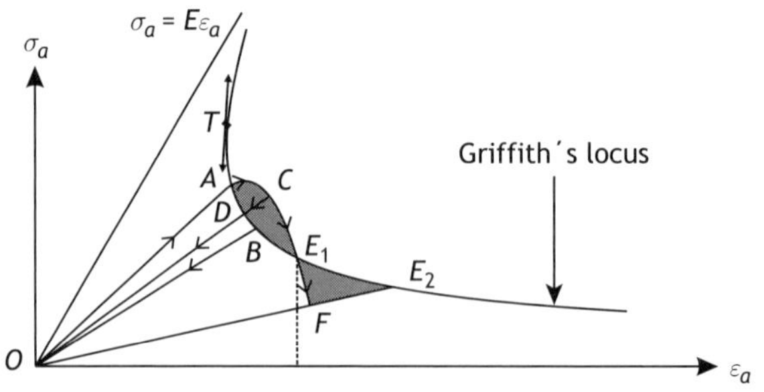

Fig. 7.11 Griffith's locus for a mode I crack.

A plot of σ_{ac} versus ε_{ac} in the stress–strain plane (fig. 7.11) is called the Griffith's locus (Berry, 1960; Cook and Hodgson, 1965).

On the one hand, for a very small crack length $2a$, the Young's modulus of the equivalent material tends to that of the matrix. Hence the Griffith's locus is asymptotic to the straight line $\sigma_a = E\varepsilon_a$. On the other hand, as the crack length approaches the sample width, the Young's modulus of the equivalent material becomes very small and the Griffith's locus is asymptotic to the horizontal axis. The point at which the Griffith's locus is tangent to a straight line parallel to the σ_a axis (point T on fig. 7.11) is found by setting $d\varepsilon_a/d\sigma_a = 0$; it corresponds to a crack length $a_c = (6L\pi)^{-1/2}$.

When a sample with a preexisting crack of length $2a$ and effective Young's modulus E_{eff} is initially loaded by a uniaxial stress σ_a, its response is linear until the straight line intersects the Griffith's locus, say at point A. Once the Griffith's locus has been reached, two cases may be considered for the loading mode.

In the first case the axial strain is increased so as to produce a quasistatic fracture propagation, δa_B, until point B is reached, at which point the sample is completely unloaded. Because the crack has extended by the amount δa_B, the Young's modulus of the equivalent material is smaller than that before fracture extension occurred. But, by the definition of brittleness, the sample remains elastic and no permanent deformation is observed after unloading.

In the second case the axial stress is increased, after point A has been reached, in such a way that the stress–strain curve reaches into the domain to the right of the Griffith's locus. Let us assume that point C is reached when the crack has grown by δa_D and that the sample is instantly unloaded in a linear manner back to zero load. The corresponding straight line OC intersects the Griffith's locus at point D and its slope corresponds to the Young's modulus for a crack length $2a + \delta a_D$.

We saw in section 7.1.1 that the area $\widehat{OADO}$ corresponds to the surface energy absorbed by the crack increment δa_D, so we conclude that the area $\widehat{ACDA}$ corresponds to the kinetic energy in the sample when the crack extension has reached δa_D.

The question now is to determine whether a new position of equilibrium may be reached or whether the sample will fail and break into two independent parts. If a new position of equilibrium exists, it corresponds to a situation for which the velocity at all points in the material is zero. Such a point is either on the Griffith's locus or to the left of it. Let us call

E_1 the point where the stress–strain curve of the loading process intersects the Griffith's locus. At this point the kinetic energy in the system is given by area $\widehat{ACE_1A}$ and therefore point E_1 cannot be a point of static equilibrium, since the velocity is not zero at all points in the material.

Let us show that if a new position of static equilibrium exists, it cannot correspond to a point on the Griffith's locus. Indeed, let F be the point of final equilibrium. If F were on the Griffith's locus, the loading process that would take the sample from point E_1 to point F would dissipate exactly the amount of energy required to create a corresponding crack increment from E_1 to F and therefore would not dissipate the kinetic energy accumulated in the sample at point E_1. Hence the final position of equilibrium, if it exists, must be within the region to the left of the Griffith's locus.

If a new position of equilibrium F exists, it corresponds to a crack length $2a + \delta a_F$, for which the effective Young's modulus is E_F. The straight line OF intersects the Griffith's locus at point E_2, and point F may be reached in two different ways. In the first loading process the sample is loaded through a dynamic process that went through C and E_1. In the second loading process, the sample is continuously loaded in a quasistatic manner, first from O to A and then from A to E_2, following the Griffith's locus until a crack extension $2a + \delta a_F$ has been reached; finally, the sample is unloaded from E_2 to F.

Because both loading processes end up with the same final crack configuration, the change in potential energy is the same for either of the two loading paths. Consequently, area $\widehat{E_1ACE_1}$ must be equal to area $\widehat{E_1E_2FE_1}$. In other words the kinetic energy that existed in the sample when point E_1 was reached has been dissipated as the surface energy required to propagate the crack from $2a + \delta a_{E_1}$ to $2a + \delta a_F$.

More generally, we may conclude that an equilibrium position exists if the kinetic energy stored in the body when the loading intersects the Griffith's locus at E_1 (the shaded area on fig. 7.11) is smaller than the area delimited between the Griffith's locus and the horizontal axis after the point E_1 on the Griffith's locus has been reached.

7.2.2 Servocontrolled testing systems

Let us now consider a real sample of rock in which multiple, randomly oriented, small cracks exist. The same reasoning as above may be conducted for the most critical crack so that it is possible to define a Griffith's locus for the sample. We may conclude from the discussion in section 7.2.1 that, once a crack starts propagating in a dynamic manner, it will stop only if the loading process is such that the system may be brought back below the Griffith's locus before the sample has completely failed.

Let us consider two different loading processes (fig. 7.12).

The first loading process is "load controlled": the load applied to the specimen may either be kept constant or slowly incremented. Then, as soon as the Griffith's locus is reached (points A or G), the crack extension velocity increases and, since it is impossible to unload the sample, the sample breaks in an unstable manner.

The second loading process is "displacement controlled": the displacement at the upper surface of the specimen may either be kept constant or slowly incremented at a constant rate. When the Griffith's locus is reached at the lower section of the locus (say at point G),

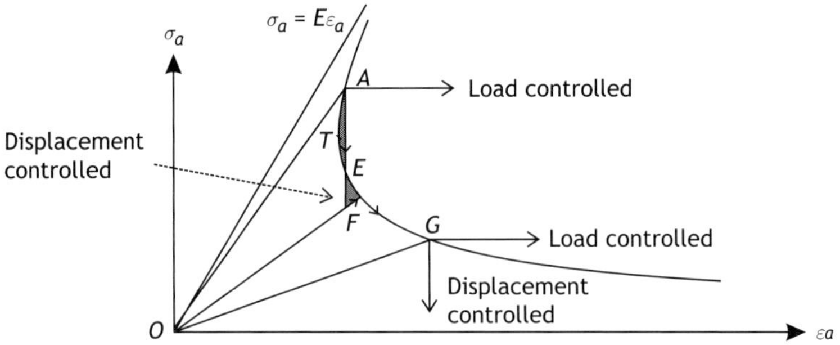

Fig. 7.12 Fracture propagation under load controlled and displacement controlled systems.

the load is progressively decreased in such a way as to keep the displacement rate constant; the crack extension then remains under control. But when the Griffith's locus is intersected in its upper part (say at point A), i.e. no preexisting crack in the sample is longer than the critical length for which the tangent to the locus is vertical, some unstable crack extension will occur with no possibility for decreasing the sample length since the displacement cannot be decreased. This leads to some instability (with the generation of kinetic energy), which stops only if a point of static equilibrium may be found below the lower part of the Griffith's locus (point F). Thereafter the crack extension becomes stable and is completely controlled by the imposed displacement rate.

This discussion illustrates the fact that the instability of fracture propagation is not an intrinsic characteristic of fracturing but depends only on the boundary conditions that accompany the fracturing process. For example, if instead of controlling either the force or the displacement, as was the case in the two above examples, a system had been set up to impose a monotonic decrease of the apparent Young's modulus of the sample (this is known as stiffness control), then it would have been possible to monitor the fracturing process of the sample completely.

Loading systems that adjust the load in such a way that a given parameter follows a precisely programmed path are provided by so-called "servo-controlled" systems (e.g. Rummel and Fairhurst, 1971). Their time of response must be of the order of a few milliseconds in order for catastrophic failure to be prevented.

7.2.3 Stress corrosion and sub-critical crack growth

Stress corrosion refers to a situation where chemically reactive fluids fill up cracks and alter the rock surface energy in such a way as to influence the critical strain energy release rate or the fracture toughness of the material (e.g. Atkinson and Meredith, 1987a).

Consider a rock loaded in such a way that the mode I stress intensity factor K_I for the most critical crack (with length $2c$) is lower than the critical value K^*_{Ic} associated with the unaltered material. For such a load the rock exhibits an elastic behavior. Assume now that a fluid circulates through the pore space and progressively alters the surface energy of the material at the tip of the most critically loaded crack, so that the critical stress intensity factor for the material at this location gets lower and lower (fig. 7.13).

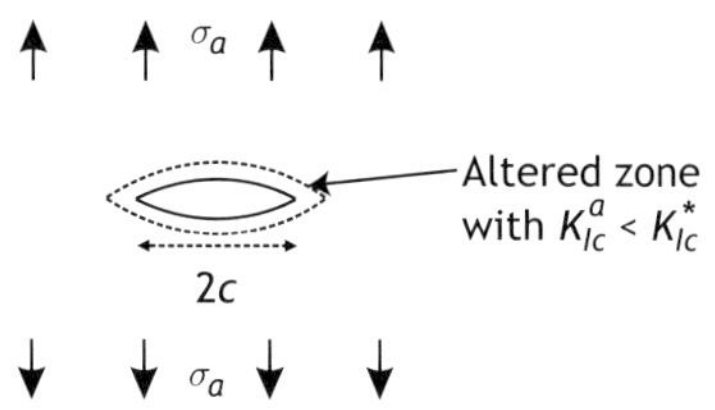

Fig. 7.13 Stress corrosion at the tip of a crack through circulation of a corroding fluid; K_{Ic}^{*} is the fracture toughness of the unaltered material while K_{Ic}^{a} is that of the altered rock.

When the fracture toughness of the altered material K_{Ic}^{a} becomes equal to the stress intensity factor at the tip of the crack, the fracture extends and reaches an unaltered zone where the fracture toughness is that of the unaltered material (K_{Ic}^{*}), so that fracture stops. In the process the crack has grown by an increment δc, and so the stress intensity factor at the tip of the crack is now slightly larger than its value before fracture extension occurred.

Through continuous fluid circulation within the pore space, alteration at the tip of the crack starts again until a new jump occurs; thus the stress intensity factor progressively increases after each new jump. When it reaches the fracture toughness of the unaltered material, if the system is under constant load conditions the fracturing process becomes unstable and leads to catastrophic failure.

Whilst the mechanism of stress corrosion involves a progressive alteration of the surface energy of the material, other mechanisms also may lead to subcritical crack growth. Two such mechanisms are gas diffusion through the crystalline lattice of the material and very slow plastic deformation.

We may note finally that the opposite effect (a strengthening effect) may be observed when the circulating fluid precipitates in the open pore space, resulting in a blunting of the crack tip with a correlative decrease in the stress intensity factor, a phenomenon referred to as crack healing (Smith and Evans, 1984).

7.3 Experimental investigations

7.3.1 Laboratory measurements

Because the propagation of fractures in rock material is important for rock fragmentation processes, such as blasting, tunnel boring, rock cutting or crushing, the International Society for Rock Mechanics suggested methods for determining the fracture toughness of rocks (Ouchterlony, 1988). However, only for mode I fracture propagation have standard testing procedures been proposed. No such standardization effort has been undertaken for mode II or mode III processes.

The standard testing procedures are indicated in fig. 7.14. They involve chevron-notched samples produced from rock cores. The advantage of a chevron-notched specimen is that

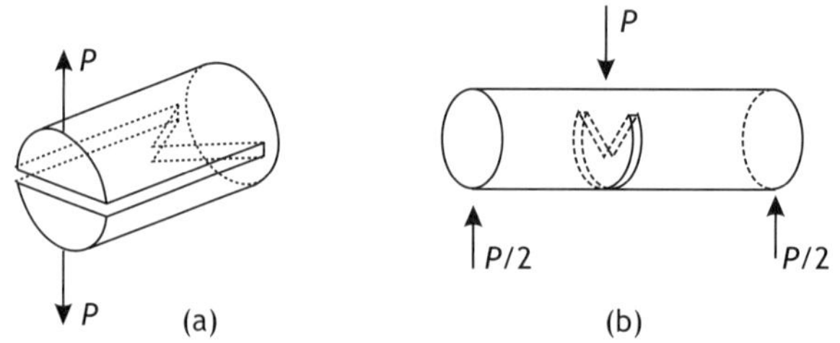

Fig. 7.14 Chevron-notched specimens for (a) short rod specimens and (b) for three-point bending tests.

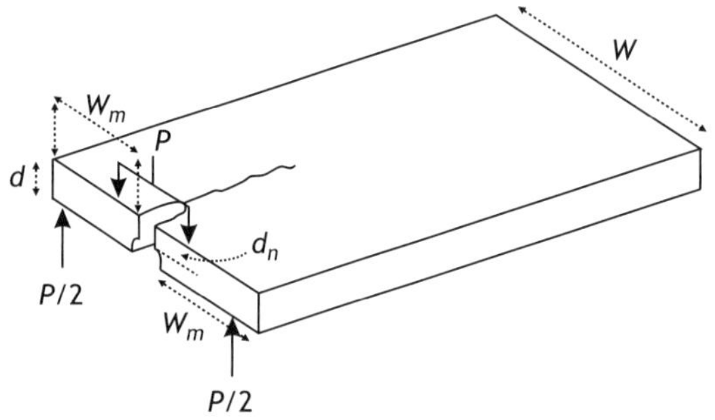

Fig. 7.15 The four-point bending configuration for sub-critical crack growth investigation.

crack growth from the apex of the triangular ligament is initially stable, so that the specimen is self-precracking (precracking is required to localize the inception of the process zone). The critical crack length for catastrophic failure is material independent and constant for a given specimen geometry. Hence there is no requirement for crack length measurement.

Another convenient testing geometry is the four-point bending plate (fig. 7.15) which provides stable fracture propagation independently of the crack length.

The stress intensity factor for the four-point bending test is (e.g. Atkinson, 1987, chapter 11)

$$K_I = PW_m \left[\frac{3(1+\nu)}{Wd^3 d_n} \right]^{1/2} \tag{7.22}$$

where P is the applied load, ν is Poisson's ratio and the other quantities are parameters that characterize the test geometry, as shown in fig. 7.15.

This testing geometry has shown itself to be effective for investigating subcritical crack growth in various corrosive environments.

A compilation of published surface energy and fracture toughness measurements were proposed by Atkinson and Meredith (1987b). For minerals, surface energy values range from 0.23 J/m^2 for calcite to 11.5 J/m^2 for quartz. For rocks, surface energy values G_c are in the range 20 to 1600 J/m^2 for sandstone (depending on the nature and quality of the cement), from 18 to 200 J/m^2 for limestone, from 80 to 500 J/m^2 for granite and between 20 and 50 J/m^2 for basalt. Reported fracture toughness values, K_{Ic}, exhibit much less dispersion and range from 0.5 MPa m$^{1/2}$ to 3.5 MPa m$^{1/2}$.

Fig. 7.16 Growth of a crack inclined to a uniaxial stress field (reproduced from Brace and Bombolakis, 1963, fig. 3, with permission from Wiley).

7.3.2 Numerical investigations on the propagation of a fracture inclined to the principal stress directions

As already mentioned, it is well established that mode I fractures propagate in their own plane and therefore perpendicular to the minimum principal stress direction (here compressive stresses are reckoned positive). Furthermore, laboratory experiments have shown that, in uniaxial compression, a crack inclined to the uniaxial compression direction propagates in a stable manner toward this direction (Brace and Bombolakis, 1963; Hoek and Bieniawski, 1965) (fig. 7.16), i.e. toward a direction perpendicular to the minimum principal stress direction.

Both Brace and Bombolakis (1963) and Hoek and Bieniawski (1965) also explored the interaction between cracks and showed that specific positions of cracks could result in an overall decrease of the apparent uniaxial compressive strength of the equivalent continuum material. However, analytical elastic solutions for these complex geometries are generally intractable and in practice numerical solutions must be sought.

With the development of fast computers, various investigations of fracture propagation in diverse loading conditions have been conducted numerically with the finite element method (e.g. Ingraffea, 1987) or the boundary element method (e.g. Cornet, 1979; Crawford and Curran, 1982; Nemat-Naser and Horii, 1982).

For example, Cornet (1979) showed that, for two-dimensional (plane strain) conditions and the critical strain energy release rate as fracture criterion, a crack inclined at 45° to a uniaxial tensile stress field propagates so as to become oriented perpendicularly to the uniaxial stress direction, i.e. perpendicularly to the minimum principal stress direction; this was observed for pure mode I fractures. Further, the fracture extension process is unstable, i.e. the fracture must be unloaded progressively in order to keep the strain energy release rate G constant and so prevent the appearance of kinetic energy and the onset of a velocity increase in the various parts of the specimen.

Cornet (1979) followed a similar approach for a uniaxial stress field σ_1 in compression. His numerical results were consistent with the experimental results obtained by Brace and Bombolakis or Hoek and Bieniawski, i.e. the fracture becomes progressively oriented parallel to the uniaxial stress direction (or perpendicularly to the minimum principal stress

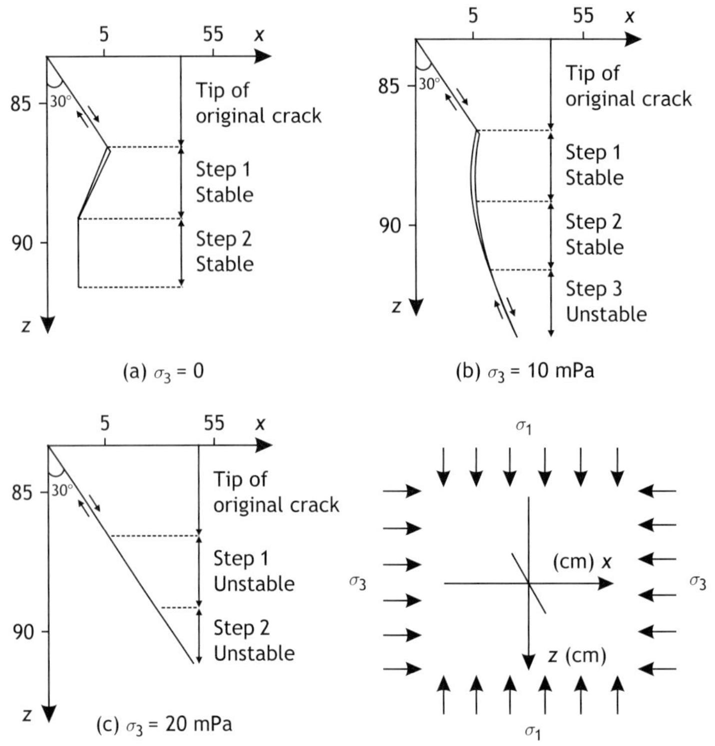

Fig. 7.17 Growth of a crack inclined to a principal stress direction in a triaxial stress field (reproduced from Cornet, 1979, fig. 4, with permission from The American Society of Mechanical Engineers).

direction): fracture propagation corresponds to a predominantly mode I mechanism. Furthermore fracture propagation is stable, i.e. the load must be progressively increased in order to keep the strain energy release rate G at a constant value.

Hence, these numerical simulations are at least qualitatively in agreement with experimental results.

When a small compressive principal stress component σ_3 is applied in a direction perpendicular to the σ_1 direction, the fracture extension geometry is progressively altered (fig. 7.17). In this model, crack surfaces are allowed either to become separated or to slide with respect to one another, but no interpenetration is allowed. When the fracture surfaces slide with respect to one another some resistive friction is mobilized, which satisfies the classical Coulomb friction law $\tau = \mu\sigma_n$ where μ is the friction coefficient, which was taken as 0.5 in this investigation.

For a compressive stress σ_1 equal to 100 MPa, a value of 10 MPa was first assumed for the minimum principal stress σ_3 (fig. 7.17(b)).

The fracture extension was found to be stable until it reaches a critical length and orientation, after which the fracture extends in an unstable manner without any mode I component (fig. 7.17(c)).

When the minimum principal stress component σ_3 was taken as 20 MPa, the fracture was found to propagate in its own plane in an unstable manner. However, for this numerical

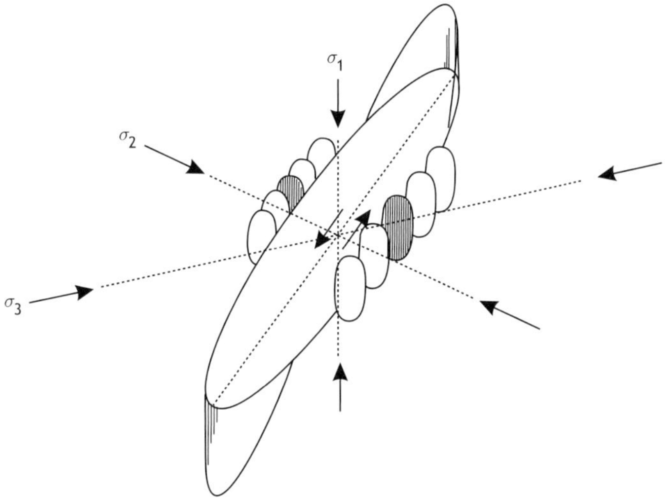

Fig. 7.18 Diagram showing the formation of tensile crack extensions associated with the shearing of a fracture that is inclined to all three principal stress directions (reproduced from Scholz, 1990, fig. 1.16, with permission from Cambridge University Press).

investigation only the relative variations of G were considered; the absolute value of G was not taken into account.

These numerical investigations were run under two-dimensional plane strain conditions, i.e. the fracture geometry was presumed to be independent of the third dimension and to extend to infinity perpendicularly to the plane of fig. 7.17. In reality, however, the fracture dimension in the third direction is finite so that, in fact, the extension of a pre-existing crack subjected to triaxial stress compressions involves combinations of mode I and mode II components in the upper and lower sections of the fracture but combined mode I and mode III components for the out-of-plane sections of the fracture (Scholz, 1990, chapter 1).

Experimental results showed that crack propagation in pure mode III results in a complex fracture extension path that involves mostly a mode I fracturing process (e.g. Palaniswamy and Knauss, 1974). This led Scholz to propose the geometry of fracture extension in compressive stress fields shown in fig. 7.18; here the fracture extension is stable, i.e. the extension of a single crack loaded in uniaxial compression cannot lead to complete unstable sample failure.

Since these simple crack extension schemes are stable, the axial load must be progressively increased in order to reach the failure of the specimen.

The growth of fractures under shear stress conditions is discussed further in section 8.1.4, where it will be seen to depend on rock type and loading conditions. In particular, the results of Lockner *et al.* (1991) showed that, for granite in triaxial compression conditions, shear cracks may extend in their own plane with a very narrow process zone (see section 8.1.4). This shows that the surface energy associated with process zone development depends on the local stress field and is not an intrinsic property.

Hence, in contrast with rock fracturing in tension, in which crack extension results in the formation of one single large fracture surface, the failure of rock in compression may involve either interactions between numerous cracks or the extension of one single fracture plane, depending on the local stress field.

7.4 Exercises

1. Why is the tensile strength of a rock much smaller than its compressive strength ?
2. Why does the measured value of tensile strength depend on the volume of the tested samples?
3. The stress intensity factor for a flat elliptical fracture of length $2a$ loaded in mode I is given by $K_{Ic} = \sigma_3^{\infty}\sqrt{2\pi a}$, where σ_3^{∞} is the far-field uniaxial tensile stress applied to the material (tensile stresses being reckoned as positive). What is the stress intensity factor for the same fracture loaded by a uniaxial compression $-\sigma_a$ but with a unform pressure P inside the fracture, with $P < -\sigma_a$?
4. Experience and numerical modeling show that a fracture inclined to a far-field uniaxial tensile stress field becomes progressively perpendicular to the stress field direction as it propagates. What happens if the far-field uniaxial stress is compressive but the fracture is subject to a uniform internal pressure P larger than the magnitude of the uniaxial compression magnitude?
5. Consider fig. 13.3, where two inflatable packers help to isolate a portion of a borehole, referred to as the interval, which is to be pressurized. The pressure in the packers is P_p whilst that in the interval is P_w, with $P_p > P_w$. The stress intensity factor associated with a mode I fracture initiated in a borehole by a pressure P_p that remains confined within the borehole is proportional to the inverse of the fracture length a: $K_I = P_p/\sqrt{\pi a}$. Both the packer pressure and the interval pressure are progressively increased until some fracturing occurs. Show that fracture initiates at the packer level but then propagates from the pressurized interval.

8 Laboratory investigations on geomaterials under compression

In chapter 1 we pointed out that a geomaterial is often a mixture of various constituents, which could be rock, soil, liquid or gas. We indicated that the definition of an equivalent continuous geomaterial implies the existence of a representative elementary volume (REV), the properties of which are assumed to be uniform throughout the volume occupied by this equivalent continuum. This implies that the volume of the body for which an equivalent geomaterial is being defined should be orders of magnitudes larger than the REV.

We also observed that rock masses always involve fracture and fault systems that exhibit fractal dimensions, so that some geomechanical properties of rock masses cannot be analyzed with the sole concept of an equivalent geomaterial. For rock masses it is often necessary to define, independently, equivalent geomaterials, equivalent fracture fields and equivalent faults.

Laboratory experimentation is typically intended to address issues related to the properties of geomaterials. It may pursue very diverse goals, whether this involves investigating a particular deformation mechanism or measuring the mechanical characteristics of an equivalent geomaterial. In the latter case the size of the tested volume must be considerably larger than that of the REV.

Experimental work on the mechanical behavior of geomaterials loaded in compression has been conducted in the laboratory for more than a century. For example, work by von Karman (1911) on limestone addressed issues related to rock plasticity whilst that by van Terzaghi (1923, 1943) investigated the plastic behavior of soils with particular attention to the role of pore pressure.

We adopt here the classical distinction between rock and soil, i.e. rocks exhibit some cohesion because of bonds that exist between their constitutive elements, whilst soils are granular materials without cement between the grains. In both rocks and soils the various constitutive elements are much smaller than the REV.

The procedure common to all these compression experiments is to load representative samples under well-controlled conditions and then to correlate displacements with the forces and pressures measured at the sample boundaries.

8.1 Laboratory testing of rocks

8.1.1 The concept of a complete stress–strain curve

During the 1960s and 1970s, a great deal of attention was given to understanding rock failure, or more generally the post-elastic behavior of rock (e.g. Hoek, 1968). In particular it was observed that in some instances rock pillars left in place to support the rooves of mined-out cavities could fail progressively, retaining some significant load-bearing capacity even though the pillars were heavily fractured. Yet in some other instances, sometimes in the same mine, the pillars would fail in an unstable manner leading to catastrophic roof collapse.

Attention turned to understanding the causes of instability in failure processes. The objective was to conduct experiments in the laboratory on rock specimens by driving them to complete collapse in a quasistatic manner, i.e. without any significant generation of kinetic energy. At the end of testing, many discontinuities had developed in the rock sample, resulting in deformations much larger than those observed normally within the elastic domain. Even though the material within the tested volume had become quite discontinuous during failure, the results were analyzed on the assumption that an equivalent continuum could be defined at all stages of the deformation process. This equivalent continuous material was characterized by a so-called "complete stress–strain curve" produced during the compression tests.

For uniaxial compression loading, the mean "axial stress" in the equivalent geomaterial is simply taken as equal to the axial force applied to the sample divided by the area of the intact sample's cross section. The axial strain in the equivalent geomaterial is taken as equal to the change in length of the sample divided by its original length, assuming complete homogeneity for the equivalent continuous material. A more accurate definition of stress magnitudes corrects for the change in cross-sectional area of the sample at the center of the specimen as determined from direct measurement (e.g. through strain gauges glued in the circumferential direction near the center of the specimen).

In chapter 7 the failure of rock in tension was shown to result from the progressive growth of flaws and microcracks. We demonstrated how the propagation dynamics of mode I fractures depends on the loading process, the crack growth being always unstable for *load controlled* loading processes. We showed how *displacement controlled* loading processes may lead to local instabilities with subsequent stable crack growth.

This concept of controlled rock failure has been generalized to rock testing under compression, and the objective of modern testing systems is to keep the rock-disintegration process stable after the peak load has been reached.

Stabilization is obtained after observing that the loading frame and the sample under testing constitute a closed system in which there are continuous energy exchanges: the loading frame becomes deformed in a reversible, linearly elastic, manner whilst the rock sample breaks progressively.

During the post-peak disintegration process, if the stiffness of the loading frame is smaller than that of the rock sample then more energy is released by the unloading of the frame caused by sample failure than can be absorbed by the failure of the rock (case 1 in fig. 8.1). The difference is transformed into kinetic energy and the failure is unstable.

When the stiffness of the loading frame is larger than that of the rock sample, less energy is released by the unloading of the frame than the energy necessary for failure to progress in the rock (case 2 in fig. 8.1). The difference must be continuously supplied to the loading frame to make the failure process stable.

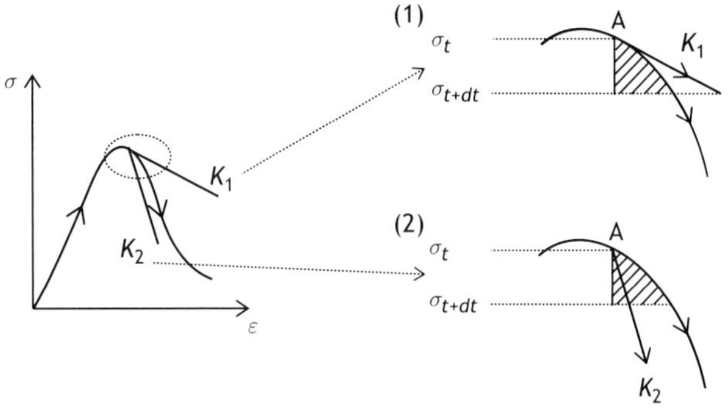

Fig. 8.1 Stiff (K_2) and soft (K_1) testing systems leading respectively to stable (2) or unstable (1) rock disintegration in compression.

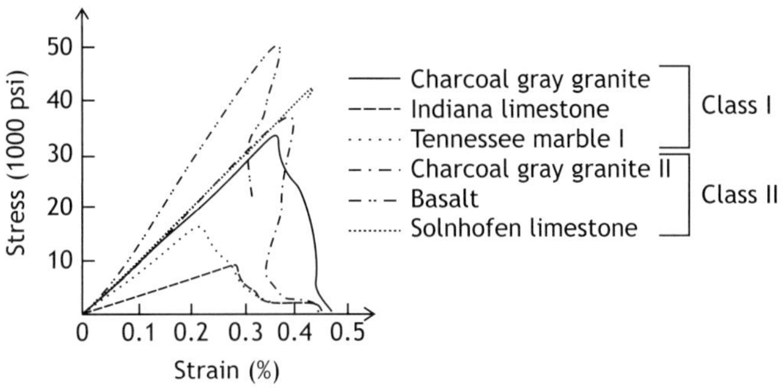

Fig. 8.2 Full stress–strain curves for different rocks (charcoal granite, Indiana limestone, Tennessee marble, basalt, Solnhofen limestone) tested under uniaxial compression; 1000 psi = 6.9 MPa, reproduced from Wawersik and Fairhurst (1970), with permission from Elsevier.

Hence, a stable system is such that exchanges of energy between the loading frame and the rock sample remain quasistatic without release of significant kinetic energy during the complete rock disintegration process.

With this in mind, Wawersik and Fairhurst (1970) developed a testing system with which they could transfer very small incremental loads from a very stiff loading frame to a rock sample by changing the temperature in the loading frame. The change in temperature produced thermal contraction or thermal expansion of the frame, depending on the sign of the temperature variation, hence loading or unloading the specimen by very small amounts. They were able to produce complete stress–strain curves for different rock types under uniaxial compression (fig. 8.2).

Later, Rummel and Fairhurst (1971) and Hudson *et al.* (1971) generalized the concept and showed how servocontrolled testing systems could be programmed to monitor failure for any testing configuration. Stabilization of the failure process is obtained by monitoring a parameter that varies monotonically during rock disintegration.

Loading of the sample is conducted so as to keep the variation rate of the selected parameter in agreement with values programmed for the testing schedule. The larger the

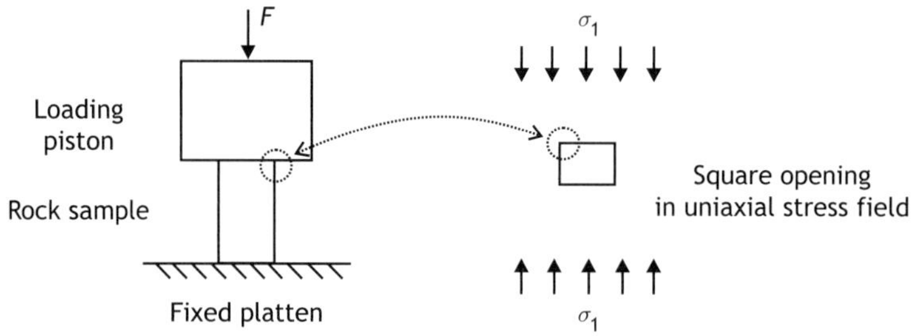

Fig. 8.3 Elastic analysis of loading conditions for uniaxial testing.

variations of this controlled parameter during rock disintegration, the more efficient the disintegration monitoring.

In the remaining part of this chapter, the discussion of the complete stress–strain curves assumes quasistatic deformation rates, except in the case of acoustic emission (section 8.1.4). Some dynamic aspects of fracture propagation are discussed in chapter 11.

8.1.2 Uniaxial compression test

Uniaxial compression tests usually are run in order to characterize the elastic response of a rock, its uniaxial compressive strength and its post-elastic behavior. In such tests the stress tensor in all parts of the sample is assumed to be simply

$$\begin{pmatrix} \sigma_1 & 0 & 0 \\ 0 & 0 & 0 \\ 0 & 0 & 0 \end{pmatrix} \tag{8.1}$$

where the axis $\hat{\boldsymbol{I}}_1$ is parallel to the axial loading direction. This implies that specific attention must be given to the loading procedure and to the sample geometry if parasitic stress fields, known as end effects, are to be avoided.

End effects

If the sample is inserted directly between the loading plattens, some stress concentration is generated in both the platten and the rock because of the discontinuity in the load applied to the platen. This stress concentration may be evaluated (Jaeger, 1969) using solutions derived from the theory of elasticity for stress distributions near square openings in infinite bodies (section 5.6.4 and fig. 5.11). Figure 8.3 shows how this elastic solution may help model the stress field in a rock specimen inserted directly between flat plattens that are made of the same material as the rock specimen.

The effect of this discontinuity is avoided by inserting between the loading platten and the rock specimen a small perforated steel disk with same cross section as that of the sample, so that the loading discontinuity is located at the interface between the loading platten and the steel disk (fig. 8.8).

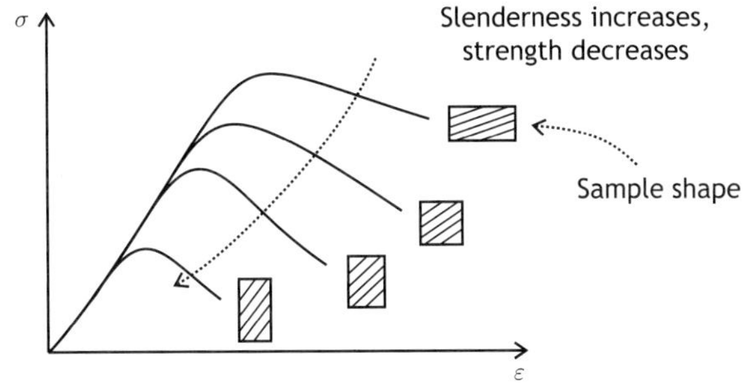

Fig. 8.4 Parasitic end effects in uniaxial rock testing affect the complete stress–strain curve for specimen aspect ratios that are too small (reproduced from Hudson *et al.*, 1972, with permission from Elsevier).

Also, because generally no slippage occurs at the steel–rock interfaces or at the disk–platten interfaces, teflon is sometimes inserted between the loading platten and the cylindrical steel plates in order to suppress the constraints on horizontal displacements at the rock specimen ends.

Further, because the Poisson's ratio of the steel plates is different from that of the rock, a parasitic stress field is generated at the interface between the loading steel disk and the rock specimen. A compressive nonzero stress component is generated within the rock in the vicinity of both ends of the specimen, in directions parallel to the interface.

The influence of these parasitic end effects on rock failure is illustrated in fig. 8.4. The figure shows complete stress–strain curves obtained for samples of the same rock with different aspect ratios (the aspect ratio $AR = l/d$ where l is the sample length and d is its diameter). End effects introduce a stress component parallel to the rock–steel interface which changes the material properties and makes it appear stronger (see section 8.1.3 on triaxial compression tests). Hence, the complete stress–strain curves are found to depend on the specimen aspect ratio when it is smaller or equal to 1 (Hudson *et al.*, 1972).

In order to limit these parasitic end effects, the sample length is set equal to two to three times the mean dimension of its cross section. With such a shape, end effects are limited to a volume with dimensions slightly less than those of the sample cross section, i.e. the stress field is indeed uniaxial within the central portion of the sample. Let us note that, when the aspect ratio is too large, the sample may fail by buckling as discussed in section 5.3.1.

Complete stress–strain curve

A typical stress–strain curve for a rock sample tested in uniaxial compression is shown in fig. 8.5. Compression is taken as positive, so that a positive axial strain implies a shortening. Four different domains are observed.

First the curve is convex upward (section OA). Then the curve becomes linear (section AB) before turning convex downward (section BC). It reaches its maximum value at C where it gets tangent to a horizontal line. This maximum value is called the uniaxial, or unconfined, compressive strength of the rock and is often referred to as σ^c. Then the load bearing capacity of the rock drops progressively to zero as the axial strain is regularly increased.

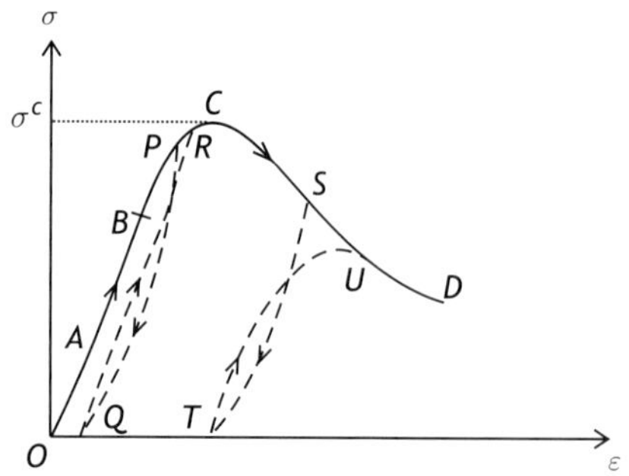

Fig. 8.5 Typical stress–strain curve for a rock loaded in uniaxial compression (reproduced from Jaeger and Cook, 1979, with permission from Wiley).

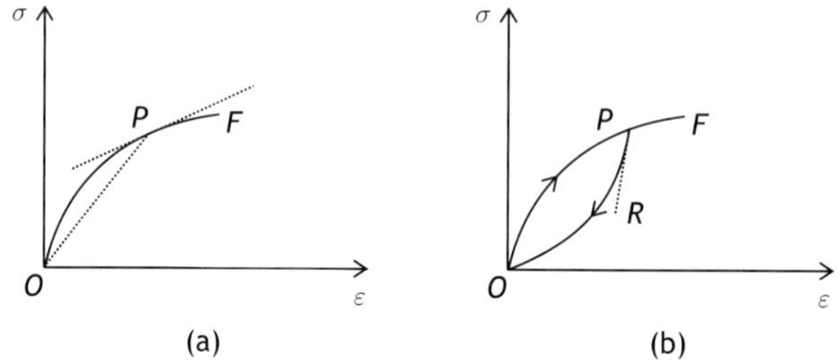

Fig. 8.6 (a) Tangent and secant modulus for rock in compression; (b) unloading modulus. For linearly elastic rocks, the tangent, secant and unloading moduli are identical.

During the first portion of the loading curve (section *OA*) in fig. 8.5, the rock gets stiffer and stiffer as the axial load increases. As shown in section 10.1.1, this stiffening of the rock is caused by the closing of microcracks, i.e. the void space left open between the two faces of pre-existing microcracks gets smaller and smaller as the uniaxial load is increased.

Also, because some microcracks are inclined with respect to the uniaxial compression direction, they will support some shear stress so that some slip may occur along these minute interfaces (Cook, 1965). However, when the sample is unloaded these interfaces do not slip back immediately, because of some frictional resistance (friction is discussed in section 10.1.2). Hence the equivalent continuum appears stiffer during unloading (curve *PQ* in fig. 8.5) than during loading (curve *QR*) and it is argued in section 9.1.1 that the observed unloading modulus may be assimilated to that of the matrix, i.e. the equivalent solid material without any microcracks.

Because of this nonlinear behavior, two elastic moduli are commonly defined for rocks loaded to a given axial load amplitude (point *P* on fig. 8.6), the secant modulus (the slope of the line *OP*) and the tangent modulus (the slope of the tangent at *P*) on fig. 8.6(a). Both moduli are smaller than the unloading modulus, the slope of *PR* in fig. 8.6(b). The departure from linearity is mostly dependent on the degree of rock heterogeneity within the REV, with a particular dependence on the shape of the pores and microcracks.

After reaching its peak value, the load-bearing capacity of the rock sample drops progressively to zero with two different possible behaviors, referred to as class I and class II by Wawersik and Fairhurst (1970) (see fig. 8.2).

With class I behavior, energy has to be continuously supplied to the rock sample for the disintegration process to proceed. For class II behavior, however, too much elastic energy is stored in the testing system (loading system plus rock sample) and energy must be progressively removed from the testing system through unloading of the specimen, while keeping the failure process developing quasistatically. This class II behavior can be observed only with a system that unloads the sample progressively, to prevent the generation of kinetic energy. This may be achieved through a servocontrolled testing system but is often achieved through a succession of loading–unloading cycles (such as the cycle *STU* in fig. 8.5); the unloading part of the cycle stabilizes the disintegration process.

Elastic limit and uniaxial compressive strength

The elastic limit corresponds to the maximum load that may be applied to the rock before the axial stress versus axial strain curve becomes convex downward. It was shown in section 7.3.2 that crack propagation in uniaxial compression is stable, i.e. the load must be increased in order to further propagate microcracks. With the development of microcracks, however, the apparent Young's modulus of the equivalent material reduces. Hence the decrease in slope of the stress–strain curve observed beyond the elastic limit but before reaching the peak load corresponds to the propagation of an increasing number of stable microcracks. The cumulated effect of their propagation leads to the observed nonlinearity.

The maximum load-bearing capacity of the sample, i.e. the load for which the axial-stress–axial-strain curve becomes horizontal, is controlled by the density of the microcracks and by their rate of growth. This implies that the post-elastic behavior of rock is strongly dependent on the deformation rate, as discussed in section 8.1.5, and that so is the uniaxial compressive strength.

Fracture geometry

The early work on rock failure in uniaxial compression was conducted with "load controlled" systems, so that post peak-load failure was always unstable. In this early work both tensile fractures parallel to the uniaxial load and shear fractures inclined to the axial load were observed.

However, with the use of stiff and servo-controlled testing systems and with proper loading procedures for avoiding end effects, it has been shown that only indirect tensile fractures (i.e. fractures parallel to the applied load and perpendicular to the minimum principal stress direction, fig. 8.7) are generated during uniaxial compression (Wawersik and Fairhurst, 1970).

These axial fractures generate small axial slabs that buckle as the load is increased. The sample's cross section is reduced progressively at its central part, away from the extremities where end effects prevent the development of these axial cracks.

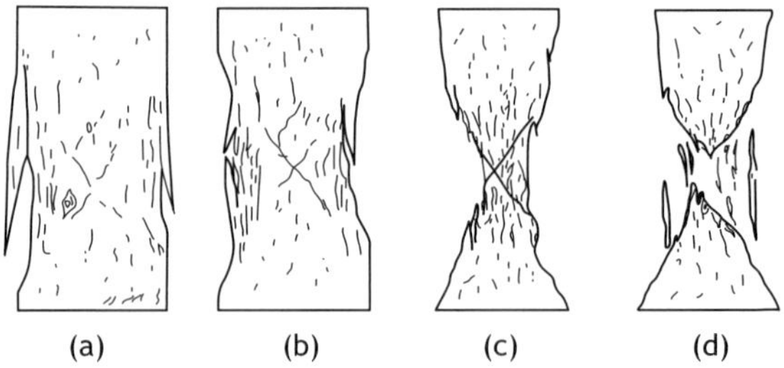

Fig. 8.7 Progressive development of failure in a cylindrical specimen of granite, loaded in uniaxial compression with a stiff testing machine (Wawersik, 1968). The axial load-bearing capacity of the sample decreases progressively from (a) 80% of peak load to (d) 0% of peak load.

As disintegration develops, the sample geometry is altered quite markedly and the definition of an equivalent continuum becomes more and more problematic because of the now ill-defined sample geometry.

8.1.3 Triaxial compression tests

Since the early work by von Karman (1911) it has been recognized that rocks tested in compression are stronger when pressure is applied in directions perpendicular to that of the axial compression.

Experimental procedures

In order to apply such a pressure during axial loading, cylindrical specimens together with their loading piston and loading base are coated with an impervious jacket. Then the sample and its loading heads are inserted within a pressure vessel, and a fluid pressure P_c is applied around the sample and the loading heads (fig. 8.8). During testing some fluid is injected or extracted from the pressure vessel, so as to keep the fluid pressure constant while the axial piston is progressively lowered inside the vessel to increase the axial load. This constant fluid pressure is referred to as the confining pressure.

Away from end effects, the stress tensor near the center of the sample is given by

$$\begin{pmatrix} \sigma_1 & 0 & 0 \\ 0 & P_c & 0 \\ 0 & 0 & P_c \end{pmatrix} \tag{8.2}$$

Even though two principal stress components are equal (and constant), these loading conditions are referred to as triaxial compression-testing conditions.

Subtracting the change in volume occupied by the loading piston inside the vessel from the volume of the confining fluid flowing in or out of the pressure vessel enables the measurement of the rock sample's volume variations (Crouch, 1970). Because the sample's axial strain is known, the change in cross-sectional area of the sample may be evaluated at all stages of the test on the assumption that the specimen remains cylindrical. Then this cross-sectional area is used to determine the applied axial stress at the center of the specimen.

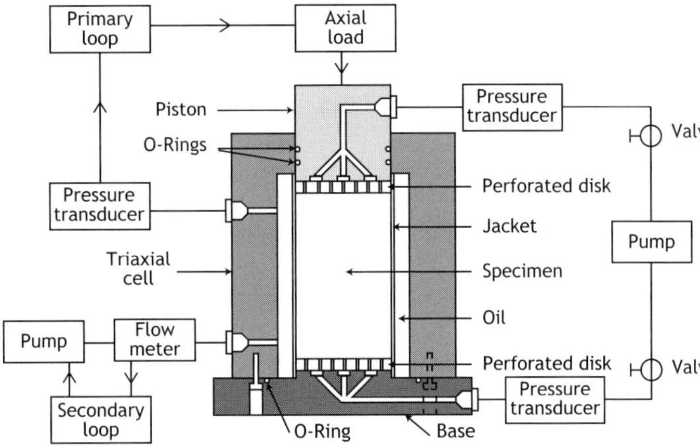

Fig. 8.8 The triaxial test: a confining pressure P_c is applied, through a secondary loop, to a cylindrical specimen loaded axially by a primary loop. This system also provides a means to monitor interstitial pore pressure inside the rock sample, as discussed in section 8.1.6.

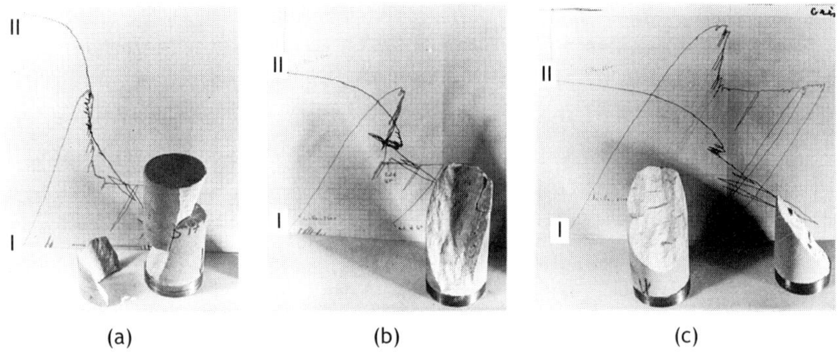

Fig. 8.9 Examples of tests for Berea sandstone under triaxial compression. Note the loading steel discs below the samples. For each test, plot I represents the axial force, along the vertical axis, versus the vertical displacement, along the horizontal axis. Plot II represents the flow of confining fluid (with vertical axis positive downward) versus the axial displacement, along the horizontal axis. The confining pressures were (a) $P_c = 3.44$ MPa; (b) $P_c = 17.4$ MPa; (c) $P_c = 44.8$ MPa.

Various parameters have been used to monitor the disintegration process of rock under triaxial compression. For example, in fig. 8.8 the flow of confining fluid monitors the loading of the rock (Cornet and Fairhurst, 1974). A servocontrolled system controls the confining pressure and generates a steady outflow. Another servocontrolled system keeps the confining pressure constant through the loading of the sample. Given that the first loop removes fluid from the cell at a steady rate, the axial piston is lowered into the cell so as to compensate the drop in pressure caused by the first loop. This double servocontrolled loop system results in both a constant confining pressure and a monotonic increase of the lateral strain (i.e. the volumetric strain minus the axial strain) of the rock sample.

Some experimental results generated with this system are shown in fig. 8.9.

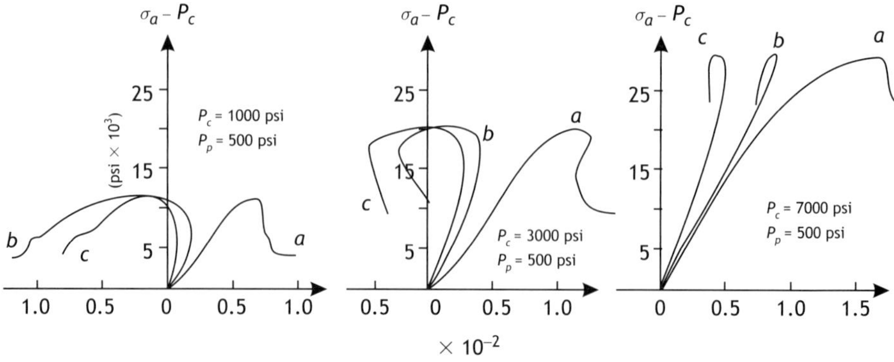

Fig. 8.10 Examples of experimental curves for Berea sandstone tested under triaxial compression. The curves *a* are plots of axial stress versus axial strain. The curves *c* are plots of axial stress versus volumetric strain. The curves *b* are plots of axial stress versus change in pore volume (discussed in section 8.1.6). The confining pressures were (left) $P_C = 3.44$ MPa (1000 psi); (middle) $P_C = 17.4$ MPa (3000 psi); (right) $P_C = 37.9$ MPa (5500 psi) (Cornet and Fairhurst, 1974).

It may be observed that for a 3.44 MPa confining pressure (shown in the leftmost picture), a class I post-elastic behavior is observed, i.e. external work must be continuously supplied to the rock for the disintegration to proceed according to the testing schedule. However, for larger confining pressures the volume variation during disintegration is not monotonic and a quasistatic disintegration process is obtained only by the use of various load–unload cycles.

We discuss in section 8.1.4 below results obtained by Lockner *et al.* (1991), who monitored the rate of acoustic emission production during triaxial tests and were able to produce experimentally a quasistatic sample disintegration process for triaxial conditions.

Volumetric strain and the elastic limit

In figs. 8.10 and 8.11 results from servo-controlled triaxial compression tests run on Berea sandstone and Indiana limestone are presented for three different effective confining pressures, 3.44 MPa (500 psi), 17.4 MPa (2500 psi) and 44.8 MPa (6500 psi). The concept of an effective confining pressure (the external confining pressure minus the internal pore pressure) relates to saturated rocks and will be discussed in section 8.1.6.

Berea sandstone is a homogeneous sandstone with volume porosity in the 15% range whilst Indiana limestone is a homogeneous limestone with porosity in a similar range of values.

The axial stress in excess of the effective confining pressure is plotted along the vertical axis (0 corresponds to the initial hydrostatic state of stress). For each confining pressure, the corresponding axial strain (curve *a*) and volumetric strain (curve *c*) are plotted along the horizontal axis. Because compressions are reckoned positive, positive axial strains correspond to sample shortening and positive volumetric strains correspond to a decrease in volume. The curves *b* correspond to changes in pore volume and are discussed in section 8.1.6 below.

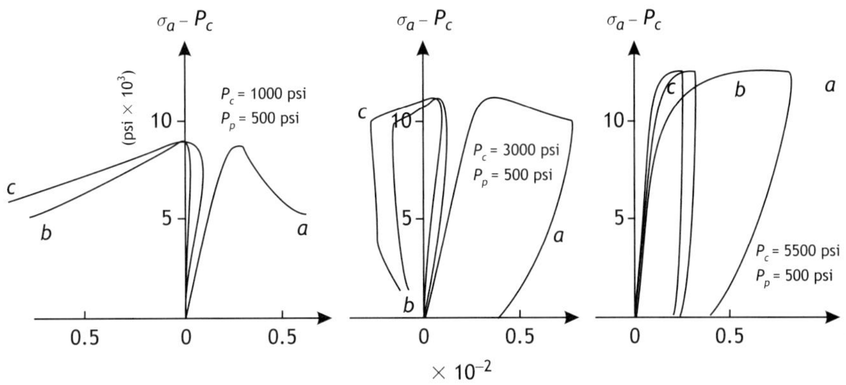

Fig. 8.11 Examples of experimental curves for Indiana sandstone tested under triaxial compression. Curves *a* are plots of axial stress versus axial strain. Curves *c* are plots of axial stress versus volumetric strain. Curves *b* are plots of axial stress versus change in pore volume (discussed in section 8.1.6). The confining pressures were (left) $P_c = 3.44$ MPa (1000 psi); (middle) $P_c = 17.4$ MPa (3000 psi); (right) $P_c = 44.8$ MPa (7000 psi) (Cornet and Fairhurst, 1974).

Initially, both types of rock exhibit an elastic response, but the mechanical behavior of Berea sandstone becomes linear only for the largest effective confining pressure. Further, for Berea sandstone loaded under a 3.44 MPa effective confining pressure, we notice a decrease in volumetric strain, i.e. the volume of the sample starts increasing (according to the sign convention), whilst, according to the axial stress axial strain curves, the elastic limit has not yet been reached.

This volume increase observed whilst the axial load is being increased is called dilatancy and was first described by Brace *et al.* (1966). It corresponds to the stable growth of microcracks in a manner somewhat similar to the indirect mode I mechanism discussed in section 7.3.2. Brace *et al.* conducted tests on granite, marble and aplite, at confining pressures up to 800 MPa. For granite they found that the onset of dilatancy is strongly time dependent: the higher the loading rate, the higher the axial stress for the onset of dilatancy. In chapter 9 we show how microcracks influence seismic velocities.

The tests of Brace *et al.* prompted detailed investigations on the seismic velocity variations that occur before the maximum load-bearing capacity of rocks is reached (Rummel, 1974; Hadley, 1975). The objective was to determine wether the monitoring of seismic velocity variations would provide a means to predict, *in situ*, rock failure at various scales. However, seismic velocities depend also on the fluid content. Furthermore, at large scales not only must microcracks be considered but also large fractures and faults. We discuss in chapter 12 how the monitoring of seismic velocity variations may help detect large-scale changes in loading conditions.

We may conclude that the elastic limit as defined from the axial stress versus axial strain curve gives an overestimate of the maximum axial load that may be applied to the rock before pre-existing flaws and microcracks start propagating in a stable manner, for the corresponding confining pressure. A more accurate definition of the elastic limit is provided by the onset of dilatancy.

When volume variations are monitored after the peak strength has been reached (Cornet and Fairhurst, 1974), for class I rocks a continuous volume increase is observed but for class II rocks the decrease in load required to maintain a stable disintegration process results in a decrease in volume of the rock sample and of the pore volume. For rocks that are saturated by an incompressible liquid, this change in pore volume induces pore pressure variations, depending on the boundary conditions, which in turn influence the failing process because of the effective stress principle (section 12.1.1).

Let us note that dilatancy is also observed for Indiana limestone samples loaded with 3.44 MPa on 17.4 MPa effective confining pressures. However, for this rock the onset of dilatancy occurs for axial loads much closer to the limit of linearity of the axial stress versus axial strain curves. Furthermore, for an effective confining pressure equal to 44.8 MPa, dilatancy is not observed; rather, a continuous decrease in volume is noticed, which gets larger and larger as the axial load is increased. This corresponds to compaction.

More generally, experience has shown that dilatancy is mostly associated with rocks that exhibit numerous sources of stress concentrations within their representative elementary volume (REV), which lead to microfracture propagation. These local stress concentrations induce local stable mode I crack growth as the axial load is increased. It is found that dilatancy disappears progressively as the confining pressure gets larger and larger; furthermore, it becomes negative, and is then called compaction, when pore collapse becomes the dominating deformation feature (section 8.1.8).

Effect of confining pressure on the complete stress–strain curves of rocks

Figures 8.10 and 8.11 give just a few examples of the diversity of the responses of rocks to a confining pressure. However, they synthesize the main features associated with an increasing confining pressure, the diversity of the responses being shown by the relative principal stress component magnitudes for which these features are observed, depending on the nature of the rock. We summarize below the four main features that have been outlined.

The first important effect of an increase in the confining pressure is that the rock sample appears stiffer, i.e. the Young's modulus (the mean value between the secant modulus and the tangent modulus) increases and Poisson's ratio decreases. Further, the axial stress versus axial strain curves become progressively more linear as the confining pressure is increased.

A second important feature associated with an increase in the confining pressure is the correlative decrease of dilatancy in the post-elastic domain.

A third consequence of an increase in the confining pressure is that the maximum load-bearing capacity of the rock sample increases.

Finally, depending on the nature of the rock, an increase in the confining pressure modifies the post-peak-load behavior significantly.

For Berea sandstone, the post-peak behavior changes from class I to class II as the confining pressure is progressively increased from 3.4 MPa to 17.4 MPa, but reverts to class I post-peak behavior when the confining pressure exceeds 30 MPa. In other words, a range of confining pressures exists for which failure is very unstable, i.e. there is more potential energy stored in the vicinity of the fracture zone than is necessary to create fresh surfaces. For this range of confining pressures, the material must be progressively unloaded to prevent the generation of kinetic energy whilst the fracturing process develops.

However, for Indiana limestone, at low confining pressures the initial class I drop in strength observed after the peak load-bearing capacity has been reached decreases less and less rapidly with increasing strain, as the confining pressure grows further. The rock becomes ductile (so that there is no loss of strength with increasing strain) for confining pressures equal to or larger than 40 MPa.

Fracture geometry and the brittle–ductile transition

Figure 8.9 presented Berea sandstone specimens after testing to produce complete stress–strain curves at three different confining pressures. It gives in outline the changes in fracture geometry that are associated with the various disintegration processes, depending on the magnitude of the effective confining pressure.

The simplest geometry is that observed for the triaxial test at a 44.8 MPa confining pressure. The specimen has sheared along a single plane. Then its deformation occurred through slippage along that plane (under a constant axial load) as a response to the imposed increase in axial strain for the equivalent continuous geomaterial.

In the test conducted under a 17.4 MPa confining pressure, a single shear surface was formed, steeper than the plane observed for the 44.8 MPa confining pressure test.

In the test conducted under a 3.44 MPa confining pressure, an axial fracture is observed; in addition conical shear zones have developed, the geometry of which is controlled by the influence of end effects at both extremities of the specimen.

For all the rocks, the confining pressure magnitude influences the geometry of the fracture surfaces associated with the disintegration process. Fractures are initially coaxial to the specimen in uniaxial compression tests (at zero confining pressure). As the confining pressure is progressively increased, however, single continuous planar fracture surfaces are generated at an angle to the maximum principal stress direction. The final fracture geometry at the end of testing, for a given confining pressure, depends on the stress field within the specimen and on the relative influence of end effects. As the confining pressure increases, the angle between the fracture plane and the maximum principal stress direction keeps increasing. After the fracture plane's inclination to the maximum principal stress direction has reached 45° (or larger values), more and more such shear planes are formed, depending on the magnitude of the confining pressure. Correlatively, no maximum value for the load-bearing capacity of the sample is observed and the maximum principal stress keeps increasing as the axial strain is increased.

This absence of a peak strength together with the formation of multiple shear planes is characteristic of a ductile behavior. The formation of a simple fracture surface associated with a significant loss of "strength", i.e. the existence of a maximum load-bearing capacity, is characteristic of a brittle behavior. Brittle behavior involves the growth of micro- and macrocracks, which correspond to material discontinuities at the scale of the

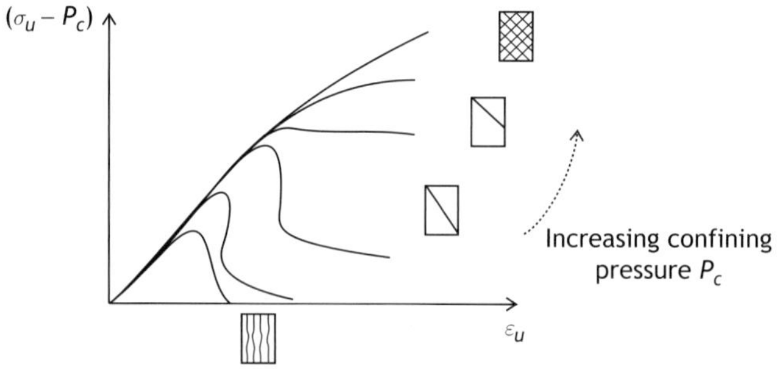

Fig. 8.12 The effect of a confining pressure on the fracture geometry (shown schematically in the rectangles) and on the differential stress (vertical axis) versus axial strain (horizontal axis) curves.

grains and crystals that make up the rock. Ductile behavior is associated with the development and the motion of dislocations in the structure of crystals at the atomic scale (e.g. Guéguen and Palciauskas, 1994). Ductility is one property that leads to the plastic behavior of a material, the other being friction along the surfaces of discontinuity. A combination of microfracture growth and friction is often referred to as cataclastic deformation. Hence plasticity involves both cataclastic and ductile deformation.

Thus, at low confining pressures and for low axial loads, the mechanical behavior of a rock is controlled mostly by elasticity and friction along free surfaces. Once the elastic limit has been reached, rock deformation involves, in addition to elasticity and friction, the growth of microcracks, but microcracks induce local stress concentrations that may lead locally to ductility. The higher the confining pressure, the larger the volumes where ductility develops. As the confining pressure continues to increase, the contribution of brittle microcrack propagation (the dissipation of potential, energy through surface energy) in the rock deformation process becomes less and less significant, whilst the deformation process is progressively dominated by the motion of dislocations, i.e. by ductility. This progressive evolution of the deformation process is called the brittle–ductile transition and is illustrated by fig. 8.12.

8.1.4 Acoustic emissions

We saw in section 7.1.1 that an unstable fracturing process involves the generation of some kinetic energy. This release of kinetic energy corresponds to an earthquake when the fracturing process involves deep geological formations. When it occurs within a rock sample and remains localized in both time and space, it is called an acoustic emission (e.g. Scholz, 1968; Lockner, 1993). Some acoustic emissions involve grain-scale failures and are efficient markers of rupture phenomena.

However, acoustic emissions may also be associated with the motion of dislocations (ductile behavior) or by slip on preexisting microcracks. Hence much can be learned from acoustic emissions, through their location, their polarization and their spectral content. For example, Schubnel *et al.* (2006) exploited measurements of acoustic velocity variations, as

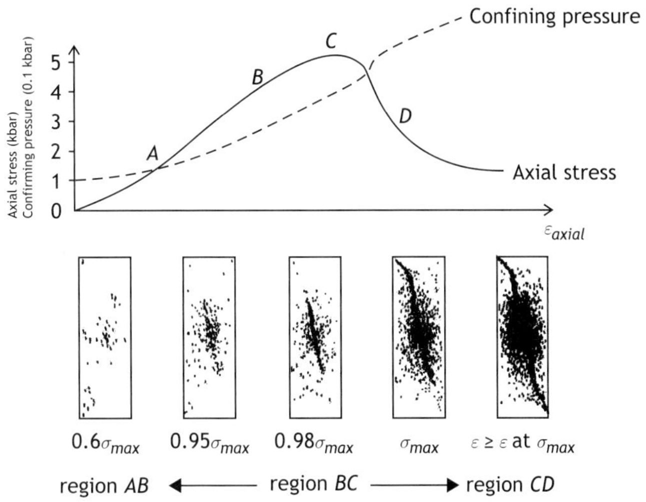

Fig. 8.13 Generation of acoustic emissions during triaxial testing (reproduced from Jaeger and Cook, 1979, with permission from Wiley).

detected from the observation of acoustic emissions, to quantify the development of failure in rock samples.

Another interesting application is the counting of acoustic emissions during rock deformation, or more precisely an analysis of the variation in rate of acoustic emission production during loading. This quantity depends on the past loading history of the rock and a consequence of this dependency, known as the Kaiser effect, is discussed in section 12.4.2, and it has led to the development of a stress determination technique.

Figure 8.13 illustrates how the mapping of acoustic emissions during loading helps characterize the mechanical behavior of rock sample as the differential stress magnitude $\sigma_a - P_c$ is increased progressively. It was produced by Hallbauer *et al.* (1973) during tests on an argillaceous quartzite tested under triaxial conditions but with a sealed pressure vessel. Because of the absence of flow in or out of the pressure vessel, the confining pressure increased regularly (the broken curve in fig. 8.13) whilst the axial piston was lowered within the pressure vessel.

During the initial linear response of the material (region *AB* of the axial stress versus axial strain curve), acoustic emissions are distributed throughout the sample. As the curve becomes nonlinear (region *BC* of the curve), the rate of acoustic emission production increases and the spatial density of events starts increasing, mostly within a limited planar volume. After the peak load has been reached (part *D* of the curve) the density of acoustic emission keeps increasing and outlines a surface that includes the planar surface already identified in the previous phase but involves also extensions apparently influenced by end effects. As this surface gets closer to the ends of the sample the load-bearing capacity of the sample drops rapidly.

Lockner *et al.* (1991) investigated the nucleation of shear fractures in granite samples in triaxial compression tests with a constant confining pressure. They were able to stabilize (i.e. render quasistatic) the shear fracture development using a servo-controlled system in which the monitoring parameter was the number of acoustic emissions observed per unit time. Because this parameter increases monotonically as the fracture grows, they obtained complete control of the shear band development.

Lockner *et al.* observed that, in all their tests, the fault plane nucleated abruptly at a point on the sample surface soon after reaching peak axial load. Then fracture evolved into a planar halfpenny-shaped crack, the growth of which was associated with acoustic emissions. These were located within a narrow process zone associated with the fracture front. For all these tests, a class II post-peak behavior was observed.

We may conclude from these observations that, depending on the nature of the rock and on the loading conditions, fresh shear surfaces may result from the progressive coalescence of many stable microcracks (Hallbauer *et al.*, 1973) or from the growth of one single fracture (Lockner *et al.*, 1991). The quantity of energy required for each of these two fracture growth processes is quite different, and this difference is the origin of the class I or class II rupture modes described by Wawersik and Fairhurst (1970). It has important consequences for the stability of the fracture development process and, at a larger scale, for the development of earthquakes, as will be discussed in chapter 11.

8.1.5 Time-dependent effects

Static and dynamic "elastic" response

Because rocks are always heterogeneous materials, if only because of pore space within them, their deformation below the elastic limit always involves some friction and an elastic deformation of the elementary rock constituents. However, friction implies the existence of shear stress thresholds in the rock (sections 2.4 and 10.1.2), so that some microcracks slip while others do not. As a result, when a rock sample is loaded in compression below its elastic limit, the stress–strain curve exhibits some level of nonlinearity, which depends on the nature of the rock but also on the loading characteristics.

This nonlinearity has led to the distinguishing of a dynamic and a static-mechanical behavior for rocks. The static behavior is observed in the laboratory for loading rates in the range 10^{-4}/s to 10^{-6}/s (Simons and Brace, 1965).

The dynamic response is that observed, for example, when one is measuring the velocity of body waves (P and S) in a rock sample for signals in the kHz to MHz frequency range. The analysis of resonances in cylindrical bars depends also on the dynamic response of the rock. Dynamic elastic constants may be derived from these velocity measurements, using equations (5.101) and (5.105), when the rock density is also known.

In addition, the range of strains observed in the dynamic regimes (10^{-7}–10^{-6}) is quite different from that observed in the static regimes (10^{-4}–10^{-3}). Thus nonlinearity effects are more pronounced for quasistatic measurements than for dynamic measurements.

Whilst discrepancies between static and dynamic elastic parameters measured at atmospheric pressure may reach several hundred percent, they are smaller than a few percent (i.e. within the precision of the measurements) when the confining pressure is larger than a certain amount that varies with rock type. For example, Simons and Brace (1965) measured

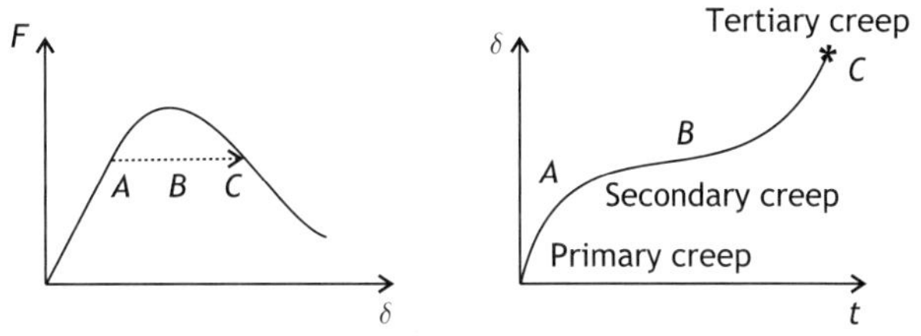

Fig. 8.14 Creep test showing the primary (*A*), secondary (*B*) and tertiary (*C*) creep phases.

a 3.64±0.1 km/s mean P wave velocity for a granite at atmospheric pressure but a 6.37±0.1 km/s velocity when the confining pressure reached 1000 MPa.

Because a confining pressure increases the normal stress that can be supported by microcracks, it diminishes the effect of friction on the overall mechanical response of the equivalent geomaterial and suppresses progressively a cause of nonlinearity. This is discussed further in section 9.1.1, where the effective medium theory is introduced.

Strain rate effects; creep and relaxation tests

When a rock is loaded beyond its elastic limit, microcrack propagation starts together with some local ductile deformation processes. Because both fracture growth and ductility depend on loading rates, so does the post-elastic rock behavior.

Three different experimental approaches are followed in the laboratory for investigating the effect of the strain rate on the post-elastic behavior of rock. Here, we refer to the macroscopic strain rate on the assumption that samples are homogeneous. We will ignore the possible local rotations that may result from the heterogeneity of the sample in the large-deformation domain.

With the first approach, complete stress–strain curves are produced at different strain rates in the range 10^{-3}/s to 10^{-6}/s. Results show that the lower the strain rate, the lower the elastic constants and the lower the maximum load-bearing capacity of the rock. However, each rock has its own strain rate dependence and there is no simple rule for modeling strain rate effects. For example, salt exhibits an elastic brittle behavior when tested at strain rates of the order of 10^{-3}/s to 10^{-4}/s but a more ductile behavior when strain rates fall below 10^{-6}/s.

With the two other approaches, either the load is kept constant and the strain is monitored (a creep test) or the strain is kept constant and the correlative variations of applied stresses are monitored (a relaxation test). A typical creep test is illustrated in fig. 8.14.

A sample is first loaded to a certain load level, larger than its elastic response but smaller than its maximum load-bearing capacity as defined for the corresponding confining pressure and strain rate. Then the load is maintained constant and the axial strain is monitored.

First, when the load is kept constant (point A), the axial strain keeps increasing but at a progressively decelerating rate. This is known as primary creep. Then (region B) the deformation process keeps on going but at a constant strain rate. This is known as secondary creep. When a critical axial strain value is reached, the creep process accelerates until complete rupture of the sample occurs with the appearance of kinetic energy (region *C*). If the deformation rate is to be kept quasistatic, the axial load must be progressively decreased until a residual value is reached that depends on the confining pressure magnitude.

If the loading is stopped before phase C is reached and the rock sample is unloaded then some residual strain is observed, which demonstrates that creep involves nonelastic deformation processes.

Creep is characterized by its strain rate, and various models have been proposed for its characterization. For example, the secondary (steady-state) creep of salt is characterized by (Dusseault and Fordham, 1993)

$$\dot{\varepsilon}_{ss} = A\left(\frac{\sigma_1 - \sigma_3}{\sigma_0}\right)^n \exp\left(\frac{-Q_0}{RT}\right) \tag{8.3}$$

where $\dot{\varepsilon}_{ss}$ is the strain rate measured in the direction of the applied maximum stress component σ_1, for a confining pressure σ_3; σ_0 is the differential stress magnitude for which the micromechanism at the origin of the creeping behavior starts operating; Q_0 is the activation energy of the corresponding mechanism, R is the universal gas constant and T is the temperature in degrees kelvin. The constants A and n are determined from the creep experiments.

This is just one of the many formulations proposed for creeping laws. More on creep and relaxation tests may be found in Dusseault and Fordham (1993).

Load–unload cycles and fatigue

Because the development of rock failure depends on the time elapsed, rock samples may be loaded beyond their elastic limit and then unloaded. This unloading brings the rock back to the elastic domain but its elastic characteristics are somewhat changed, depending on the development of the failure process when the rock was loaded beyond the elastic limit. When such loading cycles are applied repeatedly to a sample, each cycle introduces some additional nonelastic deformation so that the rock gets progressively weaker and weaker.

This process of repeated loading cycles applied beyond the elastic limit and leading progressively to the complete collapse of the rock is called fatigue loading. The number of loading cycles that may be applied before the sample fails completely depends on the amplitude of the cycles. This is clearly illustrated in fig. 8.15, produced by Haimson and Kim (1972), which presents the results of fatigue tests conducted on various samples, loaded previously in a stable manner, beyond their uniaxial compressive strength.

The results of Haimson and Kim demonstrate first that a rock sample that has been loaded in a stable manner beyond its maximum load-bearing capacity maintains a significant compressive strength. They also suggest that the number of loading cycles that leads to fatigue failure depends on the maximum load reached at each cycle. Of interest is the test that was terminated after more than 40 000 cycles without fatigue failure. This test clearly demonstrates that even though the sample had been taken beyond its peak load-bearing capacity and many microcracks were present, the sample retained a significant elastic limit and compressive strength.

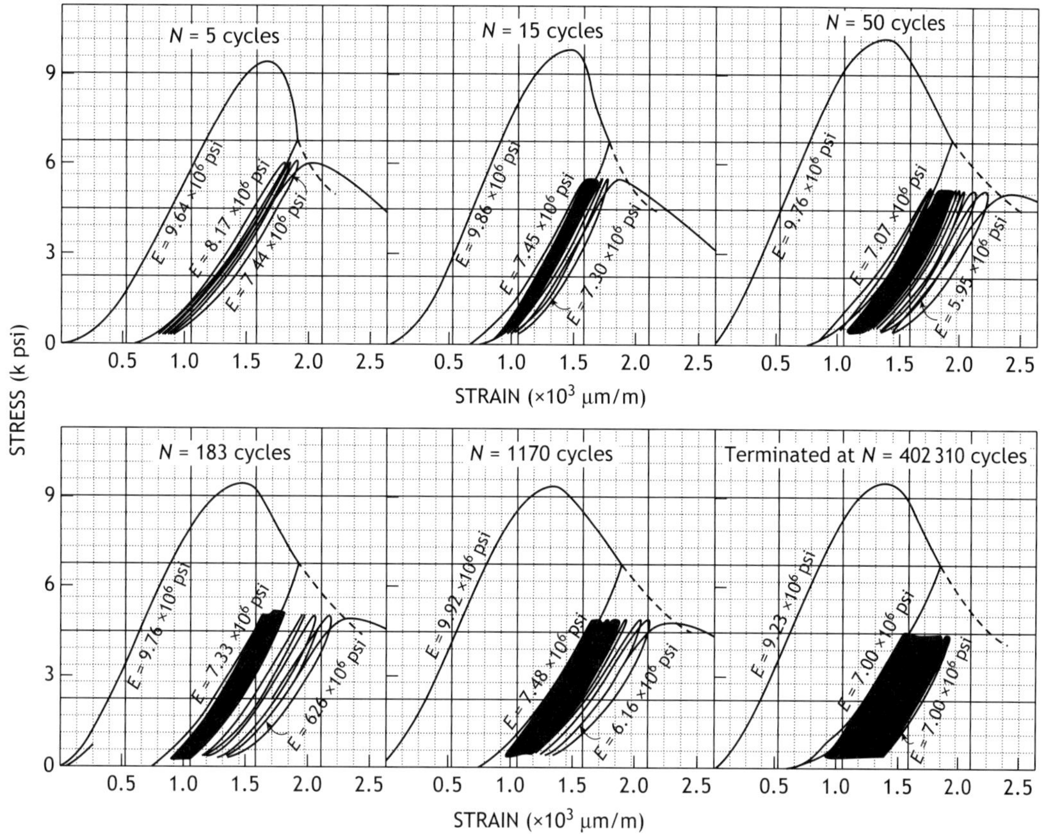

Fig. 8.15 Fatigue loading of a sample loaded beyond its peak bearing capacity (reproduced from Haimson and Kim, 1972, with permission from ASCE).

8.1.6 Influence of pore pressure and drainage conditions

When the pore space of rocks is filled with fluids, these fluids are generally under pressure. We discuss in chapter 12 various geomechanical issues related to the interaction between solids and fluids. In the present subsection, only the experimental aspects relating to laboratory compression tests conducted on rock samples saturated by a fluid under pressure are presented.

Figure 8.8 presented a triaxial apparatus that provides a means to monitor independently the externally applied loads (a confining pressure and an axial load) and the internal pore pressure that exists in the cylindrical rock sample. Note that, with this equipment, the pore pressure is only monitored at the boundary of the rock sample. Some testing equipment has been developed for hollow cylindrical specimens which insures that the pore pressure is monitored throughout the inner cylinder of a sample. However, as shown in sections 5.5 and 5.6.4, this geometry introduces nonhomogeneous stresses in the sample, so that the data interpretation remains controversial.

Triaxial compression tests are often run with two different boundary conditions on the fluid, undrained or drained. With undrained conditions, no flow is allowed in or out of the rock sample. With drained conditions, the pore pressure remains constant throughout the pore space during the duration of testing. However, because elastic and nonelastic deformation processes involve changes in pore volume, the deformation rate in drained tests must be slow enough to let the fluid flow in or out of the specimen in order to accommodate these pore volume variations. Examples of pore volume variations during triaxial compression tests on Berea sandstone and Indiana limestone are shown in figs. 8.10 and 8.11.

Experience shows that, for most rocks, strain rates slower than 10^{-5}/s are required to fulfill truly drained conditions. Then, for these slow strain rates, the stress–strain curves can be defined in terms of an effective stress $\tilde{\sigma}'$:

$$\tilde{\sigma}' = \tilde{\sigma} - P_0 \tilde{1} \tag{8.4}$$

i.e. the curves are the same (including the maximum load-bearing capacity) for all combinations of stress $\tilde{\sigma}$ and pore pressure P_0 that yield the same effective stress tensor $\tilde{\sigma}'$. This is known as the effective stress principle; it was first demonstrated for the case of soils (van Terzaghi, 1943).

In effective stress theory an effective maximum shear stress component τ_m and an effective isotropic component σ'_m are defined:

$$\tau'_m = [(\sigma_1 - P_{0)} - (\sigma_3 - P_0)] = \sigma'_1 - \sigma'_3 = \sigma_1 - \sigma_3 = \tau_m \tag{8.5}$$

$$\sigma'_m = \tfrac{1}{3}[(\sigma_1 - P_0) + (\sigma_2 - P_0) + (\sigma_3 - P_0)] = \tfrac{1}{3}(\sigma'_1 + \sigma'_2 + \sigma'_3) \tag{8.6}$$

For faster strain rates, dilatancy effects associated with the development of microcracks render heterogeneous the pore pressure distribution within the pore space. Then the interpretation of the effect of pore pressure on the stress–strain curves requires additional considerations of the fluid flow, the pore space variation rate and the solid rock deformation process, as will be discussed in chapter 12.

8.1.7 Influence of temperature

Experimental laboratory investigations on the effect of temperature on the mechanical behavior of rocks under compression have addressed two issues: the development of ductile behavior and the role of grain-scale heterogeneity.

For example, Griggs *et al.* (1960) conducted triaxial tests on various rocks (granite, peridotite, pyroxenite, basalt and marble) under various moderate to high confining pressures, at temperatures in the range 25 °C to 800 °C. For all the rocks they observed an increase in ductility as the temperature increased (e.g. fig. 8.16), i.e. the decrease in load-bearing capacity with increasing axial strain becomes more marked as the temperature increases. However, simultaneously, an increase in temperature was found to result in a decrease in the peak load-bearing capacity.

So, more generally, the higher the temperature, the more ductile the rock but, simultaneously, the weaker the rock.

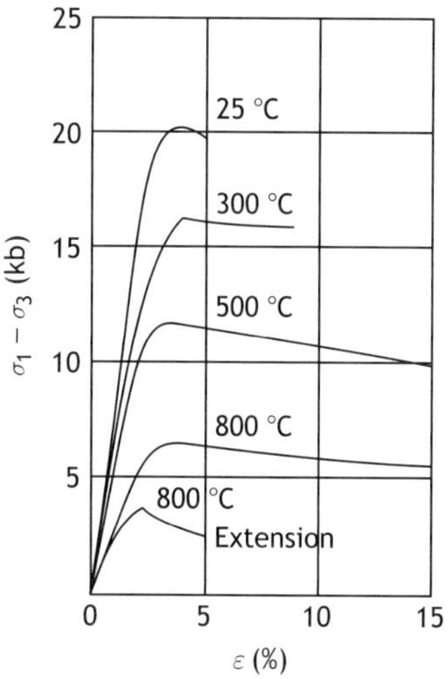

Fig. 8.16 Effect of temperature on the triaxial compression behavior of granite under 500 MPa confining pressure (reproduced from Jaeger and Cook, 1979, with permission from Wiley).

When crystalline rocks such as granite are heated up slowly (e.g. at less than 1 °C/min in Simmons and Cooper, 1978) under low confining pressures, so as to avoid macroscopic cracking linked to the thermal gradient in the rock (see section 12.5), some microcracking develops because of the mismatch between the thermal expansion coefficients of the various rock constituents.

For example, Wang *et al.* (1989) heated granite samples up to different maximum temperatures and then cooled them back down to room temperature, under low to moderate confining pressures. Microcracking was observed either through direct microscopic observations or by counting acoustic emissions or by direct P wave velocity measurements. Wang *et al.* noticed that the threshold temperature increase required for initiating acoustic emissions was $90 \pm 10\,°C$ at 7 MPa confining pressure, $115 \pm 10\,°C$ at 28 MPa confining pressure and $135 \pm 10\,°C$ at 55 MPa confining pressure.

Hence, for the four granites they tested, Wang *et al.* proposed a linear relationship between the confining pressure and the temperature variation threshold required for generating microcracks:

$$T = (85 \pm 10\,°C) + (0.9 \pm 0.4\,°C/MPa)P \tag{8.7}$$

where T is the temperature in degrees celsius and P is the confining pressure in MPa.

8.1.8 Compaction of porous rocks

The compaction of rock occurs in highly porous rocks when the pore space is reduced permanently by the relative motion of the constitutive solid grains and also by the crushing of the grains. When the rock is subjected to high enough stresses, grains may loosen up or break and then be pushed and crushed to fill part of the pore space.

In sandstones, in which the size of the pores is of the same order of magnitude as that of the grains, pore collapse results mostly from the reorganization of the grain-arrangement structure.

In materials such as chalk, in which the size of the grains may be an order of magnitude smaller than that of the pores, the pore collapse mechanism becomes very important and depends strongly on the displacement of the fluids that fill the pores.

Pore collapse may occur under purely hydrostatic macroscopic loading conditions, but from a micromechanics perspective pore collapse involves shear forces between the grains. Further, when stresses on the constitutive grains become large enough, some grain crushing may occur and this leads to a further reduction in pore space. Hence compaction may occur also because of high shear stresses.

Rock compaction may be viewed as the opposite to dilatancy. It often corresponds to a well-distributed nonelastic volume decrease within a large volume of rock. But it may also lead to localized compaction bands that develop in a direction that may be subperpendicular to the maximum principal stress direction or inclined to it, depending on the maximum shear stress magnitude.

Wong and Baud (2012) conducted an extensive review of the micromechanisms leading to compaction in porous rocks and discussed how the trace of the spherical stress component, $\sigma_m = (\sigma_1 + \sigma_2 + \sigma_3)/3$ and the maximum applied shear stress, $\tau_m = \sigma_1 - \sigma_3$ influence the onset of compaction. They call the set of values σ_m and τ_m for which compaction is observed the compactive yield C^*. The lowest hydrostatic pressure magnitude ($\tau_m = 0$) for which a permanent decrease in volume is observed is called P^*. They observed that, for dry sandstone, compaction develops depending on the values of σ_m and τ_m as normalized with respect to the hydrostatic compaction value P^*, according to the following relationship:

$$\frac{(\sigma_m/P^* - \gamma)^2}{(1-\gamma)^2} + \frac{(\tau_m/P^*)^2}{\delta^2} = 1 \tag{8.8}$$

with $\gamma = 0.5$ and δ between 0.5 and 0.7 for the various sandstones they tested. They further noted that P^* depends on both the rock porosity and the grain size. A map of the (τ_m, σ_m) values for which compaction is observed is called the *compactive yield cap*. For liquid-saturated sandstones, the same compactive yield cap is observed except that σ_m and τ_m are replaced by the effective spherical component σ'_m and the effective differential stress $\tau'_m = \tau_m$, as defined respectively by equations (8.6) and (8.5).

The modeling of compaction through plastic behavior will be discussed in section 9.2.

8.2 Laboratory testing of soil shear strength

We give the name "soils" to all uncemented clastic materials of geological origin. Their tensile strength is nil. Their mechanical behavior in compression is governed by contact forces between the grains and, just as importantly, by the flow of fluids in between the grains.

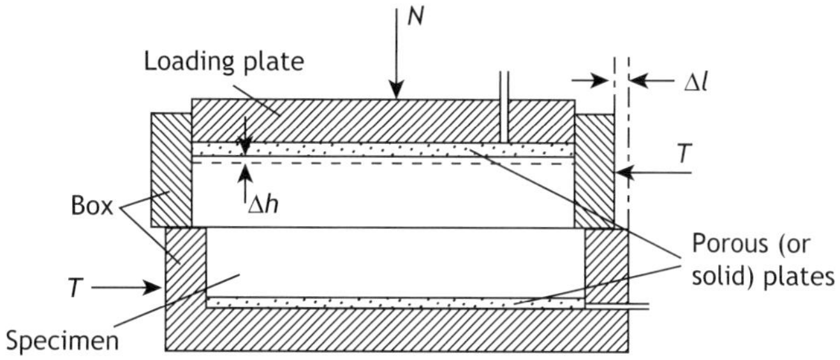

Fig. 8.17 The Casagrande direct shear box (after Craig, 1987, fig. 4.3, with permission from Taylor & Francis).

We present here some experimental aspects of the mechanical behavior of soils under shear, when all the effective principal stresses are compressive. Some aspects of consolidation theory are discussed further in chapter 12.

Determining significant characteristics of the shear strength of soils in the laboratory requires *representative* samples. This implies that neither the fluid content nor the grain structure of such a sample has been perturbed. In particular, the problem of clay swelling may become significant if proper attention has not been given to the collection, handling and storage of samples.

8.2.1 Experimental procedures

The shear strength of soils may be investigated in the laboratory with either a direct shear box or a triaxial cell.

The Casagrande direct shear box

The principle of the Casagrande direct shear box is shown in fig. 8.17. The apparatus includes upper and lower half-boxes of cylindrical or parallelepiped shape. The lower half-box is fixed and includes, at the bottom, a porous plate on which the soil specimen is set. At the top of the upper half-box surrounding the soil sample, a second porous plate is set to ensure proper drainage. On top of the upper porous plate a loading plate provides a means for applying the required normal load N. A horizontal load T is applied to the upper half-box so as to impose a controlled horizontal displacement.

During shearing, both the normal and the tangential displacements of the upper half-box are continuously monitored, together with the magnitudes of the applied normal and shear forces. The pore pressure, which is monitored at both ends of the specimen, may be maintained constant (drained conditions) or its variations may be monitored under no flow (undrained conditions).

A main shortcoming of this equipment is that the shear stress distribution within the plane where shear failure is assumed to occur is not uniform, since the horizontal force is applied at a distance from this plane. A second shortcoming is that the area of the sheared portion of the sample varies during testing.

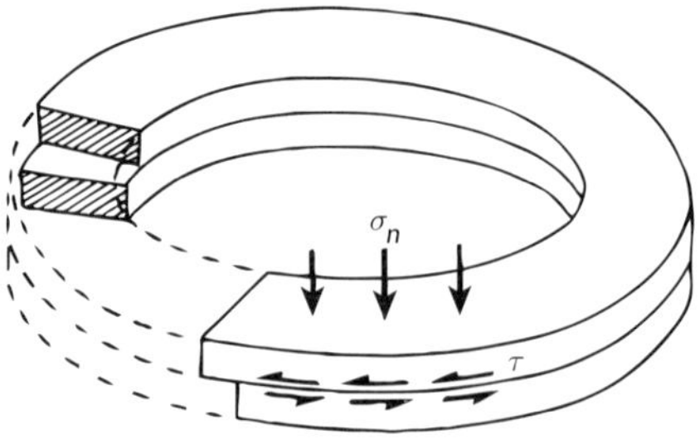

Fig. 8.18 The ring shear box (after Craig, 1987, fig. 4.23, with permission from Taylor & Francis).

The ring shear box

The ring shear box was designed to correct the shortcomings of the Casagrande shear box and to provide a way of generating large shear displacements. Its principle is shown in fig. 8.18.

An annular sample is sheared under a given normal stress within the horizontal plane through the rotation of the upper half of the sample with respect to the fixed lower half. The rotation rate must remain slow enough to allow full drainage of the sample. The shear displacement may be as large as deemed necessary, provided that there is no loss of shearing area. The only shortcoming in this apparatus is the change in shear velocity depending on the distance to the center of the ring. In order to ensure a nearly uniform velocity the inner ring diameter must be sufficiently large and the ring thickness must remain small compared with the inner ring diameter.

The triaxial cell

The triaxial cell used in soil mechanics is somewhat similar in principle to that used in rock mechanics (see fig. 8.8). However, the procedure followed for loading the specimen up to the original hydrostatic conditions may follow different paths: either the interstitial fluid in the specimen is maintained at a constant value and the sample is allowed to consolidate or no flow is allowed out of the sample and application of the hydrostatic external load results in an increase in pore pressure. So, three different testing conditions are used: an unconsolidated sample and undrained conditions during shearing; a consolidated sample and undrained conditions during shearing; a consolidated sample and drained conditions during shearing.

During testing, the sample's cross-sectional area varies and must be evaluated in order to determine the axial stress on the sample. If A_0 and V_0 are respectively the original area and original volume of the sample, the cross-sectional area A after an axial strain $\varepsilon_a = dl/l_0$ and a volumetric strain $\varepsilon_v = dV/V_0$ have occurred is

$$A = A_0 \frac{1 - \varepsilon_v}{1 - \varepsilon_a} \tag{8.9}$$

Collecting the flow of interstitial fluid during testing (either during consolidation or during shearing) provides a direct measurement of the change in pore space during the deformation process.

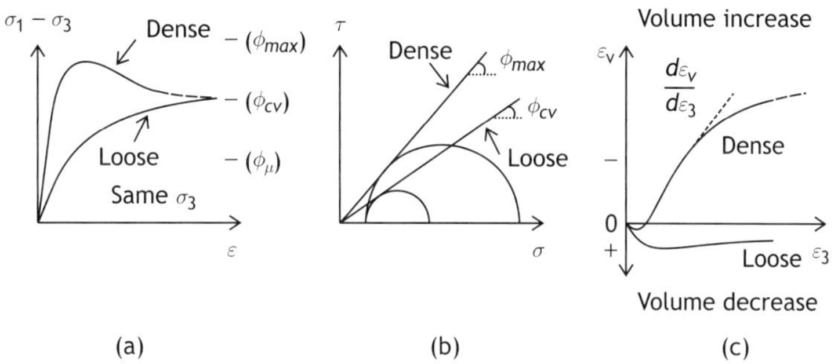

Fig. 8.19 Stress–strain curves from drained triaxial tests on sand (after Craig, 1987, fig. 4.9, with permission from Taylor & Francis).

8.2.2 The shear strength of sand

The shear strength characteristics of sand may be determined either from any shear box configuration or from drained triaxial tests. These characteristics are the same for dry and for saturated sand provided that there is no pore pressure variation during shearing. However, for wet but undersaturated sand the shear strength depends strongly on the water content. Here, only fully saturated sand is discussed.

Typical stress–strain curves for loose and packed sand are shown in fig. 8.19. For dense sand, because of the interlocking between the grains, a peak load-bearing capacity is observed. This does not occur for loose sand. Both dense and loose sand exhibit the same shear behavior after the axial strain has reached some characteristic value.

Plotting the Mohr circles that correspond to the effective stress conditions at peak load provides a means to define the angle of maximum shearing resistance ϕ_{max} for dense sand (fig. 8.19(b)). Similarly, for both dense and loose sand, a critical angle ϕ_{cv} of ultimate (sometimes called residual) strength may be defined. Typical values for the critical angle range from 25° to 35°, whilst those for the maximum angle range from 35° for uniform sand to 50° for sandy gravel.

For dense sand, a reduction in the degree of interlocking produces a volume increase associated with shear; this phenomenon is called *dilatancy*, as for the volume increase associated with the shearing of brittle rocks. No dilatancy is observed for the shearing of loose sand, which, on the contrary, exhibits some small volume decrease before the peak load is reached.

For sand, the dilatancy is characterized by the dilation angle ψ, which is defined either in terms of the maximum ($d\varepsilon_1$) and minimum ($d\varepsilon_3$) principal strain increments or in terms of the volumetric strain $d\varepsilon_v$ and the maximum shear strain $d\gamma$:

$$\sin\psi = \frac{d\varepsilon_1 + d\varepsilon_3}{d\varepsilon_1 - d\varepsilon_3} = -\frac{d\varepsilon_v}{d\gamma} \tag{8.10}$$

8.2.3 The shear strength of clay

Even more than for sand, the properties of clays depend strongly on their water content. Only fully saturated clays are considered here.

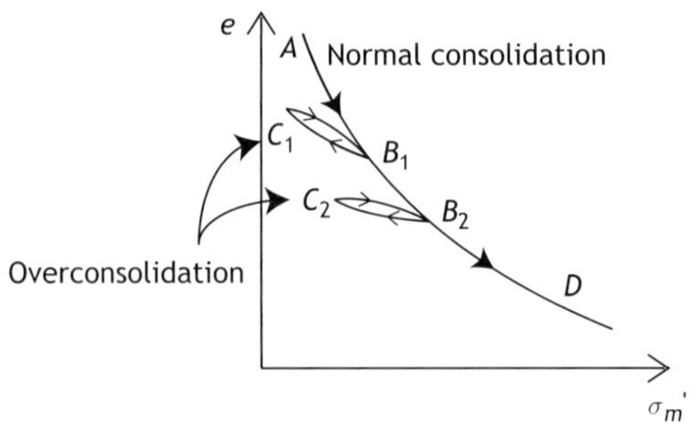

Fig. 8.20 Relationship between void ratio and the effective spherical stress component for clay.

Normally consolidated and overconsolidated clays

In a triaxial cell, if the loading under hydrostatic conditions of a saturated clay sample is conducted sufficiently slowly to let fluids flow out of the pore space then the void ratio varies with time. This is known as *isotropic consolidation*.

The void ratio after consolidation may be derived from measurements of bulk density (equation (1.9)). The relationship between void ratio and effective stress depends on the loading history of the clay. When the effective spherical stress p' acting on the sample is the maximum that has ever been applied to the sample, the clay is said to be *normally* consolidated. However, if at some time in the past an effective spherical stress larger than the present one has been applied to the sample then the clay is said to be *overconsolidated*.

The ratio of the maximum effective stress applied in the past to the present effective stress is called the overconsolidation ratio. Overconsolidation is usually the result of geological factors, e.g. erosion or the effect of past glaciations.

Figure 8.20 illustrates the relationship between, on the one hand, the void ratio e or the specific volume v and, on the other hand, the effective spherical stress component p' or the minimum effective principal stress σ_3' during consolidation.

For a normally consolidated clay, isotropic consolidation occurs when an external isotropic stress component increment is imposed on a drained clay sample (curve AB_1). Once at B_1, if the effective stress is reduced then the void ratio increases but remains smaller than the value for the normally consolidated sample under similar external loading conditions. The sample becomes overconsolidated (curve B_1C_1). Once at C_1, if the isotropic external load is increased again while the pore pressure remains constant then the void ratio decreases back to the value at B_1. Once the sample is at B_1, if the external loading is still further increased then the associated decrease in void ratio follows that of normally consolidated clay. If this unloading–loading cycle is later reproduced, say in the cycle B_2C_2 then the mean slope is reduced further.

It should be noted that the sampling of clay most often modifies the inner structure of the clay-solid assembly in such a way that the void ratio increases and the density decreases. Measuring the density of undisturbed overconsolidated clay requires special care. It may be conducted *in situ* by direct determination of the vertical stress component magnitude (see chapter 14).

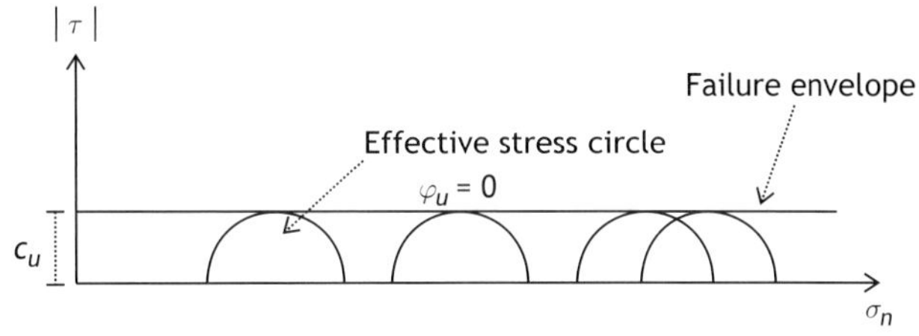

Fig. 8.21 The failure envelope for unconsolidated-undrained tests on saturated clays.

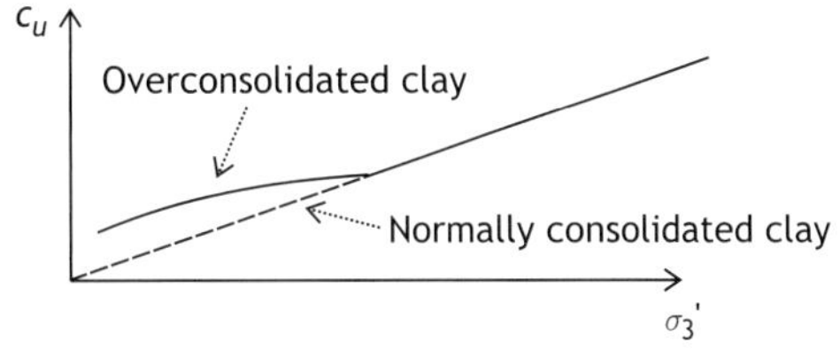

Fig. 8.22 The maximum load bearing capacity C_u of clay as a function of the effective spherical stress σ_3' in consolidated-undrained triaxial tests (after Craig, 1987, fig. 4.12, with permission from Taylor & Francis).

For unconsolidated-undrained triaxial tests on saturated clay, any increase in external isotropic load results in an equal internal pore pressure increase. Hence, the effective minimum principal stress remains constant. When triaxial tests are conducted for different confining pressures reached under undrained conditions, a Mohr circle of the total stress acting on the sample may be drawn for each test when failure initiates (fig. 8.21). The envelope to all these Mohr circles is called the failure envelope and is a horizontal straight line since failure for all tests occurs for the same effective confining pressure.

The maximum differential stress $\sigma_1 - \sigma_3$ (or $\sigma_1' - \sigma_3'$) corresponding to failure for unconsolidated-undrained conditions depends on the void ratio and on the structure of the solid grain arrangement.

During consolidated-undrained triaxial tests, consolidation results in a decrease in the void ratio and a new solid grain arrangement such that the corresponding undrained shear strength increases. When the maximum load-bearing capacity observed for consolidated-undrained triaxial tests is plotted as a function of the effective confining pressure σ_3', a straight line is obtained for normally consolidated clays (fig. 8.22). For overconsolidated clay the relationship is nonlinear as long as the clay remains overconsolidated. It becomes linear when the overconsolidation ratio reaches unity.

Sensitivity of clays

The shear strength of some clays is very sensitive to the structure of their solid-grain assembly, i.e. to the remoulding that may occur either artificially through human activity or naturally, as in landslides. The sensitivity of clay is defined as the ratio of the undrained strength in the undisturbed state and the undrained strength of the remoulded state with the same water content as the undisturbed state. Remoulding for testing purposes may be brought about by kneading or wedging. The sensitivity for most clays is within 1 to 4. When the sensitivity is somewhere between 4 and 8, the clay is described as being *sensitive* and as *extra sensitive* for a sensitivity index between 8 and 16. *Quick clays* have a sensitivity larger than 16 and as high as 100.

8.3 Failure criteria for geomaterials in compression

We saw in chapter 7 that using the concept of the mean stress in an equivalent continuum in order to define the tensile strength of rock may be quite misleading because of local stress concentrations linked to the geometry of the pore space. Further, because soils are granular materials with no cement between the grains their tensile strength is always considered to be zero.

However, for geomaterials (rocks and soils) loaded in compression, the concept of the mean stress acting in an REV has proved quite useful for investigating both their pseudo-elastic and their post-elastic behavior. We examine below the main failure (or yield) criteria applied for evaluating the compressive and shear strengths of geomaterials under compression.

8.3.1 The Tresca failure criterion

The Tresca criterion is a yield criterion commonly used in the plasticity theory of metals. It was shown to be efficient for evaluating the strength of geomaterials under compression (Salençon, 1974, p. 5). It assumes that yield occurs when the maximum differential stress $\sigma_1 - \sigma_3$ reaches a critical value:

$$\sigma_1 - \sigma_3 = 2k \tag{8.11}$$

This yield criterion is independent of both the minimum and the intermediate principal stress magnitudes and assumes the material to be isotropic. It is sometimes used to analyze the deformation of ductile rocks. It represents efficiently the yield strength of clays under unconsolidated-undrained triaxial conditions, as shown in fig. 8.21.

For the Tresca criterion, the failure envelope plotted in the Mohr plane corresponds to a horizontal straight line parallel to the horizontal axis, since all Mohr circles exhibit the same radius, $(\sigma_1 - \sigma_3)/2$.

The orientation of the failure plane is such that its normal makes the angle $2\theta = \pi/2$ with respect to the maximum principal stress orientation, i.e. it is inclined at 45° to the maximum or minimum principal stress directions and is parallel to the intermediate principal stress direction (see section 3.4.1).

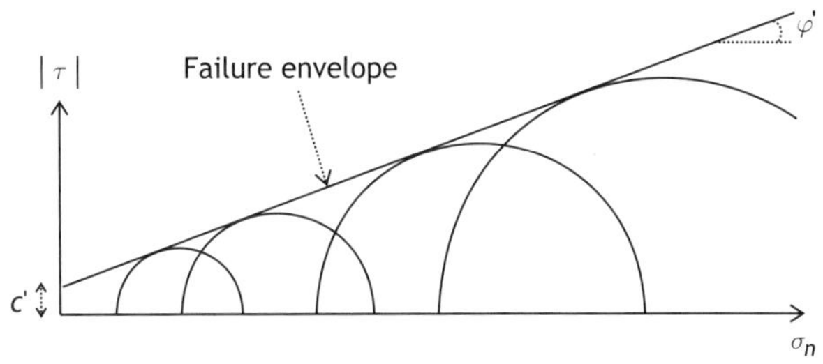

Fig. 8.23 The failure envelope for consolidated-drained saturated clays and the Coulomb failure criterion.

8.3.2 The Coulomb failure criterion

The Coulomb failure criterion is used for characterizing the onset of failure in both soils and rocks, with slightly different physical significance for its parameters depending on whether soils or rocks are under consideration. We discuss first its application to soils.

Soils

The shear strength of clay expressed in terms of effective stress components may be determined by means of either consolidated-undrained triaxial tests run with continuous monitoring of the pore pressure magnitude or consolidated-drained triaxial tests run sufficiently slowly to insure a uniform and constant pore pressure within the sample. Triaxial tests are run at various effective confining pressures. Mohr circles that correspond to the effective stress state in the sample at failure are plotted for the various effective confining pressures. They all are tangent to a common failure envelope (see fig. 8.23).

This failure envelope ($|\tau| = f(\sigma_n)$ is a straight line, with slope tan φ', which intersects the vertical axis at c'. Its equation is simply

$$|\tau| = c' + \sigma_n' \tan \varphi' = c' + (\sigma_n - P_0) \tan \varphi' \tag{8.12}$$

where c' and φ' are respectively the cohesion and friction angles for the effective stress theory and P_0 is the pore pressure. Hence, the Tresca failure criterion is equivalent to a Coulomb failure criterion with zero friction angle.

The friction angle for clay ranges between 20° and 35°. As shown in fig. 8.19, the shear strength of sand also obeys a Coulomb failure criterion with zero cohesion and friction angles ranging from 25° to 50°, depending on the grain size distribution and the shape of the grains.

For normally consolidated clays the cohesion is zero but it may reach 0.3 MPa for overconsolidated clays. It is generally somewhat smaller than the unconsolidated-undrained shear strength, which corresponds to shear under constant pore volume and may reach values as high as 1.5 MPa for very stiff clays.

Hence, whilst the unconsolidated-undrained shear strength for soils corresponds to the shear stress (defined at the macroscopic scale of the REV) required for modifying the solid grain arrangement within the REV, the cohesion defined from drained triaxial tests represents the shear stress that may be supported by the material when the effective normal stress supported by the failure plane is nil. Such loading conditions would require the effective minimum principal stress in the REV to be negative, which is impossible since the tensile strength of soils is nil. Cohesion, as defined for the Coulomb failure criterion applied in soil mechanics, is simply one of the two parameters required for characterizing the failure of the material under compression and cannot be deduced independently from the friction angle.

Rocks

As discussed in section 8.1.3, the peak load-bearing capacity of rock as measured in triaxial tests increases with the confining pressure. When the different states of stress observed at peak load for these various triaxial tests are plotted on a Mohr plane, the various Mohr circles are found to be tangent to a line (the failure envelope).

As a first approximation, when the domain of tested confining pressures remains limited, the failure envelope may be approximated by a straight line in a manner somewhat similar to that discussed for soils. Hence, in rocks, as in soils, the Coulomb failure criterion implies that failure (peak load) occurs when

$$|\tau| = S_0 + \mu\sigma_n \tag{8.13}$$

which can also be written as

$$|\tau| - \mu\sigma_n = S_0 \tag{8.14}$$

where $\mu = \tan\varphi$ is the friction coefficient and S_0 and φ are the cohesion and the internal friction angle of the rock. The criterion applies to dry triaxial tests (indicated by unprimed letters for the cohesion and the friction angle) as well as to drained triaxial tests. In the latter case the total stresses are replaced by effective stresses in a manner similar to that for the Coulomb criterion for soils.

In section 3.4.1 we showed that the normal and shear stress components supported by a plane parallel to the intermediate principal stress direction and whose normal makes an angle β to the maximum principal stress direction are

$$\sigma_n = \frac{\sigma_1 + \sigma_3}{2} + \frac{\sigma_1 - \sigma_3}{2}\cos 2\beta \tag{8.15}$$

$$\tau = -\frac{\sigma_1 - \sigma_3}{2}\sin 2\beta \tag{8.16}$$

Hence the left-hand side of equation (8.14) may be rewritten as follows:

$$|\tau| - \mu\sigma_n = \frac{\sigma_1 - \sigma_3}{2}(\sin 2\beta - \mu\cos 2\beta) - \frac{\mu}{2}(\sigma_1 + \sigma_3) \tag{8.17}$$

The plane orientation for which $|\tau| - \mu\sigma_n$ reaches its maximum value is such that

$$\tan 2\beta = \frac{-1}{\mu} \tag{8.18}$$

so that $\pi/2 \leq 2\beta \leq \pi$.

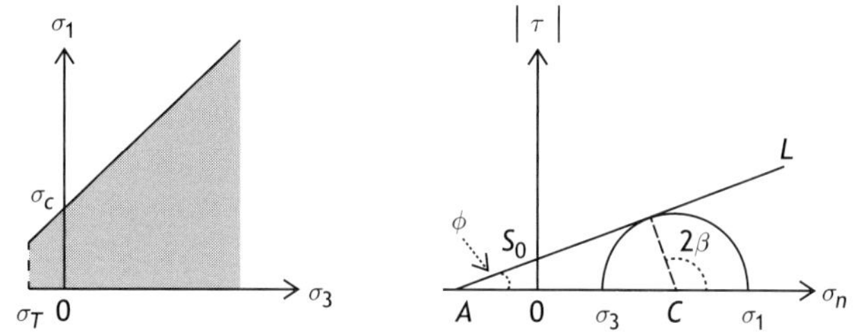

Fig. 8.24 The Coulomb failure criterion as expressed in the Mohr plane (right) and in terms of the maximum and minimum stress components (left).

However, $\sin 2\beta = (1+\mu^2)^{-1/2}$ and $\cos 2\beta = -\mu(1+\mu^2)^{-1/2}$, so the maximum value of the left-hand side of equation (8.14) is given by

$$|\tau| - \mu\sigma_n = \frac{\sigma_1 - \sigma_3}{2}(1+\mu^2)^{-1/2} - \frac{\mu}{2}(\sigma_1 + \sigma_3) \tag{8.19}$$

Failure occurs when this maximum value is equal to S_0 and, from equation (8.14), we conclude that failure occurs when

$$\sigma_1[(\mu^2+1)^{1/2} - \mu] - \sigma_3[(\mu^2+1)^{1/2} + \mu] = 2S_0 \tag{8.20}$$

Equation (8.20) expresses the Coulomb failure criterion as a function of the maximum and minimum principal stress components, for a given intrinsic friction coefficient and a given cohesion.

Let us note that, for failure under uniaxial compression, $\sigma_3 = 0$ and $\sigma_1 = \sigma^c$. So, according to the Coulomb criterion, a relationship exists between, on the one hand, the uniaxial compression strength and, on the other hand, the internal friction angle and cohesion. Further, if it is assumed that a tensile strength σ_T may be defined for the rock under consideration, then $\sigma_3 \geq \sigma_T$. Accordingly, with the convention that $\sigma_1 \geq \sigma_2 \geq \sigma_3$, the domain of values for the minimum and maximum principal stress components for which the material is not under failure lies in the shaded area shown on fig. 8.24.

Let us also observe that the orientation of the failure plane with respect to the maximum principal stress direction depends on the internal friction angle ϕ. Indeed, from equation 8.18, the values of beta at failure are

$$2\beta = \arctan(-1/\mu) + k\pi; \quad k = 1, 2, \ldots \tag{8.21}$$

or, in terms of the internal friction angle ϕ,

$$\beta = \pm(\pi/4 + \phi/2) \tag{8.22}$$

which indicates that two possible directions exist for the orientation of the shear plane (recall that β characterizes the orientation of the normal to the fracture plane with respect to the maximum principal stress component). These two directions are called conjugate directions.

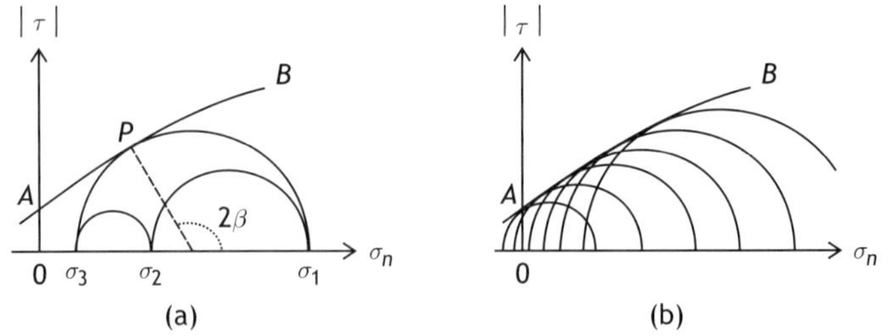

Fig. 8.25 The Mohr–Coulomb failure criterion.

The Coulomb criterion assumes that the intermediate principal stress component has no effect, since failure is assumed to occur when the largest value for $|\tau| - \mu\sigma_n$ reaches the critical value S_0. This maximum value involves only the Mohr circle with diameter $\sigma_1 - \sigma_3$ and assumes that the failure plane is parallel to the intermediate principal stress direction. The possible influence of the intermediate principal stress magnitude is discussed in section 8.3.4.

8.3.3 The Mohr–Coulomb and Hoek and Brown failure criteria

The Coulomb criterion suffers from a few drawbacks. We mentioned earlier that the failure of rock under uniaxial compressive loads does not occur along conjugate shear planes, as predicted by this failure criterion, but rather through axial splitting of the rock samples caused by indirect mode I fracture extensions. Further, for triaxial stress conditions we also mentioned that the larger the confining pressure, the more inclined is the failure plane to the maximum principal stress direction, a feature not predicted by the Coulomb criterion. Hence, the Coulomb failure criterion has been found experimentally to represent adequately the failure of rock for only limited domains of minimum principal stress magnitudes. For larger domains of minimum principal stress magnitudes, an empirical generalization was been proposed by Mohr; it is known as the Mohr–Coulomb failure criterion (e.g. Jaeger and Cook, 1979) and takes the general form

$$|\tau| = f(\sigma_n) \tag{8.23}$$

or alternatively

$$\sigma_1 = g(\sigma_3) \tag{8.24}$$

where the functions $f(\sigma_n)$ or $g(\sigma_3)$ are determined experimentally by a series of triaxial experiments conducted in a large domain of confining pressures (fig. 8.25).

Various explicit formulations have been proposed for the functions f and g. A commonly adopted formulation is that proposed by Hoek and Brown (1988), which covers a large variety of rocks. It is expressed as

$$\sigma_1 = \sigma_3 + \sigma^c \left(m\frac{\sigma_3}{\sigma^c} + 1 \right)^{1/2} \tag{8.25}$$

where m and σ^c are two fitting parameters.

For uniaxial compression tests ($\sigma_3 = 0$), equation (8.25) yields the value of σ^c for the maximum principal stress σ_1. An analysis of the published strength data led Hoek and Brown to propose the following values for m:

$5 < m < 8$ for carbonate rocks with well-developed crystal cleavage (dolomite, limestone, marble);
$4 < m < 10$ for lithified argillaceous rocks (mudstone, siltstone, shale, slate);
$15 < m < 24$ for arenaceous rocks with strong crystals and poorly developed crystal cleavage (sandstone, quartzite);
$16 < m < 19$ for fine-grained polymineralic igneous crystalline rocks (andesite, dolerite, diabase, rhyolite);
$22 < m < 33$ for coarse-grained polymineralic igneous and metamorphic rocks (amphibolite, gabbro, gneiss, granite, norite, quartz-diorite).

As in the case of the Coulomb criterion, the Mohr–Coulomb (or Hoek and Brown) criterion assumes that failure is independent of the intermediate principal stress magnitude and that the failure surface is parallel to the intermediate principal stress direction.

8.3.4 The von Mises and other polyaxial failure criteria

While the classical axisymmetrical triaxial test configuration is convenient for testing rock strength under compression, the associated condition $\sigma_2 = \sigma_3$ is not representative of many real *in situ* conditions. Since an increase in confining pressure results in an increase in strength, the question arises whether an increase in the intermediate principal stress would also result in a further increase in compressive strength.

To test this hypothesis, truly triaxial compression experiments have been run on various rocks (e.g. Chang and Haimson, 2000). A truly triaxial test is often called a polyaxial test, to differentiate it from the classical triaxial test. Polyaxial tests are run on prismatic samples so that the three principal stress components may be controlled independently.

In metals it is often assumed that the plastic yield criterion depends only on the deviatoric stress (with no effect of the spherical component) (e.g. Salençon, 1974) and the simplest yield criterion is the von Mises criterion,

$$J_2 = k^2 \quad \text{with} \quad J_2 = \tfrac{1}{6}[(\sigma_1 - \sigma_2)^2 + (\sigma_2 - \sigma_3)^2 + (\sigma_3 - \sigma_1)^2] \tag{8.26}$$

where J_2 is the second invariant of the deviatoric stress (see section 3.5).

However, for geomaterials, for which the post-elastic behavior depends on the spherical component of the stress, it has often been considered (e.g. Drucker and Pragger, 1952) that a more appropriate criterion is given by

$$(J_2)^{1/2} = a + bI_1 \tag{8.27}$$

where $I_1 = \sigma_1 + \sigma_2 + \sigma_3$ is the first invariant of the stress tensor.

Zhou (1994) proposed an extension of the Drucker–Prager criterion to be achieved by adding a quadratic term for the first stress invariant:

$$(J_2)^{1/2} = a + bI_1 + cI_1^2 \tag{8.28}$$

After considering the experimental results on cohesionless materials tested in various drained and undrained triaxial conditions, Lade (1977) proposed a yield criterion that also involved all three effective principal stress components:

$$\left(\frac{I_{\tilde{\sigma}}^3}{III_{\tilde{\sigma}}} - 27\right)\left(\frac{I_{\tilde{\sigma}}}{P_a}\right)^{m'} = \eta_1 \tag{8.29}$$

where m' and η_1 are empirical constant adjusted to fit the observations; $I_{\tilde{\sigma}}$ and $III_{\tilde{\sigma}}$ are the first and third stress invariants and P_a is the atmospheric pressure.

This failure criterion was adapted later by Ewy (1999) for cohesive materials in what is known today as the modified Lade criterion:

$$\frac{(I_{\tilde{\sigma}}^{'})^3}{III_{\tilde{\sigma}}^{'}} = 27 + \eta \tag{8.30}$$

where η and S are material constants with $I_{\tilde{\sigma}}^{'} = [(\sigma_1 + S) + (\sigma_2 + S) + (\sigma_3 + S)]$ and $III_{\tilde{\sigma}}^{'} = (\sigma_1 + S)(\sigma_2 + S)(\sigma_3 + S)$.

The parameter S is related to the material cohesion whilst η_1 is related to the friction characteristics.

Colmenares and Zoback (2002) conducted an evaluation of these polyaxial failure criteria along with some others (Wiebols and Cook, 1968). They used the published experimental results on rock failure under polyaxial stress conditions for different rock types (amphibolite, dolomite, limestone, sandstone and shale) and tested the fit between prediction and observation for the occurrence of failure. These authors concluded that the modified Lade criterion (equation (8.30)) and the Zhou criterion (equation (8.28)) as well as the Wiebols and Cook criterion were in statistically good agreement with observation. They also concluded that if only triaxial test results are available for a given material, the effect of the intermediate principal stress component on failure may be properly taken into account by the use of these polyaxial failure criteria (fig. 8.26). Parameters of either of these failure criteria are determined from appropriate triaxial tests.

8.4 Exercises

1. Why is the uniaxial compressive strength of a rock larger than its elastic limit? What is the most accurate method for determining the elastic limit of a rock?
2. Consider a rock sample loaded under uniaxial compression with a load-controlled system. Describe the deformation process observed when the load reaches the maximum load-bearing capacity of the sample.

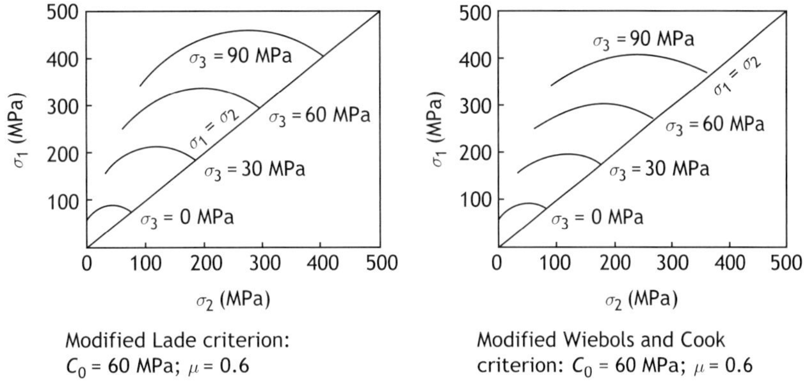

Fig. 8.26 Influence of the intermediate principal stress component on the yield strength of geomaterials (redrawn from Colmenares and Zoback, 2002, with permission from Elsevier).

3. According to Wawersik and Fairhurst (1970) the post-peak behavior of a rock sample loaded under compression beyond its maximum bearing capacity may exhibit either a class I or a class II behavior, when proper loading conditions are satisfied. What conditions must be satisfied by the loading system in order to observe a quasistatic failure process after the maximum bearing capacity of the rock sample has been reached? What is the difference between class I and class II behaviors? What is the physical mechanism that leads to a class II behavior ?
4. Servo-controlled systems provide a means to control rock failure, i.e. to keep quasistatic the failure process of a sample until its complete disintegration is reached. Explain the principle of a servocontrolled system and discuss how to use it for monitoring the failure of brittle samples loaded under triaxial compression conditions. Describe the change in pore space during the complete failure process of a class II material.
5. What is the brittle–ductile transition and what type of mechanism may cause such a transition ?
6. What is rock dilatancy? Under what conditions is it possible to observe negative dilatancy (also called contractancy)?
7. What is soil dilatancy ? What is the mechanism that leads to soil dilatancy?
8. What is an overconsolidated clay? How does the density of an overconsolidated clay compare with that of a normally consolidated clay?

9 Homogenized geomaterials

Rock masses may be viewed as structures involving both solids and fluids, with complex interfaces. They involve discontinuities at all scales, whether these correspond to solid grain limits, to fractures or to faults. Furthermore, in the vast majority of cases, rocks and rock masses are naturally under triaxial loading compression, so that significant parts of these discontinuities support some shear stress.

Different approaches are followed for geomechanical investigations on rock masses, depending on their main focus.

In a rock mechanics approach, the rock mass behavior is often dominated by the elastic response of the solid parts and by the geomechanical characteristics of the discontinuities. Fluids are generally considered only through their local pressure values.

In a soil mechanics approach, the elastic behavior of the solid particles is considered to be negligible and therefore the geomechanical behavior of the equivalent homogeneous geomaterial depends on the size distribution and shape of the particles, on the friction between the particles, on the relative content of solid, water and air and on the flow characteristics of the fluids within the pore space (generally air and water).

In a hydrogeology approach, no attention is given to the solid part and only the motion of fluids (often only liquid water) is of interest.

In all these approaches, continuum mechanics principles are applied and, depending on the discipline, attention is focused on the elastic behavior of the solids with specific attention to the interfaces, or on the plastic behavior of the equivalent homogeneous geomaterial or on the flow of the fluids within a rigid solid structure.

However, in many geomechanical problems, whether in engineering applications or in the context of earth sciences, attention must be given to a mixture of rocks, soils and fluids with explicit consideration to fractures and faults and their topology, i.e. the structure and morphology of the individual fractures and faults as well as that of their networks.

In this chapter we introduce geomechanics modeling approaches that have been derived by considering homogenized geomaterials, whether these are elastic or plastic solids or viscous fluids. The mechanical and hydraulic characteristics of fractures and faults and their influence on the hydromechanical behavior of rock masses are discussed in chapters 10 and 12.

9.1 Elastic geomaterials

We showed in section 2.4 how friction introduces an upper limit to the load that may be applied to a system of pads, springs and dashpots, a limit below which the material exhibits a true linearly reversible elastic behavior. Similarly, rock masses are normally under compression and a first objective is to describe the behavior of equivalent homogeneous geomaterials when the amplitude of the stress perturbation anywhere in the rock mass remains small compared with the existing stress level, so that elasticity theory applies.

Indeed, rock masses always exhibit a real elastic behavior, if only when one is considering elastic wave propagation. Linear elasticity has shown itself to be a very efficient modeling technique for rock engineering problems, or geoscience issues, when the short-term (days to months) transient deformation is of little importance and only the long-term (years to centuries, if not thousands of years) behavior is the main concern. It may be approached with a pseudo-elastic solution after proper consideration is given to the definition of equivalent homogeneous bodies.

Experimental investigations (see chapter 8) have shown that when a piece of rock is loaded quasistatically in compression, the axial strain versus axial stress curve is first nonlinear and then becomes linear until the so-called limit of elasticity is reached. This elastic limit corresponds to the maximum perturbation above which linearity is lost and nonreversible deformation mechanisms start developing. The modeling of some of these nonreversible deformation processes is discussed in section 9.2.

If the rock sample is unloaded before it reaches its elastic limit, the slope of the unloading stress–strain curve is steeper than that of the loading curve. However, this effect is less noticeable when some confining pressure is applied to the rock sample and may even disappear for large enough confining pressures. Furthermore, the elastic parameters that may be derived from P and S wave velocity measurements (section 5.3.5) yield different values for the elastic moduli compared with those derived from quasistatic loading conditions. The differences between the quasistatic and dynamic elastic parameters derived from field observations may reach an order of magnitude (e.g. Cayol and Cornet, 1998; Gunzburger and Magnenet, 2014). According to laboratory investigations (Simons and Brace, 1965), however, these differences disappear progressively with increasing confining pressure.

This implies that elastic parameters derived from the measurements of seismic wave velocities conducted down to a few kilometers may not be used directly for evaluating the quasistatic elastic parameters of the equivalent elastic geomaterial, but such measurements have been shown to be very helpful for investigating rock porosity or the density of microfractures, especially in sedimentary formations. This has found application in for example rock engineering, where measurements of wave propagation velocity help to identify the extent of failure zones around cavities. Furthermore, seismic wave velocitiy determination appears to be appropriate for determining the quasistatic elastic parameters of rocks at great depth (in the tens of kilometers range), where the minimum principal stress reaches values in the gigapascal domain.

In order to reconcile the quasistatic and dynamic elastic moduli, attention turned to analyzing the effect of microdiscontinuities on the elastic parameters. For example, Geertsma (1957) investigated the effect of fluid pressure decline on elastic volumetric changes in porous rocks, through application of the reciprocal theorem of elasticity (section 5.5.2). Independently, Walsh (1965a,b,c) considered a body consisting of a homogeneous isotropic elastic solid with small cavities (pores and microfractures) as a model for

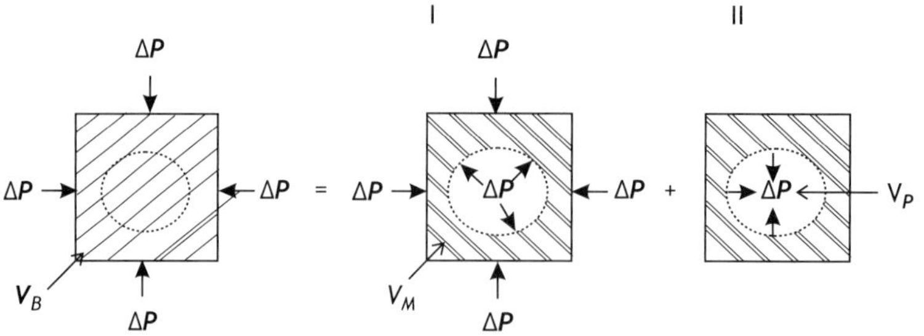

Fig. 9.1 Compressibility of a homogeneous body containing a small cavity and subjected to an external pressure variation. On the left, the volume V_B is filled with an equivalent homogeneous material B.

investigating the influence of microscopic discontinuities on the elastic behavior of rocks. His approach, which also involves the reciprocal theorem, is consistent with laboratory experimental observations.

These various derivations have been expanded during the last 20 years for the purpose of investigating body wave propagation in rock masses and also to explore the effects the extraction of fluid (hydrocarbons or water) from deep reservoirs (see chapter 12). We present below an analysis of the volume changes caused by an external hydrostatic stress component, involving the elastic compressibility, and then we discuss the effects of pores and microcracks on the other elastic parameters.

9.1.1 Effective rock compressibility

Let us consider a small part of a homogeneous isotropic elastic body which contains a single cavity of random shape and volume V_P subjected to an external pressure (the left-hand side of fig. 9.1). We call the original solid material M and the homogeneous bulk material equivalent to the solid plus the cavity B, i.e. the homogeneous material fills the volume V_B. Under an external pressure variation ΔP this small volume deforms, and the corresponding relative change of volume $\Delta V_B/V_B$ depends linearly on the applied pressure variation ΔP. The ratio of this relative volume change and the applied pressure variation defines an *effective compressibility* β_B, which is larger than that of a body with no cavity, consisting of material M with compressibility β_M:

$$\frac{\Delta V_B}{V_B} = \beta_B \, \Delta P \tag{9.1}$$

The objective is to determine this difference in compressibility.

Because the solid is linearly elastic, the effect of an external pressure may be analyzed through the superposition of two loading systems (see the right-hand side of the equation in fig. 9.1). System I involves a uniform pressure applied to both the outside boundary and the surface of the cavity, while system II involves a uniform *negative* pressure applied just to the surface of the cavity. In this analysis compressions are reckoned positive.

For system I, let us assume first that there is no cavity, i.e. that material M fills the complete volume V_B. Application of an external pressure ΔP results in a relative change in volume $\Delta V_B^I/V_B = \beta_M \Delta P$. Because the pressure is hydrostatic and uniform, it has the same value throughout the volume V_B, in particular along the border of the small cavity. Hence, if now the small cavity is present and a uniform pressure ΔP is applied on its boundary, for the solid M this is equivalent to the loading scheme without a cavity: replacing the material M within the cavity by an equivalent uniform pressure ΔP does not generate any strain in the complementary volume filled with M. Hence we conclude that, for loading system I,

$$\frac{\Delta V_B^I}{V_B} = \frac{\Delta V_M^I}{V_M} = \frac{\Delta V_P^I}{V_P} = \beta_M \Delta P \tag{9.2}$$

where ΔV_M^I and ΔV_P^I are respectively the volume variations for the volume occupied by the matrix, V_M, and by the cavity V_P.

The change in bulk volume of the original system, ΔV_B, caused by an external pressure only is the sum of the changes in volume caused by system I and by system II (see equation (9.1)):

$$\Delta V_B = \Delta V_B^I + \Delta V_B^{II} = \beta_B \, \Delta P \, V_B \tag{9.3}$$

The relative change in bulk volume caused by system II alone is given by

$$\frac{\Delta V_B^{II}}{V_B} = \frac{\Delta V_B}{V_B} - \frac{\Delta V_B^I}{V_B} = (\beta_B - \beta_M)\Delta P \tag{9.4}$$

Now we can apply the reciprocal theorem to evaluate the change in volume of the cavity, ΔV_P^{II}, caused by system II. From the reciprocal theorem we know that, given two loading systems, I and II, applied to the same linearly elastic body, the work $W_{I,II}$ done by loading system I through the displacements associated with loading system II, is equal to the work $W_{II,I}$ done by loading system II through the displacements associated with loading system I.

$$W_{I,II} = \Delta P \Delta V_B^{II} - \Delta P \Delta V_P^{II} = \Delta P \Delta V_P^I = W_{II,I} \tag{9.5}$$

The reason for the opposite signs for the terms $\Delta P \Delta V_P^{II}$ and $\Delta P \Delta V_P^I$ in equation (9.5) is that the load for each system has a different sign whilst in both cases a (negative) decrease in the cavity's volume is observed. Hence

$$\Delta V_P^{II} = \Delta V_B^{II} - \Delta V_P^I = (\beta_B - \beta_M)\Delta P V_B - \beta_M V_P \Delta P \tag{9.6}$$

This expression for the volume variations ΔV_P^I and ΔV_P^{II} is valid for a large number of small cavities, given that the stress field associated with loading system I is isotropic throughout the volume V_B, i.e. it is applicable to a porous material of porosity $n = V_P/V_B$ where V_B is the representative elementary volume (REV). Accordingly, the relative variation of the pore volume in a porous rock loaded externally by an hydrostatic pressure ΔP is

$$\frac{\Delta V_P}{V_P} = \frac{\Delta V_P^I + \Delta V_P^{II}}{V_P} = (\beta_B - \beta_M)\Delta P \frac{1}{n} \tag{9.7}$$

Now let us compute the change in solid volume $\Delta(V_B - V_P)$ caused by an external pressure variation ΔP. From equations (9.1) and (9.7) we get

$$\Delta(V_B - V_P) = \beta_B \Delta P V_B - (\beta_B - \beta_M)\Delta P\, V_B = \beta_M \Delta P\, V_B \tag{9.8}$$

which is Walsh's equation (1) (Walsh, 1965a). Differentiating and rearranging yields

$$\frac{dV_B/dp}{V_B} = \beta_M + \frac{dV_P/dp}{V_B} \tag{9.9}$$

The left-hand side of equation (9.9) corresponds to the definition of the compressibility of the bulk material (equation (9.1)) whilst the second term on the right-hand side of equation (9.9) corresponds to the rate of change of the porosity if the change in bulk volume is small compared with the total bulk volume. Adopting Walsh's convention that an increase in porosity is positive, equation (9.9) may be rewritten as

$$\beta_B = \beta_M - \frac{dn}{dp} \tag{9.10}$$

Sometimes the compressibility β_M is called the *intrinsic elastic compressibility*. More generally, rocks and rock masses are characterized by their *effective elastic constants* whilst the elastic constants of the solid matrix M, free of cavities and other discontinuities, are referred to as the *intrinsic elastic constants*. Equation (9.10) shows that the compressibility of rocks, or rock masses, does not depend on the porosity itself but on the change of porosity associated with the pore pressure variation. Accordingly, various models have been considered for investigating the pore pressure dependence of porosity variations as a function of the shape of the pores.

Of particular interest is the influence of microcracks, which have already been shown to control the strength of rocks, both in tension (chapter 7) and in compression (sections 7.3.2, 8.1.3 and 8.1.5).

9.1.2 Influence of microcracks on effective elastic constants

We assume here that the small cavity of fig. 9.1 is a flat elliptical crack of length $2c$, and we call L, b and t respectively the length, width and thickness of the small volume dv_B (fig. 9.2). Following Jaeger and Cook (1979), we first evaluate the elastic strain energy associated with this small elliptical crack. Let W_0 be the strain energy of the volume dv_B when there is no crack, and W be that of the same volume with the crack. The strain energy associated with the crack, W_c, is therefore simply

$$W_c = W_0 - W \tag{9.11}$$

We consider first the case in which the small volume dv_B is subjected to a uniaxial stress σ_n in the direction normal to the microcrack orientation and then the case in which it supports a shear stress component parallel to the microcrack orientation.

As for the compressibility evaluation, we make use of the reciprocal theorem and consider the two following loading cases. In case 1 an external uniaxial stress σ_a is applied

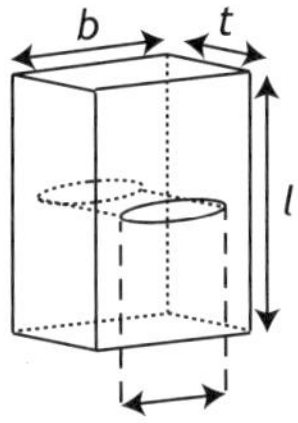

Fig. 9.2 An elliptical crack, with length 2c, in a small parallelepiped of volume dv_B with length L, width b and thickness t, consisting of homogeneous material M; σ_a is a uniaxial stress applied on equivalent material B, assumed to fill completely the small volume $dv_B = Lbt$, i.e. a volume without any crack.

to the outer boundaries of the volume dv_B that are parallel to the crack, whilst in case 2 a uniaxial stress σ_c is applied both to the outer boundary parallel to the crack and to the crack surface.

The work done by the first stress through the displacements caused by the second stress is

$$W_{12} = \sigma_a bt \frac{\sigma_c L}{E_M} \tag{9.12}$$

where E_M is the Young's modulus of the rock matrix when free of cracks, material M.

Let us call u_2 the displacement of the crack surface in the σ_a direction caused by σ_a and $t\,da$ the crack surface component normal to the σ_c direction. The work done by the second stress through the displacements associated with the first stress is

$$W_{21} = \sigma_c bt \frac{\sigma_a L}{E_B} - 2 \int_{-c}^{+c} \sigma_c u_2 t \, da \tag{9.13}$$

where E_B is the effective Young's modulus when the cavity of fig. 9.1 is a flat elliptical crack. Hence, from the reciprocal theorem we conclude that

$$\sigma_a bt \frac{\sigma_c L}{E_M} = \sigma_c bt \frac{\sigma_a L}{E_B} - 2\sigma_c t \int_{-c}^{+c} u_2 \, da \tag{9.14}$$

If we consider the case for which $\sigma_a = \sigma_c = \sigma_n$, equation (9.14) may be written as

$$\sigma_n^2 Lbt \frac{1}{2E_B} = \sigma_n^2 Lbt \frac{1}{2E_M} + \sigma_n t \int_{-c}^{+c} u_2 \, da \tag{9.15}$$

The left-hand side of equation (9.15) is the strain energy W of the volume V_B that contains the crack whilst the first term on the right-hand side of equation (9.15) is the strain energy W_0 of the same volume without the crack. Hence the strain energy associated with the crack is

$$W_c = \sigma_n t \int_{-c}^{+c} u_2 \, da \tag{9.16}$$

For a very flat elliptical crack, solutions have been published (see e.g. Jaeger and Cook, 1979) for plane stress conditions (t very small) and for plane strain conditions (t very large), as follows:

$$W_c = \pi\sigma_n^2 c^2 t/E_M \qquad \text{plane stress conditions} \tag{9.17}$$

$$W_c = \pi\sigma_n^2 c^2 t(1-\nu_M^2)/E_M \qquad \text{plane strain conditions} \tag{9.18}$$

For a penny-shaped crack (axisymmetrical conditions) of radius c,

$$W_c = 8(1-\nu_M^2)c^3\sigma_n^2/(\pi E_M) \tag{9.19}$$

Hence, for all three type of condition, plane stress, plane strain or axisymmetrical, the elastic strain energy associated with a crack under a far-field stress component σ_n differs only by a numerical factor close to 1 when $t = c$. Hereafter we will take the plane stress solution as a reasonable approximation for all three cases.

If, instead of a normal stress component σ_n, a shear component τ is applied, similar expressions are obtained in which τ replaces σ_n and, if one superposes the stress components τ and σ_n then the strain energy associated with the crack is

$$W_c = \pi c^2 t(\sigma_n^2 + \tau^2)/E_M = \pi c^2 t[\sigma_1^2\cos^2\alpha + \sigma_2^2\sin^2\alpha]/E_M \tag{9.20}$$

when the crack is inclined at an angle α to the direction σ_2 minimum principal stress (see section 3.4.1).

Returning to equation (9.15), we note that, for a flat elliptical crack,

$$\frac{1}{E_B} = \frac{1}{E_M} + 2\left(\frac{1}{\sigma_n}\right)\left(\frac{1}{bL}\right)\int_{-c}^{+c} u_2\,da \tag{9.21}$$

$$= \frac{1}{E_M} + 2\frac{W_c}{\sigma_n} = \frac{1}{E_M} + \frac{2\pi c^2\lambda}{E_M} \tag{9.22}$$

where $\lambda = t/(bLt)$ may be regarded as the crack length per unit volume. Equation (9.22) expresses the effective Young's modulus in terms of the intrinsic Young's modulus when the flat elliptical crack is oriented perpendicularly to the uniaxial stress σ_n. A very similar result is obtained for the effective bulk modulus $K_B = 1/\beta_B$ when a hydrostatic pressure is applied externally:

$$\frac{1}{K_B} = \frac{1}{K_M} + \frac{2\pi c^2\lambda}{E_M} \tag{9.23}$$

It has been shown (e.g. Walsh, 1965b) that these solutions are applicable to materials with multiple cracks provided that the distance between each crack is larger than twice the mean crack length, so that the cracks do not interact. This analysis may be generalized to include rocks in which the crack density satisfies the condition that the mean distance between cracks is larger than twice the size of the cracks. Then one may evaluate the effective Young's modulus of the rock as a function of the mean crack size per unit volume:

$$E_B = \frac{E_M}{1 + 2\pi c^2\lambda} \tag{9.24}$$

If the crack plane is inclined by an angle $(\pi/2 - \alpha)$ to the normal stress direction, then from equations (9.20) and (9.22) the effective Young's modulus is found to be

$$\frac{1}{E_B} = \frac{1}{E_M} + \cos^2\alpha \frac{2\pi c^2 \lambda}{E_M} \tag{9.25}$$

If the cracks are randomly distributed, a spherical average yields the effective Young's modulus for the equivalent geomaterial. However, since $(1/2\pi)\int_0^{\pi/2} 2\pi \sin\alpha \cos^2\alpha \, d\alpha = 1/3$ we conclude that, for materials with randomly orientated cracks distributed randomly throughout their REV, the effective Young's modulus is

$$\frac{1}{E_{eff}} = \frac{1}{E_M} + \frac{1}{3}\frac{2\pi c^2 \lambda}{E_M} \tag{9.26}$$

or

$$E_{eff} = \frac{E_M}{1 + 2\pi c^2 \lambda/3} \tag{9.27}$$

Now, Walsh (1965c) showed that the various relations that exist between the elastic constants of homogeneous materials (section 5.1) are valid for the effective elastic constants also. Hence, given that

$$\frac{\nu_M}{E_M} = \frac{1}{2}\left(\frac{1}{E_M} - \frac{1}{3K_M}\right) \tag{9.28}$$

we conclude from equations (9.24) and (9.27) that

$$\frac{\nu_{eff}}{E_{eff}} = \frac{\nu_M}{E_M} \tag{9.29}$$

Equation (9.29) shows that the ratio of the lateral strain and the axial strain is the same whether the body is cracked or homogeneous (i.e. without cracks).

Further, Walsh (1965b) showed that if the externally applied stresses are large enough then the cracks close and support some slip that "mobilizes" the friction. As suggested by the discussion in section 2.4.3, friction is not immediately reversible upon unloading, so that when an element has slipped in a given direction, dropping the applied load does not result instantaneously in a reverse slip but generates residual stresses within the material. This implies that when a rock sample is unloaded after it has supported some slip on its microcracks, the microcracks do not slip back immediately and the material appears stiffer during unloading than during loading. This was demonstrated by Walsh (1965b), who also showed that the unloading modulus of rock should theoretically be equal to that of the homogeneous matrix M.

Whilst in Walsh's work the fractures are randomly distributed, in some materials some preferential directions may exist, which results in some anisotropy of the rock. In order to quantify this effect, Kachanov (1980) introduced a crack tensor,

$$\alpha_{ij} = \frac{1}{V}\sum_{k=1}^{m(V)} (S^k)^{3/2} n_i^k n_j^k, \qquad i,j = 1,2,3 \tag{9.30}$$

where S^k is the area of the kth microcrack and $m(V)$ is the number of microcracks in the representative elementary volume V.

This discussion implies that because the elastic constants of geomaterials depend on their porous structure and on the microcrack geometry and distribution, so do the body wave velocities (for both P and S waves). In practice, though, normally the porous domains of rocks are filled with fluids, which may include liquid and gas. Hence the seismic velocities do not depend only on the structure of the pore volume; they also depend on the fluid content within the pore volume (e.g. Adelinet *et al.*, 2011).

9.2 Elementary considerations on plasticity

As already mentioned, because geomaterials or rock masses (i.e. homogeneous geomaterials plus their associated fracture fields and faults) are generally under compression, the many discontinuities that affect these bodies support some shear stress. When these shear stresses grow too large, because of some perturbation, the material response to this perturbation is no longer purely elastic.

We saw in chapter 2 that permanent deformations may be associated with various types of rheological behavior: viscoelasticity, elastoplasticity or elastovisco-plasticity. Viscoelasticity characterizes fluids, i.e. materials that may transmit elastic waves but cannot support any shear stress when they are at rest. This is the case, for example, for lavas or magmas in volcanoes. Elastoplasticity refers to materials the post-elastic behavior of which is controlled only by friction; this is often the case for a fault. Elastoviscoplasticity refers to materials that behave elastically as long as the applied load remains smaller than a critical value but that sustain permanent deformation when the load becomes larger than this critical value. In addition, for elastoviscoplastic materials the post-elastic deformation depends on the strain rate (a viscosity effect) and on the loading history, and this is the case for many geomaterials loaded beyond their elastic limit.

An important issue is the determination of whether crustal rock masses, and more generally the materials that make up the lithosphere, are elastoviscoplastic solids, i.e. whether they can support permanently some shear stress component when they are at rest, or whether in the long (or very long) term they behave as viscous fluids. This is discussed further in chapter 14, where evaluations of the stress field at various depths in different geological contexts are described.

The micromechanisms involved in the plastic behavior of geomaterials may be related to the quasistatic extension of microcracks, to friction along surfaces of discontinuity at all scales, or to perturbations of the crystalline lattice of solids. The deformation rate may be controlled by the compressibility and motion of the fluids filling the pore space, by the motion of dislocations within crystalline structures or other small-scale stable micromechanisms such as solid diffusion or pressure solution (see section 12.3). Hence, plasticity describes the deformation of geomaterials for very different domains of mean stress level and temperature.

First the yield and yield surface concepts, which describe the conditions for which irreversible deformation occurs, are introduced. Then we discuss plastic flow and associated and nonassociated flow. Finally, strain localization and bifurcation are addressed in outline.

9.2.1 Strength, yield and yield surface

In the previous chapter the inception of nonreversible deformation was referred to as occurring at the elastic limit of the material. We noted that the post-elastic domain may imply some plastic deformation, i.e. a deformation process during which the material maintains some strength and which is not therefore associated with instantaneous catastrophic failure. However, the post-elastic domain may also involve some brittle rupturing processes, the propagation of which may reach high velocities (close to that of elastic shear waves), leading to catastrophic failure of the system.

This onset of post-elastic deformation, or failure, is generally characterized by a combination of values for some of the stress tensor components, conveniently described by a function called a failure criterion. In this chapter, and thereafter, we adopt the word *yield* as referring only to the onset of plasticity as the dominant failure process, i.e. a failure process which is stable for the initial small-strain domain observed just after the onset of yielding.

On many occasions (chapters 7 and 8) we have referred to strength. We will use the word *strength* to refer to the maximum value that a given stress component may reach during a deformation process leading to large deformations, irrespective of the stability of the deformation process. For example, the tensile strength of a rock is the maximum tensile stress that may be supported by the rock before tensile rupture develops. The uniaxial compressive strength is the maximum load-bearing capacity of a volume of rock loaded in uniaxial compression. The same applies to the triaxial compressive strength.

Strength is a convenient concept for some engineering applications that require critical loading conditions to be identified. It refers to the mean value of some stress components within an REV. It does not refer to an intrinsic geomaterial characteristic, for it depends on the size of the REV as well as on many environmental parameters, such as temperature, strain rate etc., that affect the micromechanisms at the origin of failure. Strength always concerns materials that are in a state beyond yielding so that, from an engineering perspective, yield is a safer estimate of the maximum possible loading before any failure develops. Strength is often a useful concept for describing the *load levels* that control the onset of unstable failure.

In section 8.3 we examined various failure criteria derived from laboratory experiments. They refer either to yield or to strength. In the discussion below we consider only yield criteria, i.e. the set of values that must be reached by the stress components for yield to occur in the equivalent homogeneous geomaterial. In the previous chapter we pointed out that, when it occurs, the onset of irreversible volume variation is a clear marker of yield. However, yielding is not systematically associated with volume variations and, for some materials, a large plastic strain may occur without significant volume change.

A yield criterion may be formulated in terms of principal stress components. Hence, yield criteria may be represented in a three-dimensional space where each axis corresponds to the magnitude of a principal stress component. The surface that represents all combinations of principal stress components at the onset of yield defines the yield surface for the corresponding yield criterion.

We saw in section 8.3 that yield criteria may also be characterized by the relationships between shear stress and effective normal stress (e.g. the Coulomb and Mohr–Coulomb failure criteria), or by combinations of stress invariants (e.g. the von Mises, Drucker–Prager or modified Lade criteria), or by the relationship between the effective spherical stress component σ'_m and the maximum differential stress τ_m (Wong and Baud, 2012). These criteria correspond to different plastic behaviors, involving different micromechanisms that depend on the material and on the environmental conditions.

For example, for metals the yield criteria, namely the Tresca or von Mises criteria, are independent of the spherical stress component $\sigma_m = (\sigma_1 + \sigma_2 + \sigma_1)/3$. These criteria are often used to represent the onset of plasticity in crustal geomaterials at very great depth (where the spherical stress components are in the gigapascal range, with temperatures in the 500–1000 °C range, for depths reaching a few tens of kilometers).

However, for the geomaterials encountered at shallower depths, from ground surface down to 15 km depending on the geomaterial, the yield criterion does depend on the spherical component of the stress tensor. Furthermore, the spherical component has also a great influence on the compaction or on the dilatancy of the material, i.e. on its volumetric strain, so that some coupling exists between plastic flow and yield.

Yield surface in the principal stress space

We examine here some yield criteria expressed as functions of the principal stress components, first introducing the principal stress space.

The principal stress space is a three-dimensional space in which each reference axis corresponds to a principal stress magnitude. In such a space, the stress tensor components may be plotted as a point with coordinates $(\sigma_1, \sigma_2, \sigma_3)$, called a stress point. We consider only compressive stress fields (since rock masses and granular materials cannot sustain tensile stresses) and we reckon compressions as positive.

In the principal stress space, the straight line inclined equally to all axes by an angle θ_0 such that $\cos\theta_0 = 1/\sqrt{3}$ is called the *space diagonal*. All stress points that belong to the space diagonal correspond to an isotropic (also called hydrostatic) state of stress, i.e. all principal stress components are equal. Because of this property, the space diagonal is frequently called the *hydrostatic axis* or, more simply the *hydrostat*. The plane normal to the space diagonal is called the π plane (fig. 9.3). We may define three orthogonal unit vectors, $\hat{\boldsymbol{n}}_D, \hat{n}_E, \hat{n}_F$, such that $\hat{n}_D$ lies on the space diagonal, $\hat{n}_E$ lies in the π plane and is normal to the projection of the σ_1 axis onto the π plane, whilst $\hat{n}_F$ lies in the π plane and is parallel to the projection of the σ_1 axis onto the π plane:

$$\hat{n}_D = \frac{1}{\sqrt{3}}\begin{pmatrix}1\\1\\1\end{pmatrix}, \quad \hat{n}_E = \frac{1}{\sqrt{2}}\begin{pmatrix}0\\-1\\1\end{pmatrix}, \quad \hat{n}_F = \frac{1}{\sqrt{6}}\begin{pmatrix}2\\-1\\-1\end{pmatrix} \tag{9.31}$$

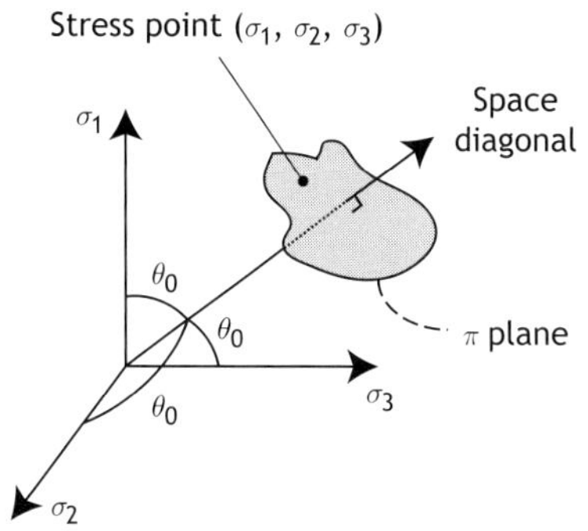

Fig. 9.3 The π plane and the space diagonal (the hydrostatic axis) in the principal stress space (redrawn from Davis and Selvadurai, 2002, with permission from Cambridge University Press).

The coordinates of the stress point $(\sigma_1, \sigma_2, \sigma_3)$ are obtained in the $(\hat{n}_D, \hat{n}_E, \hat{n}_F)$ frame of reference by computing the scalar products of the stress vector $(\sigma_1, \sigma_2, \sigma_3)$ with each unit vector $\hat{n}_D$, $\hat{n}_E$, and $\hat{n}_F$:

$$\sigma_D = \frac{1}{\sqrt{3}}(\sigma_1 + \sigma_2 + \sigma_3) \tag{9.32}$$

$$\sigma_E = \frac{1}{\sqrt{2}}(-\sigma_2 + \sigma_3) \tag{9.33}$$

$$\sigma_F = \frac{1}{\sqrt{6}}(2\sigma_1 - \sigma_2 - \sigma_3) \tag{9.34}$$

The distance of this stress point to the space diagonal is

$$(\sigma_E^2 + \sigma_F^2)^{1/2} = \sqrt{\tfrac{2}{3}}(\sigma_1^2 + \sigma_2^2 + \sigma_3^2 - \sigma_1\sigma_2 - \sigma_2\sigma_3 - \sigma_3\sigma_1)^{1/2} \tag{9.35}$$

$$= \sqrt{\tfrac{1}{3}}[(\sigma_1 - \sigma_2)^2 + (\sigma_2 - \sigma_3)^2 + (\sigma_3 - \sigma_1)^2)]^{1/2} \tag{9.36}$$

$$= \sqrt{\tfrac{2}{3}}(I_1^2 - 3I_2)^{1/2} = \sqrt{2}J_2^{1/2} \tag{9.37}$$

where I_1 and I_2 are the first and second invariants of $\tilde{\sigma}$ whilst J_2 is the second invariant of the deviatoric stress $(\tilde{\sigma} - \frac{1}{3}I_1\tilde{1})$ (see section 3.5).

Some authors (e.g. Davis and Selvadurai, 2002, chapter 3) write

$$\begin{aligned} Q &= (\sigma_1^2 + \sigma_2^2 + \sigma_3^2 - \sigma_1\sigma_2 - \sigma_2\sigma_3 - \sigma_3\sigma_1)^{1/2} \\ &= \frac{1}{\sqrt{2}}[(\sigma_1 - \sigma_2)^2 + (\sigma_2 - \sigma_3)^2 + (\sigma_3 - \sigma_1)^2)]^{1/2} \\ &= \sqrt{3}J_2^{1/2} \end{aligned} \tag{9.38}$$

and, for the spherical component of the stress tensor,

$$P \equiv \sigma_m = \tfrac{1}{3}I_1 \tag{9.39}$$

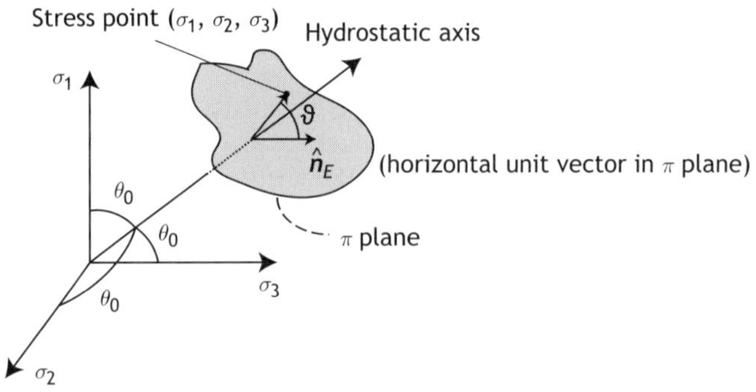

Fig. 9.4 The Lode angle in the π plane.

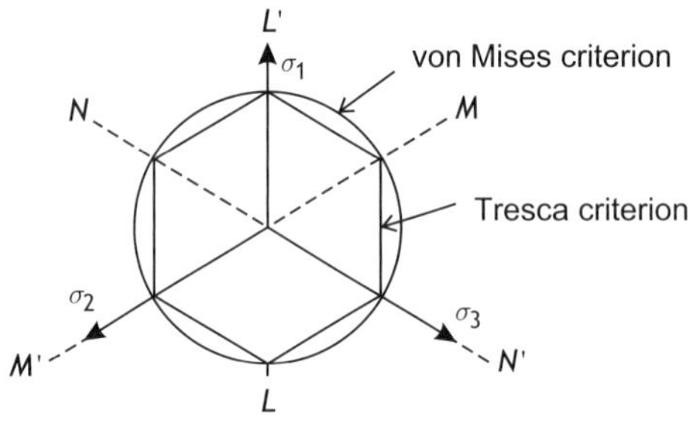

Fig. 9.5 The Tresca and von Mises yield criteria in the π plane. Note that LL', MM' and NN' are axes of symmetry because of the assumption of isotropy.

Note that, for conventional triaxial compression tests in which $\sigma_2 = \sigma_3 = P_c$ and $\sigma_1 = \sigma_a$, Q reduces to $\sigma_a - P_c$ and J_2 reduces to $\frac{1}{3}(\sigma_a - P_c)^2$, so that $(\sigma_E^2 + \sigma_F^2)^{1/2} = \sqrt{\frac{2}{3}}(\sigma_a - P_c)$.

Sometimes it is convenient to use polar coordinates in the π plane to plot the stress points at yield. For this purpose, use is made of the *Lode angle* ϑ, which defines the azimuth of the stress point direction in the π plane (with origin at the hydrostat) with respect to the $\hat{n}_E$ (horizontal) direction (fig. 9.4):

$$\tan\vartheta = \frac{\sigma_F}{\sigma_E} = \frac{2\sigma_1 - \sigma_2 - \sigma_3}{\sqrt{3}(\sigma_3 - \sigma_2)} \tag{9.40}$$

Whilst it is common to adopt the convention $\sigma_1 \geq \sigma_2 \geq \sigma_3$ for discussing the stress state at a given point, all possible combinations of principal stress magnitudes such as $\sigma_2 \geq \sigma_1 \geq \sigma_3$, $\sigma_3 \geq \sigma_1 \geq \sigma_2$, etc. must be considered when analyzing the yield at a given stress point. Further, for isotropic materials the yield criterion must satisfy some conditions of symmetry between the principal stress values. Indeed, if the set of values $(\sigma_1, \sigma_2, \sigma_3)$ corresponds to a yield situation then so does the combination $(\sigma_1, \sigma_3, \sigma_2)$. Using similar arguments for other combinations of the principal stress components one can conclude that the lines LL', MM' and NN' shown in the π plane (fig. 9.5) must correspond to axes of symmetry.

Yield criteria in the π plane

In the $(\hat{n}_D, \hat{n}_E, \hat{n}_F)$ frame of reference, the Tresca yield criterion (section 8.3) becomes

$$\sigma_1 - \sigma_3 = \frac{1}{\sqrt{2}}(-\sigma_E + \sqrt{3}\sigma_F) = 2k \tag{9.41}$$

It corresponds to a straight line inclined at 60° to the $\hat{n}_F$ direction and lying in the π plane. This criterion is independent of the hydrostatic component (the P value, (9.39)). If we adopt the classical assumption that $\sigma_1 \geq \sigma_2 \geq \sigma_3$ then only a portion of the straight line is covered by the stress condition that induces yield; but, as already mentioned, we must consider all possible states for which yield occurs, i.e. conditions in which σ_2 or σ_3 may be the largest principal stress component. Hence, for isotropic materials, the Tresca yield criterion corresponds to a hexagon in the π plane (fig. 9.5).

Because the Tresca yield criterion does not depend on the hydrostatic P value, the surface that covers all possible combinations of principal stress values for which the criterion is satisfied, i.e. the yield surface, corresponds to an infinite prism with its axis along the hydrostat and with a hexagonal cross section in the π plane. For all points inside the prism, elasticity prevails. No stress point may be found outside this prism for materials that satisfy the Tresca criterion and are under static equilibrium.

Turning to the von Mises yield criterion, we note from equation (9.38) that in the π plane it is simply written as $Q = k$. Correspondingly, equation (9.37) describes a circle that touches the Tresca criterion at the vertices of the correponding hexagon (fig. 9.5). The value of Q is independent of the hydrostatic P value, so that the yield surface for the von Mises criterion is simply an infinite cylinder, which is coaxial to the yield surface associated with the Tresca criterion.

The Coulomb failure criterion is often used to describe the triaxial compressive strength of rocks. It is also very appropriate for describing the stress values at the onset of yielding in soils, which is often referred to as their shear strength. As discussed in section 8.3.2, this criterion is expressed as

$$\tau = \mu\sigma'_n + S_0 \tag{9.42}$$

where τ and σ'_n are respectively the shear and the effective normal stress supported by the plane of failure, while μ and S_0 are material properties referred to as the friction coefficient and the cohesion. Below we do not consider explicitly the pore pressure and consider only the absolute stress values. If the pore pressure must be taken into account, similar results may be derived after replacing the principal stress components by effective principal stress components, as defined in section 8.1.6.

Taking advantage of equations (3.63) and (3.64), the Coulomb yield criterion may be expressed in terms of principal stress components:

$$\sigma_1(1 - \sin\phi) - \sigma_3(1 + \sin\phi) = 2S_0 \cos\phi \tag{9.43}$$

where ϕ is the friction angle defined by $\phi = \arctan\mu$. Equation (9.43) is independent of the intermediate principal stress σ_2, with the convention that $\sigma_1 \geq \sigma_2 \geq \sigma_3$.

Note that if $\sigma_3 = 0$ (a uniaxial compression condition) then for yield to occur the σ_1 stress component must be such that

$$\sigma_1 = 2S_0 \frac{\cos\phi}{1-\sin\phi} \equiv C_0 \tag{9.44}$$

where C_0 is the maximum stress that can be supported by the material loaded in uniaxial conditions (fig. 8.24), if it fails in shear as soils do (this expression is not valid for rocks that fail by indirect tension, as explained in chapter 8). Accordingly, the Coulomb yield criterion may be written as

$$\sigma_1 = C_0 + \sigma_3\left(\frac{1+\sin\phi}{1-\sin\phi}\right) \tag{9.45}$$

In order to obtain an expression for the Coulomb yield criterion in the π plane, we first obtain $\sigma_1, \sigma_2, \sigma_3$ as functions of $\sigma_D, \sigma_E, \sigma_F$, from equations (9.32)–(9.34):

$$\sigma_1 = \frac{1}{\sqrt{3}}\sigma_D + \sqrt{\frac{2}{3}}\sigma_F \tag{9.46}$$

$$\sigma_2 = \frac{1}{\sqrt{3}}\sigma_D - \frac{1}{\sqrt{2}}\sigma_E - \frac{1}{\sqrt{6}}\sigma_F \tag{9.47}$$

$$\sigma_3 = \frac{1}{\sqrt{3}}\sigma_D + \frac{1}{\sqrt{2}}\sigma_E - \frac{1}{\sqrt{6}}\sigma_F \tag{9.48}$$

Then, using equations (9.46) and (9.48) in (9.43), we get

$$-\sqrt{3}\sigma_E(1+\sin\phi) + \sigma_F(3-\sin\phi) = 2\sqrt{6}C_0\cos\phi + 2\sqrt{2}\sigma_D\sin\phi \tag{9.49}$$

The right-hand side of equation (9.49) depends only on σ_D, i.e. on the hydrostatic component of the stress tensor. On the left-hand side the components σ_E and σ_F are linearly related, so the intersection of the yield surface with the π plane is a straight line. Note that this straight line is inclined at an angle ψ to the $\hat{n}_F$ direction, where

$$\psi = \frac{\pi}{2} - \arctan\frac{\sqrt{3}(1+\sin\phi)}{3-\sin\phi} \tag{9.50}$$

The angle ψ is equal to 60° if $\phi = 0$, which corresponds to the Tresca criterion, but is smaller than 60° for larger values of ϕ. Now, in order to satisfy the symmetry conditions imposed by isotropy, the polygon section created by the intersection of the yield surface with the π plane (fig. 9.6) cannot correspond to an hexagon unless $\phi = 0$.

Various failure criteria for characterizing the strength of some geomaterials were mentioned in section 8.3. If we assume that these criteria are also well adapted to describing their yielding, we may compare their respective representations in the π plane (fig. 9.7).

A fundamental difference between these various yield criteria, which does not show up in their π plane representation, is the influence of the hydrostatic component. There is no influence of the hydrostatic component on either the Tresca or the von Mises criteria (owing to the cylindrical shape of the yield surface). For the Coulomb criterion, as well as for all criteria that depend on the magnitude of the spherical stress component (those pictured in fig. 9.7), the area of the corresponding polygon in the π plane depends on the

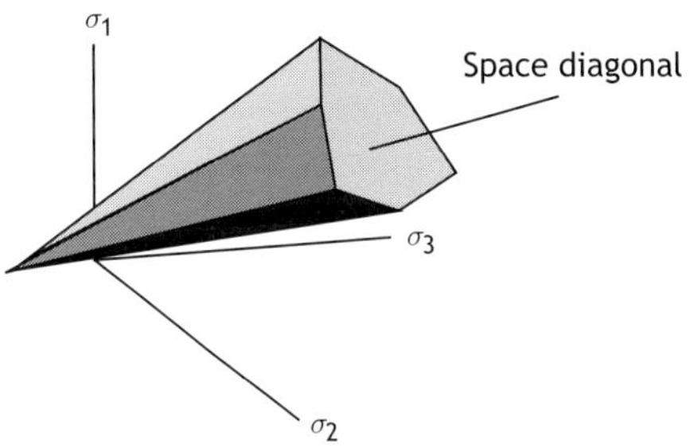

Fig. 9.6 Perspective view of the Coulomb yield surface in the principal stress space.

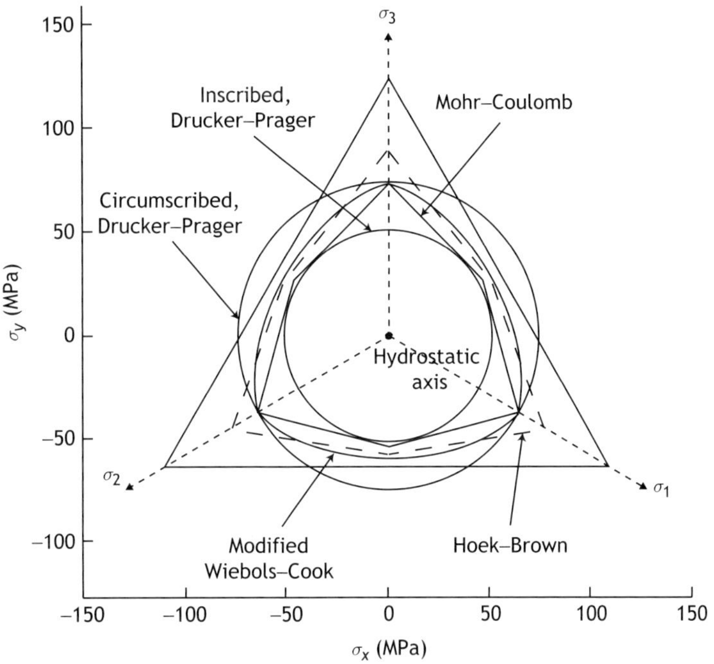

Fig. 9.7 Description in the π plane of the Mohr–Coulomb criterion, Hoek and Brown criterion, Drücker-Prager criterion as introduced in section 8.3 (redrawn from Colmenares and Zoback, 2002, with permission from Elsevier).

hydrostatic stress component. Hence the yield surface for these criteria exhibits a pyramidal shape, with the area of the polygonal cross sections normal to the hydrostat increasing with the hydrostatic component. This is a clear limitation of the Coulomb failure criterion for porous rocks resulting from compaction, as discussed in section 8.1.8.

More generally, the yield criteria for geomaterials must reflect the influence of the change in volumetric strain at the yielding point; this depends on environmental conditions such as the hydrostatic effective stress magnitude, the temperature or the loading history in cases such as the consolidation of clay.

Yield criteria in the PQ plane

A major difference between plastic deformation in metals and in geomaterials is the relative importance of the volumetric strain. Whilst for metals the change in volume associated with plasticity is negligible, for the geomaterials encountered at depths down to 5 to 10 km, the volumetric strain associated with yield is very noticeable and may have a strong influence on the yield phenomena.

We saw in section 8.1.8 that an increase in the hydrostatic pressure in a porous geomaterial leads to a permanent volume decrease when the pressure becomes sufficiently large. This implies that the yield surface as expressed in the principal stress space must intersect the hydrostat for large P values. In other words, for compactant materials the yield surface cannot extend to infinity, as e.g. with the von Mises yield criterion, but must intersect the hydrostat: for compaction to occur, the yield surface must be "capped" (i.e. it cannot extend to infinity).

Various models have been proposed for capped yield surfaces. For example, Davis and Selvadurai (2002) discussed the Cam Clay surface, which corresponds to a relationship between Q and the logarithm of the normalized magnitude of P (P and Q are defined by (9.38) and (9.39)),

$$Q + MP\left(\ln \frac{P}{P_c} - 1\right) = 0 \tag{9.51}$$

where M and P_c are material parameters. This criterion was later modified so as to exhibit an elliptical shape in the PQ plane, and is presently known as the modified Cam Clay yield surface:

$$Q^2 = M^2 P(2P_c - P) = 0 \tag{9.52}$$

where M and P_c are again material parameters.

Wong and Baud (2012) also proposed an elliptic relationship between Q and P, but with the constant adjusted to fit observation, as discussed in section 8.1.8.

Yield criteria in the PQ plane are intended to take into account the hydrostatic part of the stress tensor on yielding, i.e. the onset of volumetric strain. If we call dilatancy the nonelastic change in volume caused by the loading process, geomaterials may exhibit either dilatancy hardening or dilatancy weakening, i.e. the irreversible volumetric strain associated with plastic strain may increase or decrease the magnitudes of the principal stress components for which yielding occurs. For such materials, the yield criterion depends on the amount of plastic flow. Let us observe that, in this general definition of dilatancy, a negative dilatancy corresponds to compaction.

Hence, the analysis of plastic flow requires the definition of a yield criterion as well as a rule describing the way in which plastic flow develops and a rule describing how the dilatancy (hardening or softening) associated with plastic flow influences the yield criterion.

9.2.2 Plastic flow

When the stress components satisfy the yield criterion, some irreversible deformation occurs and therefore some plastic strain, ε^p, is added to the elastic strain ε^e associated with the elastic response of the material up to yield. Thus the total strain is the sum of the elastic and plastic strains:

$$\varepsilon = \varepsilon^e + \varepsilon^p \tag{9.53}$$

But, contrary to the elastic case, there is no one-to-one correspondence between the stress state and the plastic strain. Indeed, deformation occurs as long as the stress satisfies the yield criterion, and determination of the plastic strain requires a knowledge of the loading history. To overcome this difficulty, the plastic strain is investigated by considering the strain rate.

In what follows we consider only small plastic strains so that, following equation (4.116), the strain rate can simply be defined as

$$\dot{\varepsilon}^p_{mn} = \tfrac{1}{2}(v_{m,n} + v_{n,m}) \tag{9.54}$$

where $\boldsymbol{v}$ is the particle velocity at the point $\boldsymbol{x}$ where the yield criterion is satisfied.

If the strain rate depends only on the stress then, for isotropic materials, both the plastic strain-rate tensor and the stress tensor exhibit the same eigenvectors. Hence, in the π plane the principal directions of the plastic strain rate are the same as those of the stress, and the objective becomes to describe the functional relationship between the components of these two tensors.

Normality

It is generally assumed that for isotropic materials the principal stress directions are parallel to those of the plastic strain rate, although this does not apply when the plastic strain rate is a function of both the stress and the stressing rate. We assume below that the stress and the plastic strain rate have identical principal stress directions, so that an important characteristic of plastic flow concerns the orientation of the strain-rate principal directions with respect to the yield surface (e.g. Davis and Selvadurai, 2002).

In order to illustrate this, we follow Davis and Selvadurai (2002) in introducing the normality concept. We consider a rigid block resting on a horizontal plane with frictional contact. The mass of the block is M and the static friction coefficient μ. A force $\boldsymbol{F}$ with components F_1 and F_2 is applied to the block, its magnitude being increased in a quasistatic manner till the block starts moving (fig. 9.8(a)). Motion occurs when the driving force magnitude becomes equal to the static frictional resistance, i.e. when

$$F_1^2 + F_2^2 = (\mu M g)^2 \tag{9.55}$$

where g is the acceleration due to gravity. We note that equation (9.55) resembles a yield criterion, if all terms in the equation are normalized by the base area of the block.

The motion of the block is characterized by its velocity $\boldsymbol{v}$ (with components v_1 and v_2), which is coaxial to the force $\boldsymbol{F}$, so that

$$\frac{v_1}{v_2} = \frac{F_1}{F_2} \tag{9.56}$$

The locus of the extremity of all force vectors $\boldsymbol{F}$ that induce slip is a circle with radius $\mu M g$, and the orientation of the slip associated with a force $\boldsymbol{F}$ is normal to this circle (fig. 9.8(b)). The circle is analogous to a yield surface and the slip velocity is found to

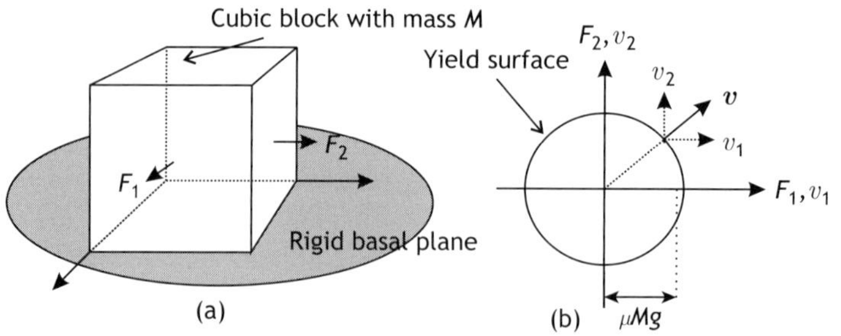

Fig. 9.8 (a) Motion of a block resisted by friction. (b) Orientation of the block's velocity (reproduced from Davis and Selvadurai, 2002, with permission from Cambridge University Press).

be normal to this yield surface. This illustrates the important concept of normality: the strain-rate vector is normal to the yield surface. The strain-rate vector is defined in a three-dimensional space in which the axes are the principal directions of the strain-rate tensor (identical to the principal stress directions, for isotropic materials) and where the coordinates are the principal values of the strain-rate tensor (the strain-rate vector is defined by the principal strain-rate components in a similar manner to the stress vector).

The rate $\dot{w}$ at which the force $\boldsymbol{F}$ works during a slip at velocity $\boldsymbol{v}$ is

$$\dot{w} = \boldsymbol{F} \cdot \boldsymbol{v} = F_1 v_1 + F_2 v_2 \tag{9.57}$$

If a given slip direction is imposed on the block, i.e. if $\boldsymbol{v}$ is prescribed, then the applied force must be coaxial to this direction. In other words, for a given force the direction of slip is that which maximizes the scalar product $\boldsymbol{F} \cdot \boldsymbol{v}$), i.e. that which maximizes $\dot{w}$.

Now assume that the friction coefficient varies with the direction of slip, with values μ_1 in the F_1 direction and μ_2 in the F_2 direction, so that slip occurs when

$$\frac{F_1^2}{\mu_1^2} + \frac{F_2^2}{\mu_2^2} = (Mg)^2 \tag{9.58}$$

The rate of work $\dot{w}_p^*$ associated with a slip (plastic deformation) may be written as

$$\dot{w}_p^* = F_1 v_1 + F_2 v_2 - \lambda \left(\frac{F_1^2}{\mu_1^2} + \frac{F_2^2}{\mu_2^2} - (Mg)^2 \right) \tag{9.59}$$

where λ is any positive scalar number and is known as a Lagrange multiplier. When slip starts, the quantity $F_1^2/\mu_1^2 + F_2^2/\mu_2^2 - (Mg)^2$ is zero and therefore equation (9.59) is precisely the rate of work.

The stationary values of $\dot{w}_p^*$ with respect to F_1 and F_2 are such that

$$\frac{\partial w_p^*}{\partial F_1} = v_1 - 2\lambda \frac{F_1}{\mu_1^2} = 0 \tag{9.60}$$

$$\frac{\partial w_p^*}{\partial F_2} = v_2 - 2\lambda \frac{F_2}{\mu_2^2} = 0 \tag{9.61}$$

so that the components of the slip velocity are given by

$$v_1 = 2\lambda \frac{F_1}{\mu_1^2} \quad \text{and} \quad v_2 = 2\lambda \frac{F_2}{\mu_2^2} \tag{9.62}$$

and the direction of slip is such that

$$\frac{v_1}{v_2} = \frac{F_1/\mu_1^2}{F_2/\mu_2^2} \tag{9.63}$$

The direction of slip is normal to the ellipse defined by equation (9.58), which is our yield criterion.

We conclude that the components of the slip rate are proportional to the components of the gradient of the yield surface, so that the condition of normality maximizes the rate of plastic work.

This result is generalized by observing that, for a given yield surface $f(\sigma_1, \sigma_2, \sigma_3)$, plastic flow occurs at a given location when the stress point at this location is situated on the yield surface. Then the rate of plastic work is

$$\dot{W}_p = \text{tr}(\tilde{\sigma}\dot{\tilde{\varepsilon}}) \tag{9.64}$$

This plastic work is maximized when the plastic strain-rate vector, the components of which are the principal values of the plastic strain-rate tensor, i.e. $\dot{\varepsilon}_1, \dot{\varepsilon}_2, \dot{\varepsilon}_3$, is normal to the yield surface $f(\sigma_1, \sigma_2, \sigma_3) = 0$.

This is known as the normality condition; it assumes the yield surface to be convex.

Associated flow rules

We discussed above how the components of the plastic strain rate may be determined from the components of the gradient of the yield surface. This relation is expressed as

$$\dot{\varepsilon}_i = \lambda \frac{\partial f(\sigma_1, \sigma_2, \sigma_3)}{\partial \sigma_i} \qquad i = 1, 2, 3 \tag{9.65}$$

Equation (9.65) insures that the plastic strain-rate vector is normal to the yield surface. It defines so-called *associated plastic flow rules*. Because of the undetermined (but positive) value of λ, equation (9.65) does not specify the magnitude of the strain rate, only its direction. Two families of associated flow rules are often considered.

Perfect plasticity

A geomaterial is said to be perfectly plastic when the plastic strain increases provided that no further change in stress occurs after yielding has started. The functional relationship between plastic strain and stress is multiply valued. Knowledge of the stress does not provide a knowledge of the strain.

Work-hardening plasticity

Work-hardening plasticity occurs when the yield surface is modified by the yielding of the material. The change in yield stress is often related to the amount of plastic work associated with the yielding process.

It may be shown (e.g. Davis and Selvadurai, 2002, chapter 4) that the normality condition implies a negative plastic strain (i.e. a volume increase) for all materials for which the yield surface expands with increasing mean stress. This applies in particular to the Coulomb yield criterion but also to the Drücker–Prager criterion cited in section 8.3. The dilatancy values observed in most geomaterials are however much smaller than those predicted by these criteria. In order to correct for this discrepancy, nonassociated flow rules need to be considered.

Nonassociated flow rules

In order to fit the dilatancy values observed for geomaterials with those predicted by the plastic flow rules, the function $f(\sigma_1, \sigma_2, \sigma_3)$ which describes the yield surface and which is taken in equation (9.65) to satisfy the normality condition is replaced by another function, called the *plastic potential function* $g(\sigma_1, \sigma_2, \sigma_3)$, such that

$$\dot{\varepsilon}_i = \lambda \frac{\partial g(\sigma_1, \sigma_2, \sigma_3)}{\partial \sigma_i}, \qquad i = 1, 2, 3 \tag{9.66}$$

Equation (9.66) is called a *nonassociated flow rule*, since the principal directions of the strain-rate tensor associated with it are *not* normal to the yield surface. With such a flow rule, the plastic strain-rate vector is normal to the surface defined by the function g, which itself is developed to fit the measured dilatancy values.

For example, for the Coulomb yield criterion, we saw in equation (9.45) that the plastic potential for the associated flow (the equation for the yield surface expressed in terms of the principal stress components) may be written as

$$\sigma_1 = C_0 + \sigma_3 \left(\frac{1 + \sin \phi}{1 - \sin \phi} \right)$$

For a nonassociated flow rule, it is written as

$$\sigma_1 = C_0 + \sigma_3 \left(\frac{1 + \sin \omega}{1 - \sin \omega} \right) \tag{9.67}$$

where ω is adjusted to fit the dilatancy observations; ω is sometimes called the dilatancy angle (e.g. Fjaer *et al.*, 2008, p. 86). When the dilatancy angle is positive, the plastic flow is dilatant and when it is negative the plastic flow is compactant.

Whether the plastic flow is associated or nonassociated, it has been supposed in the above discussion that it remains infinitesimally small. When it exceeds a critical value, the deformation process ceases to be homogeneous: some localization occurs, giving rise to shear or compaction bands. Their orientation with respect to the principal stress directions depends on the dilatancy angle.

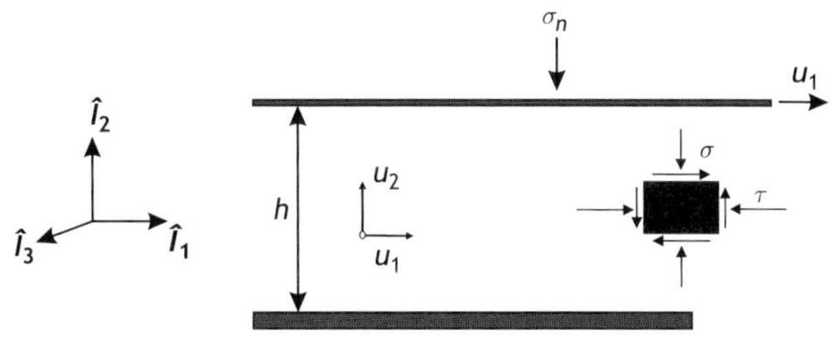

Fig. 9.9 Localization in an infinite layer of thickness h loaded by a normal stress σ_n and subjected to a uniform horizontal displacement u_1.

Localization may involve planes that are sub-perpendicular to the maximum principal stress, as for some compaction bands, or that are inclined to it by any angle smaller than 90° depending on the friction angle and on the dilatancy angles (Sulem, 2007). The analysis of this localization process is conducted by means of bifurcation theory.

9.2.3 Localization: shear bands and compaction bands

Figure 8.13, which presented the location of acoustic events during a triaxial compression test on a rock specimen, illustrated the concept of localization. During the elastic loading phase, deformation is evenly distributed throughout the specimen. After the elastic limit has been reached, however, acoustic emissions progressively localize within a zone that gets narrower and narrower as the stress–strain curve departs from linearity and the axial stress gets closer to its maximum value.

While fig. 8.13 illustrates localization during brittle failure, the same localization phenomenon is observed during plastic yielding. In order to illustrate the principle of localization associated with large plastic deformations, we will follow the discussion of Bésuelle and Rudnicki (2004).

Consider a uniform layer of thickness h subjected on its upper surface to a displacement u_1 in the $\hat{\boldsymbol{I}}_1$ direction and to a normal stress σ_n (fig. 9.9).

The reference system is such that $\hat{\boldsymbol{I}}_1$ and $\hat{\boldsymbol{I}}_3$ are horizontal and $\hat{\boldsymbol{I}}_2$ is perpendicular to the layer. The only nonzero strain components are

$$\varepsilon_{12} = u_{1,2} \qquad \text{and} \qquad \varepsilon_{22} = u_{2,2} \tag{9.68}$$

Neglecting body forces, the equilibrium conditions are

$$\sigma_{21,2} = 0 \qquad \text{and} \qquad \sigma_{22,2} = 0 \tag{9.69}$$

if the stress field is uniform throughout the layer.

Assume that the constitutive equation of the material is given by a simple relationship between the components of shear stress and shear strain,

$$\sigma_{21} = f(\varepsilon_{21}) \tag{9.70}$$

so that, when the applied stress is equal to σ_{21} an increment of shear strain $d\varepsilon_{21}$ requires an increment of shear stress $d\sigma_{21}$:

$$d\sigma_{21} = H_{tan}(\varepsilon_{21})\, d\varepsilon_{21} \tag{9.71}$$

where $H_{tan}(\varepsilon_{21}) = f'(\varepsilon_{21})$ is the slope of the tangent to the curve defined by the equation $\sigma_{21} = f(\varepsilon_{21})$.

A possible response of the material to this shear loading process is a uniform deformation so that, for a total applied displacement u_1, the uniform shear strain is u_1/h and the shear stress component is given by equation (9.70).

Hill (1958) showed that this incremental solution is unique if

$$\int_0^h \Delta d\sigma_{21}\, \Delta d\varepsilon_{21}\, dx_2 > 0 \tag{9.72}$$

where $\Delta d\sigma_{21}$ and $\Delta d\varepsilon_{21}$ denote the differences between two possible solutions for increments $d\sigma_{21}$ and $d\varepsilon_{21}$ that satisfy the equilibrium and boundary conditions.

From equation (9.71), the integrand of equation (9.72) becomes

$$\Delta d\sigma_{21}\, \Delta d\varepsilon_{21} = H_{tan}(\varepsilon_{21})\,(\Delta d\varepsilon_{21})^2 \tag{9.73}$$

The right-hand side of equation (9.73) is positive as long as $H_{tan}(\varepsilon_{21})$ is positive, and Hill's condition implies that, for such conditions, the homogeneous solution is the only possible one. However, for $H_{tan}(\varepsilon_{21}) \leq 0$ a unique solution is not warranted. The homogeneous solution is still possible if the displacement boundary condition is satisfied, but for specified shear-traction boundary conditions, quasistatic deformation cannot be achieved by increasing the shear stress component beyond the point for which $H_{tan}(\varepsilon_{21}) = 0$; inertia terms must be added to the right-hand side of equations (9.69).

Assume that the layer has been homogeneously deformed up to the shear strain $\varepsilon_{21} = \varepsilon_{21}^0$. Now we consider the case for which a further increment of deformation is associated with a nonhomogeneous solution, i.e. a case for which a band is formed such that

$$d\varepsilon_{21}^{band} = d\varepsilon_{21}^{out} + \Delta d\varepsilon_{21} \tag{9.74}$$

where $d\varepsilon_{21}^{band}$ is the increment in shear strain in the band and $d\varepsilon_{21}^{out}$ is that outside the band.

Under quasistatic loading conditions, equilibrium conditions require that

$$d\sigma_{21}^{band} = d\sigma_{21}^{out} \tag{9.75}$$

but, with the constitutive equation (9.70), equations (9.74) and (9.75) yield

$$H_{tan}(\varepsilon_{21}^0)\, d\varepsilon_{21}^{band} = H_{tan}(\varepsilon_{21}^0)\, d\varepsilon_{21}^{out} \tag{9.76}$$

where $H_{tan}(\varepsilon_{21}^0)$ is evaluated at ε_{21}^0, both inside and outside the band.

From equation (9.74) we conclude that the condition for a nonuniform shear strain to appear in the layer is

$$H_{tan}(\varepsilon_{21}^0)\, \Delta d\varepsilon_{21} = 0 \tag{9.77}$$

If $H_{tan}(\epsilon_{21}^0)$ is positive then $\Delta d\varepsilon_{21} = 0$ and the shear strain must be uniform throughout the layer, consistently with Hill's condition.

However, the shear strain may be nonuniform if $H_{tan}(\epsilon_{21}^0) = 0$, i.e. when the curve that corresponds to the function $f(\sigma_{21}, \varepsilon_{21})$ reaches its peak value. Thus, when $H_{tan}(\epsilon_{21}^0) = 0$, solutions exist for which the shear strain distribution is not uniform within the layer and,

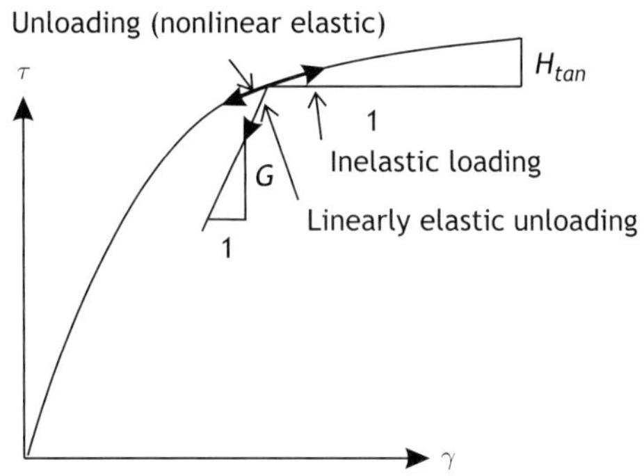

Fig. 9.10 Post-elastic loading of a geomaterial with instantaneous elastic unloading modulus G.

furthermore, in such a case no unique solution exists. Both solutions for the localization of the shear band(s) have an equal chance of occurrence.

This applies if the layer is made up of a geomaterial which is loaded beyond its elastic limit. The shear strain may be decomposed into an elastic part and a plastic part, so that

$$d\varepsilon_{21} = \frac{d\sigma_{21}}{G} + d\varepsilon_{21}^{p} \tag{9.78}$$

where G is the (incremental) elastic unloading modulus and $d\varepsilon_{21}^{p}$ is the plastic shear strain, which vanishes during unloading. If the inelastic increment of shear strain is proportional to the increment of shear stress,

$$d\varepsilon_{21}^{p} = \frac{d\sigma_{21}}{H(\varepsilon_{21}^{p})} \tag{9.79}$$

then

$$d\varepsilon_{21} = \left[\frac{1}{G} + \frac{1}{H(\varepsilon_{21}^{p})}\right] d\sigma_{21} \tag{9.80}$$

For continuous inelastic loading (fig. 9.10), the slope of the loading curve $H_{tan}(\varepsilon_{21})$ is given by

$$H_{tan}(\varepsilon_{21}) = \frac{H(\varepsilon_{21}^{p})}{1 + H(\varepsilon_{21}^{p})/G} \tag{9.81}$$

When $H(\varepsilon_{21}^{p})$ is much smaller than the elastic unloading shear modulus G then $H_{tan} \approx H(\varepsilon_{21}^{p})$.

If outside the band the material unloads elastically, equation (9.77) becomes

$$H_{tan}(\varepsilon_{21}^{0})\, \Delta d\varepsilon_{21} = [G - H_{tan}(\varepsilon_{21}^{0})]\, d\varepsilon_{21}^{out} \tag{9.82}$$

However, $H_{tan}(\varepsilon_{21}^{0})$ is smaller than or equal to G and it is initially positive. Hence, since it decreases with increasing strain, equation (9.82) is first satisfied when $H_{tan}(\varepsilon_{21}^{0}) = 0$, which requires that $d\varepsilon_{21}^{out} = 0$. This corresponds to the onset of elastic unloading of the material outside the band.

This spontaneous splitting of the deformation path is called bifurcation. It may be shown (Rudnicki and Rice, 1975) that for a material subject to associated plastic flow, the localization process occurs only after the peak stress has been reached, but for nonassociated plastic flow such shear bands may form prior to attainment of the peak stress.

9.3 Darcy flow

At the beginning of this chapter we observed that crustal geomechanics questions usually involve a mixture of solids and fluids; we have, however, concentrated our attention in the two previous sections on the mechanical behavior of the solid phases.

Now we will examine how continuum mechanics principles may be applied efficiently to the modeling of fluid flow in rock masses. Particular difficulties arise because of the complexity of the pore volume when flow occurs and because of the scarcity of data available for characterizing both the relevant pore volume and the boundary conditions on flow.

In section 1.1.1 we introduced the concept of porosity to characterize the pore space, and we pointed out that this pore space includes two parts, a nonconnected part and an interconnected part which is available for fluid flow.

Another important characteristic of the pore space is its *specific area* A_{sp}, which is defined by the ratio of the total surface area of the pore space and the total volume of the body. It varies from for example around 200 cm^2/cm^3 for sand to millions of cm^2/cm^3 for clays. It constitutes a fundamental element in the understanding of fluid–solid physicochemical interactions.

In particular, the properties of the water in the pores vary, depending on the distance to the water–solid interface. For distances smaller than 0.5 μm, water molecules are closely tied to the solid and are not free to move. This *adsorbed* water has a specific effect on materials with a very high surface area such as clays, in which the particle sizes are in the micrometer range. The adsorbed water decreases the volume of interconnected pore space available for water to flow in, and this makes clay materials quite impervious.

Sometimes the *kinematic porosity* is used to describe that part of the interconnected pore space where fluid flow may occur (e.g. de Marsily, 1981, p. 30). That part of the water which is free to flow is sometimes referred to as the *free water*. In the remaining part of this chapter, the term porosity must be understood as the kinematic porosity and the term water as the free water.

Following a similar approach to that used in the two previous sections, we will analyze the flow of fluids through porous materials by considering homogenized properties for a representative elementary volume (REV) of the material.

9.3.1 Piezometric head and seepage forces

An important question is the existence, or nonexistence, of compressible fluids in the pore space, whether air, hydrocarbon gas or water vapor. For example, in soil mechanics this

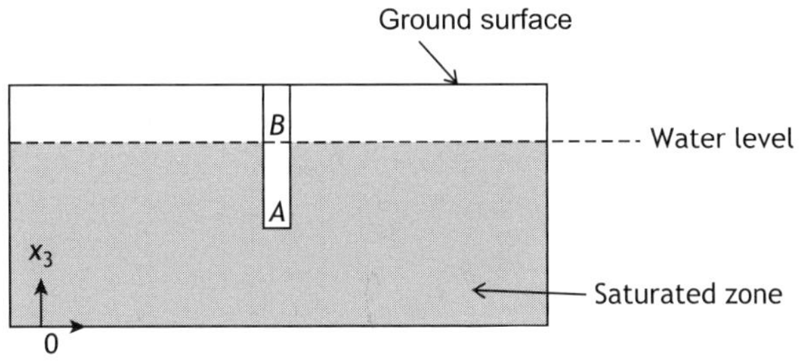

Fig. 9.11 Piezometer for measuring the hydraulic head in an aquifer at rest.

multiphase issue is characterized by either of two parameters, water content or saturation (see section 1.1.1).

An important boundary condition relevant to fluid flow in superficial rock masses (say, from the ground surface down to a few hundred meters) is the depth at which water saturation occurs, i.e. the depth below which the pore space is filled with liquid water. This is often characterized *in situ* by the water level in piezometers.

Water level and piezometric head

Applying Bernouilli's equation (6.168) to natural aquifer systems (i.e. systems saturated with liquid water), in which the only body force is gravity and for which the term $v^2/(2g)$ in equation (6.168) is negligible (owing to very slow flow velocities), we may define a *piezometric* hydraulic head h:

$$h = \frac{P}{\varrho_w g} + x_3 \tag{9.83}$$

where P is the static pressure in the water at depth x_3 (x_3 being a positive vertical coordinate with respect to an arbitrary reference height) and ϱ_w is the water density.

In order to measure the hydraulic head at any point A in an aquifer at rest within a superficial geological formation, it suffices to measure the water level at this location. This is done by drilling a vertical hole reaching below the free surface of the aquifer, placing a slotted tube within the borehole and measuring the water level B in the tube (fig. 9.11).

The vertical coordinate x_3^B of the water level in the tube yields the hydraulic head at any point A located below point B. Indeed, because the water is at rest, the hydraulic head at A is the same as at B:

$$h_A = \frac{P_A}{\varrho_w g} + x_3^A = \frac{P_B + \varrho_w g(x_3^B - x_3^A)}{\varrho_w g} + x_3^A = \frac{P_B}{\varrho_w g} + x_3^B = h^B \tag{9.84}$$

If we choose atmospheric pressure as the reference pressure (so that $P_B = 0$) then $h_A = h_B = x_3^B$. The water level in the slotted tube provides a direct measurement of the hydraulic head on the same vertical line, when the aquifer is in equilibrium. This water level is referred to sometimes as the free surface of the aquifer, or the water table or also a the *phreatic surface*. Above this surface the geological formation is not saturated.

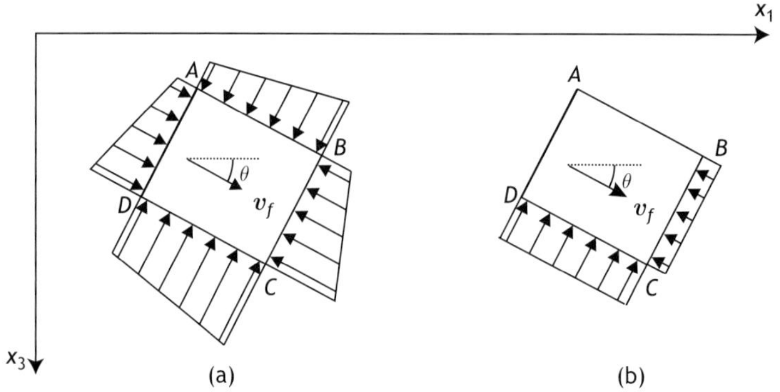

Fig. 9.12 Seepage forces associated with fluid flow in a saturated rock mass.

Below this surface the pore pressure increases as a function of both gravity and the flow conditions.

If flow occurs horizontally in the aquifer and if the hydraulic head remains constant along a vertical direction then the water level in the piezometer is independent of the depth of the piezometer. But if the flow in the aquifer is not horizontal, the hydraulic head depends on the depth reached by the piezometer in the aquifer. In such instances the free surface of the aquifer corresponds to the depth reached by the water in the piezometer only when the piezometer just reaches the saturated zone. For greater depths, the vertical direction is not a streamline and the hydraulic head varies with depth. Then, the water level in the piezometer yields a mean piezometric head that depends on the piezometer length within the aquifer.

Seepage force and seepage pressure

When water is flowing through the pore space, some hydraulic head is dissipated through viscous friction and this results in forces on the solid skeleton, called seepage forces. Seepage forces are equivalent to body forces and must be taken into account when we are considering the equilibrium of a system.

Let us consider a point P in a saturated massif in which the direction of the flow v_f is at an angle θ to the horizontal. A small cubic volume Δv with cross section $ABCD$ and side length b is centered on P with its sides parallel and perpendicular to the direction of flow (fig. 9.12(a)). The drop in hydraulic head between the faces AD and BC is Δh.

The pore water pressure varies along the boundaries of the element; it depends on both the elevation of the point under consideration and its location with respect to the flow. We can define the various values of pore pressure with respect to that at point A (P_A):

$$P_B = P_A + \varrho_w(b \sin\theta - \Delta h) \tag{9.85}$$

$$P_C = P_A + \varrho_w(b \sin\theta + b\cos\theta - \Delta h) \tag{9.86}$$

$$P_D = P_A + (\varrho_w b \cos\theta) \tag{9.87}$$

We show in fig. 9.12(b) the differences in pore pressure between sides AB and CD on the one hand and between sides AD and BC on the other hand, i.e. in the directions parallel and perpendicular to the flow direction. From (9.85)–(9.87) we have

$$P_B - P_A = P_C - P_D = \varrho_w(b \sin\theta - \Delta h) \tag{9.88}$$

$$P_D - P_A = P_C - P_B = \varrho_w b \cos\theta \tag{9.89}$$

The force on side BC is $\varrho_w(b^2 \sin\theta - b\Delta h)$ and that on side CD is $\varrho_w b^2 \cos\theta$. If there was no seepage, the value Δh would be null and the forces on sides BC and CD would be respectively $\varrho_w b^2 \sin\theta$ and $\varrho_w b^2 \cos\theta$ with resultant $\varrho_w b^2$ acting in the vertical direction. The force $J = \varrho_w \Delta h b$ represents the difference between the static and seepage cases; it is called the seepage force and acts in the direction of flow.

The quantity $i = \Delta h/b$ represents the gradient in the hydraulic head (also known as the hydraulic gradient) in the direction of flow, so the seepage force J may be expressed as follows:

$$J = bi\varrho_w b = i\varrho_w b^2 \tag{9.90}$$

The seepage pressure j is defined as the seepage force per unit area:

$$j = J/b^2 = i\varrho_w \tag{9.91}$$

Seepage pressures may destabilize a rock mass when they exceed the static stresses existing in the solid parts. These are known in the soil mechanics literature as *quick conditions* (e.g. Craig, 1987).

9.3.2 The continuity equation for flow through porous media

We saw in section 6.1.1 that the continuity equation for a fluid of density ϱ at time t may be written

$$\frac{\partial \varrho_w}{\partial t} + \operatorname{div}(\varrho_w \boldsymbol{v}) = 0$$

where $\boldsymbol{v}$ is the particle velocity at time t defined in an Eulerian frame of reference. The continuity equation expresses the fact that the change of mass within a fixed volume dv during time dt is equal to the mass that has flowed through the boundary of this same volume dv during dt.

The conservation of mass principle is applicable to fluids that flow through the pore volume. However, the local particle velocity depends on the local geometry of the pore space and therefore remains unknown. An alternative solution is to consider a mean *filtration velocity*.

Let us consider a small cubic volume dv_B of bulk material B, as defined in sections 1.1.1 or 9.1, with size equal to or larger than the REV. The area of each face is A_B and for each face there is a corresponding flux of fluid $\boldsymbol{Q}_i$, with $i = 1, 2, \ldots, 6$. The *filtration velocity* $\boldsymbol{v}_i^f$ is defined as the fluid velocity that would have been observed if the flux $\boldsymbol{Q}_i$ had been through the total area A_B of the corresponding face rather than just through the interconnected pore volume:

$$v_i^f = \frac{Q_i}{A_B} \tag{9.92}$$

thus the rate of fluid flow through the boundary of the volume dv_B is $\mathrm{div}(\varrho_w \boldsymbol{v}^f)$.

Similarly, it is possible to define a mean microscopic flow velocity by assuming that the flow occurs only through the interconnected pore space. We saw in section 1.1.1 that for isotropic materials the linear porosity l, the areal porosity f and the volume porosity n are all equal. If A^F is the total pore surface area on the cube faces then the microscopic mean velocity, also called the *seepage velocity* (Craig, 1987), is

$$v_i^m = \frac{Q_i}{A^F} = \frac{Q_i}{nA_B} = \frac{v_i^f}{n} \tag{9.93}$$

The change in fluid mass within the bulk volume dv_B during the time interval dt is $[\partial(\varrho^f n\, dv_B)/\partial t]\, dt$, where ϱ^f is the mean density of the fluids within the bulk volume dv_B. Hence, the conservation of mass for fluids flowing per unit time through a bulk volume equal to unity is

$$\mathrm{div}(\varrho_w \boldsymbol{v}^f) + \frac{\partial}{\partial t}(\varrho_w n) = 0 \tag{9.94}$$

The continuity equation for porous materials, as expressed by equation (9.94), has been formally derived for incompressible fluids such as water, as well as for compressible fluids such as a mixture of liquid and gas when the flow is steady. The validity of equation (9.94) has also been demonstrated for compressible fluids when the flow is nonsteady and the porosity is variable (e.g. de Marsily, 1981, chapter 3).

9.3.3 Darcy's law and the permeability tensor

Generally the hydraulic head is nonuniform in rock masses and thus some flow occurs. Around the year 1856, Darcy established an empirical relationship between the filtration velocity and the hydraulic gradient.

Darcy's experiment

Consider a vertical cylindrical container of cross section S with a porous base. The cylinder is filled with sand up to a height L and with water up to a height h_A (fig. 9.13).

Darcy observed that the rate of water flow, Q, through the sand is directly proportional to the loss of hydraulic head $(h_A - h_C)/L$, where h_A and h_C are respectively the hydraulic heads at A and C:

$$Q = KS\frac{\Delta h}{L} \tag{9.95}$$

where K is a constant, called the *permeability*, which depends on the porous material through which the flow is occurring. When both sides of equation (9.95) are divided by the area S we get

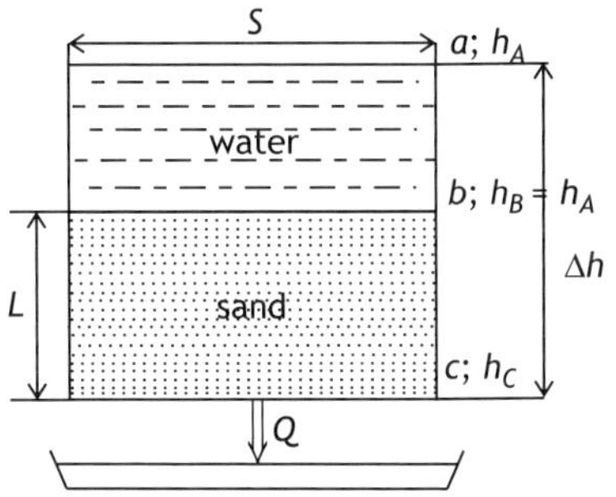

Fig. 9.13 Darcy's experiment for measuring the permeability of a sand.

$$v^f = Ki \tag{9.96}$$

where i is the *hydraulic gradient*. Equation (9.96) is the simplest form of Darcy's law. It has been found experimentally that the constant K varies as the inverse of the dynamic fluid viscosity μ_d.

Further, from the Navier–Stokes equation we know that the causes of fluid flow are pressure gradients and gravity forces, so that the general form of Darcy's law is

$$\boldsymbol{v}^f = -\frac{k}{\mu_d}(\text{grad}\, P + \varrho_w g \delta_{i3} \hat{\boldsymbol{I}}_i), \qquad i = 1, 2, 3 \tag{9.97}$$

where μ_d is the dynamic viscosity (see (6.151)). Darcy's law is valid for both compressible and incompressible fluids (de Marsily, 1981). Here k is the *intrinsic permeability* of the material. In the example above, we took k to be a scalar, which implied that the permeability is isotropic. More generally, however, the permeability must correspond to a second-order tensor $\tilde{k}$, if the various components of the filtration velocity vector are linearly dependent on the components of the pressure gradient vector. For example, a sedimentary sand exhibits a stronger permeability in a direction parallel to the sedimentary layers than in the direction perpendicular to them. Similarly, a material with fractures randomly distributed in all directions, but under a triaxial compressive stress field, exhibits very different permeability characteristics depending on the direction of flow: the fractures perpendicular to the maximum principal stress component are much less hydraulically conductive than those perpendicular to the minimum principal stress component, with intermediate values for fractures perpendicular to the intermediate principal stress.

Since $\tilde{k}$ is a second-order tensor, it is characterized either by six components in any frame of reference or by the three Euler angles that characterize the principal permeability directions and the corresponding three principal permeability values.

For isotropic materials the principal values are equal and any direction of space is a principal direction, so that the intrinsic permeability reduces to a scalar k. The dimension of the intrinsic permeability k is homogeneous to an area:

$$[k] = \frac{[Q][\mu_d]}{[A][PL^{-1}]} = \frac{[L^3T^{-1}][ML^{-1}T^{-1}]}{[L^2][ML^{-2}T^{-2}]} = [L^2] \tag{9.98}$$

Traditionally, however, the intrinsic permeability is expressed in darcy units. One darcy is the permeability of a porous material for which a pressure head of 1 atmosphere (760 mm of mercury) produces a flow of 1 cm^3/s through an area equal to 1 cm^2 of fluid with viscosity 1 centipoise. In petroleum engineering, permeabilities are measured in millidarcy.

In hydrogeology, water is considered incompressible and Darcy's law is then written as

$$\boldsymbol{v}^f = -\frac{k}{\mu_d} \operatorname{grad}(P + \varrho_w g x_3) \tag{9.99}$$

since the hydraulic gradient, $\operatorname{grad} h = -\Delta h/L$, is negative when the x_3 axis is positive upward. The hydraulic head is $h = (P/\varrho_w g) + x_3$, so that

$$\boldsymbol{v}^f = -\frac{k\varrho_w g}{\mu_d} \operatorname{grad} h \tag{9.100}$$

from which we note, using (9.95), that

$$K = \frac{k\varrho_w g}{\mu_d} \tag{9.101}$$

The constant K is homogeneous to a velocity (i.e. the units are m/s). In the same way as for k, K is represented by a two-dimensional tensor for materials with anisotropic permeabilities. Because K depends on the fluid viscosity, it aslo depends strongly on temperature. For example, if measurements of K are conducted at 20 °C, 10 °C and 2 °C, the values measured at 10 °C and 2 °C are respectively 77% and 56% of that measured at 20 °C.

Permeability, hydraulic conductivity and transmissivity

The role of fractures on the flow of fluids in rock masses is discussed further in the next chapter (section 10.2), for the presence of fractures raises specific questions about the representative elementary volume (REV). Let us note here that the hydraulic conductivity has the same definition as the permeability, but this term is normally used in the context of fracture networks whilst the term permeability is reserved for porous systems. As stated in the first chapter, pores have dimensions of the same order of magnitude in all directions whereas fractures exhibit a dimension two to three orders of magnitude smaller in one direction than in the two perpendicular directions.

Often flow occurs within formations which exhibit a uniform finite extension in one direction, which we call the layer thickness; an example is flow in the layers of a sedimentary system (fig. 9.14). If Q is the total flow rate through that layer and l is the extension of the layer in the direction perpendicular to the figure, the rate of flow through a volume with unit extension in the direction perpendicular to fig. 9.14 is

$$\frac{Q}{l} = \int_0^e \boldsymbol{v}^f \cdot \boldsymbol{n}\, dz = \int_0^e v_x^f\, dz \tag{9.102}$$

where z is the axis perpendicular to the layer and x is the axis parallel to it. Assume that the permeability is anisotropic, with one principal direction parallel to z, and consider the hydraulic head $\operatorname{grad} h$ in the direction of flow, x; then

$$v_x^f = -\tilde{K} \operatorname{grad} h \tag{9.103}$$

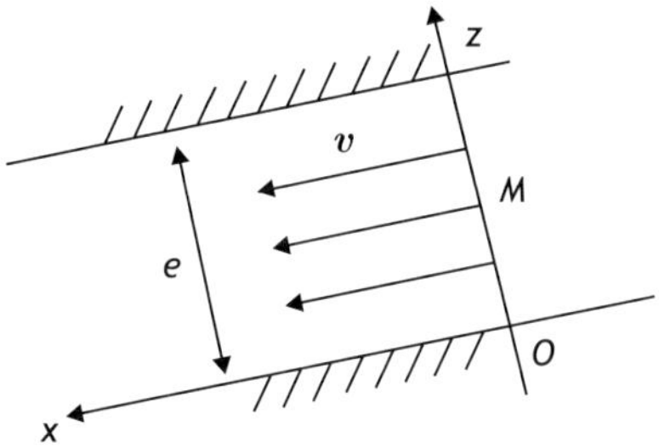

Fig. 9.14 Darcy flow through a sedimentary layer with thickness *e*.

If the hydraulic gradient is independent of x then

$$\frac{Q}{l} = -\text{grad}\, h \int_0^e \tilde{K}\, dz \tag{9.104}$$

The tensor $\tilde{T} = \int_0^e \tilde{K}\, dz$ is called the *transmissivity* of the formation. If the permeability is isotropic then the transmissivity is the scalar $T = eK$. Its dimensions are expressed in m^2/s.

Permeability measurements

The concept of permeability refers to stationary flow, and permeability measurements should theoretically be conducted only under stationary conditions. Experience has shown, however, that in most cases the concept is also valid for transitory conditions, since the corrective term that involve changes in filtration velocity is negligible.

The reliability of permeability measurements in the laboratory depends on the representativity of samples. This is of particular significance for soils and more generally for granular materials, the structure of which is often perturbed during sampling. In addition, when the permeability involves fractures, important issues are raised concerning scale effects, as will be discussed in the next chapter. Hence various methods have been developed for conducting *in situ* measurements in unperturbed conditions at very different scales.

For situations in which the concept of a representative elementary volume is meaningful, various laboratory testing techniques have been developed. The most classical technique is the longitudinal permeameter test. Its configuration is identical to the triaxial test configuration discussed in section 8.1.3 for testing pore pressure effects. With this configuration the material may be tested under various effective-stress conditions, given that a stationary axial flux is imposed on the specimen. The head loss that occurs through the specimen for a corresponding flow rate depends directly on the material's permeability. The technique lends itself to the investigation of various effects related either to the solid (stress-field considerations) or to the fluid (compressibility, viscosity, temperature effects).

For *in situ* methods we may cite the Dupuit method, used for sands and gravels, which involves pumping water into one borehole that reaches the bottom of the strata of concern while monitoring the change in water level in neighboring wells (fig. 9.15).

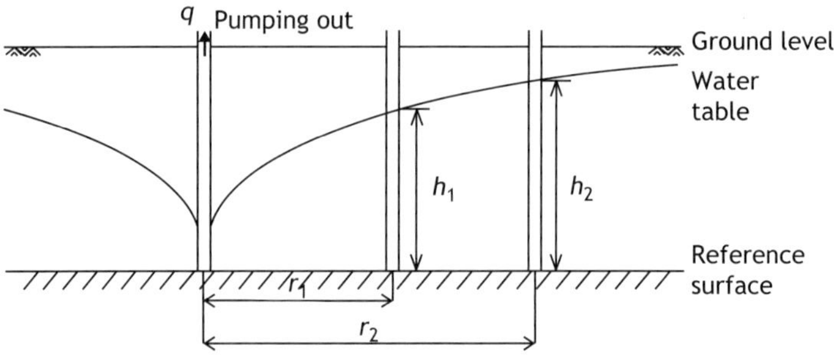

Fig. 9.15 *In situ* measurement of permeability with a stationary pumping test.

The hydraulic gradient at any distance r from the well is assumed to be constant with depth and is equal to the slope of the water table:

$$i_r = \frac{dh}{dr} \tag{9.105}$$

where h is the height of the water table at radius r from the well. At the distance r, the area of the surface through which flow takes place is $2\pi rh$ so that, applying Darcy's law, we have

$$q = 2\pi krh\frac{dh}{dr} \tag{9.106}$$

which yields, on integrating,

$$q\int_{r_1}^{r_2} \frac{dr}{r} = 2\pi k \int_{h_1}^{h_2} h\,dh \tag{9.107}$$

or

(9.108)

$$q\ln\frac{r_2}{r_1} = \pi k(h_2^2 - h_1^2) \tag{9.109}$$

so that finally

$$k = \frac{2.3\,q\log(r_2/r_1)}{\pi(h_2^2 - h_1^2)} \tag{9.110}$$

Typical permeability values range from 10^{-2} m/s for coarse gravels to 10^{-10}m/s for clays. The permeability depends strongly on the particle-size distribution. For well-graded granular materials the permeability is much larger than for granular materials with a large distribution of particle sizes.

Transient methods have also been proposed for measuring *in situ* the permeability of sands (Papadopoulos and Cooper, 1967), using draw-down tests in wells.

As discussed at the beginning of this chapter, the geometry of pores, and more importantly that of fractures, depends strongly on the state of stress. However, the permeability, or hydraulic conductivity, of a geological structure depends on the geometry of the pore space, so some relationship must exist between the permeability and the effective stress

$\tilde{\sigma}' = \tilde{\sigma} - P\tilde{1}$, the more so when the density of fractures increases. For this reason, when the variations in pore pressure caused by fluid flow become significant, the modeling of flow through permeable materials must couple the descriptions of the solid response and of the fluid. This may be of particular significance for nonsteady flow conditions. Hydromechanical coupling and hydrothermomechanical coupling are discussed further in chapter 12.

9.4 Exercises and further reading

9.4.1 Exercises

1. What is the π plane in the theory of plasticity? Demonstrate that in the π plane, the Lode angle ϑ is such that

$$\tan\vartheta = \frac{2\sigma_1 - \sigma_2 - \sigma_3}{\sqrt{3}\,(\sigma_3 - \sigma_2)} \tag{9.111}$$

where σ_1, σ_2 and σ_3 are the principal stress components, with the convention that

$$\sigma_1 \geq \sigma_2 \geq \sigma_3$$

2. Represent the stress conditions implied by the von Mises plastic criterion in the *PQ* plane.
3. Why cannot the von Mises criterion be used for analyzing the plastic deformation associated with compaction or dilatancy?
4. Under what conditions is the Coulomb yield criterion compatible with *associated* plastic flow?
5. What are the units of the intrinsic permeability k? What are those of the permeability K used in hydrogeology, and those of the transmissivity?

9.4.2 Further reading

On the concept of the elastic crust

Burov, E. B., 2010. The equivalent elastic thickness (T_e), seismicity and the long term rheology of the lithosphere: time to burn-out "crème brulée"? Insights from large-scale geodynamic modeling. *Tectonophys.*, **484**, 4–26.

Jackson, J., 2002. Strength of the continental lithosphere: time to abandon the Jelly Sandwich? *GSA Today,* September, 4–10.

Watts, A. B., and Burov, E., 2003. Lithospheric strength and its relationship to the elastic and seismogenic layer thickness. *Earth Planet. Sci. Lett.*, **213**, 113–131.

On plasticity

Davis, R. O., and Selvadurai, A. P. S., 2002. *Plasticity and Geomechanics.* Cambridge University Press, 287 pp.

Sulem J., and Vardoulakis, I. G., 2004. *Bifurcation Analysis in Geomechanics.* CRC Press, 484 pp.

On hydrogeology

de Marsily, G., 1986. *Quantitative Hydrogeology.* Academic Press, 464 pp.

Hiscock, K. M., and Bense, V. F., 2014. *Hydrogeology: Principles and Practice, Second Edition.* Wiley, Blackwell, 520 pp.

10 Fractures and faults

The concept of fracture discussed in this chapter applies to structures for which the dimension in one direction is three to six orders of magnitudes smaller than the dimensions in the two perpendicular directions. These structures are often referred to as *joints*, in the rock mechanics literature. In the geological literature, however, joints refer sometimes only to mode I fractures (no tangential displacement has occurred along the fracture). The word "fracture", as used below, does not imply any restriction on the direction of relative displacements that occur across these surfaces of discontinuity.

Fractures are assimilated to sub-planar structures, which include solid bridges together with the interconnected pore space in between the bridges (see section 1.3.1). Generally fractures involve both solid and fluid phases and here these phases are assumed not to interact except through the effective stress concept already introduced (section 8.1.6). Issues relating to solid–fluid interactions are discussed in chapter 12.

We will separate, in a somewhat artificial manner, the concept of fracture from that of microcracks. Microcracks never exceed a few centimeters and are dealt with implicitly through the definition of the appropriate mechanical or hydraulic characteristics for an equivalent continuous geomaterial (chapter 9). In chapter 7, on fracture mechanics, we outlined the influence of microcracks as local sources of stress concentration. But we emphasize that microcracks are not fractures: they remain of microscopic scale, with simple geometry. Sometimes their surface is assumed to be stress free (Kachanov *et al.*, 2010). Fractures are always larger than the decimeter scale, so their roughness cannot be ignored. They are dealt with explicitly through their mechanical and hydraulic characteristics. Furthermore, attention must be given not only to the individual fracture properties but also to the corresponding fracture network characteristics (see the discussion of fracture fields in section 1.3.3). In this chapter we consider only quasistatic loading processes. Dynamic issues are discussed in chapters 7, 11 and 12.

Although faults result from the growth of fractures, their properties cannot be obtained by simply scaling up fracture characteristics. As mentioned in section 1.3.4, a fault includes a core made of a gouge, with heavily fractured (i.e. cataclastic) zones on both sides of the fault core; these give the fault its specificities. The mechanical and hydraulic characteristics of faults are discussed in section 10.3.

10.1 Mechanical properties of fractures

Fractures can be assimilated to very thin sub-planar objects that include solid contacts and pores, which are usually filled with fluids. Like every structure involving solids, these objects exhibit an elastic and a post-elastic domain. The truly elastic domain is typically that corresponding to wave propagation, i.e. the domain of very small strains (in the 10^{-7}–10^{-6} range); considerable work has been devoted to understanding the influence of fractures on wave propagation (e.g. Kachanov *et al.*, 2010).

For quasistatic loading conditions, because the mechanics of rock masses is significantly influenced, and sometimes completely dominated, by that of the fractures they contain, a great deal of attention has been given to characterizing the mechanical behavior of fractures (e.g. Barton and Stephansson, 1990; Myer *et al.*, 1995). In this section we concentrate on the response of single fractures to quasistatic loading processes, for strains larger than 10^{-5}, and very often larger than 10^{-4}.

In chapter 8 we introduced various testing procedures and equipments for investigating, in the laboratory, the mechanical properties of solid geomaterials (the triaxial cell, Casagrande shear box and ring shear box). Implicit in these testing procedures is the assumption that the tested specimens are representative of the material under consideration. In other words, the assumption that the concept of a representative elementary volume (REV) applies to the material is implicit. Also implicit is the assumption that the sampling technique does not alter the properties to be tested in the laboratory.

However, we saw in section 1.3.1 that the roughness of fractures is self-affine and may be characterized by the Hurst exponent (Schmittbuhl *et al.*, 1995). This implies that, given that the surface roughness characteristics are scale dependent, whether we are considering the central line average (CLA), the mean squares value (MSV) or the root mean square average (RMS) of elevations, the test results for natural fractures are dependent on the size of the tested surface (e.g. Bandis *et al.*, 1983; Scholz, 1990). In addition, the fracture properties of interest are often altered during the sampling operation so that the fracture characteristics measured in the laboratory are only partially representative of *in situ* properties.

Hence, one of the difficulties in characterizing fracture properties is the up-scaling of laboratory experimental results to field conditions, and, in the practice of rock mass reconnaissance, laboratory tests are often complemented by *in situ* tests. The International Society for Rock Mechanics (ISRM) has assembled a set of suggested procedures for testing fracture properties, both in the laboratory and in the field (e.g. Brown, 1981) and the interested reader is referred to these well-established procedures (http://www.isrm.net).

10.1.1 Stiffness and compliance of a fracture

Like any surface in a solid, a fracture always supports some stress components, so that some displacements occur at the fracture surfaces when these stress components vary. The fracture can be assimilated to a thin volume of soft material, and the stress components under consideration are those of the stress vector supported by the fracture plane (with components σ_n and $|\boldsymbol{\tau}|$) whilst the displacements (with components d_n and d_s) are those between two points A and B located at some distance from the upper and lower surfaces of the fracture (fig. 10.1).

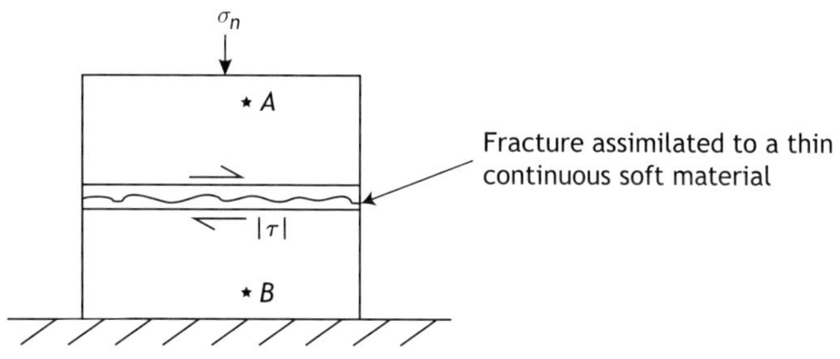

Fig. 10.1 Test for measuring the stiffness of a fracture.

By analogy with the Hooke elastic spring introduced in chapter 2, the stiffness of the fracture is defined by the matrix $[S]$ that relates the components of the stress vector to those of the displacement vector (Goodman *et al.*, 1968):

$$\begin{pmatrix} \sigma_n \\ |\tau| \end{pmatrix} = \begin{pmatrix} S_{nn} & S_{ns} \\ S_{sn} & S_{ss} \end{pmatrix} \begin{pmatrix} d_n \\ d_s \end{pmatrix} \tag{10.1}$$

It is often assumed that $S_{ns} = S_{sn} = 0$, i.e. that the stiffness matrix is diagonal. The dimensions of its components are homogeneous to pressures over displacements and so are expressed in MPa/m. The inverse matrix $[C] = [S]^{-1}$ is called the *compliance matrix* of the fracture.

Fractures are generally assumed to exhibit zero tensile strength, even though some may in fact exhibit a nonzero tensile strength because of local cementation. Hence, the stiffness of fractures is defined for compressive stress fields. Bandis *et al.* (1983) described laboratory investigations on the stiffness of fractures under compression. The principle of their tests is now described.

Normal stiffness

A fracture, together with some intact rock on both its sides is loaded under a normal compression σ_n in a direct shear testing machine (fig. 10.1). Displacements d_t are measured between the points A and B located in the intact rock on each side of the fracture, on a line normal to the fracture plane, and a curve (σ_n, d_t) is plotted. A sample of totally intact rock with similar geometry is also loaded in compression and the corresponding stress–displacement curve (σ_n, d_i) is produced. This curve is then subtracted from the curve obtained with the fractured specimen in order to produce a curve for the displacement associated with the squeezing of the fracture for the corresponding applied normal stress (curve σ_n, d_n), as shown in fig. 10.2. The quantity $d_n = d_t - d_i$ is called the fracture closure.

Various stiffness curves were produced by Bandis *et al.*, depending on whether the fracture surfaces were originally interlocked (intact fracture surfaces) or mismatched (the two sides of the fracture are displaced relative to one another by 0.5–1.0 mm). The pore

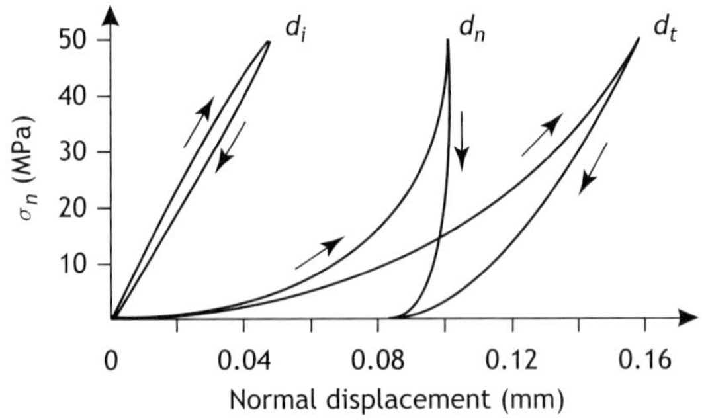

Fig. 10.2 The normal stiffness of the fracture (curve σ_n, d_n) is obtained after subtracting the response of the intact rock (curve σ_n, d_i) from that of the fractured rock (curve σ_n, d_t). Redrawn from Bandis *et al.*, 1983, with permission from Elsevier.

space within mismatched fracture surfaces is much larger than that of interlocked, mated, fractures, so that nonlinearity is much stronger for mismatched fracture surfaces.

The first unloading of the fracture exhibits a strong residual displacement, which indicates that some unrecoverable compaction has occurred within the fracture volume during the loading process. However, if such loading–unloading cycles are repeated a number of times, the nonlinear curves become reproducible without cumulative residual displacements between cycles: the fracture response to normal loading cycles becomes elastic but remains nonlinear.

Such tests have been conducted, for axial loads approaching about one-third of the uniaxial compressive strength of the intact rock, on various kinds of rocks (slate, dolerite, limestone, siltstone), with Young's moduli ranging from 78 GPa for the dolerite to 24 GPa for the sandstone. The area of actual contact between the surfaces ranged from 40% to 70% of the total area, so that some residual aperture was always present even for the largest normal stress values (up to 55 MPa).

Bandis *et al.* (1983) observed that the reproducible stress–fracture-closure relationship obtained for mated (interlocked) fractures may be approximated by a hyperbolic function:

$$\sigma_n = \frac{d_n}{\alpha - \beta d_n} \tag{10.2}$$

where α and β are constants such that when $\sigma_n \rightarrow \infty$, $\alpha/\beta \rightarrow d_m$, where d_m is the maximum closure of the fracture.

Thus the normal stiffness of the mated fracture, is

$$S_{nn}^m = \frac{\partial \sigma_n}{\partial d_n} = \frac{1}{\alpha\left(1 - \beta d_n/\alpha\right)^2} \tag{10.3}$$

If the specific stiffness S_{nn}^s at zero normal stress is defined by

$$S_{nn}^s = \frac{1}{\alpha} \tag{10.4}$$

then the mated-fracture normal stiffness under any normal load is given by

$$S_{nn}^m = \frac{S_{nn}^s}{\left(1 - d_n/d_m\right)^2} \tag{10.5}$$

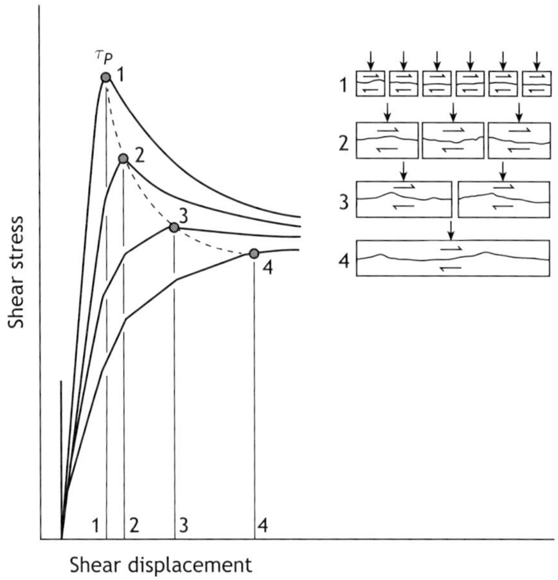

Fig. 10.3 The initial shear stiffness of a fracture, as tested in the laboratory, is strongly scale dependent. Reproduced from Bandis *et al.*, 1981, with permission from Elsevier.

For uncorrelated fracture surfaces, Bandis *et al.* (1983) observed that they could fit their data with a logarithmic function, so that

$$\ln \sigma_n = C + J d_n \tag{10.6}$$

Then the uncorrelated-fracture normal stiffness becomes

$$S^u_{nn} = \frac{\partial \sigma_n}{\partial d_n} = J\sigma_n \tag{10.7}$$

Various analytical models have been proposed for evaluating the normal stiffness of fractures from the roughness characteristics of the surfaces, with some hypotheses on the mechanical response of the asperities to normal loads (e.g. Cook, 1993).

Shear stiffness

Bandis *et al.* (1983) confirmed previous observations that the shear stress versus shear displacement reversible relationship for fractures may be approximated by a hyperbolic function, as long as the shear stress remains smaller than the shear strength of the fracture: thus

$$|\boldsymbol{\tau}| = \frac{d_s}{m + n d_s} \tag{10.8}$$

where d_s is the shear displacement between two points located on each side of the fracture, whilst m and n are fitting parameters. The constant m is the inverse of the initial shear stiffness S^i_{ss} and n is the inverse of the horizontal asymptotic value for the hyperbolic $|\boldsymbol{\tau}|$ versus d_s curve (fig. 10.3); S^i_{ss} is the initial slope of the tangent to the $|\boldsymbol{\tau}|$ versus d_s curve.

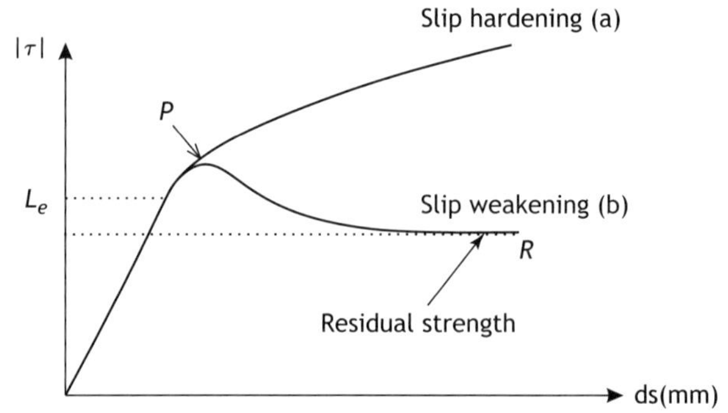

Fig. 10.4 Stress–displacement curves produced during direct shear testing. (a) The shear stress increases with displacement (slip hardening); (b) the shear stress reaches a peak value (slip weakening) and then decreases to a value called the residual strength.

Experience shows that the shear stiffness of a fracture depends on the magnitude of the normal stress acting on the fracture, as will be discussed in section 10.1.4.

Furthermore, it must be kept in mind that experimental data derived from laboratory tests are strongly size dependent: the larger the sample size, the lower the observed shear stiffness, for a given normal stress (see fig. 10.3). Various empirical approaches have been proposed for extrapolating laboratory test results to site conditions, as discussed in the proceedings of a dedicated symposium on rock joints (Barton and Stephansson, 1990).

10.1.2 Friction

When the shear stress component is progressively increased beyond the elastic limit, some irreversible shear motion occurs along a fracture, as shown on fig. 10.4.

Beyond the elastic limit (the point L_e in fig. 10.4), in some instances the slope of the curve $(|\tau|, d_s)$ decreases but stabilizes progressively to a positive value, so that the shear stress must be continuously increased in order to generate further displacement. In such instances the friction is said to be slip hardening (fig. 10.4(a)). More often, the shear stress reaches a maximum value, after which it decreases as slip occurs. In such instances the friction is said to be slip weakening (fig. 10.4(b)).

Further, depending on whether the normal stress magnitude is low or high, the slip motion respectively is or is not associated with a normal displacement. We address first the latter case, when the normal stress is large enough to prevent any significant normal displacement during shear motion; this is the situation for normal stresses in the range 5–150 MPa.

The quasistatic approach

Extensive work on rock friction was conducted during the 1960s and 1970s, with some confusion about the results because of the lack of a clear definition of the various

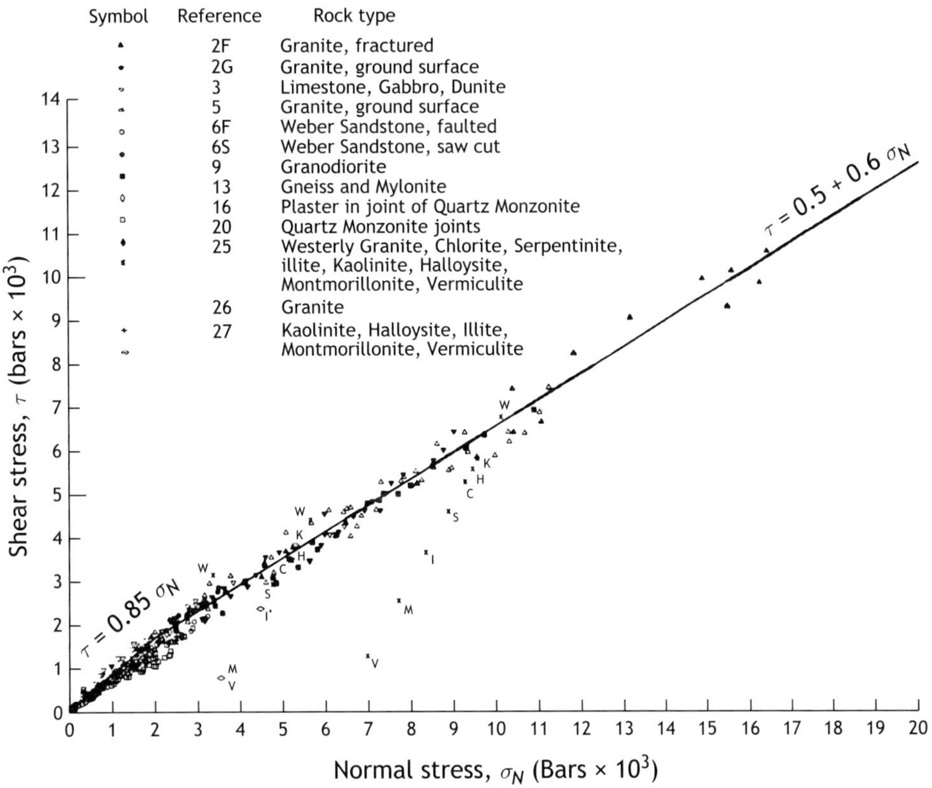

Fig. 10.5 Maximum friction data (reproduced from Byerlee, 1978, with permission from Springer).

quantities under consideration (Byerlee, 1978). Byerlee proposed to specify whether the shear stress value under consideration concerns the elastic limit L_e, the peak value (the point P in fig. 10.4, referred to as the maximum friction) or the residual value after some shear displacement has taken place (the point R in fig. 10.4, referred to as the residual friction). He conducted a compilation of all the results published in the literature (fig. 10.5) concerning the maximum friction, for various rock types tested in dry conditions, and concluded that

$$\text{for} \quad 5\,\text{MPa} < \sigma_n < 20\,\text{MPa}, \qquad \tau_p = 0.85\sigma_n \tag{10.9}$$

$$\text{for} \quad \sigma_n > 20\,\text{MPa}, \qquad \tau_p = 0.5 + 0.6\sigma_n \tag{10.10}$$

Byerlee observed that equations (10.9) and (10.10) are valid for very smooth surfaces as well as for interlocked surfaces, and for very diverse rock types. These two equations are commonly referred to, in the geophysics literature, as *Byerlee's friction law*. It is often generalized to wet conditions by introducing a coefficient μ and a cohesion C. The coefficient $\mu = \tau/\sigma_n$ is generally termed the friction coefficient while the value $\varphi = \arctan \mu$ is the angle of friction for the fracture surface. However, by analogy with the Coulomb yield

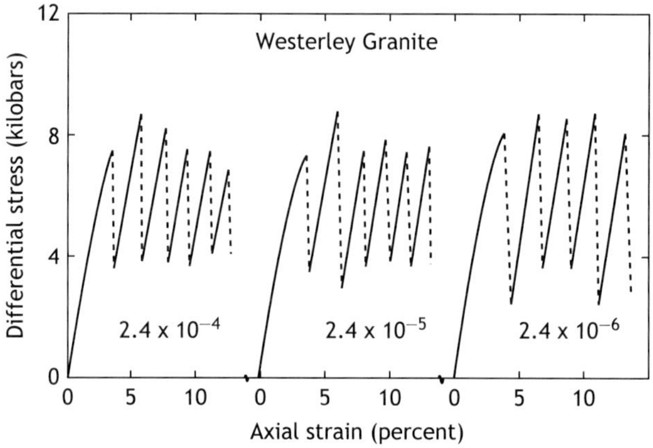

Fig. 10.6 Stick–slip motion during triaxial tests on the ground surface of westerly granite (reproduced from Byerlee and Brace, 1968, with permission from Wiley).

criterion, it is common to call the multiplying factor of σ_n the "friction coefficient", when the friction law includes a constant term as in equation (10.10).

The rate and state approach

Many experiments on friction (e.g. Byerlee and Brace, 1968) have established the fact that, for certain kinds of rocks and surface roughness, motion occurs in a stick–slip manner, as shown on fig. 10.6. Once the peak shear strength of the tested fracture surface has been reached, slip starts with some strong acceleration. The slipping block velocity becomes larger than that of the shear loading system and then stops, whilst the loading system keeps moving at its own constant velocity. Hence, after the block has stopped it is loaded again, until the peak shear stress reaches the shear strength of the surface and unstable motion starts again, and so on.

It was suggested in section 2.5.1 that stick–slip motion may be caused by differences between the values of the friction coefficient for static and dynamic conditions, although these differences remain of second order as compared with the maximum static shear strength.

The concept of a dynamic friction coefficient has been explored by many authors, and a good synthesis of this work was given by Segall (2010, chapter 11), which we will follow.

Slip-weakening friction

In particular, it has been found that the stick–slip phenomenon depends on the stiffness of the testing system, in a manner somewhat similar to failure development in intact rocks, as already discussed in section 8.1.1.

Let us consider the model introduced in section 2.5.1, in which the friction is equal to μ_s as long as the block is static but decreases with slip amplitude ($\partial\mu/\partial\delta < 0$) as soon as motion starts. The motion of the block starts when the force exerted by a spring of

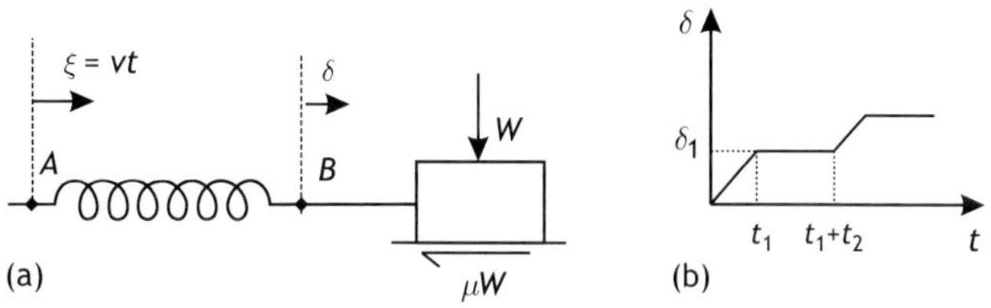

Fig. 10.7 The stick–slip principle revisited.

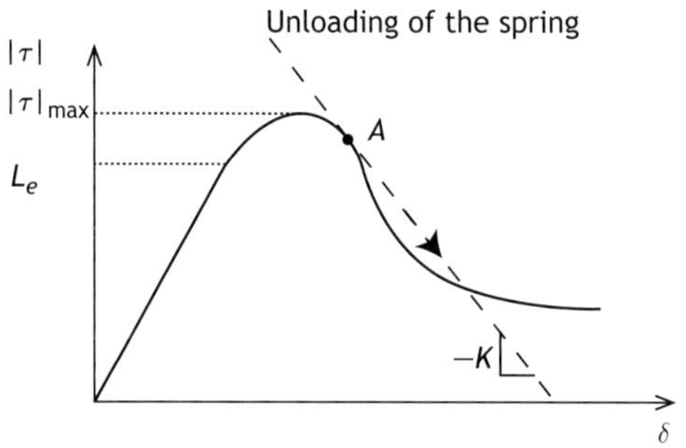

Fig. 10.8 Instability caused by the stiffness of the spring.

stiffness k equals the frictional resistance on the block's horizontal surface:

$$k(vt - \delta) = \tau = \mu_s \sigma_n \tag{10.11}$$

where σ_n is the normal stress acting on the contact area at which friction τ develops. During motion the spring unloads and simultaneously the frictional resistance drops (fig. 10.8).

If the frictional resistance decreases faster than the load exerted by the spring, it becomes smaller than the force acting on the block and so the block accelerates. But when the unloading of the spring reduces the spring force by an amount larger than the loss in frictional resistance, the friction is stable. Thus the friction is

$$\text{unstable,} \qquad \text{if } \sigma_n \frac{\partial \mu}{\partial \delta} < -k < 0 \tag{10.12}$$

$$\text{stable,} \qquad \text{if } -k < \sigma_n \frac{\partial \mu}{\partial \delta} \tag{10.13}$$

Note that, in the unstable configuration, once the block starts it accelerates and acquires some kinetic energy, but at the same time the driving force decreases, as discussed in section 2.5.1. When the driving force drops to zero, the velocity of the block is nonzero and so there is some overshoot, and the block stops at a point where the frictional resistance is larger than the spring force.

Observe that the stability of the slipping process depends on the stiffness of the testing system, on the applied normal stress and on the frictional weakening rate $\partial\mu/\partial\delta$. The larger the normal stress, the less stable the system for a given system stiffness, as verified experimentally in the laboratory by Dieterich (1978).

This simple model implies that, as soon as the block moves, the frictional resistance starts decreasing. But experiments have shown that the motion of the block may be stable initially and becomes unstable only after some displacement has taken place. Furthermore, in some conditions fractures may heal after they have slipped, i.e. their frictional properties depend on the environmental conditions. Hence, further refinement of the friction law is necessary.

Velocity-weakening friction

Since the frictional resistance varies with motion, the question arises whether it depends on the slip rate.

Let us assume now that the friction is a decreasing function of the slip rate alone, for our one-dimensional model of fig. 10.7. When motion initiates, there is an exact balance between the spring driving force and the frictional resistance, as expressed in equation (10.11). At time t, the velocity of the block is v_0 and the drop in frictional resistance associated with motion $\Delta\delta$ is

$$\left.\frac{\partial \tau}{\partial v}\right|_{v_0} \Delta v = -k\Delta\delta$$

or

$$\left.\frac{\partial \tau}{\partial v}\right|_{v_0} \left(\frac{d\Delta v}{dt}\right) = -k\Delta v \tag{10.14}$$

Integration of equation (10.14) over a duration Δt leads to

$$\Delta v = \exp\left(\left.\frac{-k\Delta t}{\partial \tau / \partial v}\right|_{v_0}\right) \tag{10.15}$$

which suggests that the velocity perturbation should grow exponentially, independently of the system stiffness, a feature which is not physically acceptable, as already pointed out. This has led to the development of the rate and state friction theory outlined below.

Rate and state friction

Experimental observations, pioneered by Dieterich (1979) and Ruina (1983), have demonstrated the effects of velocity on friction. This is well illustrated by velocity-stepping tests conducted in steady-state conditions, as shown on fig. 10.9.

For slip at a constant velocity, a unique steady-state response is observed. Associated with a step increase in velocity (velocity increase a, in fig. 10.9), the friction instantaneously increases but then exponentially decreases as displacement takes place. It reaches progressively a value lower than the steady-state value observed for the original slower velocity (drop b in the friction coefficient, in fig. 10.9). When the velocity drops back to its original value, the friction coefficient initially decreases but then progressively increases back to its initial steady-state value.

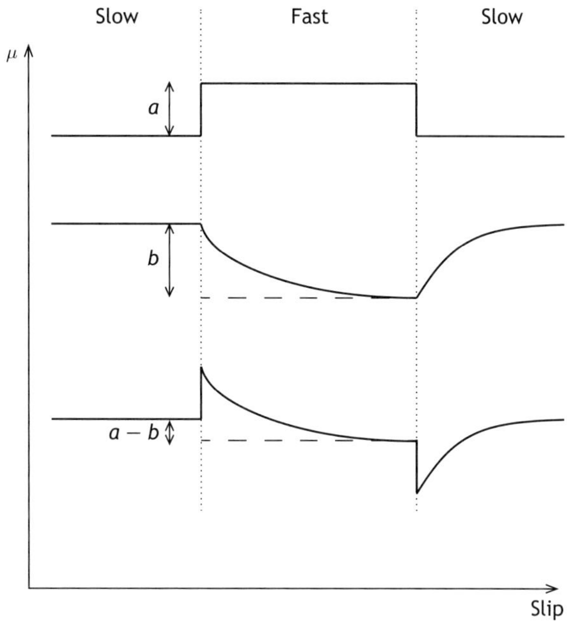

Fig. 10.9 Schematic illustration of a velocity-stepping test and its modeling by the single-state-variable constitutive equation (10.17). The vertical axis corresponds to the friction coefficient whilst the horizontal axis corresponds to the accumulated amount of slip (see the text for the significance of the parameters *a* and *b*). Reproduced from Scholz, 1990, with permission from Cambridge University Press.

This instantaneous, synchronous, increase of friction (a on fig. 10.9) with an increase of slip velocity is referred to as direct velocity strengthening ($\partial\tau/\partial v > 0$). Experiments show that the instantaneous friction increase varies as the logarithm of the slip velocity variation.

The subsequent gradual variation in friction occurs over one or more characteristic displacement scales d_c that depend on the surface roughness or on the presence of filling material (gouge) in between the slipping surfaces.

The final steady-state frictional dependence on slip velocity, $\partial\tau_{ss}/\partial v$, can be either positive or negative (the quantity $a - b$ in fig. 10.9).

This may be summarized by stating that the friction coefficient depends on the current slip velocity magnitude v and on one or more state variables that characterize the state of the slipping surfaces:

$$\tau = \sigma_n \mu(v, state) \tag{10.16}$$

Equation (10.16) corresponds to a so-called rate and state friction law. The most commonly used form of equation (10.16) is

$$\tau = \sigma_n \left(\mu_0 + a \ln \frac{v}{v_0} + b \ln \frac{\theta}{\theta_0} \right) \tag{10.17}$$

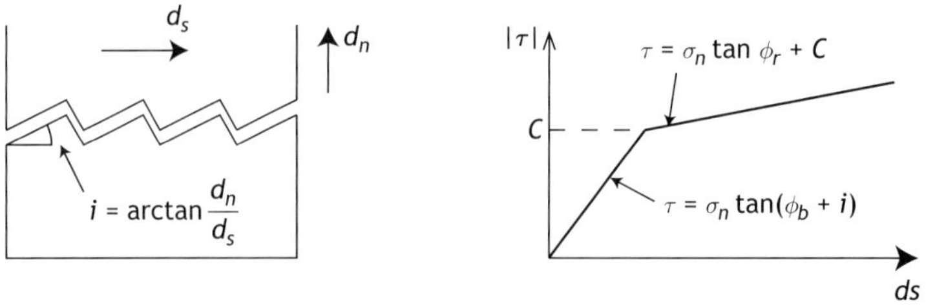

Fig. 10.10 (a) Modeling fractures as sawtooth structures; (b) bilinear shear strength.

where μ_0 is the steady-state friction coefficient at velocity v_0, θ is a so-called state variable that characterizes the state of the surface and θ_0 is the value of θ for a steady velocity v_0.

Scholz (1990) observed that if $a - b < 0$ then the frictional behavior is velocity weakening and the slip is unstable, while if $a - b > 0$ then the slip behavior is velocity strengthening and slip is stable.

10.1.3 Shear strength at low normal stress and fracture dilatancy

When the normal stress acting on a fracture is low, say smaller than 5 MPa, the peak shear strength depends strongly on the roughness of the fracture surfaces.

Patton (1966) proposed taking into account this roughness by modeling fractures as sawtooth-like structures (fig. 10.10(a)), so that the shear strength of a fracture may be modeled by a bilinear relationship (fig. 10.10(b)):

$$\text{for } \sigma_n < \sigma_T, \qquad \tau_p = \sigma_n \tan(\phi_b + i) \tag{10.18}$$

$$\text{for } \sigma_n > \sigma_T, \qquad \tau_p = \sigma_n \tan \phi_r + C \tag{10.19}$$

where i is the angle between the inclined slipping sawtooth surface and the mean fracture plane, ϕ_b is the friction angle associated with the longer inclined part of the sawtooth surfaces, whilst ϕ_r is the residual friction angle for the mean fracture surface after shearing of the teeth and is such that $\phi_r < \phi_b + i$.

This model assumes that, for low normal stresses, slip occurs along the inclined part of the sawtooth surface whilst, for higher normal stress, the asperities do not slip but generate a cohesion C for the fracture surface. Once this cohesion has been broken, some residual friction controls the shear resistance of the mean slipping surface.

The angle i is called the fracture dilatancy angle, or more simply the dilatancy angle. It corresponds to a normal displacement associated with the riding on the roughness of the fracture surfaces. A more general definition of the fracture dilatancy angle is given by

$$i = \arctan(d_n/d_s) \tag{10.20}$$

where d_n is the normal displacement observed when the shear displacement along the fracture is equal to d_s.

In chapters 8 and 9 we introduced the concept of dilatancy to describe the volume increase associated with shear stresses, whether the micromechanics involve the growth of mode I microcracks or the motion of grains in granular materials. In this book we will

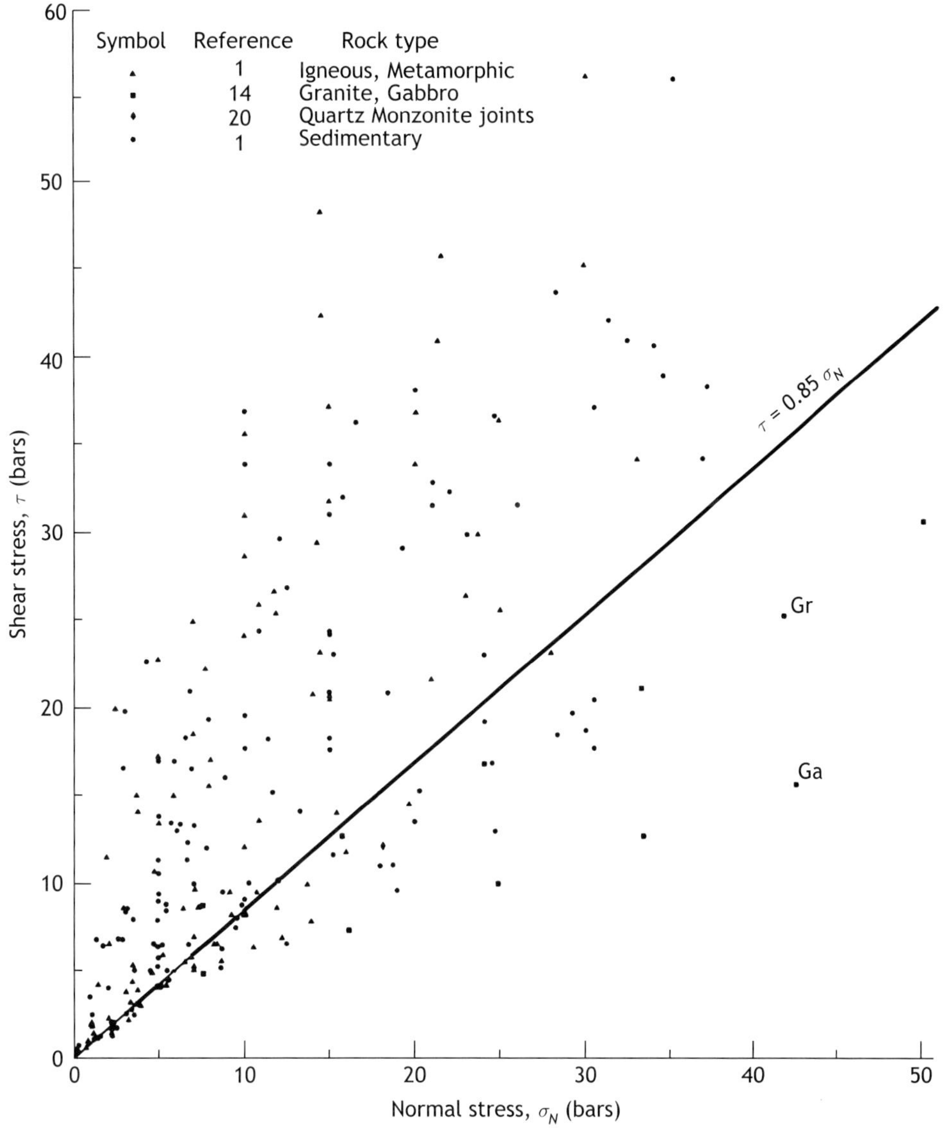

Fig. 10.11 Peak shear strength of unfilled natural fractures (reproduced from Barton, 1976, with permission from Elsevier).

reserve the term "dilatancy" for the bulk volume increase associated with shear stresses and we will use the term "fracture dilatancy" when referring to an increase in volume caused by the shearing of preexisting fractures. Note that dilatancy may occur only for nonlinear elastic materials (Stuart, 1974); it is incompatible with a linear elastic response.

Barton (1976) observed that, under low normal loads, the peak shear strength of fractures, i.e. the maximum shear stress that may be applied to a fracture under a given normal load, varies widely (fig. 10.11). He conducted a series of direct shear tests on a variety of artificial tension fractures generated in various brittle materials. For each test he measured the peak fracture dilatancy angle i_p and plotted the value of the *total friction angle*

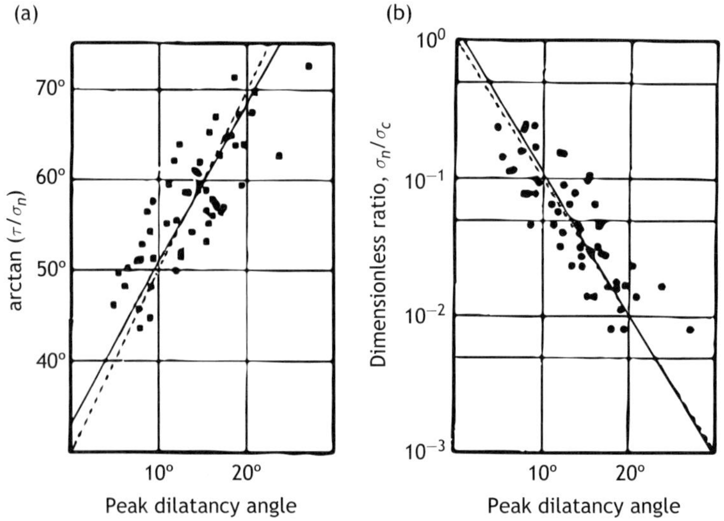

Fig. 10.12 (a) Plot of total friction angle ϕ_t versus peak fracture dilatancy angle i_p; (b) plot of the ratio of the uniaxial compressive strength and the applied normal stress versus the peak fracture dilatancy angle (reproduced from Barton, 1976, with permission from Elsevier; Barton used the term "dilatation" rather than "dilatancy").

$\phi_t = \arctan(\tau_p/\sigma_n)$ as a function of i_p (fig. 10.12(a)). Indeed, he noted that the experimental evidence shows that the peak fracture dilatancy angle is observed when the peak shear strength τ_p is reached. On a different diagram (fig. 10.12(b)) he plotted the peak fracture dilatancy angle versus the ratio of the applied normal stress σ_n and the uniaxial compressive strength σ_c of the material. He determined the best linear least squares fit to the data (the solid lines) and then changed the slopes slightly so as to fit the following equations (the dotted lines):

$$\frac{\tau_p}{\sigma_n} = \tan(2i_p + 30°) \tag{10.21}$$

$$i_p = 10 \log_{10}\left(\frac{\sigma^c}{\sigma_n}\right) \tag{10.22}$$

Then he could determine the maximum shear strength for any given normal load after eliminating the peak fracture dilatancy angle i_p between equations (10.21) and (10.22):

$$\tau_p = \sigma_n \tan\left[20 \log_{10}\left(\frac{\sigma^c}{\sigma_n}\right) + 30°\right] \tag{10.23}$$

Note that, on the one hand, for very smooth fractures the peak strength should be related to the normal stress through the basic static friction angle ϕ_b, so that, as σ_n gets closer to σ^c, equation (10.23) should tend to

$$\tau_p = \sigma_n \tan \phi_b \tag{10.24}$$

On the other hand, for very rough surfaces, Barton observed that the coefficient 20 in equation (10.23) fits the observations satisfactorily. Hence, he proposed a generalization of equation (10.23) to represent the shear strength of all fracture surfaces:

$$\tau_p = \sigma_n \tan\left[JRC \log_{10}\left(\frac{JCS}{\sigma_n}\right) + \phi_b\right] \tag{10.25}$$

where *JRC*, the joint roughness coefficient, is a parameter that characterizes the fracture roughness whilst *JCS*, the joint wall compressive strength, represents the uniaxial compressive strength of the fracture walls.

Equation (10.25) is known as Barton's peak shear strength criterion. It takes into account implicitly the fracture dilatancy associated with the roughness of the surface, through the parameter *JRC*; *JRC* is defined empirically and varies from 0, for very smooth fractures, up to 20 for very rough fractures, according to a scale proposed by Barton. The joint wall compressive strength *JCS* takes into account the breaking of asperities. It is equal to the unconfined compressive strength σ^c of the fracture wall when the fracture is unaltered, but it may be as low as $\sigma^c/4$ if the fracture is weathered; this parameter may be determined with normalized Schmidt hammer tests (Barton *et al.*, 1985).

Bandis *et al.* (1981) proposed a simple tilt test to determine the *JRC* parameter for natural fractures (fig. 10.13). Individual blocks with through-going fractures are slowly

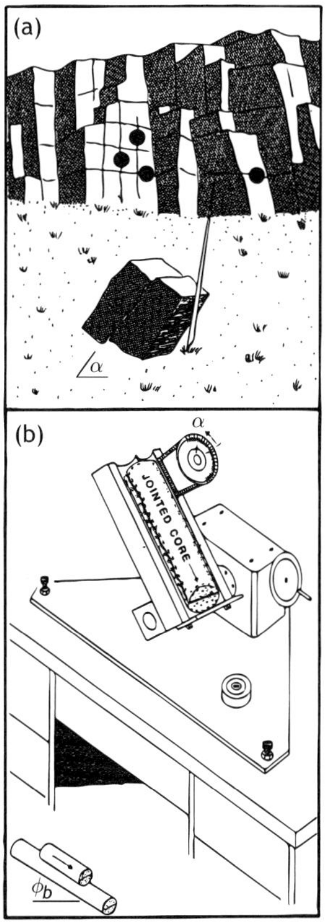

Fig. 10.13 Determination of the *JRC* coefficient from tilt tests (reproduced from Barton *et al.*, 1985, with permission from Elsevier).

tilted up to the inclination α at which an upper block slips on a lower block. The normal stress σ_{n0} supported by the fracture when slip starts is directly determined from the mass of the slipping block and the tilt angle α. The parameter is given by

$$JRC = \frac{\alpha - \phi_b}{\log_{10}(JCS/\sigma_{n0})} \tag{10.26}$$

Grasseli and Egger (2003) showed, for laboratory experiments an equivalence between the empirical *JRC* parameter and the roughness characteristics that are accessible to direct optical measurements. Xie and Pariseau (1995) discussed the relationships between Barton's *JRC* scale and the fractal attributes of roughness.

10.1.4 Empirical constitutive equations for fractures

The above discussion indicates that, after a few cycles, the cyclic shear or normal loading of fractures under compression exhibits a reversible behavior that may be characterized by a normal stiffness and a shear stiffness. Both involve nonlinear functions, of the normal stress acting on the surface, which depend on the size of the fracture. When the shear stress exceeds a critical value, we have identified two different behaviors depending on the normal stress magnitude.

When the normal stress is larger than 5 MPa

The change in volume associated with the shear is negligible and the peak shear strength is characterized by a linear relationship between the normal stress and the peak shear stress, $\tau_p = \mu\sigma_n + C$, called Byerlee's law in the geophysics literature and the Coulomb friction law in the engineering literature.

When the fracture is filled by a fluid under pressure P the normal stress must be replaced by the effective normal stress, so that the peak shear strength is now given by

$$\tau_p = \mu(\sigma_n - P) + C \tag{10.27}$$

However, once slip starts, the friction coefficient μ varies and its variation is often represented by a rate and state friction law, which depends in particular on environmental factors, including temperature, weathering etc. A commonly used rate and state equation for a fracture filled by a fluid under pressure P is

$$\tau = (\sigma_n - P)\left(\mu_0 + a \ln \frac{v}{v_0} + b \ln \frac{\theta}{\theta_0}\right) \tag{10.28}$$

where τ is the shear stress supported by the fracture when the slip velocity is v, and θ is a so-called state variable that characterizes the wear of the surface. The rate and state slip equation implies that either slip hardening occurs, leading to stable slip when $b - a > 0$, or slip weakening takes place, leading to slip instability when $b - a < 0$.

We will discuss in section 11.2 examples where the same slipping surface involves large stable slipping patches together with localized slip zones where the slip is unstable and leads to repeated seismic signals from the same source (also called an asperity). Such a behavior may be modeled by assuming that the parameters a and b in equation (10.28) vary with the location within the shearing surface.

Results from both natural fault monitoring (Nadeau and McEvilly, 1999) and geothermal energy development projects (Cornet, 2012) suggest that the sizes of the unstable slip zones increase with depth, i.e. with the normal stress magnitude. Given that these unstable slipping phenomena generate seismic waves and that the magnitude of these waves depends on the size of the unstable slip zone (see section 11.2), this is consistent with the observation that earthquakes (mechanical vibrations felt at ground surface) are only observed for source depths larger than a few kilometers. For shallower depths the size of the unstable slip zones is small, so that the amplitudes of the corresponding seismic waves are also small and disappear in the local seismic noise before reaching the ground surface.

When the normal stress is smaller than 5 MPa

The slip motion implies some riding on the asperities of the fracture surface, resulting in fracture dilatancy. For these low effective-normal-stress conditions, a description of the fracture mechanical behavior must involve the fracture stiffness matrix, the shear strength as a function of the effective normal stress and the shear displacement and, finally, the fracture dilatancy, with its dependence on the effective normal stress and the amount of slip displacement; all these quantities depend on the fracture size.

Figure 10.3 indicates that the shear stress versus shear displacement curve may exhibit a range of shapes, from the bilinear peak-residual curve observed for small-size samples to a smooth hyperbolic response for large fractures. Further, the shear-stress–shear-displacement relationship depends on the normal stress acting on the fracture, as shown in fig. 10.14. Similarly, the higher the normal stress, the smaller the dilatancy angle for a given shear displacement (fig. 10.15).

In order to take into account the effect of the fracture size on the mechanical behavior of a fracture, Barton *et al.* (1985) pointed out that numerous observations from laboratory experiments suggest that the peak shear strength is generally reached when the shear displacement δ_p has reached about 1% of the length L of the fracture ($L = 10$ cm). From equation (10.25), if a fluid fills the fracture and is under a constant pressure P, the corresponding peak friction angle ϕ' is given by

$$\phi' = JRC \log_{10}(JCS/\sigma_n') + \phi_r \tag{10.29}$$

After the peak shear strength has been reached the roughness is progressively destroyed, as shear displacement takes place, until it reaches a stable residual value. Barton *et al.* generalized equation (10.29) to any post-peak shear displacement:

$$\phi'(\text{mob}) = JRC(\text{mob}) \log_{10}(JCS/\sigma_n') + \phi_r \tag{10.30}$$

where $\phi'(\text{mob})$ and $JRC(\text{mob})$ are respectively the mobilized drained (see sections 8.1.6 and 12.1.1) friction angle and the mobilized roughness for the corresponding shear displacement d_s. They observed that the dimensionless ratios $JRC(\text{mob})/JRC(\text{peak})$ and $d_s/d_{s\ peak}$ vary during shear in an almost identical manner, for a wide range of normal stresses and for a wide variety of fracture surfaces. This observation led them to propose a

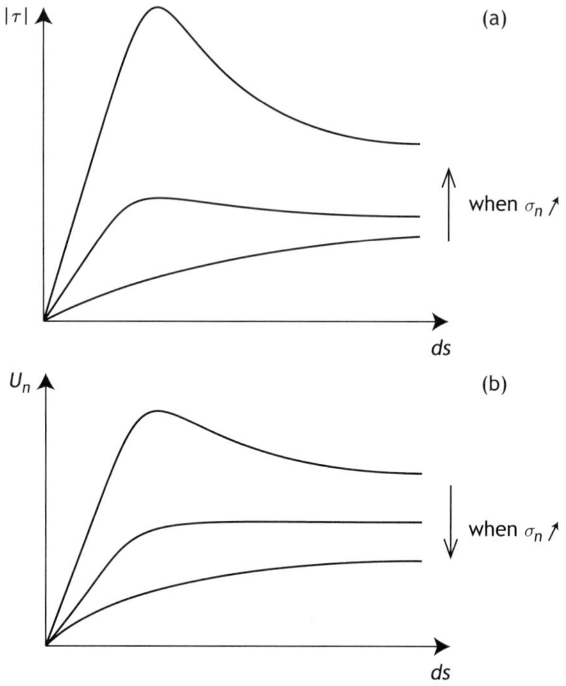

Fig. 10.14 Influence of the normal stress magnitude on (a) the shear stress $|\boldsymbol{\tau}|$ versus shear displacement d_s curve and (b) the normal displacement U_n versus shear displacement d_s curves for preexisting fractures.

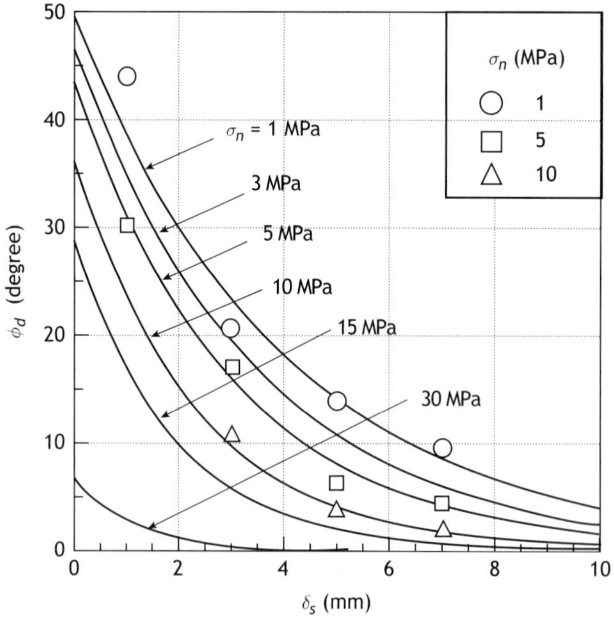

Fig. 10.15 Influence of the normal stress magnitude on the maximum dilatancy angle ϕ_d versus shear displacement δ_s curve (reproduced from Matsuki *et al.*, 2010, with permission from Elsevier).

parametric characterization of the shear strength and dilatancy for fractures up to 1 m long and for normal stresses smaller than about 10 MPa and shear displacements reaching a few percent of the fracture length. They pointed out that these empirical relationships apply also to reverse shear tests, as is for example the case during seismic loading (e.g. Hutson and Dowding, 1990; Huang *et al.*, 1993).

This discussion indicates that a description of the mechanical behavior of fractures remains somewhat empirical. We have reported propositions formulated by Barton's group in the early 1980s, for they capture quite effectively the phenomenology of the shearing process. However, various alternative models have been proposed for numerical modeling, as discussed for example in the ISRM rock joints symposia (e.g. Barton and Stephansson, 1990; Myer *et al.*, 1995).

As an example of an alternative numerical model, we present below the "continuously yielding joint" model of Cundall and Lemos (1990), which exhibits the following main characteristics.

1. The shear stress versus shear displacement curve always tends to a "target" shear strength for the fracture, i.e. the instantaneous gradient of the curve depends directly on the difference between strength and stress.
2. The target shear strength decreases continuously as a function of the accumulated irreversible displacement.
3. The dilatancy angle is to be taken as the difference between the apparent friction angle (determined by the current shear stress and normal stress) and the residual angle.

The response to normal loading is expressed incrementally:

$$\Delta\sigma_n = k_n \Delta u_n \tag{10.31}$$

where the normal stiffness k_n is given by

$$k_n = a_n \sigma_n^{e_n} \tag{10.32}$$

with a_n and e_n model parameters intended to provide the required increase in normal stiffness with normal stress.

For shear loading, the model displays an irreversible (plastic) behavior from the onset of shearing (see fig. 10.16). The shear stress increment is calculated as

$$\Delta\tau = F k_s \Delta u_s \tag{10.33}$$

where the shear stiffness k_s also increases with the normal stress component:

$$k_s = a_s \sigma_n^{e_s} \tag{10.34}$$

In equation (10.33) the factor F that modifies the shear modulus to $F k_s$ depends on the distance between the current shear stress versus shear displacement curve and a target strength curve (see fig. 10.16):

$$F = \frac{1 - \tau/\tau_m}{1 - r} \tag{10.35}$$

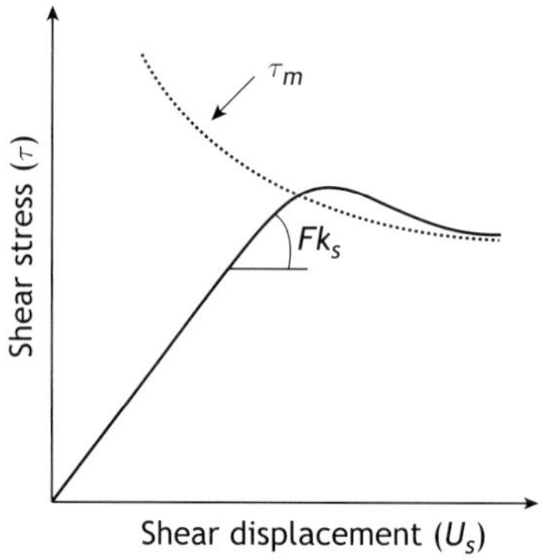

Fig. 10.16 The continuously yielding joint model: the shear stress versus shear displacement curve and the target shear strength τ_m.

where the factor r accounts for the behavior observed upon unloading (or upon reloading after unloading, during cyclic loading). It is set equal to τ/τ_m whenever there is a loading reversal. The target strength is defined as

$$\tau_m = \sigma_n \tan \phi_m \operatorname{sign}(\Delta u_s) \tag{10.36}$$

The parameter ϕ_m corresponds to the friction angle that would apply at the maximum dilation angle. As damage accumulates with shearing, this friction angle decreases continuously, according to

$$\Delta\phi_m = \frac{-1}{R(\phi_m - \phi)\Delta u_s^p} \tag{10.37}$$

where the plastic shear displacement is defined as

$$\Delta u_s^p = (1 - F)|\Delta u_s| \tag{10.38}$$

and ϕ is the basic friction angle of the rock surface; R is a material parameter which expresses the fracture roughness.

Accordingly, this model, uses six parameters to fit the observed fracture mechanical behavior, $a_n, e_n, a_s, e_s, R, \phi_m^i$, where ϕ_m^i is the fracture friction angle before any shear displacement has taken place.

10.2 Hydraulic properties of fractures

We saw in chapter 1 that rock masses involve solids and fluids and that the motion of fluids is localized within the interconnected pore space. Then we proceeded by distinguishing pores, which exhibit dimensions of the same order of magnitude in all directions, and fractures, which are characterized by their specific shape. This led us in section 9.3.3 to distinguish two kinds of permeability: a primary permeability associated with flow through the interconnected pores and a secondary permeability, often called the hydraulic conductivity, associated with flow through the fracture network. We concentrate here on modeling the flow of incompressible Newtonian fluids through a single fracture, assuming that there the flow has no effect on the fracture geometry.

The simplest model of flow through a fracture is that introduced in section 6.3, where we considered a unidirectional flow between two parallel plates. Using mass conservation (equation (6.36)) and the Navier–Stokes equation (6.157), we observed that the flow rate Q per unit width of fracture is given by

$$Q = \int_0^b v_1 \, dx_3 = -\frac{p_{,1}}{2\mu_d}\left[\frac{x_3^3}{3} - \frac{bx_3^2}{2}\right]_0^b \tag{10.39}$$

Thus

$$Q = -\frac{b^3}{12\mu_d} p_{,1} \tag{10.40}$$

where b is the distance between the two parallel plates (representing the fracture thickness), $p_{,1}$ is the pressure gradient in the direction of flow and μ_d is the dynamic fluid viscosity. For a rectangular fracture of width w, the flow rate is

$$Q = -\frac{w b^3}{12\mu_d} p_{,1} = \frac{T}{\mu_d} p_{,1} \tag{10.41}$$

where the quantity $T = wb^3/12$ is known as the fracture transmissivity. Equation (10.41) is often referred to as the *cubic law*.

Note that equation (10.41) is valid only for laminar flow in one direction when the plate surfaces are smooth. We saw, however, that fractures under compression exhibit solid contacts, which control the solid mechanics behavior of the fracture, and a complex pore space, in which the flow cannot be characterized by the simplified Poiseuille model: for flow between rough surfaces, some advective terms must be kept in the Navier–Stokes equation, which render it nonlinear. Many authors have discussed this issue (e.g. Barton and Stephansson, 1990). In particular, it has been shown that, because of fracture roughness, flow has a tendency to concentrate in local channels (Tsang and Tsang, 1989; Pyrak-Nolte and Morris, 2000). An example of a model adopted to account for the effect of roughness on the hydromechanical behavior of a fracture is shown in fig. 10.17. Channeling is of central concern when one is considering heat exchange phenomena or solid–fluid chemical interactions (e.g. Neuville *et al.*, 2011).

In many instances, however, the cubic law remains an acceptable approximation, especially for flow in large rock masses, in which the effect of the fracture network structure dominates effects resulting from the details of the flow in individual fractures. We present below results of Witherspoon *et al.* (1980), who examined the validity of the cubic law for the flow of liquid water in a single fracture.

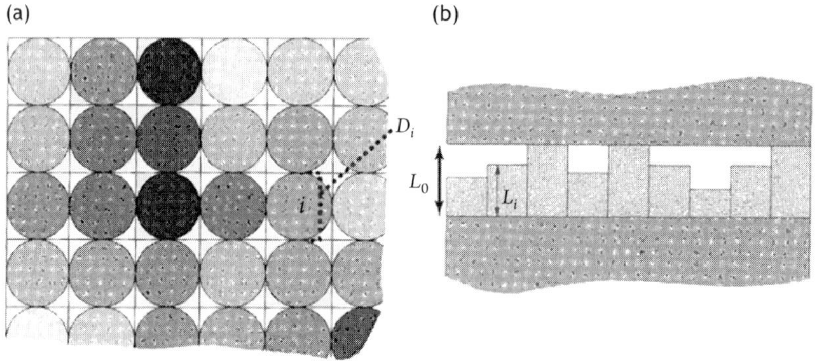

Fig. 10.17 Flow through a rough fracture under load. The irregular shape of the flow path and its dependence on the effective normal load is modeled by parallel cylinders of equal diameter D but irregular length L_i: (a) plan view, (b) cross section of the fracture. The stress field is transmitted through the solid contacts whilst the flow occurs through the empty spaces (reproduced from Pyrak-Nolte and Morris, 2000, with permission from Elsevier).

Witherspoon *et al.* observed that, by analogy with Darcy's law, for steady isothermal flow conditions the water flux per unit drop in piezometric head, $Q/\Delta h$, may be written as

$$\frac{Q}{\Delta h} = C(2b)^3 \tag{10.42}$$

where $2b$ is the fracture hydraulic aperture and C is a constant given by

$$C = \frac{2\pi}{\ln(r_e/r_b)} \frac{\varrho g}{12\mu_d} \quad \text{for radial flow,} \tag{10.43}$$

$$C = \frac{w}{L} \frac{\varrho g}{12\mu_d} \quad \text{for straight flow,} \tag{10.44}$$

For the case of radial flow, r_e is the outer diameter of the planar circular fracture and r_b is its inner diameter of the circle (e.g. a borehole cross section), through which the flow enters the fracture. For the straight-flow case, L is the fracture length.

Witherspoon *et al.* ran flow experiments through both mechanically open fractures, i.e. two fracture surfaces with no contact between them, and mechanically closed fractures, i.e. two fracture surfaces in contact with each other through the peaks of the two rough surfaces. The mechanically closed fracture was also tested under normal load conditions as high as 20 MPa. A so-called fracture "thickness" (equivalent to the fracture hydraulic aperture discussed below) was defined, following a procedure somewhat similar to that described in section 10.1.1. By comparison with the displacements observed for a similar parallelepipedic volume of intact rock under similar loading conditions, a fracture closure is defined for a variety of normal stress conditions. Hence, for mechanically closed fractures, an *aperture* may be defined from the measured fracture closure and from a residual aperture defined by a curve fitting procedure. These equivalent fracture apertures, which we will call *hydraulic apertures*, ranged from 250 μm to 4 μm; here, the smaller value is the residual hydraulic aperture measured under 20 MPa normal load.

The cubic law, expressed as $Q/\Delta h = C(2b)^3$, has been found to be valid whether the fractures are mechanically open (no contact point between the surfaces on each side of the fracture) or whether they are closed and under normal load, provided that a corrective factor f is applied to the equivalent fracture aperture determined from the fracture closure. This corrective factor, defined by replacing the constant C in the cubic law by C/f, has been found to range from 1.04, when the blocks just touch each other, to 1.65 for very high normal loads. A 1.65 correction of the equivalent hydraulic aperture corresponds to a 40% reduction in flow rate compared with that between two parallel, nontouching, fracture surfaces.

Interestingly, Barton *et al.* (1985) challenged the hydraulic aperture measurements of Witherspoon *et al.* because they were not direct fracture opening measurements but were derived from indirect fracture closure observations. They proposed some empirical relationships based on the *JRC* coefficient introduced in the previous section and derived by another curve fitting procedure.

Clearly, the concept of a fracture hydraulic aperture is convenient for evaluating the flow in fractured rock masses. However, no unique method exists for determining these hydraulic apertures for single fractures. A common practice is to conduct in situ hydraulic tests in boreholes. An inflatable straddle packer is set on a fracture intersected by a borehole (see section 13.2). Then, after conducting an injection test on the fracture, a cubic law is fitted to observations. By repeating the test on a number of fractures from the same fracture set, the hydraulic conductivity of single fractures from the corresponding fracture set may be characterized by an expected value and a confidence level, assuming a gaussian law of error distribution.

Such tests assume no disturbance of the fracture properties upon testing; however, we showed in fig. 10.15, from Matsuki *et al.* (2010), how a shear displacement influences the dilatancy angle of a fracture under some normal stress. These authors found that this change in dilatancy results in an increase in the fracture transmissivity, an effect which depends on the orientation of the shear direction with respect to that of the flow direction. The observation that the hydraulic transmissivity of fractures varies with the shear displacement is the concept applied during hydraulic stimulation, as further discussed in section 12.4.

It should be kept in mind that these fracture hydraulic apertures are not true mechanical apertures. They should not be confused with openings sometimes observed *in situ* at the intersection between fracture planes and boreholes, by various logging techniques. Such logging observations apply to only a very local property of the fracture and cannot be extrapolated to the complete fracture surface.

10.3 Mechanical and hydraulic characteristics of faults

Faults are *not* large fractures, as already stated. They always exhibit a complex structure that changes with time depending on the loading process, whether through active tectonics or because of mechanical readjustments associated with erosion, glaciations, or any other mechanism (e.g. Turpeinen *et al.*, 2008). Even when no active mechanical loading is applied to them, their properties may vary with time because of fluid circulation. We will introduce some basic concepts of structural geology and then outline some consequences for the hydromechanical characteristics of faults.

10.3.1 Faults and fault growth

The word "fault" is derived from an eighteenth century mining term describing a surface across which coal layers were offset (Twiss and Moores, 1992, chapter 4). Hence, faults always correspond to shear displacements, and these displacements may range from a few meters to a few hundred kilometers. Accordingly, their thickness may range from tens of centimeters to a few kilometers, and we will see that this concept of a fault thickness requires some specific attention. Like fractures, faults are characterized by the fact that one of their dimensions is several orders of magnitude smaller than their dimensions in the two perpendicular directions.

Depending on their shear displacement, faults are classified into three categories. When the fault exhibits a dip the upper block is called the hanging wall whilst the lower block is called the footwall; such faults are called dip-slip faults. If the hanging wall rides up the footwall, the dip-slip fault is called a thrust or reverse fault (fig. 10.18(b)), whilst if it rides down the fault it is called a normal fault (fig. 10.18(a)). When the fault is vertical, it is called a strike-slip fault. A strike-slip fault is said to be right-lateral when the motion is associated with a clockwise relative motion (fig. 10.18(d)), and left-lateral if the relative motion is counterclockwise (fig. 10.18(c)). However in many cases, the fault's relative displacement involves both a dip-slip component and a strike-slip component so that faults may be, for example, sinistral-normal or dextral-normal.

Another possible way of describing fault motion consists in characterizing the direction of slip in the fault plane by the direction cosines of a unit vector contained in the fault plane and parallel to the slip direction, as mentioned in section 1.3.1.

Anderson (1905) pioneered the application of mechanical concepts to analyze the origin of faults and suggested that one should apply the Coulomb failure criterion to relate fault formation to the existing stress field. Today, the classification into normal, reverse and strike-slip faulting is sometimes called Andersonian fault theory. The basic approach is

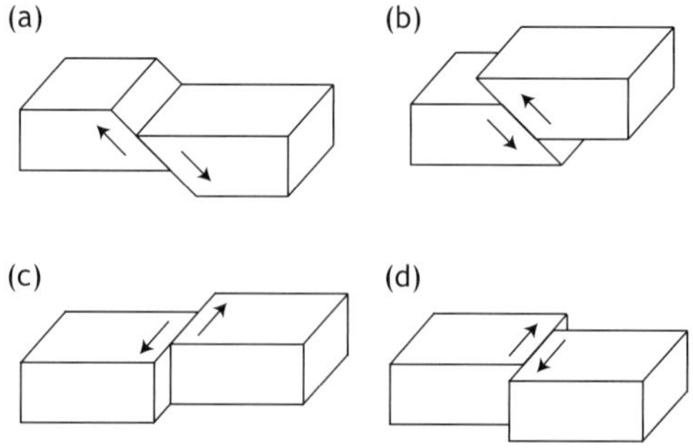

Fig. 10.18 The different fault types of structural geology. (a) Normal faulting; (b) thrust or reverse faulting; (c) left-lateral and (d) right-lateral strike-slip faulting. In (a) and (b) the right-hand block is the hanging wall and the left-hand block is the footwall.

that, during faulting, one principal stress component, σ_v, is taken as vertical, so that the two other principal stress components, σ_H and σ_h, are horizontal, σ_H being the larger.

According to the Coulomb failure criterion discussed in section 8.3.2, normal faulting occurs when the minimum principal stress, σ_3, is horizontal and the maximum principal stress, σ_1, is vertical. Then the intermediate principal stress, σ_2, is horizontal and the fault is parallel to the $\sigma_H = \sigma_2$ direction, with a dip equal to $\pi/2 - \phi$, where ϕ is the friction angle for the fault. If the vertical principal stress is the minimum principal value then one of the horizontal stress components is the maximum stress component, so that failure occurs by reverse faulting, the fault direction being parallel to the minimum horizontal principal stress direction $\sigma_h = \sigma_2$. Finally, if the vertical stress component is the intermediate principal stress $\sigma_v = \sigma_2$, then a vertical strike-slip fault is formed and is inclined an angle ϕ to the $\sigma_H = \sigma_1$ direction.

To the first order, Anderson' classification has been shown to be very helpful, but this theory leaves many questions unanswered. Indeed, we saw in section 9.2 that plastic localization depends on the flow rule that characterizes plasticity. For example, Sulem (2007) showed that, because of the contractant properties of the clay that makes up the fault core of the 10 km long Aigion fault in the Corinth rift zone, in Greece, this active normal fault is oriented perpendicularly to the local minimum principal stress direction, a feature not consistent with Anderson's faulting theory. Similarly, direct stress measurements conducted in the vicinity of the San Andreas fault, in California, suggest that locally the observed principal stress components imply a resolved shear stress on the fault plane that is contrary to the known direction of slip along the fault (Shamir and Zoback, 1992); this is still a subject of active debate (Scholz and Saucier, 1993).

These difficulties are encountered because faults are not simple planar structures but exhibit a complex morphology that involves mechanical constituents with very diverse rheologies, as already stated in section 1.3.4. This complex morphology reflects in turn the long evolution process that transformed a network of simple small-scale fractures into the present-day large-scale fault structure.

For example, as shown in section 8.1.3, when a specimen is loaded under triaxial compression beyond its elastic limit, some axial microcracking develops but it is not clear yet how these microcracks progressively yield the observed macroscopic shear failure surface. This localization process is not explained by Coulomb theory. Furthermore, as shown by Lockner (1993), for some values of the confining pressure shear fractures have been found to propagate in their own plane without preliminary development of an extensive process zone (see the discussion in section 8.1.4). The mechanism for the development of a process zone for a fracturing process under triaxial compression remains to be properly modeled.

Further, detailed field work has shown that fault outcrops involve numerous branches at different scales and that fault growth is generally not symmetrical with respect to the center of the fault. For example Manighetti *et al.* (2001) investigated the growth of normal faults in the Afar region of eastern Africa, whilst Davis *et al.* (2005) investigated the growth of thrust faults in New Zealand. Both concluded that there is a patchy, nonsymmetrical, growth of fault systems as shown in fig. 10.19. This patchiness of fault growth is linked to the fact that it occurs through seismic activity, each patch corresponding to an earthquake.

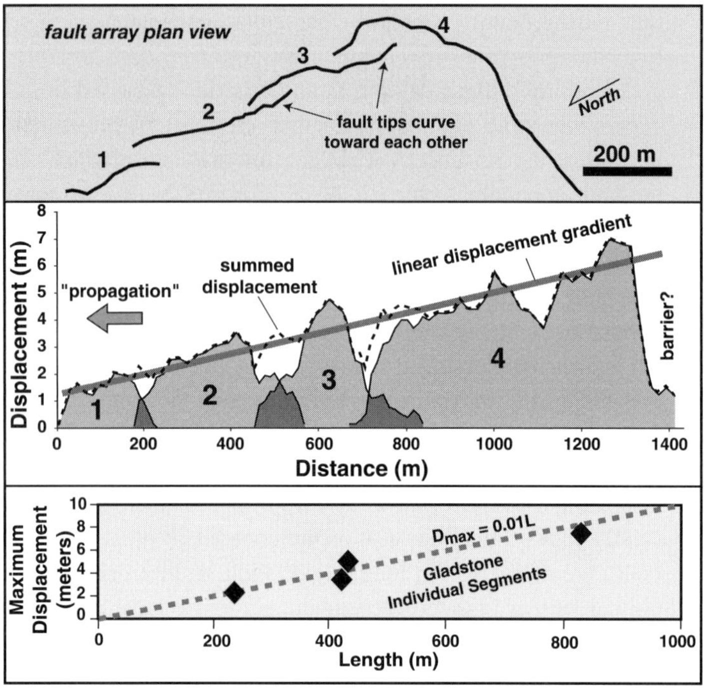

Fig. 10.19 Distribution of fault displacements along the various segments of a thrust fault. The fault growth is discontinuous in time and linked to its seismic activity, resulting in a complex distribution of shear displacement along the fault. Yet the maximum shear displacement seems to be related to the total fault length (reproduced from Davis *et al.*, 2005, with permission from Elsevier).

But then these large slips are accompanied by multiple smaller-scale readjustments, some of which are associated with the many aftershocks observed after the main shock.

Another source of heterogeneity in fault morphology is the nature of the geomaterials that are faulting. For example, consider a dip-slip fault developing in a sedimentary system comprising a series of horizontal brittle and plastic geomaterials. When the slip gets large enough, it results in complete separation of the brittle layers but in the smearing of the plastic layers along the fault plane (fig. 10.20). The mechanics of clay smearing was discussed for example by Egholm *et al.* (2008), who evaluated the sealing capacity of clay smear in soft sediments. They concluded that the thickness of the clay smear is a function of the initial sedimentary bed's thickness and of the friction-angle contrast. We mention here that the thickness and continuity of this plastic layer strongly influence the hydraulic characteristics of the fault, as will be discussed in the next subsection.

10.3.2 Discussion on the hydromechanical characteristics of faults

Because of their developmental history faults are complex structures, the mechanics of which involves elastic, plastic and fluid geomaterials as well as various fracture networks. The properties of these diverse structural elements depend on the magnitude

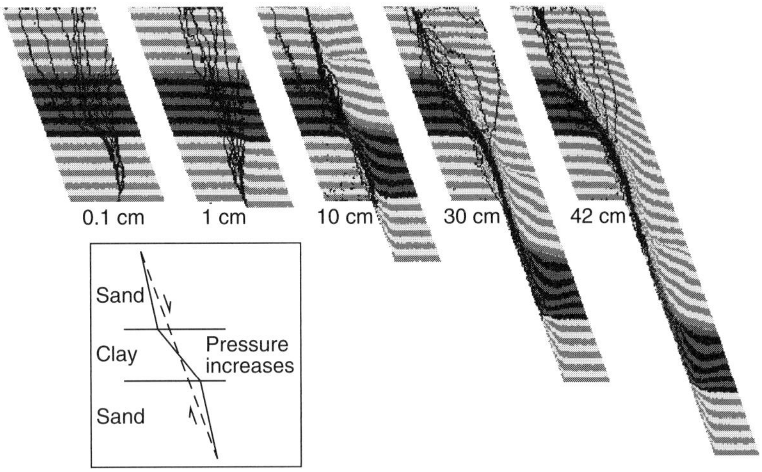

Fig. 10.20 The smearing of clay along dip-slip faults in sedimentary formations (reproduced from Egholm *et al.*, 2008, with permission from the Geological Society of America).

of the effective minimum principal stress, $\sigma_3 - P_p$, as well as on temperature; thus their hydromechanical characteristics vary with depth. Furthermore, a key factor in understanding the hydromechanical characteristics of faults is the coupling between their various hydrothermochemomechanical properties, which depends on the deformation rate.

Describing the mechanics of faulting is beyond the scope of this book, and further reading on this topic is suggested at the end of this chapter. We only outline here-after some important fault characteristics.

It is often convenient to differentiate active faults from dead faults, i.e. faults that have not experienced any shear motion within the last 10^4 to 10^5 years. The evaluation of such time scales is often provided by the shearing of well-dated recent sediments.

Dead faults

Dead faults are important, for they modify locally the elastic response of rock masses and often correspond to zones of weakness. They may also influence the local fluid flow and correlatively the heat flux characteristics.

Dead faults are often modeled by equivalent homogenized geomaterials that fill up the fault volume. Their elastic properties are characterized by planar anisotropy, i.e. the Young's modulus in the direction normal to the fault plane is taken as 10% to 20% smaller than its value in directions parallel to the fault plane. Similarly, the shear modulus for shear strains within the fault plane is taken as 10% to 20% smaller than that for shearing in directions perpendicular to the plane.

A fault's shear strength is often considered to be well approximated by a Coulomb failure criterion but, if the fault core corresponds to a plastic fine-grained granular geomaterial, the yield criterion must take into account the appropriate dilatancy, positive or negative, for the gouge material.

Very often, faults exhibit locally a very high fluid transmissivity, which may lead to local fluid convection. Convection corresponds to fluid flow driven by the density contrasts associated with different temperatures (see chapter 12). This is the case, for example, for the heat anomaly observed in the upper Rhine graben (Pribnow and Schellschmidt, 2000; Guillou-Frottier *et al.*, 2013), a region where exploitation of this geothermal energy is under development.

In dead faults, fluid circulation may result in local material alterations that give rise to occasional instabilities.

Active faults

Active faults are the site of shear displacements that are occurring presently (within the current 10 000 years). This shear motion occurs generally through occasional large earthquakes that mobilize slip surfaces, the size of which is related to the earthquake magnitude (see section 11.2). The time lapse between two large earthquakes may vary from tens of years to thousands of years. It is sometimes called the recurrence time, although it is not periodic. These large earthquakes are followed by some slow motion for periods of time that also depend on the earthquake's magnitude (Scholz, 1990, chapter 4). They are associated with large-scale stress variations, which result in large-scale pore pressure variations that are partly the source of the slow deformation observed after large earthquakes (Brodsky *et al.*, 2003).

Some active faults, however, exhibit locally continuous slow motion that prevents the occurrence of a large earthquake. This is the case, for example, for the strike-slip Philippine fault on Leyte island (Duquesnoy *et al.*, 1994), where a yearly slip motion close to 3 cm/yr has been determined through repeated GPS measurements. On this island no earthquakes larger than magnitude 5 have ever been recorded. However, further to the north, the same fault is the site of occasional magnitude-7 earthquakes. This variability in the seismic behavior of the fault seems to be related to the temperature distribution along it, the creeping section being located in an area of high temperatures (300 °C at 2000 m depth).

Interestingly, the Philippine fault zone, in the section where it creeps, has been found to correspond to a hydraulic barrier in the direction normal to the fault (Prioul *et al.*, 2000). Similarly, the active normal Aigion fault in the Corinth Rift has also been found to constitute a hydraulic barrier at the location where clay is smeared along the fault (Cornet *et al.*, 2004).

Hence active faults may correspond to flow barriers in the direction normal to the fault and to preferential flow conduits in directions parallel to the fault.

This has been confirmed by results from the Nojima-fault drilling project, in Japan. Two boreholes were drilled through this strike-slip fault after the magnitude-7.2 1995 Kobe earthquake. They intersected the slip zone associated with the earthquake at 624 m and 1140 m depths respectively. Lockner *et al.* (2009) conducted permeability measurements on cores collected at various depths and obtained the permeability profile shown in fig. 10.21. This profile clearly indicates that the gouge of the fault core corresponds to a low permeability zone (10 microdarcy) whilst on both sides of the core highly conductive zones are observed (with millidarcy permeability). It should be noted that the local

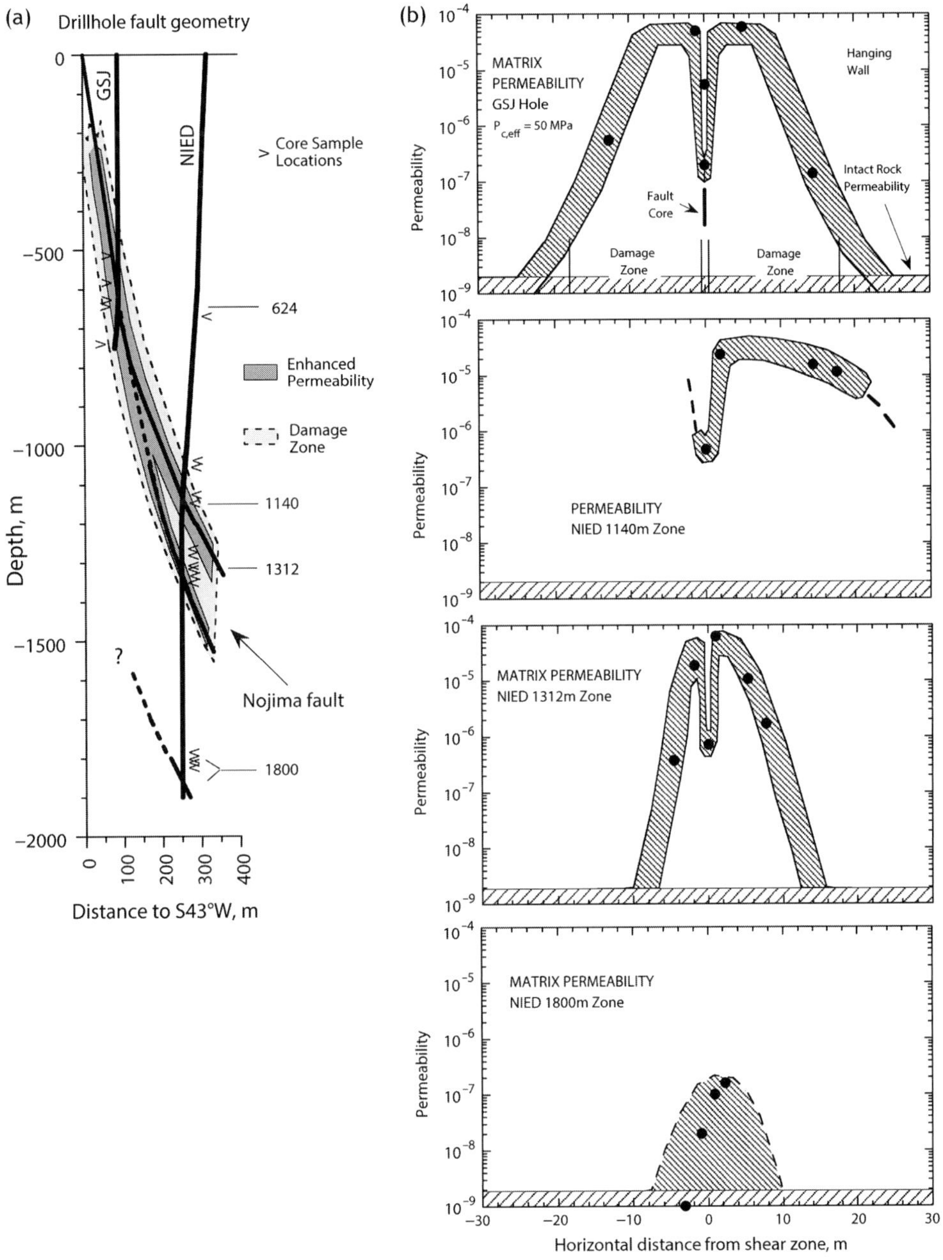

Fig. 10.21 Permeability variations in darcy through the Nojima fault as measured after the magnitude-7.2 Kobe earthquake through measurements on cores. (a) Geometry of the fault and location of the cores; (b) results from the various cores (top, the GSI hole; three lower pictures, the NIED hole). Redrawn from Lockner *et al.*, 2009, with permission from Springer.

intact rock is a highly impervious granodiorite (10 nanodarcy permeability), so that the fault corresponds to a highly anisotropic flow zone, with a high hydraulic conductivity in the direction parallel to the fault but a low conductivity in the direction perpendicular to it. Boullier *et al.* (2004) discussed how the coupling between solid and fluid during the dynamic rupture process associated with the earthquake may have led to the increase in hydraulic conductivity around the slip zones.

We mention also the permeability measurements conducted on faults in the Dead Sea rift system by Gabay *et al.* (2014). Gabay *et al.* have identified a five-order-of-magnitude difference between the permeability measured parallel to the fault ($\sim 3.5 \times 10^{-10}$ m^2) and that measured normal to the fault (10^{-15} m^2).

10.4 Further reading

Cook, N. G. W., Goodman, R. E., Myer, L. R., and Tsang, C. F. (eds.), 1996. *Fractured and Jointed Rock Masses.* CRC Press, 786 pp.

Gudmunsson, A., 2011. *Rock Fractures in Geological Processes.* Cambridge University Press, 570 pp.

Handy, M. P., Hirth, G., and Hovius, N. (eds.), 2005. *Tectonic Faults, Agent of Change of a Dynamic Earth.* MIT Press, 446 pp.

Scholz, C., 2002. *The Mechanics of Earthquakes and Faulting.* Cambridge University Press, 471 pp.

11 Elements of seismology

We saw in section 10.1.2 that slip on pre-existing fractures may be characterized by rate and state friction laws which imply that slip may be stable or unstable, depending on the loading conditions and material properties. Further, we saw in section 10.3 that unstable slip associated with fault activity may lead to earthquakes of very variable sizes.

In section 8.1.4 we discussed how acoustic emissions are generally observed in rock samples loaded under compression. These high-frequency signals (the frequencies are generally larger than 100 000 Hz) are generated by local instabilities often associated with microcrack growth; these instabilities, depending on the loading conditions, may remain localized or may lead progressively to the complete failure of the rock specimen in a stable or unstable manner.

All these instabilities generate mechanical waves that propagate through the medium surrounding the source. It is important to determine how instabilities start, grow and stop and how this process is related to the associated mechanical wave field.

The object of seismology is to better understand the sources of mechanical waves in geomaterials and to investigate the propagation of these waves through the earth. However, although there is no theoretical restriction on the frequency range of interest, traditionally signals of seismological interest have frequencies ranging from a few fractions of a millihertz to a few tens of hertz, for crustal seismology, or kilohertz for the seismicity induced by human activity (i.e. mining and reservoir engineering). Longer periods are usually investigated in geodetics and higher frequencies in rock physics.

Because wave propagation depends on the mechanical characteristics of the material through which the waves propagate, the analysis of seismic waves may be used to investigate the various structural elements that have influenced the wave train on its travel from the source to the receiver. For example, analysis of the seismic signals generated by large earthquakes has helped refine our understanding of the deeper structure of the earth. Analysis of the mechanical waves observed in other planets, and even stars, is one of the few means available to understand the structure of these celestial objects.

In order to exploit the information that may be retrieved from mechanical waves for characterizing the medium through which the waves have propagated, artificial sources have been developed for specific needs, such as the exploration of hydrocarbon reservoirs or the monitoring of the development of unstable zones in underground workings. The words "seismics" and "seismic engineering" often refer to the analysis of waves generated by engineered vibratory sources.

The objective of this chapter is to introduce elementary concepts of wave propagation and natural sources of vibration. These elements of seismology are more and more often applied to various engineering problems such as mitigating man-made seismicity. They also help us to understand present-day natural deformation processes, as will be discussed in chapter 14, and have proved to be helpfull in gaining an efficient understanding of deep fluid–solid interactions; they also shed light on the relationship between seismic and non-seismic deformation processes. These concepts may be used to validate, or invalidate, numerical models of rock mass perturbations induced by engineering undertakings.

11.1 Seismic waves

11.1.1 Body waves

The seismic wave equation

A seismic wave corresponds to a particle motion which varies with time t and depends on the particle location $\boldsymbol{x}$, and for which the corresponding strains are infinitesimal (generally smaller than 10^{-6}). Although earthquakes may generate large displacements, sometimes in the multimeter range, the term "seismic wave" is understood here as referring to elastic motions observed at large distances from the source. Let $\boldsymbol{u}(\boldsymbol{x}, t)$ be the vector function that describes such a seismic motion.

We saw in section 5.2 that $\boldsymbol{u}(\boldsymbol{x}, t)$ must satisfy Navier's equation if the motion is that of a particle in an isotropic, homogeneous, linearly elastic body, with Lamé elastic coefficients λ_L and G. In cartesian coordinates Navier's equation is written as

$$(\lambda_L + G)u_{i,ij} + Gu_{j,ii} = \varrho \ddot{u}_j, \qquad i,j = 1,2,3 \tag{11.1}$$

since no body force is involved in this motion. In vector form, Navier's equation is

$$(\lambda_L + G)\nabla(\nabla \cdot \boldsymbol{u}) + G\,\nabla^2 \boldsymbol{u} = \varrho \ddot{\boldsymbol{u}} \tag{11.2}$$

where

$$\nabla^2 u_j = \frac{\partial^2 u_j}{\partial x_1^2} + \frac{\partial^2 u_j}{\partial x_2^2} + \frac{\partial^2 u_j}{\partial x_3^2} = u_{j,ii}$$

is the Laplacian of u_j.

Taking advantage of the vector identity (see the appendix)

$$\nabla^2 \boldsymbol{u} = \nabla(\nabla \cdot \boldsymbol{u}) - \nabla \wedge (\nabla \wedge \boldsymbol{u}) \tag{11.3}$$

equation (11.2) becomes

$$(\lambda_L + 2G)\nabla(\nabla \cdot \boldsymbol{u}) - G\,\nabla \wedge (\nabla \wedge \boldsymbol{u}) = \varrho \ddot{\boldsymbol{u}} \tag{11.4}$$

However, according to Helmholtz' theorem (e.g. Fung, 1965, chapter 8), any analytic vector field $\boldsymbol{u}(\boldsymbol{x}, t)$ may be expressed in terms of a scalar potential Φ and a vector potential $\boldsymbol{\Upsilon}$:

$$\boldsymbol{u}(\boldsymbol{x},t) = \nabla\Phi(\boldsymbol{x},t) + \nabla \wedge \boldsymbol{\Upsilon}(\boldsymbol{x},t) \tag{11.5}$$

Thus the displacement is the sum of the gradient of the scalar potential $\Phi(\boldsymbol{x},t)$ and the curl (rot) of the vector potential $\boldsymbol{\Upsilon}(\boldsymbol{x},t)$, and we have the properties $\nabla \wedge (\nabla\Phi) = 0$ and $\nabla \cdot (\nabla \wedge \boldsymbol{\Upsilon}) = 0$. Inserting equation (11.5) into equation (11.4) we obtain

$$(\lambda_L + 2G)\nabla(\nabla^2\Phi) - G\,\nabla \wedge \nabla \wedge (\nabla \wedge \boldsymbol{\Upsilon}) = \varrho\frac{\partial^2}{\partial t^2}(\nabla\Phi + \nabla \wedge \boldsymbol{\Upsilon}) \tag{11.6}$$

However, the divergence of the curl is zero, $(\nabla \cdot (\nabla\wedge) = 0)$, so that

$$\nabla \wedge \nabla \wedge (\nabla \wedge \boldsymbol{\Upsilon}) = -\nabla^2(\nabla \wedge \boldsymbol{\Upsilon}) + \nabla[\nabla \cdot (\nabla \wedge \boldsymbol{\Upsilon})] = -\nabla^2(\nabla \wedge \boldsymbol{\Upsilon}) \tag{11.7}$$

which, after substitution into equation (11.4), leads to

$$\nabla\left[(\lambda_L + 2G)\nabla^2\Phi - \varrho\frac{\partial^2\Phi}{\partial t^2}\right] = -\nabla \wedge \left[G\nabla^2\boldsymbol{\Upsilon} - \varrho\frac{\partial^2\boldsymbol{\Upsilon}}{\partial t^2}\right] \tag{11.8}$$

given that the elastic constants are uniform throughout the body where waves are propagating.

A solution to equation (11.8) is obtained when both terms in brackets are zero. This implies that the scalar potential must satisfy

$$\nabla^2\Phi = \frac{1}{C_P^2}\frac{\partial^2\Phi}{\partial t^2} \tag{11.9}$$

where $C_P = [(\lambda_L + 2G)/\varrho]^{1/2}$. Similarly the vector potential must satisfy

$$\nabla^2\boldsymbol{\Upsilon} = \frac{1}{C_S^2}\frac{\partial^2\boldsymbol{\Upsilon}}{\partial t^2} \tag{11.10}$$

where $C_S = (G/\varrho)^{1/2}$.

Equations (11.9) and (11.10) are the seismic wave equations. They are valid only in homogeneous, isotropic, linearly elastic materials. They have been derived in cartesian coordinates but they may be expressed also in spherical coordinates, which is useful when considering seismic wave propagation in the earth (e.g. Stein and Wysession, 2003, chapter 3).

For spherical waves, the area of the wave front, $4\pi\rho^2$, increases with the square of the distance ρ to the origin. Hence, if we assume the conservation of energy during spherical wave propagation, the energy per unit wave front area decays as the inverse of the squared distance to the source, $1/\rho^2$. This decay is referred to as *geometric spreading*. Because the energy is proportional to the square of the amplitude of the wave, geometric spreading implies that the body wave amplitude decays as $1/\rho$.

Let us observe that, for distances far enough from the source, spherical waves may be assimilated to plane waves, and this is a very common hypothesis made in observational seismology.

P and S waves

Equation (11.9) describes how the scalar field Φ propagates in a three-dimensional space. It represents a homogeneous wave equation with no forcing function. In order to introduce the source of this wave field, it is necessary to introduce a corresponding force in the equation of motion, so that it becomes

$$\nabla^2\Phi - \frac{1}{C_P^2}\frac{\partial^2\Phi}{\partial t^2} = f(\boldsymbol{x}, t) \tag{11.11}$$

where $f(\boldsymbol{x}, t)$ is a forcing function, analogous in its principle to the body force considered for the law of motion in continuum materials (section 6.1.2).

The solution to equation (11.9) in terms of harmonic plane waves is written as

$$\Phi(\boldsymbol{x}, t) = A\exp[i(\omega t \pm \boldsymbol{k}\cdot\boldsymbol{x})] = A\exp[i(\omega t \pm k_1x_1 \pm k_2x_2 \pm k_3x_3)] \tag{11.12}$$

where i is the imaginary number in complex variable theory ($i^2 = -1$), $\boldsymbol{x}$ is the position vector and $\boldsymbol{k}$ ($\boldsymbol{k} = k_j\hat{\boldsymbol{I}}_j; j = 1, 2, 3$) is known as the wave vector or wave number vector. This solution describes a plane wave propagating with velocity C_P in the direction given by the wave vector $\boldsymbol{k}$.

Hence the magnitude of the wave vector, $|\boldsymbol{k}|$, gives the spatial frequency of the wave and the direction of the wave vector gives the direction of wave propagation. Wave fronts correspond to surfaces, which, at any time, exhibit the same constant phase $\omega t \pm \boldsymbol{k}\cdot\boldsymbol{x}$; they are perpendicular to the direction of wave propagation.

For example, let us consider a wave propagating in the x_3 direction. The scalar potential is

$$\Phi(x_3, t) = A\exp[i(\omega t - kx_3)] \tag{11.13}$$

and the corresponding displacement is

$$\boldsymbol{u}(x_3, t) = \nabla\Phi(x_3, t) = (0, 0, -ik)A\exp[i(\omega t - kx_3)] \tag{11.14}$$

with no component in the x_1 and x_2 directions. The corresponding volumetric strain is

$$\varepsilon_v = \nabla\cdot\boldsymbol{u}(x_3, t) = -k^2A\exp[i(\omega t - kx_3)] \tag{11.15}$$

which shows that the wave is associated with a finite volumetric variation. As the wave propagates, the material is alternately compressed and expanded in the direction of wave propagation. This is the compressional P wave discussed in section 5.3.5 (see fig. 11.1).

When the first motion associated with the P wave occurs in the direction of wave propagation it is called compressional. When this first motion is in the opposite direction of wave propagation it is called dilatational.

Let us turn now to the vector potential $\boldsymbol{\Upsilon}$; a similar type of solution may be found. Its components must be solutions of

$$\nabla^2\Upsilon_j = \frac{1}{C_S^2}\frac{\partial^2\Upsilon_j}{\partial t^2}, \qquad j = 1, 2, 3 \tag{11.16}$$

for which the solution is given by

$$\boldsymbol{\Upsilon}(\boldsymbol{x}, t) = \boldsymbol{B}\exp[i(\omega t \pm \boldsymbol{k}\cdot\boldsymbol{x})] \tag{11.17}$$

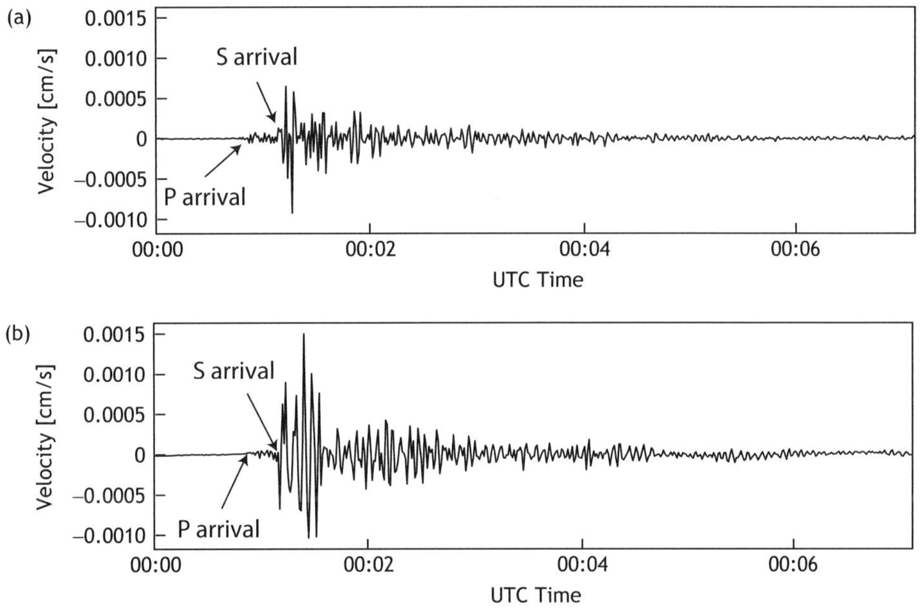

Fig. 11.1 Example of seismic signal showing the P waves and S waves arrivals for a small earthquake located about 4 km from the receiver. (a) The vertical component of the particle velocity, (b) one of its two horizontal components. Time is in seconds.

where $\boldsymbol{B}$ is a constant vector.

For a wave propagating only in the x_3 direction, the vector potential is

$$\boldsymbol{\Upsilon}(x_3, t) = (B_1, B_2, B_3)\exp[i(\omega t - kx_3)] \tag{11.18}$$

and the resulting displacement field is given by the curl of this vector potential:

$$\boldsymbol{u}(x_3, t) = \nabla \wedge \boldsymbol{\Upsilon}(x_3, t) = (ikA_2, -ikA_1, 0)\exp[i(\omega t - kx_3)] \tag{11.19}$$

There is no component of displacement in the x_3 direction: the only displacement associated with the wave is normal to the wave propagation direction. It corresponds to the shear waves (S waves) discussed in section 5.3.5 (see fig. 11.1). The corresponding volumetric change, which is the divergence of the displacement, is zero. Shear waves propagating in linearly elastic media induce no volume variation.

The component of $\boldsymbol{\Upsilon}(x_3, t)$ in the direction of wave propagation, x_3, has no effect on the displacement field since the curl operator discards it. Only A_1 and A_2 contribute to the displacement. Because each component of the displacement depends on only one of these components, shear waves can have two independent polarizations. Interestingly, in anisotropic materials the shear wave velocity depends on the direction of polarization. This leads to different S wave arrivals, a phenomenon known as shear wave splitting and will be discussed further in section 13.5.

In seismology the x_3 axis is generally taken as vertical and the vertical x_1 x_3 plane is oriented in such a way as to include the source and the receiver. Hence plane waves

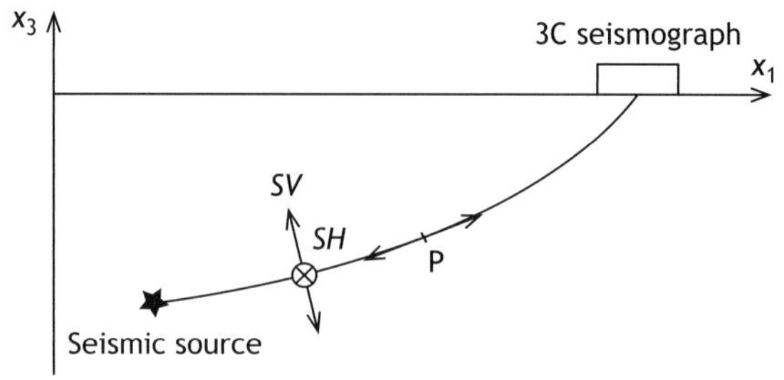

Fig. 11.2 P and S body waves. The P waves are compressional and the S waves (SV and SH) are pure shear. The SV waves vibrate in the vertical plane perpendicular to the direction of wave propagation. The SH waves vibrate in the horizontal plane. Body waves become plane waves at long distances from their source.

traveling directly from the source to the receiver propagate in the x_1x_3 plane. Shear waves polarized in the vertical x_1x_3 plane are called SV waves, whilst those that are polarized parallel to the x_2 direction, in the horizontal plane, are called SH waves (fig. 11.2). It will be shown in section 11.1.2 that, at horizontal interfaces between different materials, the P and SV waves interact with each other, whilst the SH waves remain separate.

In homogeneous isotropic materials, the determination of the polarization direction of the P waves provides a means to determine the orientation of the straight line that joins the source to the receiver (also called a seismic ray). The distance d between the source and the receiver may be retrieved from the first arrival times for P and S waves, t_P and t_S, respectively:

$$d = C_P(t_P - t_0) = C_S(t_S - t_0) \tag{11.20}$$

where t_0 is the time when the source event occurred. Eliminating t_0 in equation (11.20) yields the distance between the source and the receiver:

$$d = \frac{C_P C_S}{C_P - C_S}(t_S - t_P) \tag{11.21}$$

When Poisson's ratio for the body is close to 0.25 the P wave velocity C_P is close to $\sqrt{3}C_S$, so that equation (11.21) becomes $d \simeq 2.37C_S(t_S - t_P) \simeq 1.37C_P(t_S - t_P)$.

Radiated energy

Plane waves correspond to a deformation field and as such carry energy, which is referred to as the radiated energy. It may be decomposed into kinetic energy and strain energy.

Let us consider first an SH wave propagating in the x_3 direction, with displacement in the x_2 direction:

$$u_2 = B\cos(\omega t - kx_3) \tag{11.22}$$

For a geomaterial with density ϱ, the kinetic energy associated with the motion of the particles in a volume V is

$$K_{SH}(V) = \tfrac{1}{2} \int_V \varrho \left(\frac{\partial u_2}{\partial t} \right)^2 dV \tag{11.23}$$

or, for a unit wave front averaged over a wavelength λ,

$$\begin{aligned} K_{SH}(\lambda) &= \tfrac{1}{2} \varrho B^2 \omega^2 \int_0^\lambda \sin^2 (\omega t - kx_3)\, dx_3 \\ &= \frac{1}{2\lambda} \varrho B^2 \omega^2 \frac{\lambda}{2} = \tfrac{1}{4} B^2 \omega^2 \varrho \end{aligned} \tag{11.24}$$

The elastic strain energy is (see section 5.4.1)

$$We_{SH} = \tfrac{1}{2} \int_V \sigma_{ij} \varepsilon_{ij}\, dV \tag{11.25}$$

However, the only nonzero strain components are

$$\varepsilon_{32} = \varepsilon_{23} = \tfrac{1}{2}(u_{2,3}) = \tfrac{1}{2} Bk \sin(\omega t - kx_3) \tag{11.26}$$

and so the only nonzero stress components are

$$\sigma_{32} = \sigma_{23} = GBk \sin(\omega t - kx_3) \tag{11.27}$$

The strain energy per unit area of wave front averaged over a wavelength in the propagation direction is

$$\begin{aligned} We_{SH}(\lambda) &= \frac{1}{2\lambda} \int_0^\lambda GB^2 k^2 \sin^2(\omega t - kx_3)\, dx_3 \\ &= \tfrac{1}{4} GB^2 k^2 = \tfrac{1}{4} B^2 \omega^2 \varrho \end{aligned} \tag{11.28}$$

since $G = C_{SH}^2 \varrho$ and $\omega = kC_{SH}$.

Hence, the kinetic energy and the strain energy per unit area of wave front, averaged over a wavelength, are equal, so that the total energy averaged over a wavelength is

$$E_{SH}(\lambda) = \tfrac{1}{2} B^2 \omega^2 \varrho \tag{11.29}$$

For waves with a given amplitude, those with high frequencies transport more energy than those with low frequencies.

Similarly, we may consider a P wave characterized, by the scalar potential

$$\Phi(x_3, t) = A \exp[i(\omega t \pm kx_3)] \tag{11.30}$$

so that the displacements is

$$\boldsymbol{u}(x_3, t) = \nabla \Phi(x_3, t) = (0, 0, -ik) A \exp[i(\omega t \pm kx_3)] \tag{11.31}$$

with real part

$$Ak \sin(\omega t - kx_3) \tag{11.32}$$

The kinetic energy per unit wave front averaged over a wavelength is

$$K_P(\lambda) = \frac{1}{2\lambda}\varrho A^2 k^2 \omega^2 \int_0^\lambda \cos^2(\omega t - kx_3)\, dx_3 = \tfrac{1}{4}A^2\omega^2 k^2 \varrho \tag{11.33}$$

The only nonzero stress component is

$$\sigma_{33} = (\lambda_L + 2G)\varepsilon_{33} = \varrho C_P^2 \varepsilon_{33} \tag{11.34}$$

Thus the strain energy per unit wave front averaged over the wavelength λ is

$$We_P(\lambda) = \frac{1}{2\lambda}\int_0^\lambda \varrho C_P^2 A^2 k^4 \cos^2(\omega t - kx_3)\, dx_3 = \tfrac{1}{4}A^2 k^2 \varrho \tag{11.35}$$

and the total energy averaged over the wavelength λ is

$$E_P(\lambda) = K_P(\lambda) + We_P(\lambda) = \tfrac{1}{2}A^2\omega^2 k^2 \varrho \tag{11.36}$$

This expression is the same as that obtained for SH waves, except for the factor k^2 which differs because the constant A, introduced in equation (11.30), refers to the scalar potential whilst the constant B in the expression for SH waves is related directly to the displacement amplitude.

Because of energy conservation, when seismic waves travel from a stiff geomaterial to a softer one, i.e. to a material with lower elastic constant and lower wave velocity, their amplitude increases. Practically speaking, this implies that seismic signals recorded on soft geomaterials are stronger than those recorded on stiff geomaterials and that damage caused by earthquakes is more extensive for structures built on soft ground than for those built on stiff rocks, as long as the response of the geomaterial is truly elastic.

Attenuation and quality factor

Because of geometrical spreading and energy conservation, we saw that when spherical waves propagate from a point source the energy per unit front wave decreases as $1/\rho^2$, if ρ is the distance to the source (not to be confused with ϱ the material density). This is called *geometric attenuation.*

In addition to this purely geometrical effect, which is independent of frequency, the amplitude of waves decreases because geomaterials are not purely elastic. At each cycle some energy is lost in heat. When this loss remains small, it is characterized by an elastic quality factor Q, which is defined by

$$\frac{4\pi}{Q} = -\frac{\delta We}{We} \tag{11.37}$$

where We is the strain energy. Two different quality factors are defined, Q_K and Q_G. The quality factor Q_K concerns changes in volume, i.e. it refers to the modulus of incompressibility K introduced in chapter 5, $K = (3\lambda_L + 2G)/3$. The quality factor Q_G concerns the shear motion and involves the shear modulus G. Because shear waves involve only shear motion, the associated quality factor Q_S is equal to Q_G. However, for P waves, which involve both the elastic constants λ_L and G, the corresponding quality factor Q_P depends on both Q_K and Q_G. It may be shown that

$$\frac{Q_P}{Q_S} \leq \frac{3(\lambda_L + 2G)}{4G} \tag{11.38}$$

The limit is reached when the energy losses associated with Q_K become negligible compared with the energy losses associated with Q_G, which is generally the case. For deep geomaterials, λ_L and G are of the same order of magnitude; the ratio Q_P/Q_S is close to 2.25.

Let us consider a P wave propagating in the x_1 direction:

$$u_1 = A(x_1)f\left(t - \frac{x_1}{C_P}\right), \qquad u_2 = u_3 = 0 \tag{11.39}$$

The kinetic energy per unit volume is

$$K_P(x_1,t) = \tfrac{1}{2}\varrho v_1^2(x_1,t), \quad \text{where } v_1(x_1,t) = A(x_1)\frac{\partial}{\partial t}\left[f\left(t - \frac{x_1}{C_P}\right)\right] \tag{11.40}$$

Given that the kinetic energy is equal to the elastic strain energy, the total energy per unit volume, which is the sum of the kinetic energy and the strain energy, is

$$W_P = K_P + We_P = \varrho A^2(x_1)\left[\frac{\partial}{\partial t}f\left(t - \frac{x_1}{C_P}\right)\right]^2 \tag{11.41}$$

At time t, during the time interval dt the quantity of energy inside a volume $C_P dS\,dt$ that passes through the surface dS normal to the direction of wave propagation is

$$dW_P = C_P\,dS\,dt\,We_P = C_P\,dS\,dt\,\varrho A^2(x_1)\left[\frac{\partial}{\partial t}f\left(t - \frac{x_1}{C_P}\right)\right]^2 \tag{11.42}$$

The flux of seismic energy associated with this P wave is

$$\varphi(x_1,t) = C_P\varrho A^2(x_1)\left[\frac{\partial}{\partial t}f\left(t - \frac{x_1}{C_P}\right)\right]^2 \tag{11.43}$$

The quantity $I_P \equiv \varrho C_P$ is the *seismic impedance* for P waves, and the flux of seismic energy may be written as a function of the seismic impedance:

$$\varphi(x_1,t) = I_P A^2(x_1)\left[\frac{\partial}{\partial t}f\left(t - \frac{x_1}{C_P}\right)\right]^2 \tag{11.44}$$

In order to evaluate the loss of energy over each cycle, let us assume that $f(x_1,t)$ is sinusoidal:

$$f(x_1,t) = \sin\left[\omega\left(t - \frac{x_1}{C_P}\right)\right] = \sin(\omega t - kx_1) \tag{11.45}$$

The seismic energy flux is

$$\varphi(x_1,t) = I_P A^2(x_1)\left[\frac{\partial}{\partial t}\sin(\omega t - kx_1)\right]^2 = I_P A^2(x_1)\omega^2\cos^2(\omega t - kx_1) \tag{11.46}$$

For a very thin slice, of thickness δx_1, in the direction of wave propagation, the difference between the entering and exiting seismic flux is

$$\varphi(x_1,t) - \varphi(x_1+\delta x_1,t) = -\frac{\partial}{\partial x_1}\varphi(x_1,t)\delta x_1 \tag{11.47}$$

The integral of this difference over a period T, where $T = 2\pi/\omega$, gives the energy loss per cycle:

$$\delta W_P(x_1) = \int_{t-T/2}^{t+T/2} \frac{\partial \varphi(x_1,t)}{\partial x_1}\delta x_1 dt \tag{11.48}$$

or, after deriving the energy flux,

$$\begin{aligned}\frac{\delta W_P(x_1)}{\delta x_1} &= 2I_P\omega^2 A\frac{dA(x_1)}{dx_1}\int_{t-T/2}^{t+T/2}\cos^2(\omega t - kx_1)\,dt \\ &- I_P\omega^2A^2k\int_{t-T/2}^{t+T/2}\sin 2(\omega t - kx_1)\,dt\end{aligned} \tag{11.49}$$

But the integral of $\cos^2$ term is equal to $T/2$ and the second integral is zero, so that

$$\frac{\delta W_P(x_1)}{\delta x_1} = 2I_P\omega^2 A\frac{dA(x_1)}{dx_1}\frac{T}{2} = I_P\omega^2AT\frac{dA}{dx_1} \tag{11.50}$$

Further, the mean elastic strain energy in the slice of thickness δx_1, during one cycle, is

$$\begin{aligned}We_P(T) &= \frac{1}{T}\int_{t-T/2}^{t+T/2} We\delta x_1\,dt = \frac{\varrho}{2T}\int_{t-T/2}^{t+T/2}\left[\frac{\partial}{\partial t}A\,\sin(\omega t - kx_1)\right]^2\delta x_1 dt \\ &= \frac{\varrho\omega^2A^2}{2T}\int_{t-T/2}^{t+T/2}\cos^2(\omega t - kx_1)\,\delta x_1 dt = \frac{\varrho\omega^2A^2}{4}\delta x_1\end{aligned} \tag{11.51}$$

The ratio of the energy lost during one cycle versus the total energy associated with the corresponding cycle depends on the quality factor, as expressed by equation (11.37):

$$\frac{\delta We}{We} = \frac{4I_P\omega^2AT(dA/dx_1)\delta x_1}{\varrho\omega^2A^2\delta x_1} = 4C_PT\frac{1}{A}\frac{dA}{dx_1} = \frac{-4\pi}{Q_P} \tag{11.52}$$

from which we get

$$\frac{dA}{dx_1} = -\frac{4\pi}{Q_P}\frac{A(x_1)}{4C_PT} = -\frac{\omega}{Q_P}\frac{A(x_1)}{2C_P} \tag{11.53}$$

Integrating between 0 and x_1 yields

$$A(x_1) = A(0)\exp\left(-\frac{\omega}{2C_PQ_P}x_1\right) \tag{11.54}$$

Thus P waves are progressively attenuated as they propagate in the x_1 direction. This attenuation is proportional to ω/C_P and inversely proportional to the wavelength $2\pi C_P/\omega$. It is called the *physical attenuation*. The higher the frequency, the stronger the physical attenuation.

Hence, for receivers further away from a source, the associated seismic signal becomes more and more difficult to identify within the background seismic noise. However,

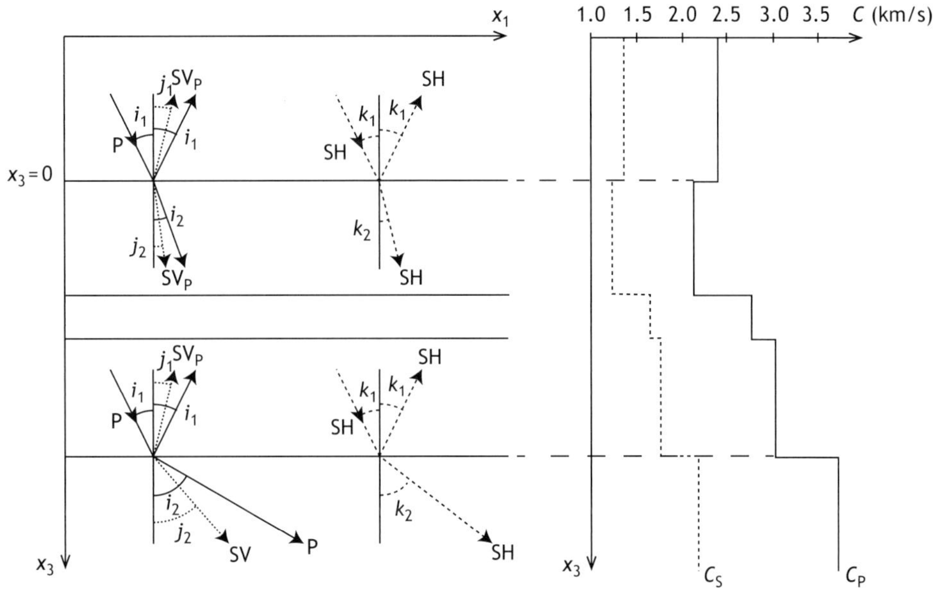

Fig. 11.3 A layered half space with a plane wave propagating in the x_1x_3 plane. The right-hand diagram indicates the vertical variation of the P and S wave velocities.

various numerical techniques have been developed for identifying a seismic signal at great distances from the source, either by summing records from multiple receivers (stacking) or through various signal-processing techniques (e.g. Stein and Wysession, 2003).

Had we considered a shear wave, we would have obtained the same results, but with C_P and Q_P replaced by C_S and Q_S.

11.1.2 Refraction, reflection, diffraction

In the previous subsection we considered the propagation of mechanical waves in a homogeneous infinite body. In many geomechanical applications, however, the medium in which seismic waves propagate may be assimilated to a horizontally layered medium, where the elastic coefficients, density and Q factors vary from one layer to another. Further, when the depth range considered for the analysis is small compared with the radius of the earth, the problem can be modeled as a *horizontally layered half space* with seismic velocities, densities and Q factors characterized by constants that vary as step functions of depth only. Also, for distances far enough from the source, the planar approximation is valid so that the problem of spherical wave propagation reduces to that of plane wave propagation in a horizontally layered medium (fig. 11.3).

With such a model, the wave propagation in each layer is solved according to the homogeneous formulation introduced in the previous section, but, in addition, attention must be given to describing wave propagation through the various layer interfaces.

Plane waves at interfaces

A bending of wave fronts is observed as a plane wave goes from one medium to another. This is most conveniently investigated through plane wave potentials and is known as Snell's law in seismology.

Plane wave potentials for layered geomaterials

Consider a plane wave propagating in a medium which involves two different materials in contact along a horizontal plane (fig. 11.3) and let the planar wave propagate in the x_1x_3 plane. As shown in the previous subsection the displacement field may be decomposed into a scalar potential Φ that describes P waves and a vector potential field $\boldsymbol{\Upsilon}$ that describes S waves. In order to separate the SV and SH waves, the vector potential $\boldsymbol{\Upsilon}$ is decomposed into two terms,

$$\boldsymbol{\Upsilon}(x_1,x_3,t) = \boldsymbol{\Psi}(x_1,x_3,t) + \nabla \wedge \boldsymbol{\chi}(x_1,x_3,t) \tag{11.55}$$

so that the displacement may be written

$$\boldsymbol{u}(x_1,x_3,t) = \nabla\Phi(x_1,x_3,t) + \nabla \wedge \boldsymbol{\Psi}(x_1,x_3,t) + \nabla \wedge \nabla \wedge \boldsymbol{\chi}(x_1,x_3,t) \tag{11.56}$$

The vector potentials are chosen such that their components are

$$\begin{aligned} &\text{for } \boldsymbol{\Psi}(x_1,x_3,t), \qquad [0, \Psi(x_1,x_3,t), 0] \\ &\text{for } \boldsymbol{\chi}(x_1,x_3,t), \qquad [0, \chi(x_1,x_3,t), 0] \end{aligned} \tag{11.57}$$

Hence each component of the displacement vector is described as a function of the partial derivatives of a scalar potential:

$$\text{for P waves,} \qquad \nabla\Phi(x_1,x_3,t) = (\Phi_{,1}, 0, \Phi_{,3}) \tag{11.58}$$

$$\text{for SV waves,} \qquad \nabla \wedge \Psi(x_1,x_3,t) = (-\Psi_{,3}, 0, \Psi_{,1}) \tag{11.59}$$

$$\text{for SH waves,} \qquad \nabla \wedge \nabla \wedge \chi(x_1,x_3,t) = [0, -(\chi_{,11} + \chi_{,33}), 0] \tag{11.60}$$

The P and SV waves contribute to the x_1 and x_3 components of the displacement whilst the SH waves contribute only to the x_2 component:

$$u_1(x_1,x_3,t) = \Phi_{,1} - \Psi_{,3} \tag{11.61}$$

$$u_3(x_1,x_3,t) = \Phi_{,3} - \Psi_{,1} \tag{11.62}$$

$$u_2(x_1,x_3,t) = -(\chi_{,11} + \chi_{,33}) = -\nabla^2\chi(x_1,x_3,t) \tag{11.63}$$

This demonstrates our previous proposition, that the P and SV waves are coupled whilst the SH waves are independent. This coupling and decoupling persists when the waves cross interfaces between two media with their interfaces perpendicular to the x_1x_3 plane. The boundary conditions at the interfaces constrain the displacements and the stress vectors $\boldsymbol{t}$. Because the interfaces are horizontal, the stress vector acting on the interface is

$$\boldsymbol{t} = \tilde{\sigma}\boldsymbol{n} \qquad \text{with components } \sigma_{31}, \sigma_{32}, \sigma_{33} \tag{11.64}$$

For the P and SV waves, the displacement components x_1 and x_3 are nonzero, so the associated stress tensor components σ_{31} and σ_{33} are also nonzero. For these waves u_2, and therefore σ_{23}, is zero. The opposite is true for the SH waves, which are associated with zero displacements in the x_1x_3 plane, so that the only corresponding nonzero-stress components are $\sigma_{32} = \sigma_{23}$. At the interface, the P and SV waves do not interact with the SH waves. However, P and SV are coupled so that, at the interface, the P waves generate SV waves and conversely the SV waves generate P waves.

Angle of incidence

We now consider harmonic plane P and SV waves, so that the corresponding Φ and Ψ potentials are

$$\text{for P waves,} \quad \Phi(x_1, x_3, t) = A\exp[i(\omega t - k_{x_1}x_1 \pm k_{x_3P}x_3)] \tag{11.65}$$

$$\text{for SV waves,} \quad \Psi(x_1, x_3, t) = B\exp[i(\omega t - k_{x_1}x_1 \pm k_{x_3SV}x_3)] \tag{11.66}$$

The wave vector gives the direction of wave propagation, and therefore there is no motion in the horizontal x_2 direction. The direction of propagation in the x_1x_3 plane is given by the k_{x_1} and k_{x_3P} terms. The components of the wave vectors satisfy the conditions

$$|k_P|^2 = k_{x_1}^2 + k_{x_3P}^2 = \omega^2/C_P^2 \tag{11.67}$$

$$|k_{SV}|^2 = k_{x_1}^2 + k_{x_3SV}^2 = \omega^2/C_{SV}^2 \tag{11.68}$$

where C_P and C_{SV} are respectively the P wave and SV wave velocities in the corresponding medium.

However, the direction of wave propagation may also be characterized by its angle of incidence, which is the angle that the wave vector makes with the vertical direction. Because the wave vectors for the P and SV waves are different, we use i to refer to the P wave incidence and j to refer to the SV wave incidence, so that

$$\sin i = k_{x_1}/(k_{x_1}^2 + k_{x_3P}^2)^{0.5} = k_{x_1}/|k_P| \tag{11.69}$$

$$\sin j = k_{x_1}/(k_{x_1}^2 + k_{x_3SV}^2)^{0.5} = k_{x_1}/|k_{SV}| \tag{11.70}$$

Hence the direction of wave propagation depends on the velocity in the medium. When a wave propagates in a medium in which the elastic properties and the density change continuously, the wave propagation direction varies accordingly and the envelope of these successive orientations is called the ray path.

Snell's law

Let us consider now the relation between the angles of incidence for transmitted and reflected P and SV waves at an interface between two different solids. The interface, located in the plane $x_3 = 0$, separates medium 1 from medium 2; P and SV waves propagate respectively at velocities C_{P_1} and C_{SV_1} in medium 1 (the top medium in fig. 11.3), which are taken to be larger than the propagation velocities C_{P_2} and C_{SV_2} in medium 2 (the second medium from the top in fig. 11.3). A P wave incident from medium 1 generates reflected and transmitted P waves; in addition, it is partly converted to both a reflected SV wave and a transmitted SV wave. Each of these waves may be described by a potential.

In medium 1 we thus observe a downgoing P wave and an upgoing, reflected, P wave, together represented by the potential

$$\Phi(x_1, x_3, t) = A_1 \exp[i(\omega t - k_{x_1}x_1 - k_{x_3P_1}x_3)] \\ + A_2 \exp[i(\omega t - k_{x_1}x_1 + k_{x_3P_1}x_3)] \tag{11.71}$$

as well as an upgoing, reflected, SV wave, represented by the potential

$$\Psi(x_1, x_3, t) = B_2 \exp[i(\omega t - k_{x_1}x_1 + k_{x_3SV_1}x_3)] \tag{11.72}$$

The direction of propagation for each wave is given by the components of the wave vector $\boldsymbol{k}$. For example, the reflected P wave with amplitude A_2 and the reflected SV wave with amplitude B_2 travel in the $+x_1$ and $-x_3$ directions, respectively.

Similarly, potentials may be defined for the downgoing, transmitted, P and SV waves in medium 2:

transmitted P wave: $$\Phi^{tr}(x_1, x_3, t) = A' \exp[i(\omega t - k_{x1}x_1 - k_{x_3P_2}x_3) \tag{11.73}$$

transmitted SV wave: $$\Psi^{tr}(x_1, x_3, t) = B' \exp[i(\omega t - k_{x1}x_1 - k_{x_3SV_2}x_3) \tag{11.74}$$

The boundary conditions at the interface $x_3 = 0$ are that the components of the displacement and stress vectors must be continuous. For the displacement and stress vectors to be continuous for all $\boldsymbol{x}$ and t, the term $\omega t - k_{x_1}x_1$ must be the same for each of the potentials. Thus, the horizontal component k_{x_1} of the wave number vector must be the same for each wave, which implies that ω/k_{x_1} is also constant so that along the interface all the waves travel at the same velocity and stay in phase. This implies that

$$\frac{c_{P_1}}{\sin i_1} = \frac{c_{SH_1}}{\sin j_1} = \frac{c_{P_2}}{\sin i_2} = \frac{c_{SH_2}}{\sin j_2} \tag{11.75}$$

which is *Snell's law*.

The incident and reflected P waves have the same angle i_1 to the normal. The transmitted P and SH waves change direction by a factor that depends on the respective velocities in the two media. This change in direction upon transmission in a medium with a different velocity is called *refraction*.

The angle to the normal j_1 of the reflected SV wave in medium 1 is such that

$$\sin j_1 = \sin i_1 \frac{c_{SV_1}}{c_{P_1}} \tag{11.76}$$

so that, since $c_{SV_1} < c_{P_1}$, the reflected SV ray is closer to the vertical than the P ray in the same medium.

The angle to the normal of the refracted P wave depends on the ratio of the P wave velocities in the two media:

$$\sin i_2 = \sin i_1 \frac{c_{P_2}}{c_{P_1}} \tag{11.77}$$

If the velocity in medium 2 were faster than in medium 1, the P wave would travel more horizontally in medium 2 than in medium 1.

A similar approach may be followed for SH waves crossing an interface. They give rise to a reflected SH wave and a refracted SH wave (fig. 11.3) such that

$$\frac{C_{SH_1}}{\sin j_1} = \frac{C_{SH_2}}{\sin j_2} \tag{11.78}$$

where C_{SH_1} and C_{SH_2} are respectively the SH wave velocities in mediums 1 and 2 whilst j_1 and j_2 are the angles of incidence in mediums 1 and 2.

When a seismic wave crosses an interface, not only does its direction change according to the velocity contrast between the media on either side of the interface but also its amplitude. This amplitude variation is described by the *transmission coefficients*. These transmission coefficients depend not only on the velocity contrast but also on the differences between the elastic constants of the two materials, and they may be derived from the scalar and vector potentials used for describing wave propagation. This is beyond the scope of this chapter and the interested reader is referred to books on seismology (e.g. Stein and Wysession, 2003).

Critical angle and wave guides

Let us consider a P wave crossing a horizontal interface between two mediums. We saw that when the velocity C_{P2} in medium 2 is larger than that in medium 1 (C_{P1}), the refracted P wave is closer to horizontal than the P wave was in medium 1. A critical incidence angle i_c exists for which the refracted wave is horizontal. This occurs when $i_2 = \pi/2$, or when

$$\sin i_c / C_{P_1} = 1/C_{P_2}, \quad \text{which occurs when } \sin i_c = C_{P_1}/C_{P_2} \tag{11.79}$$

For incident angles $i_1 > i_c$, no refraction of P wave is possible. However, depending on the SV velocity value in medium 2 compared with that of the P velocity in medium 1, an SV wave may still be refracted.

Generally, velocities increase with depth so that a downgoing wave has a tendency to become more and more horizontal as it propagates away from the source. Eventually the ray gets horizontal, then turns upward and eventually reaches the surface. Ray paths in layered media may be modeled with Snell's law, so that analysis of wave arrivals may be used to determine both the thicknesses of and the velocities in the various layers. This is the basic principle applied in seismic prospecting.

Another interesting consequence of Snell's law is the existence of trapped waves when the seismic source is located in a layer bounded by two other layers with higher velocities. Rays that have incidence angles beyond the *critical angle* keep being reflected at the upper and lower interfaces, so that most of the seismic energy is trapped in the soft layer which acts as a wave guide. This is observed sometimes in faults, because they are generally much softer than the surrounding geomaterials (fig. 11.4).

11.1.3 Surface waves

For sufficiently large earthquakes, the associated seismic signals are observed all over the earth. The corresponding wave trains observed at very large distances from the source are

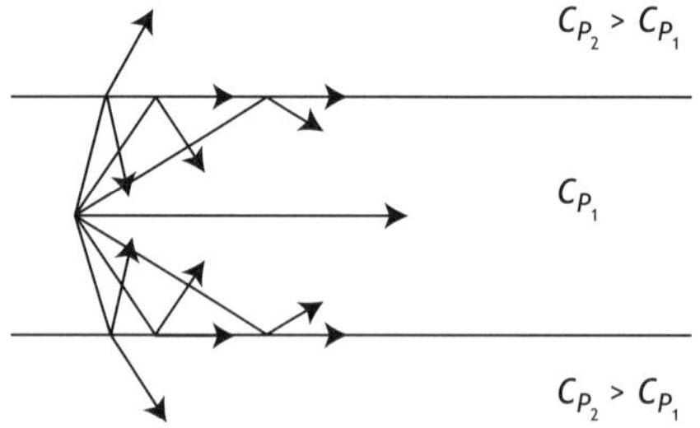

Fig. 11.4 Trapped waves in low velocity layers that constitute wave guides.

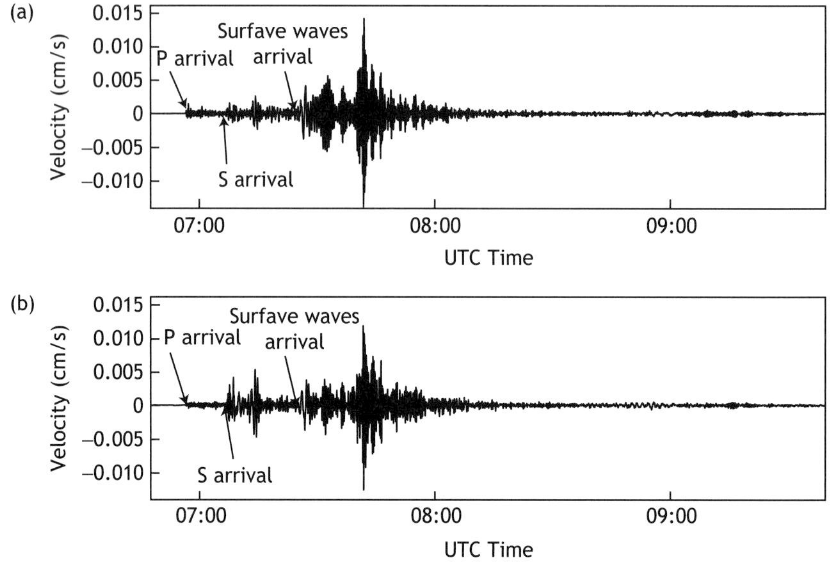

Fig. 11.5 P, S and surface waves associated with a teleseismic signal from a magnitude 7.9 earthquake that occurred more than 10 000 km away from a receiver located in the Gulf of Corinth (Greece). The upper signal shows the vertical component of particle velocity; the lower signal shows one of its horizontal components. Time is in hours.

called *teleseisms*. Seismograms of teleseisms, i.e. the records of teleseisms, are dominated by the large, longer-period, waves that arrive after the P and S body waves (fig. 11.5). These are known as *surface* waves.

Surface waves correspond to a particular case of a guided wave when one interface is the free surface whilst a positive velocity gradient is observed as one moves below the free surface.

The first surface waves to arrive after the S waves are the *Love* waves, polarized in the x_2 direction (similarly to SH waves). They are formed through the constructive interference of repeated SH wave reflections at the ground surface. Their amplitude exhibits an exponential decay with depth.

Love waves are followed by *Rayleigh* waves, which are polarized in the x_1x_3 plane and correspond to a combination of P and SV motions. They travel at velocities $C_R \approx 0.92C_S$.

These surface waves have their energy concentrated near the surface of the earth. As a result of geometrical spreading, their energy spreads two dimensionally and therefore decays as ρ^{-1}. This is to be compared to the geometrical spreading of P and S body waves, which occurs over a volume and therefore is proportional to ρ^{-2}. Hence surface waves travel much longer distances than body waves. They may circle the earth's surface many times and give rise to resonances, called the normal modes of the earth. Surface waves have yielded important information on the structure of the lithosphere and the asthenosphere.

Another example of surface waves is given by the *Stoneley* waves observed in boreholes during sonic logs; they are also known as tube waves. Sonic logs correspond to the plots of P and S wave velocities along a borehole filled with mud or water. During sonic logs a source generates P and S waves in the geomaterial surrounding the borehole, which propagate along the borehole and are recorded at different receivers located some distance from the source.

Moving the tool along the borehole provides a means to obtain continuous measurements of P and S wave velocities along the borehole. Such a plot is called a sonic log because of the high frequencies used generally with such tools (tens of kilo-Hertz). Interestingly, after the P and S waves have been observed, some further energy arrives at the recording sensors. There late arrivals of energy correspond to Stoneley waves and are somewhat reminiscent of Rayleigh waves. They may be viewed also as guided waves. The attenuation of Stoneley waves is sometimes used to investigate the hydraulic conductivity of fractures (Paillet and Cheng, 1991).

Huygens' principle and diffraction

In a few applications, linear ray theory derived from Snell's law fails to explain what is observed. For example, within the earth, seismic body waves "bend" around the earth's core and are observed at places not accessible according to Snell's law. Also, in seismic profiles "bright spots" are sometimes observed, i.e. locations within the volume through which waves are propagating that transmit more energy than would be expected from simple linear ray theory.

These phenomena may be explained by Huygens' principle, which assimilates wave fronts to a set of point sources that give rise to circular wave fronts. These circular wave fronts interfere locally positively to give rise to another wave front and everywhere else destructively. Each source is known as a diffractor.

An important point is that diffraction effects depends on the ratio of the wavelength and the size of the interface through which the wave propagates (Stein and Wysession, 2003).

11.2 Kinematics of earthquake sources

Earthquakes are generated by unstable motions that occur along faults or fractures and very often may be assimilated to displacement discontinuities.

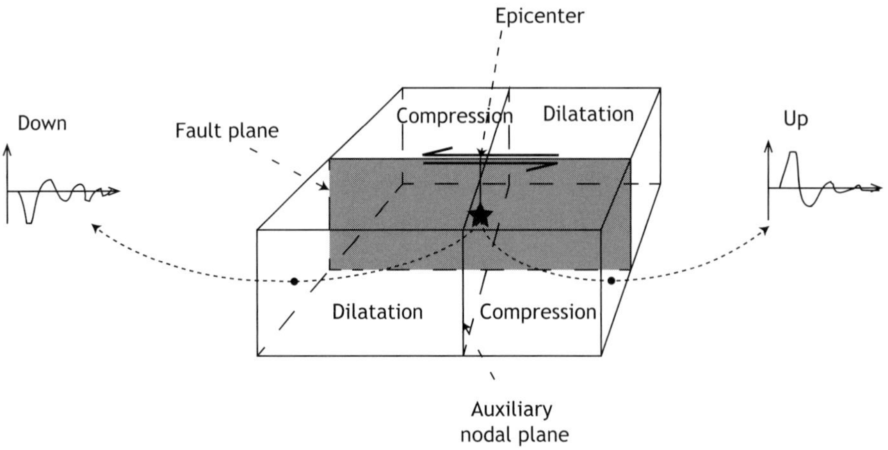

Fig. 11.6 Observation of P wave polarities at stations distributed in various azimuths with respect to the earthquake source (redrawn from Stein and Wysession, 2003, with pemission from Wiley).

One of the objective of seismology is to extract from the radiated seismic waves the kinematic characteristics of these displacement discontinuities: the dip and azimuth of the plane where the motion took place; the orientation of slip motion in the plane; the shape, size and amplitude distribution of the displacement discontinuity; the time history of the growth of the displacement discontinuity.

11.2.1 Focal plane solutions and focal mechanisms

The polarity of the first P waves arrivals, as observed at various stations after an earthquake, has been found to follow a regular pattern which depends on the station location with respect to the fault plane where the earthquake originated. For earthquakes distant enough from the station, so that rays propagate near verticality when they reach the station, the vertical component of the first motion is directed upward at some stations and downward at other locations (fig. 11.6).

Burridge and Knopoff (1964) have demonstrated that the radiation pattern of waves associated with a dislocation (i.e. an instantaneous, uniform, displacement discontinuity) is equivalent to that generated by a double couple (fig. 11.7) known as the body forces equivalent to the fault slip. One couple corresponds to the shear stress drop (assumed to be uniform) that occurred on the dislocation surface. The second couple corresponds to a fictitious stress drop on a fictitious dislocation of identical shape as the actual dislocation, but located in the plane perpendicular to the actual fault plane and intersecting the actual dislocation at its center.

Hence, two planes are identified: the actual slip plane and an auxiliary plane perpendicular to it. These two planes constitute the nodal planes of the focal mechanism of the earthquake: they correspond to the focal plane solution. Note that the unit vector parallel to the shear displacement in one nodal plane is the unit normal to the other nodal plane.

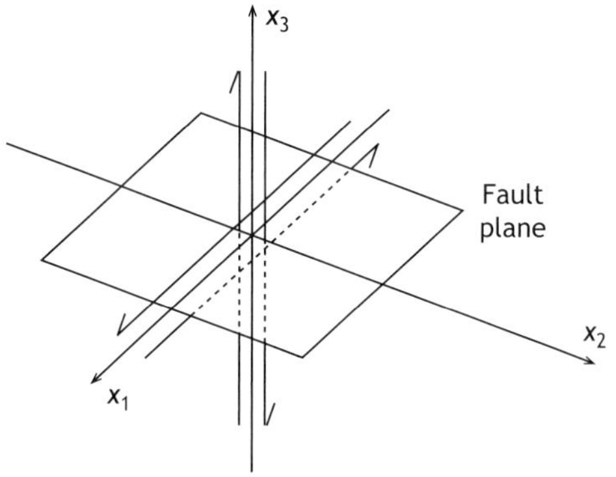

Fig. 11.7 A double couple is equivalent to a dislocation (indicated as fault plane; the x_1x_3 plane is the second nodal plane). The x_2 axis is known as the null axis. The moments of the two couples cancel, so the net torque is zero.

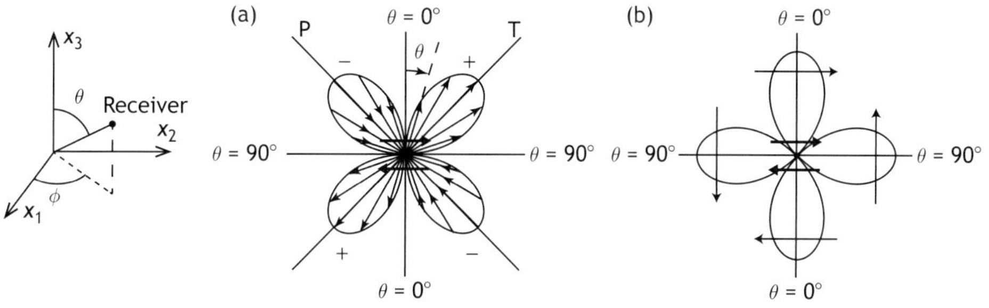

Fig. 11.8 Radiation pattern in the x_1x_3 plane (a) for P waves and (b) for S waves.

We conclude that a focal mechanism is characterized by three angles, e.g. the dip and azimuth of the fault plane and the pitch of the slip vector in the fault plane. This focal-mechanism representation assumes that the seismic source is assimilated to a point and this in turn implies that the receiver is sufficiently far from the source.

Let us consider the radiation field in spherical coordinates (fig. 11.8(a)), with θ measured from the vertical x_3 axis and ϕ measured in the horizontal x_1x_2 plane. The x_1 axis is parallel to the shear motion in the fault plane and the x_2 axis coincides with the intersection of the nodal planes (also called the null axis).

It may be shown that far from the source, in a homogeneous body, the radial motion component associated with P waves is

$$u_r = \frac{1}{4\pi \varrho C_P^3 R} \dot{M}\left(t - \frac{R}{C_P}\right) \sin 2\theta \cos\phi \tag{11.80}$$

which indicates that the radial motion depends on three factors (Burridge and Knopoff, 1964).

The first factor implies that the wave amplitude decays as the inverse of the distance to the source, R. The second factor indicates that the amplitude depends on a source time function, characterized by the seismic-moment rate function. It is the time derivative of the seismic-moment function

$$M(t) = GD(t)S(t) \tag{11.81}$$

which describes the faulting process in terms of the elastic shear modulus G of the material, the slip $D(t)$ and the slipping area $S(t)$. The quantity M_0,

$$M_0 = G\bar{D}S \tag{11.82}$$

called the *seismic moment* of the earthquake, is considered to be an effective and reliable characterization of the corresponding earthquake; $\bar{D}$ is the average slip over the area S and represents the amplitude of the equivalent displacement discontinuity. We will discuss in section 11.2.4 how the three scalars M_0, $\bar{D}$ and S may be estimated from the spectrum of the far-field (at some distance from the source) seismic displacement time function.

The third factor of equation (11.80) describes the P wave radiation pattern. The radial displacement is zero for points located in the fault plane ($\theta = \pi/2$) or in the auxiliary nodal plane ($\phi = \pi/2$). It is maximum in the x_1x_3 plane, along axes oriented 45° from the x_1 and x_3 axes. These two directions are called the P and T axes of the focal mechanism; P refers to the direction of maximum dilation and T to that of maximum compression. The P and T axes correspond to the maximum and minimum principal stress directions of the quasistatic stress drop associated with the seismic rupture (see section 11.2.4).

Similarly, if u_θ and u_ϕ are the components of the far-field shear displacement in a spherical frame of reference ($u_\rho = 0$ for shear waves), it may be shown that

$$u_\theta = \frac{1}{4\pi\varrho C_S^3 r}\dot{M}\left(t - \frac{R}{C_S}\right)\cos 2\theta \cos\phi \tag{11.83}$$

$$u_\phi = \frac{1}{4\pi\varrho C_S^3 r}\dot{M}\left(t - \frac{R}{C_S}\right)(-\cos 2\theta \sin\phi) \tag{11.84}$$

These S waves are maximum in the x_1 and the x_3 directions and zero for points located on the null axis.

Hence, the amplitude of the radiated seismic waves at a given point depends on the point's relative location with respect to the fault plane. Since P waves are the first to arrive at a given location, the polarity of the first arrivals may be used to identify the orientation of both nodal planes, provided that polarity observations are available at many different locations. These data are used for determining the slip motion direction in either nodal plane, given that the slip motion in one plane is parallel to the normal to the other plane.

However, the polarity observations are conducted at stations distant from the source, and we saw from Snell's law that, because of the changes in velocity in the mediums separating the source from the observation points, seismic rays are not straight lines. Hence, in order to take advantage of the polarity data from the various observation points, it is necessary to trace the various ray trajectories back to the point source. The direction of a ray, at its source (assimilated to a point), is called the *take-off angle*.

Consider a sphere of unit radius centered on the point source. The rays, at their origin, intersect this sphere at points which may be plotted on a Schmidt stereographic projection

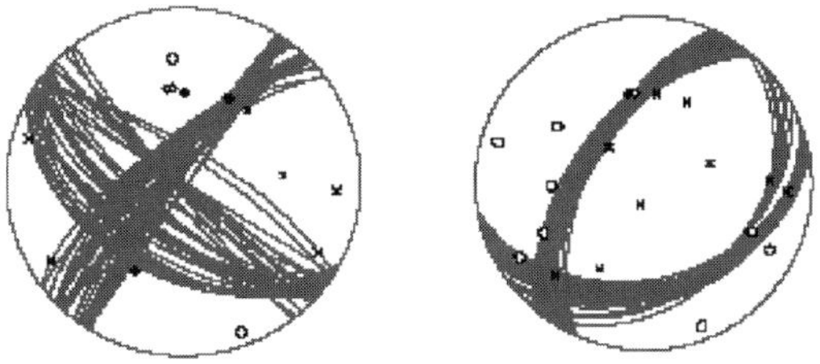

Fig. 11.9 Focal plane solution for an aftershock of the Sierentz magnitude-4.8 1980 earthquake: (left) the solution is constrained because the data comes from only a few stations, so that one plane is poorly defined; (right) an example of a well-constrained mechanism due to a large number of well-distributed stations (reproduced from Maury *et al.*, 2013, with permission from the Geological Society of France).

(see section 1.3.1). In crustal seismology, because the seismic stations are very distant from the source, the seismic rays that reach the seismic stations first are those that propagate downward. Hence only points in the lower hemisphere are of interest. However, for the monitoring of man-made seismicity it is more common to consider intersections with the upper hemisphere. The observed polarity data for the rays are shown at the corresponding points on the stereographic projection (with small circles for compression and crosses for dilatation in fig. 11.9). Statistical analysis of these stereograms can identify the dip and azimuth of both nodal planes and therefore the slip direction for both the fault plane and the auxiliary nodal plane.

Figure 11.9 indicates that the uncertainty on the determination of the dip and azimuth of the nodal planes depends on the distribution of seismic stations with respect to the fault plane geometry. In theory, focal plane solutions are characterized by three angles: the dip and azimuth of either nodal plane and the pitch of the slip direction in the same plane. Generally, however, the location of the points where polarity data are available is not evenly distributed, so that one nodal plane is better constrained than the other. Accordingly, focal plane solutions are characterized by eight numbers: the expected values for the dip and azimuth associated with each nodal plane and numbers characterizing the uncertainty associated with each of these four values (the standard deviation or extent of the domain of acceptable values for a given confidence level).

Another plot commonly used for characterizing source mechanisms shows only the expected nodal plane orientations and indicates which of the quadrants are compressional and which are dilatational. Examples of such plots are given in fig. 11.10 for the various fault types discussed in section 10.3. These simplified "beach ball" representations of earthquakes are usually taken to represent the focal mechanisms for the corresponding earthquakes. They do not address the issue of uncertainty in the nodal plane orientation determination, in contrast with focal plane solutions.

Let us note that analysis of the polarity data associated with the first P wave arrivals provides a means to identify the two nodal planes but from this information it is not possible to identify which of the two planes is the actual fault plane. This information may be retrieved from additional data, such as for example the location of the aftershocks generally associated with earthquakes of a significant size or the locations of multiplets, as will be

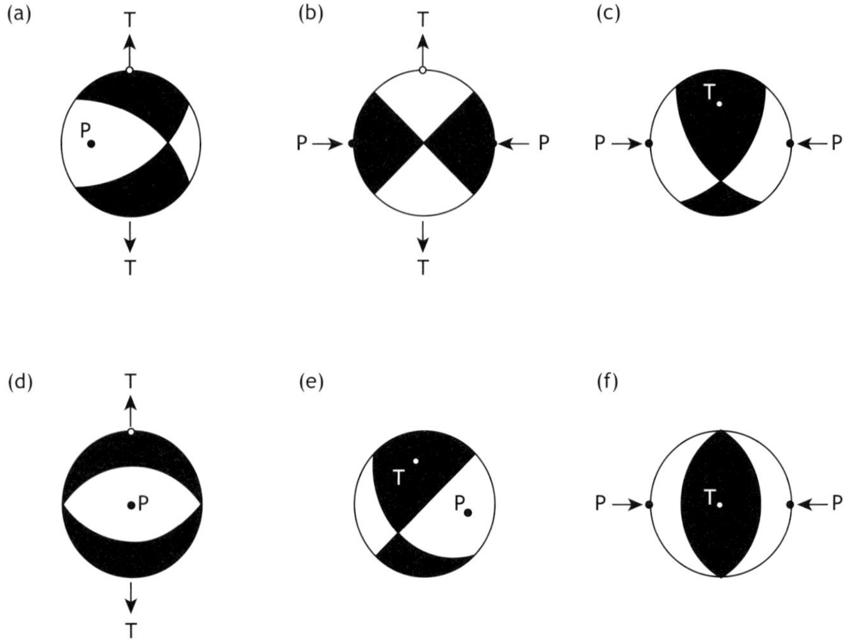

Fig. 11.10 Focal mechanisms for different fault types. The black and white correspond respectively to compressional and dilatational P wave arrivals panels (a), (c) and (e) correspond to various combined dip-slip and strike-slip motions; (b) shows pure senestral strike-slip faulting on a vertical plane oriented N135°E or dextral strike-slip faulting on a plane oriented N45°E; (d) pure normal faulting on a north or south dipping plane; (f) pure thrust faulting on an east or west dipping plane.

discussed in section 11.2.3 below. For large earthquakes, the surface waves observed at teleseismic distances (i.e. more than 1000 km, or 30°, from the source) also provide useful complementary information (see e.g. Stein and Wysession, 2003).

11.2.2 Seismic moment tensor

The double couple used to model the radiation pattern of a pure shear displacement discontinuity implies a set of forces that auto-equilibrate themselves. It exhibits no resultant torque and precludes any change in volume. Another set of forces that equilibrates itself and does not exhibit any resultant torque is a simple couple, in which the two forces are opposed in direction and are aligned on the same straight line (fig. 11.11).

It may be shown that the far-field radiation pattern of any complex seismic source may be modeled by a linear superposition of nine different couples (e.g. Aki and Richards, 1980; Julian *et al*, 1998). Three couples (M_{11}, M_{22}, M_{33}) correspond to simple couples oriented in mutually perpendicular directions. The other six couples are double couples acting in the planes defined by any two of the three axes associated with the simple couples. These nine couples constitute the components of the seismic moment tensor $\tilde{M}$:

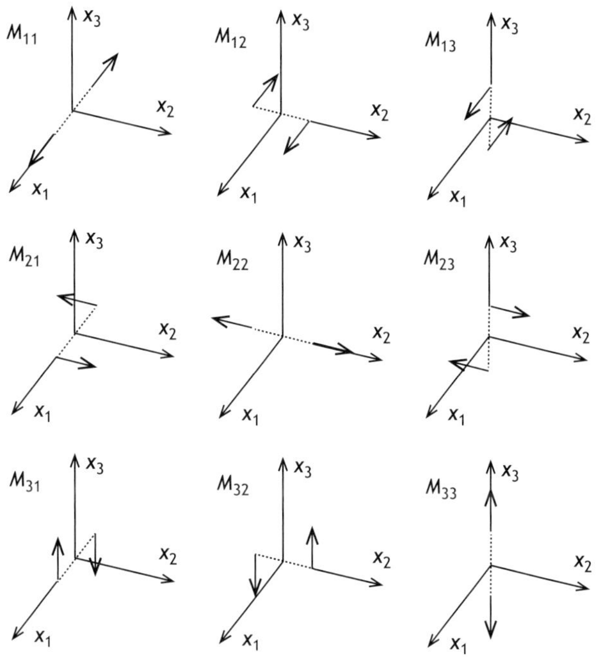

Fig. 11.11 Elementary couples used to fit the radiation pattern of complex sources (reproduced from Stein and Wysession, 2003, chapter 4, with permission from Wiley).

$$\begin{pmatrix} M_{11} & M_{12} & M_{13} \\ M_{21} & M_{22} & M_{23} \\ M_{31} & M_{32} & M_{33} \end{pmatrix} \tag{11.85}$$

The couples M_{ij} and M_{ji} must balance in order for there to be no resultant torque. Hence, the seismic moment tensor is symmetrical.

With this notation, the seismic moment tensor for the double couple considered in the previous section is simply

$$\begin{pmatrix} 0 & M_0 & 0 \\ M_0 & 0 & 0 \\ 0 & 0 & 0 \end{pmatrix} = M_0 \begin{pmatrix} 0 & 1 & 0 \\ 1 & 0 & 0 \\ 0 & 0 & 0 \end{pmatrix} \tag{11.86}$$

The matrix that represents the seismic moment for a double couple takes the simple form shown in equation (11.86) because it is expressed in the $\hat{\boldsymbol{n}}, \hat{\boldsymbol{s}}, \hat{\boldsymbol{n}} \wedge \hat{\boldsymbol{s}}$ frame of reference, where $\hat{\boldsymbol{n}}$ is the unit normal to one nodal plane and $\hat{\boldsymbol{s}}$ is the slip vector in the same nodal plane (or the unit normal to the other model plane). By expressing the components of the normal and slip vectors in an arbitrary frame of reference $(\hat{\boldsymbol{I}}_i; i = 1, 2, 3)$, $\hat{\boldsymbol{n}} = n_i \hat{\boldsymbol{I}}_i$ and $\hat{\boldsymbol{s}} = s_j \hat{\boldsymbol{I}}_j$, equation (11.86) may also be given in the $(\hat{\boldsymbol{I}}_i; i = 1, 2, 3)$ frame of reference:

$$\tilde{M} = M_0 \begin{pmatrix} 2n_1 s_1 & n_1 s_2 + n_2 s_1 & n_1 s_3 + n_3 s_1 \\ n_2 s_1 + n_1 s_2 & 2n_2 s_2 & n_2 s_3 + n_3 s_2 \\ n_3 s_1 + n_1 s_3 & n_3 s_2 + n_2 s_3 & 2n_3 s_3 \end{pmatrix} \tag{11.87}$$

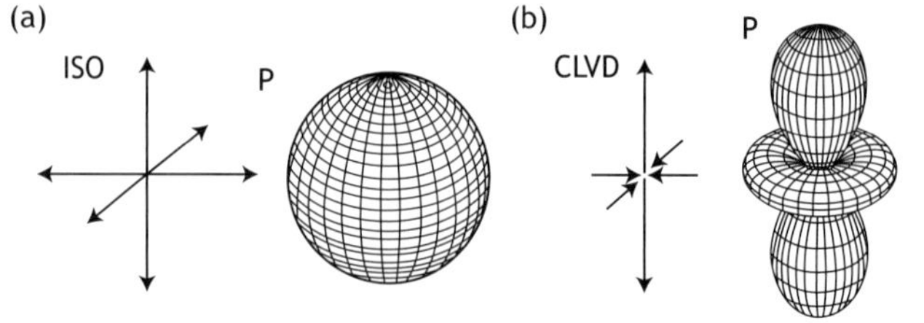

Fig. 11.12 Three-dimensional radiation diagram for (a) an explosion (ISO) and (b) a compensated linear vector dipole (CLVD) (reproduced from Sileny and Milev, 2008, with permission from Elsevier).

where the components of the matrix have been obtained using the rules for changing the frame of reference (see the appendix).

Observing that the unit vector $\hat{\boldsymbol{t}}$ in the T axis direction is given by $\hat{\boldsymbol{t}} = \hat{\boldsymbol{n}} + \hat{\boldsymbol{s}}$, we note that $\tilde{M}\hat{\boldsymbol{t}} = M_0\hat{\boldsymbol{t}}$. Similarly, if $\hat{\boldsymbol{p}}$ is the unit vector in the P axis direction then $\tilde{M}\hat{\boldsymbol{p}} = -M_0\hat{\boldsymbol{p}}$. Hence the vectors $\hat{\boldsymbol{t}}$ and $\hat{\boldsymbol{p}}$ are found to be eigenvectors of the seismic moment tensor, with eigenvalues M_0 and $-M_0$. The third eigenvector is the unit vector along the null axis, and the corresponding eigenvalue is zero. Hence the trace of the tensor is zero, which implies that there is no change in volume during the slip motion associated with this double couple.

Isotropic seismic tensor

Let us consider now the moment tensor defined by

$$\begin{pmatrix} P & 0 & 0 \\ 0 & P & 0 \\ 0 & 0 & P \end{pmatrix} \tag{11.88}$$

Its trace is $3P$ and therefore it is associated with a change in volume, either positive (for an explosion) or negative (for an implosion), with the convention that compressions are positive.

The corresponding radiation diagram is illustrated in fig. 11.12.

The compensated linear vector dipole (CLVD) and the tensile crack model

Compensated linear vector dipoles (CLVDs) are another example of seismic sources that imply no change in volume. They are characterized by components of the type

$$\begin{pmatrix} -T & 0 & 0 \\ 0 & T/2 & 0 \\ 0 & 0 & T/2 \end{pmatrix} \tag{11.89}$$

so their trace is null. Their radiation diagram is illustrated in fig. 11.12. It may be shown (e.g. Aki and Richards, 1980) that the radiation pattern generated by tensile cracks can be modeled by seismic moment tensors that combine an isotropic component for the volume change and a CLVD component for the tensile opening:

$$\begin{pmatrix} P & 0 & 0 \\ 0 & P & 0 \\ 0 & 0 & P+2T \end{pmatrix} = \begin{pmatrix} P' & 0 & 0 \\ 0 & P' & 0 \\ 0 & 0 & P' \end{pmatrix} + \begin{pmatrix} -2T/3 & 0 & 0 \\ 0 & -2T/3 & 0 \\ 0 & 0 & 4T/3 \end{pmatrix} \quad (11.90)$$

where $P' = P + 2T/3$. The trace of the tensor is $3P + 2T$. It corresponds to an increase in volume and has been proposed for interpreting the radiation patterns associated with hydraulic fractures, whether of natural origin, such as volcanic dykes (Aki and Richards, 1980), or man-made, such as hydraulic fractures conducted for the stimulation of hydrocarbon or geothermal reservoirs (Sileny *et al.*, 2009).

11.2.3 Seismic source location determination

We have seen that the determination of focal plane solutions from the polarity of first arrivals, and more generally that of seismic moment tensors from the modeling of body wave radiation diagrams, requires the velocity structure of the medium and the location of the event to be known accurately. Furthermore, the determination of event locations also requires an accurate evaluation of the seismic velocity field.

Recent developments, however, have provided a means to obtain very precise relative event locations without a full understanding of the overall velocity structure of the medium. These relative locations have proved helpful for characterizing a seismically active structure, as well as possible variations with time of its velocity field.

The objective of this subsection is to introduce the principles of the various event location procedures, but the techniques themselves are not presented.

Determinations based on P and S wave arrival time picking

In a homogeneous body the elastic parameters and the density are uniform throughout the volume under consideration, and so also are the P and S wave velocities. Consequently, in homogeneous bodies, body waves propagate along linear seismic rays from the seismic source to the various recording sensors. The picking (detection) of P or S wave arrival times at four different recording sensors provides enough information to identify the three components of the source location $\boldsymbol{x}^S$ with respect to that of the four recording sensors $\boldsymbol{x}^{rs_i}; i = 1, \ldots, 4$, and the time t_0 at which the source generated the corresponding waves. This is done by solving a set of nonlinear equations. For the case of P waves we have

$$t_P^{rs_i} = t_0 + [(x_1^{rs_i} - x_1^S)^2 + (x_2^{rs_i} - x_2^S)^2 + (x_3^{rs_i} - x_3^S)^2]^{1/2}/C_P, \qquad i = 1, \ldots, 4 \quad (11.91)$$

However, arrival time pickings are not exact and errors in time picking directly influence the location determination. This situation is typical for physical measurements: a large number of measurements is required to correct, statistically, the errors associated with each individual measurement when these errors can be assumed to obey a Gaussian law. This is particularly relevant for automatic time-picking procedures (e.g. Baer and Kradolfer, 1987; Earle and Shearer, 1994).

It may be observed that a set of four stations with arrival time picking for both P and S waves yields eight equations with four unknowns, which may be solved by a least squares method (see section 13.1.2, data reduction). More generally, when N stations yield M first P wave arrivals and Q first S wave arrivals, a set of $M + Q$ equations can be constructed with only four unknowns. This redundancy can be used to advantage for improving the location determination accuracy.

Such a location procedure is applicable when a source is not too distant from the receivers, so that the medium where rays propagate may be assumed to be largely homogeneous. Generally, however, the medium exhibits some spatial variations in the seismic velocities and ray tracing is then compulsory for the location procedure.

Codes have been developed for various cases and today are freely available on the internet, e.g. the code HYPO71 (http://jclahr.com/science/software/hypo71/8517.html) or the code HYPOINVERSE-2000, which can be downloaded from the US Geological Survey website. Both codes work on the assumption that the velocity field is known.

Recently codes have been developed that solve simultaneously for the event location and the velocity field determination. The medium is discretized in sub-volumes, in which the velocity field is assumed to be constant but unknown. When first-arrival-time data for a large number of events are obtained at a large number of seismic receivers, which are well distributed around the event sources location, a solution may be obtained that minimizes the differences between observed and computed arrival times (e.g. Thurber and Eberhardt-Phillips, 1999; Thurber *et al.*, 2003). Hence, arrival times are used to conduct, simultaneously with the source location determination, a *seismic tomography* of the medium. A seismic tomography characterizes structural volumes within the medium under consideration through their seismic velocity characteristics. The method is known as local earthquake tomography (LET). Codes for conducting simultaneously source location determination and the tomography of the medium are available on the internet.

Determinations based on waveform cross-correlation

When two seismic events are very close to each other, the seismic rays between either source and a given recording station are very close to each other. They may be considered as being parallel and therefore travel in on the same velocity field. Further, it is well established that the seismic sources of small earthquakes may involve the same instability phenomenon, so that the radiation patterns of these repeating sources are very similar (Poupinet *et al.*, 1984; Frechet *et al.*, 1989; Nadeau *et al.*, 1994). Such repeating events with similar source mechanisms are called *doublets* when the same source is repeated only once, *triplets* when it is repeated twice times or *multiplets* for a larger number of repeats. The similarity between the signals generated by such repeating sources may be used for very precise relative location determination.

If the first event of a doublet occurs at time t_1 and the second event at time t_2, and if $\boldsymbol{r}$ is the vector between both sources whilst $\boldsymbol{n}$ is the normalized vector of the ray leaving the sources and arriving at station i (fig. 11.13), the difference in P wave arrival times τ_i at station i, for the two sources is

$$\tau_i = \frac{\boldsymbol{r} \cdot \boldsymbol{n}}{C_P + t_2 - t_1} \tag{11.92}$$

or, in terms of the relative coordinates of event 2 with respect to those of event 1,

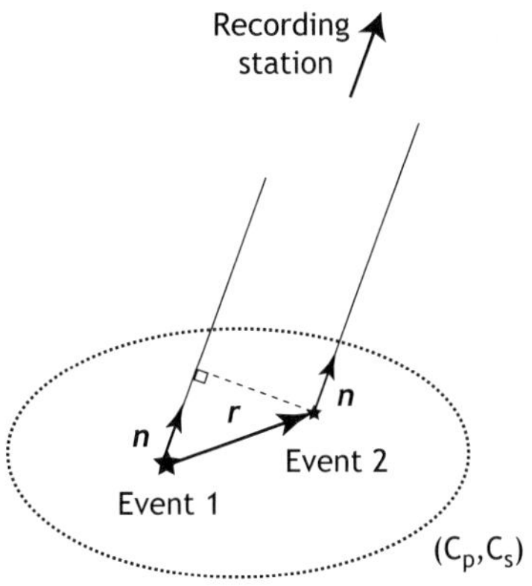

Fig. 11.13 The relative locations of two repeating events, called a doublet.

$$\tau_i = \frac{1}{C_P}(X_1 \cos\theta \sin\varphi + X_2 \sin\theta \sin\varphi + X_3 \cos\varphi) + \delta t \tag{11.93}$$

where $\delta t = t_2 - t_1$ and φ and θ are respectively the inclination and the azimuth of the unit vector $\boldsymbol{n}$ parallel to the ray direction.

The time lapse δt may be evaluated by comparing the P and S wave arrival times at the same station or by comparing the P (or S) wave arrival times for the same events at different stations. The time lapse τ_i caused by the relative difference in location $\boldsymbol{r}$ of the two events is determined by cross-correlation between the signals, hence the name *waveform cross-correlation* (WCC) given to this method. The time resolution may be as small as 1/10 to 1/20 of the period associated with the corner frequency of the signals (see the next section for a definition of the corner frequency).

This relative relocation procedure may be used to identify the slip vector direction at the source and therefore which nodal plane of a focal mechanism is the slipping plane (e.g. Rietbrock *et al.*, 1996). The WCC technique has also been applied to the detection of very small time variations in seismic velocities (Zhang and Thurber, 2003), a procedure which has shown itself to be efficient for investigating *in situ* fluid percolation processes (Calo *et al.*, 2011).

11.2.4 Elementary considerations on source spectra

We saw in section 11.2.1 how the polarity of the first P wave arrivals provides information on the dip and azimuth of the nodal planes associated with a double couple and therefore on the slip vector direction associated with the corresponding nodal planes. But these polarities do not yield any information on the amplitude of the slip vector nor on its time history. We very briefly discuss below how waveform modeling may yield further information on the kinematics of seismic sources.

The seismogram observed on a seismometer results from a combination of three actions. First, some vibratory motion is generated at the seismic source. Then this seismic signal propagates through the medium between the source and the seismometer. Finally the seismic motion is recorded by the seismometer.

Each operation has effects that depend on the frequency of the seismic waves. A convenient method for analyzing a seismic signal $u(t)$ is through its Fourier transform $U(\omega)$, where

$$u(t) = \frac{1}{2\pi}\int_{-\infty}^{\infty} U(\omega)e^{i\omega t}d\omega, \qquad U(\omega) = \int_{-\infty}^{\infty} u(t)e^{-i\omega t}dt \tag{11.94}$$

The essence of this approach is that the time function $u(t)$ that represents the seismic motion may be viewed as a sum of elementary harmonic waves that are generated throughout the duration of the slip motion at the source. Hence a seismic motion may be described either by its time function or by the fourier transform (FT) of its time function.

The combined effects of the three actions, source + transmission + reception, may be modeled as a convolution of four terms:

$$u(t) = x(t) * e(t) * q(t) * i(t) \tag{11.95}$$

if we recall that the convolution of two functions is written

$$s(t) = w(t) * s(t) = \int_{-\infty}^{\infty} w(t-\tau)s(\tau)\ d\tau \tag{11.96}$$

and call $x(t)$ the source time function and $i(t)$ the response function of the seismometer. The function $e(t)$ represents the combined effects of elastic refractions and reflections that affect the waves during their travel between the source and the seismometer, whilst $q(t)$ represents anelastic effects during the same travel.

Modeling the source time function of earthquakes still is an active domain of research. However, the solution proposed by Brune (1970) for modeling displacements has proved acceptable for it fits many observations. In particular, after applying Brune's source model (Brune, 1991), Hanks and Thatcher (1972) observed that the spectrum of the far-field shear displacement generated by an earthquake, $\Omega(f)$, may be approximated by a constant long-period level Ω_0, a spectral corner frequency f_0 and a high-frequency spectral asymptote $f^{-\gamma}$.

On a log-log plot of the far-field shear displacement spectrum, corrected for instrument response and attenuation, Brune's model reduces to two straight lines that intersect each other at the *corner frequency* f_0. Such plots may be used to evaluate the seismic moment M_0 and the mean source radius r:

$$M_0 = 4\pi\varrho C_S^3\Omega_0 R \tag{11.97}$$

$$r = \frac{2.34C_S}{2\pi f_0} \tag{11.98}$$

where R is the distance between the source and the observing station. With this seismic source model, the seismic moment provides a means to evaluate an equivalent amount of slip, which may be interpreted as an equivalent uniform *static stress drop* $\Delta\tau = \tau_i - \tau_f$, where τ_i is the initial uniform shear stress supported by the fault before the earthquake and τ_f is the uniform shear stress that is presumed to be supported by the fault after the earthquake:

$$\Delta\tau = \frac{7}{16}\frac{M_0}{r^3} \tag{11.99}$$

Alternative source models have been proposed for modeling the waveforms generated by seismic sources (e.g. Madariaga, 1977). They all confirm the important result that a log-log plot of the frequency spectrum of the displacement field generated by an earthquake, as observed at distant stations, is bilinear and provides a way to evaluate the seismic moment and the source dimension. However, for different source models, different values are obtained for the equivalent uniform amount of slip motion and for the corresponding static stress drop, although the orders of magnitude are the same for all models.

An important issue for seismic source modeling is determining whether the earthquake released the complete amount of shear stress that was initially supported by the fault or whether a significant amount of shear is still supported by the fault in the area where the slip occurred.

Earthquakes correspond to an energy release, some of which is transformed into heat, some of which is transformed into surface energy and some of which is transformed into kinetic energy. Furthermore, part of this kinetic energy gives rise to the far-field seismic waves. So, clearly some relationship must exist between the radiated seismic energy and the total energy released by the earthquake, but this relationship is still being investigated. It is generally recognized that the radiated seismic energy is only a few per cent of the total energy released by the earthquake. Some authors consider that it may be as small as a few thousandths. The reader interested in this topic is referred to the proceedings of a workshop that was focused on precisely this discussion (Abercrombie *et al.*, 2006).

11.3 Scaling of seismic events

Because the amount of energy released by an earthquake cannot be measured, various alternative procedures have been proposed for scaling earthquakes, depending of the objective of the scaling process.

The scaling of seismic events was first initiated in Italy during the eighteenth century. It relied on a quantification of the extent of the damage associated with these events. This led to the concept of seismic intensity and the definition of intensity scales that are used for various earthquake engineering applications, i.e. for the design of structures that are earthquake resistant. However, with the development of seismic recording systems, observations on the far-field seismic wave amplitudes generated by earthquakes have led to the definition of magnitude scales.

Empirical relationships have been proposed between these various scaling procedures, so that a rough evaluation of the total energy radiated by earthquakes over a given area for a given time interval may be produced. These empirical relationships have proved very useful for constructing maps characterizing seismic hazard.

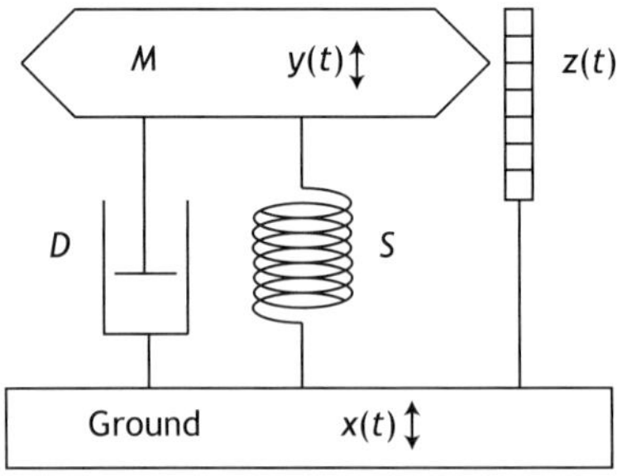

Fig. 11.14 Principle of a seismometer based on inertia.

First we introduce the sensors that have been developed for monitoring seismic vibrations in various conditions and then we discuss intensity and magnitude scales.

11.3.1 Seismometry

The objective of seismometry is to measure vibratory ground displacements. Various sensors have been designed for this purpose, the physical principles of which depend on the domain of application. Traditionally, seismology exploits records produced by seismometers, which are inertial instruments. But other sensors exist. We list below sensors used in the various domains of ground vibration monitoring.

Seismometer

The principle of a seismometer is based on inertia. The simplest model of an inertial seismometer consists of a mass M and spring system with stiffness S, to which is associated a damping system with viscosity D (fig. 11.14). The mass, usually called the internal mass, can move relative to the instrument frame, which is fixed to the ground. A seismometer records either the motion of the mass with respect to the frame or the force necessary to keep the mass from moving with respect to the frame.

Let $x(t)$ be the time-dependent ground motion, whilst the motion of the mass is $y(t)$. The motion of the mass with respect to ground is $z(t) = y(t) - x(t)$. Any force $f(t)$ acting on the mass generates an acceleration of the mass, $\ddot{y}(t)$, such that

$$M\ddot{y}(t) = f(t) - Sz(t) - D\dot{z}(t) \tag{11.100}$$

so that the relationship between the ground motion and the recorded motion of the mass (the seismogram) with respect to the ground is

$$M\ddot{z}(t) + D\dot{z}(t) + Sz(t) = f(t) - M\ddot{x}(t) \tag{11.101}$$

Equation (11.101) shows that an acceleration of the ground, $\ddot{x}(t)$, has the same effect as an external force $f(t) = -M\ddot{x}(t)$ acting on the mass in the absence of ground acceleration. Hence, recording the force required to keep the mass from moving with respect to ground is the same as measuring the ground motion; this is the principle of modern feedback seismometers.

For mechanical seismometers, any movement of the ground moves the frame. Initially, because of inertia, the mass does not move and therefore the initial motion recorded by the system is that of the frame with respect to the mass. Then, through the force exerted by the spring and the damping system on the mass, the mass moves according to a known, well-calibrated, response. The eigen frequency (natural frequency) ω_0 of such systems depends on the relative values of the mass M and the spring stiffness S: $\omega_0 = \sqrt{S/M}$.

The seismometer record represents the relative motion between the mass and the frame of reference. Because the motion of the mass is precisely defined by its response function, the motion of the frame may be identified. Hence, as already mentioned, the motion of the ground can be identified after deconvolution of the seismometer response from the seismometer record.

In modern feedback instruments, the force necessary to keep the mass from moving with respect to the reference frame is measured continuously and the record of the force amplitude variation with time constitutes the seismic record.

Usually, seismometers measure ground displacement in three orthogonal directions, one of which is vertical. However, simpler instruments are sometime undirectional, in which case the monitored component is usually parallel to the vertical direction, given that seismic rays from distant earthquakes arrive at the station parallel to the vertical direction.

Today, so-called broadband seismometers may record motions with frequencies ranging from 500 Hz (0.002 seconds per cycle) to 0.001 Hz (1000 seconds per cycle) with a dynamic range reaching 40 db. Classical short-period seismometers have eigenfrequencies equal to a few hertz and can record signals from a few hertz to a hundred hertz. In this frequency range, seismometers are constituted by three geophones (see below) mounted orthogonally.

The sensitivity of seismometers is in the 1500 V/m to 20 000 V/m range.

Geophone

A geophone is a passive analog (rather than digital) device that converts ground velocity into electrical voltage. A geophone operates on the principle of inertia and comprises a spring-mounted magnetic mass moving within a wire coil to generate an electrical signal. The eigenfrequencies of geophones are typically in the 10 Hz range. They are constrained to motion in one dimension. However, in some devices, three moving coils are mounted orthogonally to provide complete three-dimensional monitoring.

Geophones operate **only in the air**, i.e. either directly on ground or inserted in watertight containers for borehole usage. The typical sensitivity of geophones is 30 V s/m.

Accelerometer

An accelerometer measures the change in force experienced by a mass held at rest in the frame of reference of the device. Conceptually it behaves as a damped mass on a spring. When the accelerometer experiences an acceleration, the spring is able to accelerate the mass at the same rate as the casing. In commercial devices, piezoelectric or capacitive components are used to convert the acceleration into a voltage. Piezoelectric accelerometers rely on piezoceramics (components generating an electrical output signal upon mechanical excitation and, conversely, generating a mechanical pulse upon electrical excitation) and are referred to as microelectromechanical systems (MEMS). Their performance varies in a very wide range, depending on the application. They are unmatched in terms of their upper frequency range (megaHz).

Strong-motion station or accelerograph

Strong-motion stations, also called accelerographs, record the motion generated by earthquakes close to the source, where instruments with high sensitivity, such as seismometers or geophones, saturate and are therefore useless. An accelerograph contains one or three components, each component corresponding to an accelerometer (usually of the MEMS type). It operates often on a triggering mode, i.e. recording is initiated only when the signal generated by the accelerometer is larger than a certain threshold value. They are used to measure peak ground acceleration (PGA) or peak ground velocity (PGV); these are the maximum acceleration or velocity observed at a given location, for a given earthquake. They are deployed not only on ground but also at various floors in buildings. Hence they are essential for characterizing the response of structures to strong earthquakes, and they help to produce so-called *response spectra* for buildings; a response spectrum is the envelope of the maximum observed acceleration (or velocity) for a series of damped oscillators with various frequencies.

Hydrophone

A hydrophone is a microphone designed to work underwater for recording sounds that propagate in water. The typical frequency range of hydrophones ranges from tens of hertz to tens of kilohertz. They use piezoelectric components to convert pressures to electrical signals. They are frequently used in arrays to sum (or stack) the signals recorded at precisely located points in order to amplify some frequencies whilst filtering others. For frequencies lower than a few hertz, sensors that record pressure variations with time are often referred to as dynamic pressure transducers.

Strainmeter, extensometer, tiltmeter

A strainmeter measures changes in length, as does an extensometer, over frequencies ranging from a few hertz (the domain of seismology) to periods that may reach many years (the domain of geodesy). Usually the words strainmeter or extensometer are associated with the physical principle of the measurement, but the term used varies from one field to another. Tiltmeters operate in the same frequency range as strainmeters or extensometers but detect changes in angle, for example the inclination of the normal to the ground surface at a given location.

Usually the resolution (the detection of a relative variation in the measurement) of strainmeters as used in geophysical applications reaches 10^{-7}, while that of tiltmeters may be as fine as 10^{-10}. Because such measurements are very sensitive to atmospheric pressure and temperature variations, they are conducted in specially isolated locations, such as deep caverns, or vaults or boreholes.

11.3.2 Intensity and magnitude

The intensity describes the effects of earthquakes as observed at ground surface, whilst the magnitude relies on far-field seismic wave recording. Both may be used, through empirical relationships, to obtain a rough evaluation of the seismic energy radiated by an earthquake, but these scaling procedures are not intended to address the issue of the total energy released by the earthquake. Indeed, the total released energy involves various dissipation processes, including for example heat generation or heat dissipation through fluid circulation, that are still open to debate.

Intensity scales and peak ground motion

According to Musson and Cecic (2011), the intensity of an earthquake can be defined as a classification of the strength of ground shaking at any place during an earthquake, as estimated from observed effects (http://bib.telegrafenberg.de/publizieren/vertrieb/nmsop/). Thus intensity refers to macroseismic effects, i.e. effects that are evaluated without the use of instruments. It usually involves a scale expressed in Roman numerals to stress the fact that the values are only descriptive and not quantitative.

Two different intensity scales have been defined, the Modified Mercalli (MM) scale and the MSK scale (from the name of its authors, Medvedev, Sponheurer and Karnik). They may be considered as being similar for all practical purposes. The following is an abbreviated description of the 12 levels of the Modified Mercalli intensity scale:

I. Not felt except by a very few under especially favorable conditions.
II. Felt only by a few persons at rest, especially on upper floors of buildings.
III. Felt quite noticeably by persons indoors, especially on upper floors of buildings. Many people do not recognize it as an earthquake. Standing motor cars may rock slightly. The vibrations are similar to the passing of a truck. The duration can be estimated.
IV. Felt indoors by many, outdoors by few during the day. At night, some awakened. Dishes, windows, doors disturbed; walls make cracking sound. Sensation like heavy truck striking building. Standing motor cars rock noticeably.
V. Felt by nearly everyone; many awakened. Some dishes, windows broken. Unstable objects overturned. Pendulum clocks may stop.
VI. Felt by all, many frightened. Some heavy furniture moved; a few instances of fallen plaster. Damage slight.
VII. Damage negligible in buildings of good design and construction; slight to moderate in well-built ordinary structures; considerable damage in poorly built or badly designed structures; some chimneys broken.
VIII. Damage slight in specially designed structures; considerable damage in ordinary substantial buildings with partial collapse. Damage great in poorly built structures. Fall of chimneys, factory stacks, columns, monuments, walls. Heavy furniture overturned.
IX. Damage considerable in specially designed structures; well-designed frame structures thrown out of plumb. Damage great in substantial buildings, with partial collapse. Buildings shifted off foundations.

X. Some well-built wooden structures destroyed; most masonry and frame structures destroyed with foundations. Railway lines bent.
XI. Few, if any (masonry) structures remain standing. Bridges destroyed. Railway lines substantially bent.
XII. Damage total. Lines of sight and level are distorted. Objects thrown into the air.

Damage is caused by the forces associated with the seismic acceleration field. Hence intensity maps may be expressed in terms of variables characteristic of the peak ground motion (i.e. the peak ground acceleration or peak ground velocity) when the dynamic response of the structures involved in the intensity evaluation is known. These estimated peak ground motions may be converted in turn into magnitudes. For example, Gutenberg and Richter (1942) suggested that a satisfactory relationship between the intensity I and the peak ground acceleration a is given by

$$\log a = \frac{I}{3} - \frac{1}{2} \tag{11.102}$$

where the peak acceleration a is expressed in gals (cm/s^2) and the intensity is expressed in levels of the modified Mercalli scale.

This procedure is particularly useful for regions where historical reports on earthquakes preceded the development of efficient seismic recording instrumentation. Such a translation of intensity scales into magnitude scales helps broaden the time domain for which seismic data are available, from tens of years for instrumental data to hundreds of years, or even thousands of years, for intensity maps (e.g. Grünthal and Wahhlström, 2012).

The various definitions of magnitude

The concept of the magnitude of an earthquake relies on the measurement of wave amplitudes on seismograms. It is assumed that the wave amplitude reflects the earthquake size once the signal has been corrected for the sensor response, for the geometric spreading effect and for attenuation. Magnitude scales have the form

$$M = \log\frac{A}{T} + F(h, D) + C \tag{11.103}$$

where A is the maximum amplitude of the signal at the dominant period T; F is a function that corrects for the depth h of the hypocenter (the location of the initial source of the earthquake) and the distance D between the epicenter (the projection at ground level of the hypocenter) and the recording seismometer; C is a regional correction factor. Magnitude scales are logarithmic, so that an increase by one unit implies a ten-fold increase in amplitude.

At regional distances, S waves generally exhibit the largest amplitude and the magnitude evaluated from the S wave magnitudes is called M_L, or the local magnitude. For teleseisms, the surface waves exhibit the largest magnitude and the magnitude derived from them is called M_S. The generally accepted definition of the surface wave magnitude is

$$M_S = \log\left(\frac{A}{T}\right)_{max} + 1.66 \log \Delta + 3.3 \qquad (11.104)$$

where A is the maximum surface wave amplitude and T is its period, whilst Λ is the distance in degrees between the recording station and the epicenter.

When body waves rather than surface waves are used to evaluate the magnitude, it is denoted m_B, M_B or m_b. Body wave magnitudes may be converted to surface wave magnitudes with the IASPEI (International Association of Seismology and Physics of the earth Interior) recommended relationship

$$m_b = 0.56M_S + 2.9$$

.

However, we saw in section 11.2.1 that earthquakes generally correspond to displacement discontinuities that are kinematically analogous to double couples. These double couples are characterized by their seismic moment, and seismic moments may be directly evaluated from the spectral analysis of seismic signals. Hence yet another magnitude scale has been defined (Kanamori, 1977) that keeps the logarithmic scaling but introduces a variable directly related to the double couple characteristics. It is called the moment magnitude M_w and is related to the seismic moment M_0 by

$$M_w = \frac{\log M_0}{1.5} - 10.73 \qquad (11.105)$$

where M_0 is expressed in ergs (dyne cm). The M_w magnitude is equivalent to the M_S magnitude for earthquakes with M_w magnitudes smaller than 8.2. For larger magnitudes the M_S magnitude saturates and should not be used.

For small to very small seismic sources, the magnitude is sometimes defined by the total signal duration. These local magnitude scales are then tied to body wave magnitudes or moment magnitude scales by comparing the values obtained for the largest events.

11.3.3 Empirical scaling relationships

Kanamori and Anderson (1975) observed that the relationship between the average offset, $\bar{D}$, the average shear stress drop, $\Delta\tau$ and the strain energy release, Δw, published in the literature for various static dislocation models may be given as

$$\Delta\tau = CG(\bar{D}/\bar{L}) \qquad (11.106)$$

where G is the elastic shear modulus, C is a nondimensional shape factor close to unity and $\bar{D}/\bar{L} \equiv \Delta\bar{e}$ is a representative strain variation. If S is the dislocation area, the change is strain energy is

$$\Delta w = S\bar{D}\bar{\tau} \tag{11.107}$$

where $\bar{\tau} = (\tau_i + \tau_f)/2$ is the mean shear stress, with τ_i and τ_f respectively the initial and final shear stress, i.e. the shear stress supported by the slipping surface before and after slip has occurred.

The static moment M_0 is given by

$$M_0 = GS\bar{D} = (G/\bar{\tau})\Delta w \tag{11.108}$$

whilst the radiated seismic energy is taken as a fraction η of the strain energy:

$$E_S = \eta\Delta w \tag{11.109}$$

Given Haskell's solution (1964) for the wave field radiated from a dislocation for which the dislocation rise time is a linear ramp function, Kanamori and Anderson observed that various useful relationships may be derived between moment and source area, moment and magnitude, magnitude and energy or magnitude and fault area. After testing these relationships for a large number of earthquakes, they came up with the following proposed relationship:

$$M_S \approx \frac{2}{3}\log M_0 \approx 2\log L, \qquad \text{for } M_S \geq 6 \tag{11.110}$$

where M_S is the magnitude calculated from the surface wave amplitude, M_0 is the seismic moment and L is the fault rupture length. For small earthquakes they found the scaling rules

$$M_S \approx \log M_0 \approx 3\log L \approx \log E_S \tag{11.111}$$

whilst for very large earthquakes they observed that more appropriate scaling rules were

$$M_S \approx \frac{1}{3}\log M_0 \approx \log L \approx \frac{1}{3}\log E_S \tag{11.112}$$

These empirical relationships also suggest that $M_S \approx \log S$, where S is the ruptured fault area.

Hence, for example, a magnitude-4 earthquake is likely to be associated with a 1 km long fault whilst a magnitude-6 earthquake corresponds to the rupture of a 10 km long fault, and a magnitude 7 earthquake to a 30 km long fault. This shows that the dynamics of magnitude-7 and larger earthquakes depends on the thickness of the elastic crust (usually somewhere between 20 and 30 km). Hence the growth of the rupture front of a very large earthquake is a two-dimensional problem which affects lithospheric-scale structures, whilst that of smaller events is indeed a three-dimensional problem.

The Gutenberg–Richter magnitude–frequency relation

The number of earthquakes that occur yearly in a given region varies with magnitude; there are few events of large magnitude and many more of small magnitude. The relationship between the number N of events with magnitude larger than a certain value M_S and this threshold value M_S, known as the Gutenberg–Richter relationship, is

$$\log N = a - bM_S \tag{11.113}$$

For example, when considering the aftershocks observed after a main shock, it is generally found that b is close to 1. But, for swarms that correspond to a seismic activity with a very wide range of magnitudes and no main shock, the b coefficient is generally much larger than 1. This difference in b coefficients is sometimes attributed to the volume of the seismically active region, with b values close to 1 for fault-type ruptures but larger b values (1.5 to 2) for more diffuse deformation processes, such as those with fluid diffusion, as discussed in chapter 12.

The Gutenberg–Richter relationship applies also at very large scales. For example, every year there are only one or two earthquakes in the world with magnitude larger than 8, but about 10 with magnitude larger than 7 and more than 100 with magnitude larger than 6.

The parameter a is sometimes used to evaluate the detection threshold of a seismic network. Indeed, for small earthquakes that occur far from a recording station the signal will be too small to be observed on the station record. The resolving power of a station is defined by the minimum magnitude that may be detected by a station; it depends on both the event magnitude and the distance between the event and the station. The resolving power of a network is arbitrarily defined by the minimum magnitude that may be detected by a given number of stations of the seismic network (normally four, since a minimum of four stations is required for locating an event using the first P wave arrival times). Hence the minimum magnitude that fits the Gutenberg–Richter relationship for earthquakes detected with a given network defines the detection power of the network. Some events with lower magnitudes are not detected, so that the total number of events is no longer representative of the local deformation process.

11.3.4 Seismic and aseismic motions

We have just seen that small events may not be detected by a given seismic network but that the sensitivity of seismic sensors varies with the signal frequency. Hence, two possibilities may occur for seismic signals not to be detected. First, even though their frequency content may be within the optimum response bandwidth of the sensor, the wave amplitude is too small to be separated from the background noise or, second, the frequency bandwidth of the signal is outside the detection domain of the sensor and the signal again goes undetected.

An example of the bandwidth effect on signal detection by a short-period seismic monitoring network was published by Bernard *et al.* (2004). In fig. 11.15 it may be observed that the complete signal duration for the event of interest lasts about 1000 seconds, during which much higher frequencies are observed for about 20 seconds. This signal is interpreted as a low-velocity slip motion, during which an asperity was ruptured

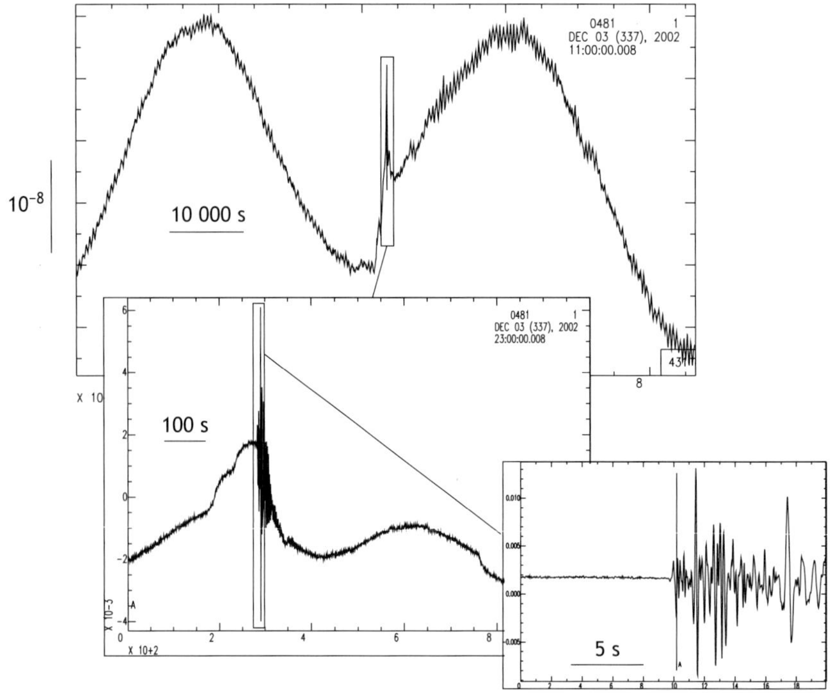

Fig. 11.15 Seismic signal recorded with a broadband borehole dilatometer. The time scale in the figures changes from 10 000 s, to 100 s, to 5 s (reproduced from Bernard *et al.*, 2004, with permission from Elsevier Masson).

giving rise to the high-frequency signal. Only the high-frequency signal was detected by the local short-period seismic monitoring network.

Examples where such slow motion occurs and generates occasional high-frequency signals have also been reported for hydraulic stimulations (e.g. Bourouis and Bernard, 2007). The slow expansion of the slipping surface gives rise to high-frequency multiplet sources that highlight the large-scale motion and provide a means to determine the fracture growth velocity (see the discussion in section 12.4.2).

These examples illustrate the fact that motion may occur at some low velocity over large patches of the slipping surface and yet go undetected by classical seismic monitoring systems. Such motions are termed nonseismic, aseismic or creeping motions. Although they go undetected, such slow motions influence the local stress field significantly and may progressively load the non-slipping parts of the surface, termed asperities (Cornet, 2012).

Such nonseismic motions are well known for some large faults, such as for example the 1200 km long Philippine Fault, which creeps by about 3 cm/yr on Leyte Island (Duquesnoy *et al.*, 1994) without any significant earthquake in the region of creep. Yet the same fault ruptures with magnitude-7 earthquakes further to the north or to the south of the region of creep.

This illustrates the necessity to deploy recording systems with a very broad band for large-scale *in situ* regional deformation investigations as well as for monitoring manmade perturbations.

11.4 Further reading

Aki, K., and Richards, P. G., 2002. *Quantitative Seismology.* University Science Books, 700 pp.

Stein, S., and Wysession, M., 2003. *An Introduction to Seismology, Earthquakes, and earth Structure.* Blackwell, 498 pp.

Udias, A., Madariaga, R., and Buforn E., 2014. *Source Mechanisms of Earthquakes: Theory and Practice.* Cambridge University Press, 300 pp.

12 Elements of solid–fluid interactions

Generally geomaterials involve both solid and fluid phases, and in some situations the mechanics of the fluids and that of the solids may be considered independently. Often, however, fluids and solids do interact, so specific attention must be given to these interactions.

Another term used for describing solid–fluid interactions is the word *coupling*, and we refer to *hydromechanical coupling* when considerations on only fluid and solid mechanics are involved. We refer to *hydrothermomechanical coupling* when we are investigating the combined effect of temperature, fluid pressure and stress on the deformation process of multiphase geomaterials and reserve the term *thermomechanical coupling* for purely temperature effects in the mechanics of solids.

First, we introduce linear theories that have been proposed for hydromechanical coupling and for thermomechanical coupling. Then, after a very schematic introduction to solid–fluid physicochemical interactions, we address some nonlinear issues related to hydromechanical coupling in the context of brittle geomaterials, namely hydraulic fracturing and hydraulic stimulation.

We follow this with a discussion on fluid-induced seismicity and its implications for understanding better fluid migration in the brittle crust.

The chapter ends with considerations on two different aspects of thermal fracturing.

12.1 Linear hydromechanical coupling

12.1.1 Terzaghi's effective stress concept

We introduced the concept of effective stress in connection with the development of failure in saturated geomaterials under compression (chapters 8–10). When a volume of geomaterial B, consisting of a solid phase S and a fluid phase F, is loaded under compression, the fluid phase supports a *mean pressure* P_p whilst the solid phase supports a *mean*, or *effective stress* $\tilde{\sigma}'$, so that the *total stress* $\tilde{\sigma}$ supported by the equivalent geomaterial is

$$\tilde{\sigma} = \tilde{\sigma}' + P_p \tilde{1} \tag{12.1}$$

In equation (12.1) compressions are reckoned positive. This effective stress definition immediately raises the question how to define the mean pressure in the fluid and mean

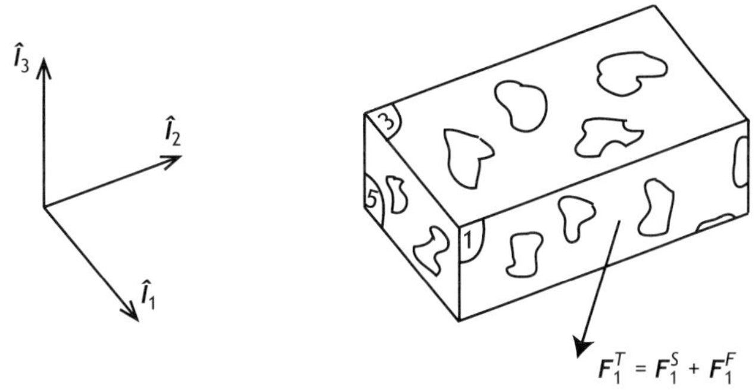

Fig. 12.1 Representative elementary volume for the definition of effective stresses.

stress in the solid. We assume here that the pore space is fully interconnected and filled completely by fluid F (thus we have saturated material).

Let us consider a representative elementary volume (REV) in the shape of a parallelepiped of this multiphase (solid + liquid) geomaterial (fig. 12.1).

The faces of the parallelepiped are numbered 1 through 6, so that faces 1 and 4, 2 and 5, 3 and 6 are respectively parallel to one another. Let us call $\boldsymbol{F}_i^S$ the resultant force supported by the solid through face i and $\boldsymbol{F}_i^F$ the resultant force supported by the fluid through the same face. The total force supported by face i is thus

$$\boldsymbol{F}_i^T = \boldsymbol{F}_i^S + \boldsymbol{F}_i^F \tag{12.2}$$

It is implicit that $\boldsymbol{F}_i^T = \boldsymbol{F}_{i+3}^T;\ i = 1, 2, 3$, so that the system is at rest. When the fluid is at rest, it can support only a pressure (there can be no shear stress in the fluid), and the mean pressure supported by the fluid through face i is

$$P^F = |\boldsymbol{F}_i^F|/A_F = |\boldsymbol{F}_{i+3}^F|/A_F| = |\boldsymbol{F}_i^F|/(fA_T) \tag{12.3}$$

where A_F is the total area of the pores through face i, f is the areal porosity of face i and A_T is the total area of face i (see section 1.1.1). Similarly, we may define *mean stress* components

$$\sigma_{ij}^S = F_{ij}^S/A_S = F_{ij}^S/[(1-f)A_T] \tag{12.4}$$

so that component j; $j = 1, 2, 3$, of the total force $\boldsymbol{F}_i^T$ supported by face i is

$$F_{ij}^T = [fP^F\delta_{ij} + (1-f)\sigma_{ij}^S]A_T \tag{12.5}$$

It has been suggested that equations (12.3) and (12.4) should be used to define the mean pressure in the fluid, and the mean stress components in the solid, within the REV. However, this averaging process depends on the porosity of the REV, which itself depends on both the fluid pressure and the stress field. The results of many laboratory experiments (e.g. Handin *et al.*, 1963; Brace and Martin, 1968; Cornet and Fairhurst, 1974) suggest that the failure of porous rocks under compression does not depend on these average stresses but rather on the effective stresses, as defined by

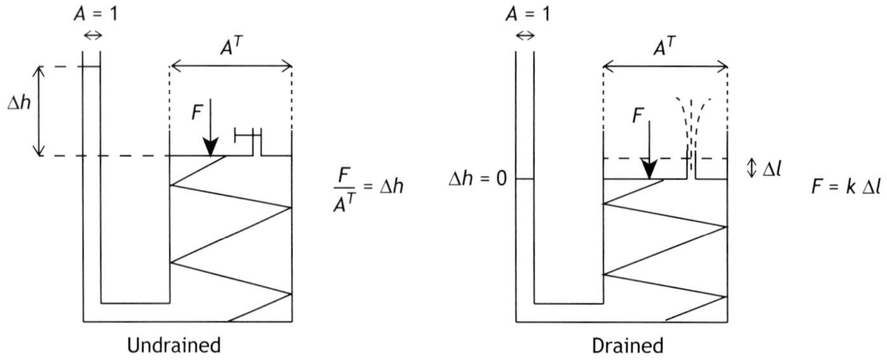

Fig. 12.2 Schematic representation of the role of drainage conditions on soil deformation. A force F is supported by the fluid when no flow is allowed to leave the system as it is applied. If fluid is allowed to flow out of the system, the force is taken up by the spring and the corresponding vertical displacement depends on the spring stiffness. The deformation rate depends then on the flow rate.

$$\sigma'_{ij} = F^S_{ij}/A_T \qquad \text{and} \qquad P_p = |\boldsymbol{F}^F_i|/A_T \tag{12.6}$$

as in equation (12.1). The use of equation (12.6) presumes that the solid and the fluid phases coexist simultaneously, in the same proportion, at all points within the REV, i.e. the REV is assumed to be filled with an equivalent bulk material B, as defined in section 9.1.1 or fig. 7.10.

This reasoning was initially proposed by van Terzaghi (1943), who analyzed the shear strength of saturated soils and demonstrated the effect of the drainage conditions on this shear strength, following the loading scheme of fig. 12.2.

The figure illustrates how a change in total stress affects both the pore pressure and the mean stress in the solid, depending on the drainage conditions.

In the absence of drainage (fig. 12.2, left) the increment in external stress is completely balanced by the increment in the fluid pressure. When drainage takes place, however (fig. 12.2, right), the fluid pressure drops and the increase in external load is progressively transferred to the solid. This is called *consolidation* in soil mechanics. It corresponds to a progressive, irreversible, reduction of pore volume associated with the drainage of fluid (e.g. Craig, 1987). A somewhat similar, irreversible, deformation mechanism occurs in hydrocarbon reservoirs as hydrocarbons are pumped out of the reservoirs (e.g. Segall, 2010; Wang, 2000). In some locations (e.g. the North Sea), consolidation has resulted in vertical motions reaching a few meters.

We saw in chapter 7 that the brittle behavior of geomaterials is governed by the growth of fissures. This growth depends on local stress concentrations, which are not fully described by the concept of the mean effective stress in a solid.

Yet we showed in section 9.1 that, when local stress concentrations are not an important factor, elasticity theory may be adapted to define elastic constants for a so-called *effective medium*, which corresponds to a continuous material that is equivalent to the solid matrix together with the pore space (material B of fig. 9.1). The applicability of effective medium elasticity to fracture propagation analysis requires that the length scale of the fault of interest is much larger than that of the microcracks and pores that make up the pore space of the geomaterial. This proposition was applied when we were determining the effect of the process zone on fracture propagation in section 7.1.3.

For such an effective material, any change in total stress results in an elastic, reversible, variation in the pore volume. However, if the pore volume varies, this affects the fluid pressure in a way that depends on the drainage conditions. This hydromechanical type of coupling was investigated initially by Biot (1941, 1956a) and the theory that developed from this early work is known as linear poroelasticity theory (e.g. Rice and Cleary, 1976; Coussy, 1991; Detournay and Cheng, 1993; Wang, 2000; Guéguen *et al.*, 2003).

12.1.2 Linear poroelasticity

The original method introduced by Biot for analyzing the elastic response of isotropic porous materials was to add two variables to the classical elastic theory, namely the pore pressure P_p, as defined by equation (12.6), and the change in fluid volume per unit bulk volume, ζ, also referred to as the change in fluid content (according to our notation of section 9.1.1, $\zeta = \delta V_F/V_B$); ζ is generally taken positive for a gain in fluid content and so it decreases if fluid is taken out of the bulk material.

If the fluid mass content is defined as the ratio of the mass M^F of fluid F in an undeformed REV volume and the REV volume V_B then the increment in fluid content as defined by Biot is

$$\zeta = \frac{\delta M_F}{\varrho_{F_0}} \tag{12.7}$$

where ϱ_{F_0} is the fluid density in the reference state and δM_F is the change in the mass of fluid during the loading of a unit bulk volume. Note that Biot's definition of fluid content is not to be confused with the classical soil mechanics definition of water content, w, defined as the ratio of the mass of water and the mass of solid ($w = M_w/M_S$, see chapter 1).

Biot assumed that the bulk volumetric strain ε_{v_B} and the fluid content ζ are linear functions of the spherical total stress component $\sigma_m = (\sigma_{11} + \sigma_{22} + \sigma_{33})/3$ and of the pore pressure P_p:

$$\varepsilon_{v_B} = a_{11}\sigma_m + a_{12}P_p \tag{12.8}$$

$$\zeta = a_{21}\sigma_m + a_{22}P_p \tag{12.9}$$

The physical meaning of the parameters a_{ij} may be seen from the following:

$$a_{11} = \left.\frac{\delta\varepsilon_{v_B}}{\delta\sigma_m}\right|_{\delta P_p=0} \equiv \frac{1}{K_B} \tag{12.10}$$

$$a_{12} = \left.\frac{\delta\varepsilon_{v_B}}{\delta P_p}\right|_{\delta\sigma_m=0} \equiv \frac{1}{H} \tag{12.11}$$

$$a_{21} = \left.\frac{\delta\zeta}{\delta\sigma_m}\right|_{\delta P_p=0} \equiv \frac{1}{H_1} \tag{12.12}$$

$$a_{22} = \left.\frac{\delta \zeta}{\delta P_p}\right|_{\delta \sigma_m = 0} \equiv \frac{1}{R} \tag{12.13}$$

The constant K_B corresponds to the bulk modulus of material B when the pore pressure P_p is maintained constant, i.e. when the bulk material deforms in drained conditions under changes in applied total stress σ_m. The inverse value, $1/K_B$, is the bulk material compressibility β_B, introduced in section 9.1.

The constant $1/H$ describes how the bulk volume V_B changes when the pore pressure varies while the external spherical stress component is maintained constant. It is called the poroelastic expansion coefficient.

The constant $1/R$ describes how the fluid content changes when the pore pressure varies while the total external spherical stress component is maintained constant. It is called the unconstrained specific storage coefficient and is sometimes referred to as S_{σ_m} (Wang, 2000).

Biot assumed the existence of a potential energy density, U, defined by

$$U = \tfrac{1}{2}(\sigma_m \varepsilon_{v_B} + P_p \zeta) \tag{12.14}$$

Since

$$\frac{\partial^2 U}{\partial \zeta \, \partial \varepsilon_{v_B}} = \frac{\partial^2 U}{\partial \varepsilon_{v_B} \, \partial \zeta} \tag{12.15}$$

he showed that

$$a_{12} = \frac{1}{H} = a_{21} = \frac{1}{H_1} \tag{12.16}$$

so that

$$\varepsilon_{v_B} = \frac{\sigma_m}{K_B} + \frac{P_p}{H} \tag{12.17}$$

$$\zeta = \frac{\sigma_m}{H} + \frac{P_p}{R} \tag{12.18}$$

Constitutive equations of linear poroelasticity

Two kinematic variables are used in poroelasticity, namely the displacement field $\boldsymbol{u}$ that describes the motion of solid and the filtration velocity field $\boldsymbol{v}^f$, which describes the motion of the fluid with respect to the solid (see section 9.3). For small strains of the solid the classical definition used in elasticity applies:

$$\varepsilon_{ij} = \frac{1}{2}(u_{i,j} + u_{j,i}); \qquad \varepsilon_{v_B} = \varepsilon_{ii} = \operatorname{tr} \tilde{\varepsilon} \tag{12.19}$$

Extensions are often considered as positive (Detournay and Cheng, 1993; Wang, 2000), which requires that stresses in compression be reckoned negative. Note that, in the classical convention of soil mechanics pressures are taken as positive, as stated when we were discussing equation (12.1).

As pointed out in section 4.3.4, the small-strain tensor is the same whether it is expressed in a Lagrangian frame or an Eulerian frame of reference. So, we will select an Eulerian frame since it is more convenient for expressing fluid flow. Consequently, if v_i^f is the rate of discharge of fluid through the unit area of a surface normal to direction x_i, the rate of change in fluid content within a unit volume of material B is (Detournay and Cheng, 1993)

$$\frac{\partial \zeta}{\partial t} = -v_{i,i}^f \tag{12.20}$$

Equation (12.20) results from mass conservation, introduced in section 6.1.1.

An alternative micromechanical expression for the increment of fluid content is the difference between the fractional change in pore volume and the fractional change in fluid volume (Berryman, 1992):

$$\zeta = \frac{\delta V_P - \delta V_F}{V_B} \tag{12.21}$$

For incompressible fluids (for which $\delta V_F = 0$), the increment in fluid content corresponds to the pore volume variation divided by the reference volume V_B. If the pore volume is incompressible (i.e. $\delta V_p = 0$), an added fluid volume would result only from fluid compressibility.

Combining equations (12.10)–(12.13) with the linear elastic behavior of the effective body B yields, in the principal-stress frame of reference,

$$\varepsilon_1 = \frac{\sigma_1}{E} - \frac{\nu}{E}(\sigma_2 + \sigma_3) + \frac{P_p}{3H} \tag{12.22}$$

$$\varepsilon_2 = \frac{\sigma_2}{E} - \frac{\nu}{E}(\sigma_1 + \sigma_3) + \frac{P_p}{3H} \tag{12.23}$$

$$\varepsilon_3 = \frac{\sigma_3}{E} - \frac{\nu}{E}(\sigma_1 + \sigma_2) + \frac{P_p}{3H} \tag{12.24}$$

$$\zeta = \frac{1}{3H}(\sigma_1 + \sigma_2 + \sigma_3) + \frac{P_p}{R} \tag{12.25}$$

Summing equations (12.22)–(12.24) yields equation (12.8).

Equations (12.22)–(12.25) depend implicitly on time when $v_{i,i}^f \neq 0$, i.e. when the fluid moves with respect to the solid. However, for undrained conditions (no flow during the instantaneous elastic response, i.e. $\zeta = 0$) or fully drained conditions (loading occurs very slowly so that the fluid pressure remains constant, i.e. $\delta P_p = 0$, or the fluid pressure returns to its original value when the loading perturbation is in place.), these equations do not imply any time dependence.

When there is no time dependence, reference must be made to whether the deformation process takes place under drained or undrained conditions. Hence the bulk material elastic parameters, in equations (12.22)–(12.24), must be correspondingly defined for drained or for undrained conditions.

We note that $E_B, \nu_B, \lambda_B, G_B$ and K_B (respectively Young's modulus, the Poisson ratio, the first Lamé coefficient, the second Lamé coefficient, also called the shear or Coulomb modulus, and the bulk modulus) are the elastic parameters that characterize the behavior of effective bulk material B under fully drained conditions (P_p is independent of time).

When flow occurs, the filtration velocity is described by Darcy's law (see section 9.3.3):

$$v_i^f = -K_f(P_{p,i} - \varrho_F g_i) \tag{12.26}$$

where K_f is the geomaterial permeability, $P_{p,i}$ is the fluid pressure gradient in the i direction, ϱ_F is the fluid density and g_i is the component of gravity in the i direction.

This set of equations is completed by the law of motion for the geomaterial B:

$$\operatorname{div}\tilde{\sigma} + \varrho_B \mathbf{g} = \varrho_B \ddot{\mathbf{u}} \tag{12.27}$$

For seismic wave propagation, the frequency range of the solid motion implies alternating particle velocities that are generally too fast to be associated with any significant large-scale fluid flow, so that undrained conditions prevail in this type of problem (e.g. Biot, 1956b, 1962). However, for lower frequencies, as in the case of those associated with earth tides, the periodic straining of the solid results in periodic water level variations in piezometric wells. These water level fluctuations can be used to characterize the fluid diffusivity (see later) of the geological formations surrounding the well (e.g. Segall, 2010).

The question of the relative velocity between solid and fluid is also fundamental for understanding the stability of fracture development, particularly for shear fractures since unstable shear may lead to earthquakes. This remains a topic of active research and will not be explored further here. It involves nonlinear poroelasticity, i.e. poroelastic parameters that depend upon both the total stress and the pore pressure for the microcracked materials under consideration.

Biot's coefficients revisited

In addition to the parameters K_B (the bulk modulus measured under drained conditions), H ($1/H$ is the poroelastic expansion coefficient measured under constant total stress) and R ($1/R \equiv S_{\sigma_m}$ is the specific storage coefficient measured under constant total stress), which have already been defined, Biot and Willis (1957) introduced the parameter $\alpha = K_B/H$ to describe the ratio of the volume of fluid added to the storage capacity (see below) and the corresponding change in bulk volume, when the pore pressure remains constant.

Whilst S_{σ_m} refers to the *unconstrained* specific storage coefficient, the *constrained* specific storage coefficient, or specific storage coefficient at constant volumetric strain, $S_{\varepsilon v}$, is defined as follows:

$$S_{\varepsilon v_B} = \left.\frac{\delta\zeta}{\delta P_p}\right|_{\varepsilon v_B = 0} \equiv \frac{1}{M} \tag{12.28}$$

where ε_{v_B} is the bulk volumetric strain. From equations (12.17) and (12.18) we get

$$\zeta = \frac{K_B}{H}\varepsilon_{v_B} + \left(\frac{1}{R} - \frac{K_B}{H^2}\right)P_p \tag{12.29}$$

so that

$$S_{\varepsilon v_B} = S_{\sigma_m} - \frac{K_B}{H^2} \tag{12.30}$$

Making use of the coefficient α, we may write

$$\zeta = \alpha \varepsilon_{v_B} + \frac{P_p}{M} \tag{12.31}$$

It is often convenient to define the rate of change in pore pressure when the bulk material is loaded under undrained conditions, i.e. when $\zeta = 0$. This is expressed by Skempton's B coefficient (Skempton, 1954):

$$B \equiv -\frac{\delta P_p}{\delta \sigma_m} \tag{12.32}$$

which may be expressed in terms of Biot's coefficients R and H:

$$B = \frac{R}{H} \tag{12.33}$$

Note that equations (12.17) and (12.18) may be rewritten as

$$\varepsilon_{v_B} = \frac{1}{K_B}(\sigma_m + \alpha P_p) \tag{12.34}$$

$$\zeta = \frac{\alpha}{K_B}\sigma_m + \frac{\alpha}{K_B B} P_p \tag{12.35}$$

so that

$$S_{\sigma_m} = \frac{1}{R} = \frac{\alpha}{K_B B} \tag{12.36}$$

The quantity $\sigma'_m \equiv \sigma_m + \alpha P_p$ is sometimes called the *effective mean stress*. However, often the term *effective stress* refers to the quantity $\sigma'_{ij} \equiv \sigma_{ij} - P_p \delta_{ij}$ as originally defined by van Terzaghi (1943), with the convention that compressions are reckoned positive. Terzaghi's effective stress concept is useful for describing the role of pore pressure during the failure of geomaterials under compression, provided that proper attention is given to pore pressure variations induced by dilatancy (Brace and Martin, 1968) or contractancy (Cornet and Fairhurst, 1974) associated with the failure process.

Hence, in order to avoid any confusion between these two definitions of effective stress we propose to refer explicitly to *Terzaghi's effective stress*, $\sigma'_{ij} \equiv \sigma_{ij} - P_p \delta_{ij}$, with compressions taken as positive, when we are considering failure problems, and to *Biot's effective stress*, $\sigma'_m \equiv \sigma_m + \alpha P_p$, with compressions taken as negative, when we are dealing with poroelasticity issues.

Further considerations on compressibility

Considerations on the compressibility, which refers to the volume variation associated with a *pressure* variation, need to include the type of bulk material B, the fluid phase F that fills the interconnected pore space, the solid phase S and also the interconnected pore volume V_p. The *pressure* variation may refer to a change in the externally applied mean stress $\delta\sigma_m$ or to a variation in the pore pressure δP_p, or both.

When the mean stress variation $\delta\sigma_m$ is equal to the pore pressure variation, this particular loading condition is sometimes referred to as an "unjacketed" loading condition, by analogy with laboratory testing procedures (Wang, 2000). This corresponds to the situation described by equation (9.2) and discussed in section 9.1.1. It involves the solid matrix compressibility β_S, also called the unjacketed material compressibility. In section 9.1.1 we called it the *intrinsic elastic compressibility*, and, from equation (12.34), it is related to K_B and Biot's parameter α as follows:

$$\beta_S = \frac{1}{K_B} - \frac{\alpha}{K_B} \tag{12.37}$$

so that

$$\alpha = 1 - \frac{K_B}{K_S} \tag{12.38}$$

We note that when the solid matrix compressibility $\beta_S = 1/K_S$ is much smaller than the bulk material compressibility $\beta_B = 1/K_B$, Biot's parameter α is very close to unity, so that Biot's effective stress becomes equivalent to Terzaghi's effective stress.

Note that material M as defined in section 9.1.1 and material S as defined in sections 1.1.1 and 12.1.1 are not exactly the same. Indeed, whilst both refer to the solid part of the REV, in section 9.1.1 the solid phase is assumed to consist of a single homogeneous constituent (the solid matrix M), whilst in sections 1.1.1 and 12.1.1 the solid phase consists of many constituents, some of which may be much "softer" than others. For such a heterogeneous solid phase, the stress tensor within the various solid grains is not itself hydrostatic even though the load applied to the REV may be hydrostatic.

The *drained pore compressibility* is defined as the fractional change in pore volume when the mean external stress σ_m varies whilst the pore pressure P_p remains constant:

$$\frac{1}{K_{V_p}} \equiv -\frac{1}{V_p}\frac{\delta V_p}{\delta \sigma_m}\bigg|_{\delta P_p=0} \tag{12.39}$$

For a fully saturated REV of bulk material, the volume porosity n may be defined as a ratio of the interconnected pore volume or the fluid volume and the REV volume V_B:

$$n = \frac{V_p}{V_B} = \frac{V_F}{V_B} \tag{12.40}$$

Thus from equation (12.21), the increment in fluid content ζ is

$$\zeta = n\frac{\delta V_p}{V_p} - n\frac{\delta V_F}{V_F} \tag{12.41}$$

However, the change in fluid volume is related to the change in fluid pressure through the fluid compressibility:

$$\frac{1}{K_F} \equiv -\frac{1}{V_F}\frac{\delta V_F}{\delta P_p} \tag{12.42}$$

so that, from equation (12.35) we obtain

$$\frac{\delta V_p}{V_p} = -\frac{\alpha}{nK_B}\sigma_m + \frac{1}{n}\left(\frac{\alpha}{K_B B_{sk}} - \frac{n}{K_F}\right)P_p \tag{12.43}$$

Equation (12.43) expresses the relative change in pore volume when both the external mean stress σ_m and the pore pressure P_p vary. So, for linear poroelasticity in drained conditions, i.e. for constant pore pressure and no dependence of the poroelastic parameters on the loading conditions, we get

$$\frac{1}{K_{V_p}} = \left.\frac{\alpha}{nK_B}\right|_{\delta P_p=0} \tag{12.44}$$

After Rice and Cleary (1976), Wang (2000) defined an *unjacketed pore compressibility*, $1/K_{V_{p}unj}$, when referring to the pore volume variation observed if both the external mean stress σ_m and the pore pressure P_p vary by the same amount (these were the loading conditions in equation (9.2)):

$$\frac{1}{K_{V_{p}unj}} \equiv -\frac{1}{V_p}\frac{\delta V_p}{\delta P_p} \tag{12.45}$$

After Geertsma (1957), we noted in section 9.1.1 that for homogeneous single-constituent solid materials these unjacketed loading conditions lead to

$$\frac{\delta V_p}{V_p} = \frac{\delta V_M}{V_M} = \frac{\delta V_B}{V_B} = \frac{1}{K_S} = \frac{1}{K_{V_{p}unj}} \tag{12.46}$$

In the general case, i.e., for multiconstituent solids, Brown and Korringa (1975) and Rice and Cleary (1976) observed that these two quantities, $1/K_S$ and $1/K_{V_{p}unj}$, are different, especially for clay-rich materials. If we write $P_d = \sigma_m - P_p$, equation (12.43) becomes

$$\frac{\delta V_p}{V_p} = -\frac{\alpha}{nK_B}P_d + \frac{1}{n}\left(\frac{\alpha}{K_B B_{sk}} - \frac{n}{K_F} - \frac{\alpha}{K_B}\right)P_p \tag{12.47}$$

Applying the above definition of the unjacketed pore compressibility to equation (12.47), we obtain

$$\frac{1}{K_{V_{p}unj}} = -\frac{1}{n}\left(\frac{\alpha}{K_B B_{sk}} - \frac{n}{K_F} - \frac{\alpha}{K_B}\right) \tag{12.48}$$

$$= -\frac{1}{n}\left[\left(\frac{1}{K_B} - \frac{1}{K_S}\right)\left(\frac{1}{B_{sk}} - 1\right) - \frac{n}{K_F}\right] \tag{12.49}$$

from which we may derive Skempton's B coefficient:

$$B = \frac{1/K_B - 1/K_S}{1/K_B - 1/K_S + n(1/K_F - 1/K_{V_{p}unj})} \tag{12.50}$$

Hence, for geomaterials the solid phase of which is made up of constituents with similar elastic properties (such as sand and sandstone), $1/K_{V_{p}unj} \simeq 1/K_S$ so that

$$B \simeq \frac{1/K_B - 1/K_S}{1/K_B - 1/K_S + n(1/K_F - 1/K_S)} \tag{12.51}$$

Let us note that if the solid phase compressibility is negligible compared with that of the bulk material, we then have for Skempton's B coefficient

$$B \simeq \frac{\beta_B}{\beta_B + n\beta_F} \tag{12.52}$$

which is the original result in Skempton (1954).

Storage capacity and hydraulic diffusivity

The *storage capacity* is a measure of the amount of fluid that must be added to, or removed from, an REV to affect the fluid pressure. It depends on the mechanical constraints applied to the solid. We have already introduced the unconstrained specific storage coefficient, $S_{\sigma_m} = 1/R$, in equation (12.13) and the constrained specific storage coefficient, $S_{\varepsilon v_B} = 1/M$, in equation (12.28). Using equation (12.36) in (12.48) we obtain

$$S_{\sigma_m} = \frac{1}{K_B} - \frac{1}{K_s} + n\left(\frac{1}{K_F} - \frac{1}{K_{V_{punj}}}\right) \tag{12.53}$$

so that, for incompressible solid constituents ($1/K_S = 1/K_{V_{punj}} = 0$), the unconstrained storage capacity is

$$S_{\sigma_m} = \frac{1}{K_B} + \frac{n}{K_F} \tag{12.54}$$

In hydrogeology it is a common procedure to consider the uniaxial specific storage, which corresponds to the storage capacity observed when no horizontal strain is allowed for the solid whilst the vertical stress is supposed to be constant.

The hydraulic diffusivity governs the diffusion of fluid in a rock mass. It depends on both the permeability and the storage capacity. Hence the hydraulic diffusivity is defined for unconstrained (D_{σ_m}), constrained ($D_{\varepsilon v}$) or uniaxially constrained (D_{uniax}) conditions, by the following equations (Wang, 2000):

$$D_{\sigma_m} \equiv \frac{k_f}{\mu_d S_{\sigma_m}} \tag{12.55}$$

$$D_{\varepsilon v_B} \equiv \frac{k_f}{\mu_d S_{\varepsilon v_B}} \tag{12.56}$$

$$D_{uniax} \equiv \frac{k_f}{\mu_d S_{uniax}} \tag{12.57}$$

where k_f is the intrinsic permeability of the rock mass and μ_d is the dynamic fluid viscosity.

Constitutive equations of linear poroelasticity

Because shear stresses do not induce any volume change for a linearly elastic material, they have no effect on the fluid content or pore pressure. Hence equations (12.22)–(12.25), which have been given in the principal-direction (eigenvector) frame of reference, may be easily generalized to any frame of reference:

$$\varepsilon_{11} = \frac{\sigma_{11}}{E_B} - \frac{\nu_B}{E_B}(\sigma_{22} + \sigma_{33}) + \frac{\alpha}{3K_B}P_p \tag{12.58}$$

$$\varepsilon_{22} = \frac{\sigma_{22}}{E_B} - \frac{\nu_B}{E_B}(\sigma_{11} + \sigma_{33}) + \frac{\alpha}{3K_B}P_p \tag{12.59}$$

$$\varepsilon_{33} = \frac{\sigma_{33}}{E_B} - \frac{\nu_B}{E_B}(\sigma_{11} + \sigma_{22}) + \frac{\alpha}{3K_B}P_p \tag{12.60}$$

$$\zeta = \frac{1}{3H}(\sigma_{11} + \sigma_{22} + \sigma_{33}) + \frac{P_p}{R} \tag{12.61}$$

$$\varepsilon_{12} = \frac{1 + \nu_B}{E_B \sigma_{12}} \tag{12.62}$$

$$\varepsilon_{23} = \frac{1 + \nu_B}{E_B \sigma_{23}} \tag{12.63}$$

$$\varepsilon_{31} = \frac{1 + \nu_B}{E_B \sigma_{31}} \tag{12.64}$$

or more generally

$$\varepsilon_{ij} = \frac{1 + \nu_B}{E_B}\sigma_{ij} - \frac{\nu_B}{E_B}\sigma_{kk}\delta_{ij} + \frac{\alpha}{3K_B}P_p\delta_{ij}, \qquad i,j = 1,2,3 \tag{12.65}$$

which is referred to as the *pure-compliance* formulation (see (10.1)). Indeed, the strain and the increment of fluid content, which are dimensionless, are functions of the stress and pore pressure, which both have units of force per unit area.

As in the case of standard linear elasticity, a *pure-stiffness* formulation may be derived. It relates the stresses and pore pressure to the strains and the increment of fluid content and involves the elastic constants of the undrained effective bulk material B. In addition we may define a *mixed-stiffness* formulation, which relates the stress and fluid content increment to the strain and pore pressure, or a *mixed-compliance* formulation, which relates the strain and pore pressure to the stress and fluid content increment.

It may be shown (e.g. Wang, 2000) that the mixed-stiffness formulation leads to

$$\sigma_{ij} = 2G_B\varepsilon_{ij} + \lambda_B\varepsilon_{kk}\delta_{ij} - \alpha P_p\delta_{ij}, \qquad i,j = 1,2,3 \tag{12.66}$$

$$\zeta = \alpha\, \varepsilon_{kk} + \frac{\alpha}{K_{B_u}B} \tag{12.67}$$

where K_{B_u} is the undrained bulk modulus of the effective bulk material. Recall that in equation (12.66) compressions (and therefore the pore pressure) are reckoned as negative.

Stresses induced by radial fluid flow in a hollow cylinder

As an application of linear poroelasticity theory, let us consider the infinite thick hollow cylinder problem already addressed in section 5.3.4 (fig. 12.3). We will follow the derivation of Fjaer *et al.* (2008), and we adopt here the convention that compressions as reckoned as positive.

The inner radius is R_1 and the external radius is R_2. The material is porous, so that flow occurs through the pore space when the external pressure P_2 is not equal to the internal pressure P_1. We showed in section 5.3.4 that for this axisymmetrical plane strain problem the equilibrium conditions reduce to

$$\sigma_{\rho\rho,\rho} + \frac{\sigma_{\rho\rho} - \sigma_{\theta\theta}}{\rho} = 0 \tag{12.68}$$

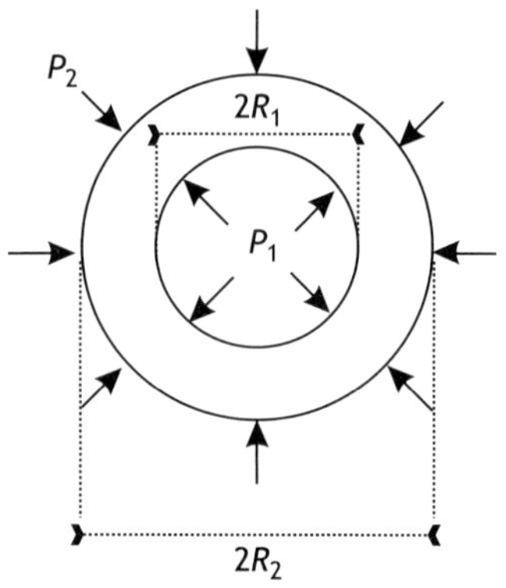

Fig. 12.3 Infinite hollow porous cylinder, loaded with an internal fluid pressure P_1 and an external fluid pressure P_2, with its axis in the z direction.

Using equation (12.66) and the definition of a small strain (see section 4.3.8), equation (12.68) becomes

$$u_{\rho,\rho\rho} + \frac{u_{\rho,\rho}}{\rho} - \frac{u_\rho}{\rho^2} + \frac{\alpha}{\lambda_B + 2G_B} P_{p,\rho} = 0 \tag{12.69}$$

which may be integrated to yield

$$\frac{1}{\rho}(\rho u_\rho)_{,\rho} + \frac{\alpha}{\lambda_B + 2G_B} P_p = 2C_1 \tag{12.70}$$

where C_1 is an integration constant. The solution of equation (12.70) is

$$u_\rho = C_1 \rho + \frac{C_2}{\rho} - \frac{\alpha}{\lambda_B + 2G_B} \frac{1}{\rho} \int_{R_1}^{\rho} \rho' P_p \, d\rho' \tag{12.71}$$

where C_2 is a second integration constant. Following the derivation given in section 5.3.4 we conclude that

$$C_1 = \frac{P_2 R_2^2 - P_1 R_1^2}{2(\lambda_B + G_B)(R_2^2 - R_1^2)}, \qquad C_2 = \frac{(P_2 - P_1) R_2^2 R_1^2}{2G_B(R_2^2 - R_1^2)} \tag{12.72}$$

The stress field associated with the displacement field (12.71) is obtained upon applying the definition of small strain in cylindrical coordinates and the mixed-stiffness poroelasticity constitutive equation (12.66). The components $\sigma_{\rho\rho}$, $\sigma_{\theta\theta}$ are principal components so we note them with only one subscript as σ_ρ, σ_θ.

If $\Delta P_p(\rho) = P_p(\rho) - P_2$, we obtain

$$\sigma_\rho = \frac{R_2^2(\rho^2 - R_1^2)P_2 + R_1^2(R_2^2 - \rho^2)P_1}{\rho^2(R_2^2 - R_1^2)} \tag{12.73}$$

$$+ \frac{2\eta_B}{\rho^2} \left(\int_{R_1}^{\rho} \rho' \Delta P_p(\rho') \, d\rho' - \frac{\rho^2 - R_1^2}{R_2^2 - R_1^2} \int_{R_1}^{R^2} \rho' \Delta P_p(\rho') \, d\rho' \right) \tag{12.74}$$

$$\sigma_\theta = \frac{R_2^2(\rho^2 + R_1^2)P_2 - R_1^2(R_2^2 + \rho^2)P_1}{\rho^2(R_2^2 - R_1^2)} \tag{12.75}$$

$$- \frac{\eta_B}{\rho^2}\left(\int_{R_1}^{\rho} \rho'\Delta P_p(\rho')d\rho' - \rho^2\Delta P_p(\rho) + \frac{\rho^2 + R_1^2}{R_2^2 - R_1^2}\int_{R_1}^{R^2} \rho'\Delta P_p(\rho')d\rho'\right) \tag{12.76}$$

where the poroelastic stress coefficient η_B is defined by

$$\eta_B = \frac{G_B}{\lambda_B + 2G_B}\alpha = \frac{1 - 2\nu_B}{2(1 - \nu_B)}\alpha \tag{12.77}$$

If the external radius R_2 is much larger than the internal radius R_1, the stress field associated with the inner pressure P_1 and the far-field pressure P_2 is given by

$$\sigma_\rho = \left(1 - \frac{R_1^2}{\rho^2}\right)P_2 + \frac{R_1^2}{\rho^2}P_1 + \frac{2\eta_B}{\rho^2}\int_{R_1}^{\rho} \rho'\Delta P_p(\rho')\,d\rho' \tag{12.78}$$

$$\sigma_\theta = \left(1 + \frac{R_1^2}{\rho^2}\right)P_2 - \frac{R_1^2}{\rho^2}P_1 - \frac{2\eta_B}{\rho^2}\int_{R_1}^{\rho} \rho'\Delta P_p(\rho')\,d\rho' + 2\eta_B\Delta P_p(\rho) \tag{12.79}$$

so that at the wall of the inner cylinder ($\rho = R_1$), the radial and tangential stress components are

$$\sigma_\rho = P_1 \tag{12.80}$$

$$\sigma_\theta = 2P_2 - P_1 - 2\eta_B(P_2 - P_1) \tag{12.81}$$

When the far-field pressure P_2 is zero, the circumferential stress component becomes $\sigma_\theta = (2\eta_B - 1)P_1$, which is always negative since $0 < \eta_B < 0.5$.

If one recalls that α represents the ratio of the volume of fluid added to the storage capacity and the change in bulk volume under the constraint that the pore pressure remains constant, we observe that α is zero for a material without any pore space. For such a material $\sigma_\theta = -P_1$, which is the solution, obtained with classical linear elasticity for purely solid materials in section 5.3.4.

If we assume now that Biot's coefficient α is equal to unity and that Poisson's ratio for the bulk material is equal to 0.25 then $\eta_B = 1/3$, so that $\sigma_\theta = -P_1/3$; this shows that fluid percolation in the rock may decrease by a factor 2/3 the *hoop stress* (i.e. the tangential stress component $\sigma_{\theta\theta}$) at the borehole wall.

12.2 Linear thermomechanical coupling

Most substances when heated expand (however, water dilates when it cools from 4 °C to 0 °C). If this thermal expansion is prevented from taking place, for some reason, then *thermal stresses* appear in the material if its temperature varies. Hence thermomechanics issues appear when the temperature in a given such body, or set of bodies, is not uniform.

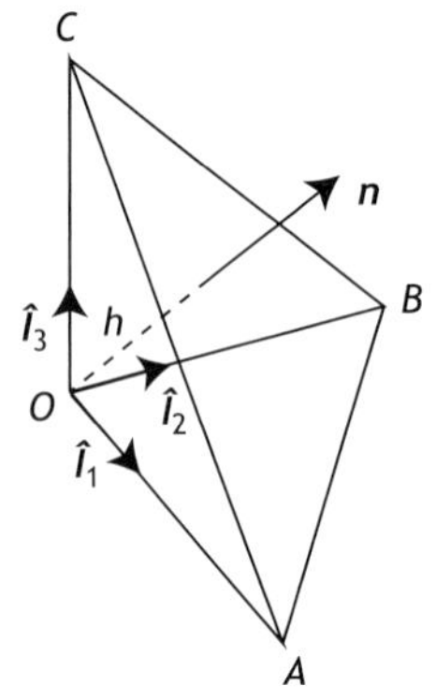

Fig. 12.4 Heat flux through a surface of random orientation.

Two problems must be addressed when the temperature is not uniform in a body: that of heat transfer between domains where the temperature differs and that of the thermal stresses that appear (Berest and Weber, 1988). We first address the issue of heat transfer schematically and then discuss the effects associated with thermal stresses in the context of linear thermoelasticity.

12.2.1 Heat transfer

When different parts of a body are at different temperatures, heat will flow from the hotter parts to the cooler parts. This transfer of heat may occur by three different processes: conduction, convection and radiation.

For heat transfers by conduction, the heat passes directly through the matter of the body. For heat transfer by convection, heat is transferred through the motion of fluid matter. For heat transfer by radiation, heat is transferred directly by electromagnetic waves.

Because geomaterials generally involve solid and fluid phases, both conduction and convection are important processes for heat transfer in crustal geomechanics, but the role of radiation is usually negligible. Hence heat transfer occurs by conduction in the solid part and by convection through fluid circulation. In addition, in nonsaturated systems (with coexistence of solid, liquid and vapor phases) attention must be given to the transfer of heat between the liquid and the vapor phases when both the pressure and temperature of the fluid vary.

Heat conduction in an isotropic solid

Let us consider a solid in which heat is flowing but in which no heat is generated. The rate at which heat is transferred across any surface S, at a point O, per unit perpendicular area and per unit time is called the flux of heat at that point across the surface. It is often denoted $\boldsymbol{q}$. It may be shown (Carslaw and Jaeger, 1986) that the flux of heat is continuous at the surface separating two media. Let us show that if the heat flux is given for each of three mutually perpendicular planes meeting at point O, namely q_1, q_2 and q_3, these values may be used to determine the heat flux through a surface of any orientation passing through O.

We consider the tetrahedron $OABC$ shown in fig. 12.4. The direction cosines of the normal to face ABC with respect to axes $\hat{\boldsymbol{I}}_1$, $\hat{\boldsymbol{I}}_2$ and $\hat{\boldsymbol{I}}_3$, are respectively n_1, n_2 and n_3. If ω is the area of triangle ABC, those of faces OAC, OBC and OAB are respectively $n_1\omega$, $n_2\omega$

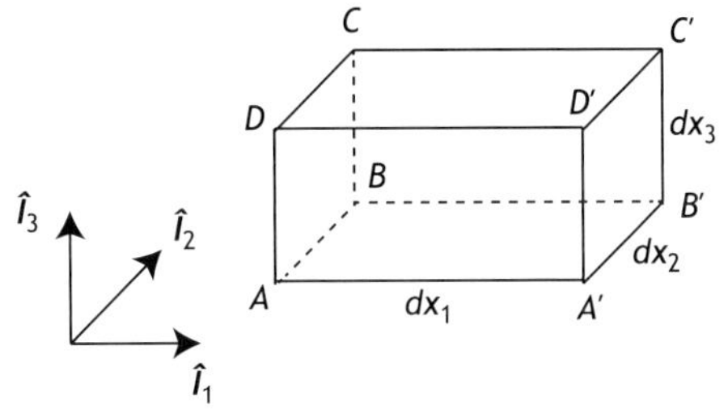

Fig. 12.5 Heat flux through a small parallelepipedic volume.

and $n_3\omega$. If $\boldsymbol{q}$ is the heat flux through area ω, the heat gained by the tetrahedron because of the heat flux through the various faces is

$$(n_1q_1 + n_2q_2 + n_3q_3 - |\boldsymbol{q}|)\omega = \tfrac{1}{3}h\omega\varrho\kappa_c\frac{\partial\theta}{\partial t} \tag{12.82}$$

where h is the distance between point O and the surface ABC, ϱ is the density of the material, κ_c is its specific heat and $\partial\theta/\partial t$ is the change in temperature per unit time.

When the tetrahedron volume shrinks to zero, the right-hand side of equation (12.82) becomes zero and equation (12.82) yields

$$|\boldsymbol{q}| = n_1q_1 + n_2q_2 + n_3q_3 \tag{12.83}$$

The values q_1, q_2 and q_3 are the components of the heat flux vector $\boldsymbol{q}$ at O. The heat flux magnitude at point O is simply $|\boldsymbol{q}| = \sqrt{q_1^2 + q_2^2 + q_3^2}$.

The temperature is characterized at any point $\boldsymbol{x}$ by the function $\theta(\boldsymbol{x}, t)$ and isothermals correspond to surfaces for which, at a given time t, the temperature is the same at all points of the surface. No two isothermals may intersect each other since the temperature is unique at any given point and any given time.

Experimental results show that

$$q_1 = -K_h\theta_{,x_1} \tag{12.84}$$

or more generally

$$\boldsymbol{q} = -K_h \operatorname{grad} \theta \tag{12.85}$$

where K_h is known as the thermal conductivity of the material. Equation (12.85) is known as *Fourier's law*.

Let us consider the case of a small solid parallelepiped $ABCDA'B'C'D'$ in which the temperature is a continuous function of spatial coordinates and time (fig. 12.5). Faces $ABCD$ and $A'B'C'D'$ are perpendicular to the x_1 direction and such that $AA' = dx_1$, $AB = dx_2$, $AD = dx_3$, so that the parallelepiped volume dv is equal to $dx_1\,dx_2\,dx_3$. We assume first that no heat is generated within this small volume.

The total rate of gain of heat in the small volume dv because of the heat flux into dv generates a change in temperature such that

$$- \operatorname{div} \boldsymbol{q}\, dv + \varrho C_h \frac{\partial \theta}{\partial t} dv = 0 \tag{12.86}$$

where C_h is the heat capacity of the material at temperature θ. Inserting equation (12.85) into equation (12.86), we obtain

$$\theta_{,11} + \theta_{,22} + \theta_{,33} - \frac{1}{\kappa_h} \frac{\partial \theta}{\partial t} = 0$$

or

$$\kappa_h \nabla^2 \theta = \frac{\partial \theta}{\partial t} \tag{12.87}$$

where $\kappa_h = K_h/(\varrho C_h)$ is known as the *heat diffusivity* of the material. Equation (12.87) is known as the *heat conduction equation*. Carslaw and Jaeger (1986) published solutions to this equation for a large variety of problems.

For steady conditions (no temperature variation with time, i.e. $\partial\theta/\partial t = 0$) equation (12.86) becomes

$$\nabla^2 \theta = 0 \tag{12.88}$$

If a heat source produces heat at a rate $A(\boldsymbol{x}, t)$ within the material, the heat conduction equation becomes

$$\nabla^2 \theta - \frac{1}{\kappa_h} \frac{\partial \theta}{\partial t} = -A(\boldsymbol{x}, t) \tag{12.89}$$

Steady conditions are generally assumed when one is considering the temperature in the crust. The natural heat flux that is observed at most places on earth varies between 20 $\mathrm{mW/m^2}$ and 40 $\mathrm{mW/m^2}$, with temperature gradients varying from 20 °C/km to 40 °C/km, depending on the thermal conductivity. The source of this heat flux is twofold. Some comes from the earth core through convection in the mantle, but most is associated with the radiogenic decay that occurs in the crust.

Solutions to heat conduction problems require both that the initial temperature distribution be known and that the boundary conditions be prescribed, whether by the known temperature variations (or absence of variation) at the boundary of the volume under consideration or by an imposed heat flux.

For a heat flux imposed through a surface of normal $\boldsymbol{n}$ and unit area, we may write

$$q_n = -K_h \operatorname{grad} \theta \cdot \boldsymbol{n} \tag{12.90}$$

For the contact between two different materials, we have continuity of temperature at the contact and equality between normal heat fluxes through the surface:

$$K_{h1} \operatorname{grad} \theta_1 \cdot \boldsymbol{n} = K_{h2} \operatorname{grad} \theta_2 \cdot \boldsymbol{n} \tag{12.91}$$

where grad θ_1 is the temperature gradient in medium 1 and grad θ_2 that in medium 2. However, often the contact between mediums involves a complex interface, and so the transfer of heat is characterized by a specific function. This is of particular relevance when fluid circulates along the interface.

Heat transfers through fluid transport

The transfer of heat associated with the motion of fluids is referred to as heat transfer by convection. The frame of reference used to investigate convection is Eulerian (see section 4.2) and the fluid velocity is $\boldsymbol{v}^f$. The change in the quantity of heat contained within a small fixed volume dv includes both the heat relating to the intrinsic temperature variation within dv and the heat associated with the flux of matter through dv. Hence, the temperature variation per unit fluid volume is given by

$$\kappa_h \nabla^2 \theta = \frac{\partial \theta}{\partial t} + \boldsymbol{v}^f \operatorname{grad} \theta \tag{12.92}$$

For rock masses comprising both solids and fluids and in which the temperature is heterogeneous, the evaluation of temperature variations at any given point requires a description of the heat transfer associated with conduction, $\boldsymbol{q}$, for the solid part (described by equation (12.85)) and the heat transfer associated with convection of the fluid part, $\boldsymbol{q}'$, where

$$\boldsymbol{q}' = \varrho C_h \boldsymbol{v}^f \theta \tag{12.93}$$

An example of a vertical temperature profile is shown in fig. 12.6. It represents the temperature variation between ground surface and a depth of 5 km at Soultz-sous-forêts, in eastern France. The upper linear part of the profile corresponds to heat conduction, whilst the quasi-absence of temperature variation observed between 1500 m and 3500 m corresponds to a domain where convection occurs. Such a thermal profile implies high temperatures at relatively shallow depths and has led to local geothermal energy exploitation.

It must be kept in mind that the temperature influences strongly both the density and the viscosity of fluids. Further, fluids are present in rocks both in the pore space and in the fracture network. The heat transfer associated with fluid motion in the pore space is taken into account through the fluid filtration velocity (see section 9.3). For the heat transfer associated with the flow through the fracture system, first the fracture pattern must be described and then the distribution of the flow velocity through the fracture system must be evaluated.

Modeling heat transfer in rock masses requires appropriate numerical methods. The interested reader is referred to the book *Coupled Thermo–Hydro Mechanical Processes of Fractured Media*, edited by Stephansson, Jing and Tsang (Stephansson *et al.*, 1996), which presents various applications of this aspect of rock mechanics, with special attention to the long-term disposal of nuclear waste.

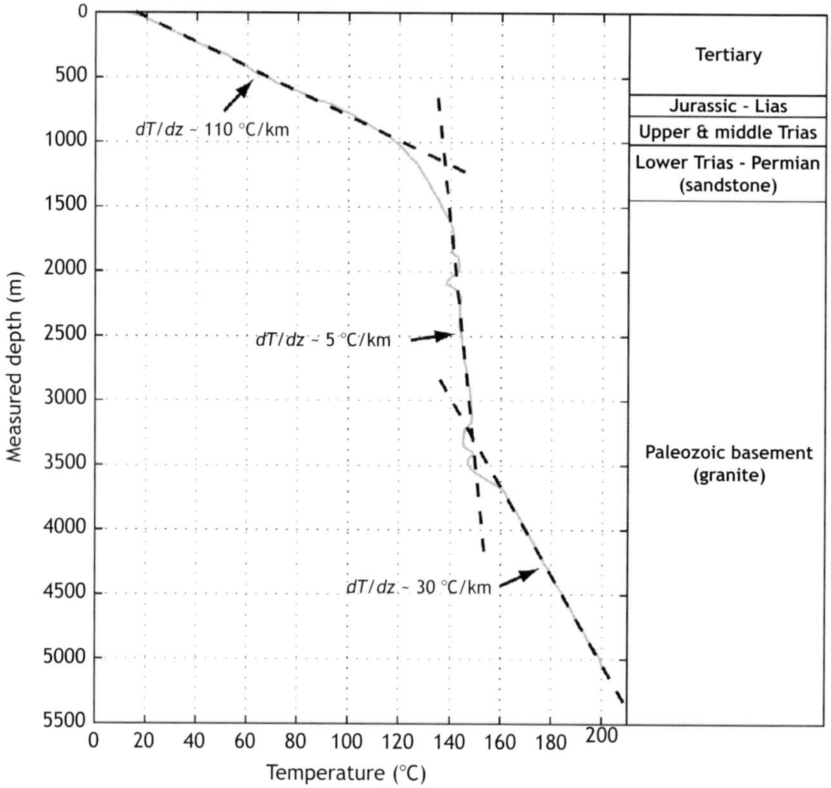

Fig. 12.6 Vertical thermal profile at Soultz-sous-forêts, eastern France. The high-temperature gradient observed in the sedimentary cover above 1000 m is caused by a deep convective system in the fractured granite down to 3500 m. Below 3500 m the vertical temperature profile is controlled by heat conduction in the granite (reproduced from Genter *et al.*, 2010, with permission from Elsevier Masson).

12.2.2 Linear thermoelasticity

Thermal stresses

As already pointed out, for all materials temperature variations are associated with expansion or contraction. Let us consider a parallelepipedic volume of homogeneous isotropic rock with length l_1, width l_2 and thickness l_3. A uniform temperature variation ΔT throughout the parallelepiped generates thermal expansion in all directions. The three changes in length along the axes are

$$\delta l_i = \alpha_T \Delta T l_i, \qquad i = 1, 2, 3$$

where α_T is the *linear thermal expansion coefficient* of the rock. When compressions are reckoned negative, increases in length are positive for positive thermal expansion coefficients. The three principal strain components associated with temperature variations are

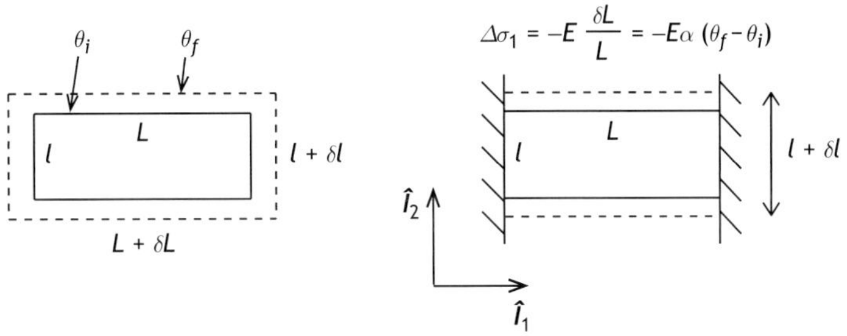

Fig. 12.7 Left: volumetric expansion of a material occurs when its temperature increases from θ_i to θ_f. Right: thermal stresses are generated when the material's thermal expansion is prevented in some direction.

$$\varepsilon_i = \frac{\delta l_i}{l_i} = \alpha_T \delta T, \qquad i = 1, 2, 3 \tag{12.94}$$

and the volumetric strain is $\varepsilon_v = 3\alpha_T \Delta T = \alpha_v \Delta T$, where α_v is the volumetric thermal expansion coefficient.

Now let us assume that in direction 1 expansion is completely prevented, whilst the rock may dilate in the other directions (fig. 12.7). This requires that an axial stress component $\Delta\sigma_1 = -E\varepsilon_1$ be applied in the $\hat{\boldsymbol{I}}_1$ direction (E is Young's modulus for the rock).

If, in addition, expansion in the $\hat{I}_2$ and $\hat{I}_3$ directions is also prevented then the stress in the $\hat{I}_1$ direction must be

$$\Delta\sigma_1 = \frac{E}{1 - 2\nu} \alpha_T \, \Delta T = 3K\alpha_T \, \Delta T \tag{12.95}$$

where ν is Poisson's ratio for the rock and K is its bulk modulus. The same holds for directions $\hat{I}_2$ and $\hat{I}_3$ as well.

Hence we observe that for isotropic materials thermal stresses are also isotropic, i.e. they do not exhibit any shear component whatever the frame of reference.

Stress–strain relations and Navier's equation for linear thermoelasticity

For isotropic linearly elastic rocks, the strain and stress components resulting from different loading processes are simply summed when all the loading processes are applied simultaneously. Hence the constitutive equations for linear thermoelasticity are simply

$$\sigma_{ij} = \lambda \varepsilon_v \delta_{ij} + 2G\varepsilon_{ij} - (3\lambda + 2G)\alpha_T \Delta T \delta_{ij}, \qquad i,j = 1, 2, 3 \tag{12.96}$$

where λ and G are the Lamé coefficients and ε_v is the volumetric strain.

Following the same derivation as in section 5.2, we observe that Navier's equation, which combines the conditions of equilibrium with the stress–strain equation and the small-strain definition, becomes

$$(\lambda + G)u_{i,ij} + Gu_{j,ii} - (3\lambda + 2G)\alpha_T T_{,j} = 0, \qquad i,j = 1, 2, 3 \tag{12.97}$$

Let us note that, when the three stress–strain equations described by equation (12.96) are summed we obtain the volumetric strain,

$$\varepsilon_v = \frac{\sigma_{kk}}{K_t} + \alpha_T \Delta T \tag{12.98}$$

where K_T is the isothermal bulk modulus of the rock. It is similar to the poroelasticity equation (12.17), but the constant α_T replaces the constant $1/H$ and the temperature replacing the pore pressure. Further, we note that Fourier's equation describing the heat flux is similar to Darcy's equation describing the fluid flow.

In fact it has been shown (Biot, 1941; Rice and Cleary, 1976) that there is a complete mathematical analogy between thermoelasticity and poroelasticity. Hence, the solutions obtained in poroelasticity are readily applied to problems in thermoelasticity with analogous initial and boundary conditions.

Stresses induced by a radial heat flux in a hollow cylinder

By analogy with the problem of the internally and externally pressurized infinite hollow cylinder (fig. 12.3 and section 12.1.2), we obtain the circumferential (also called hoop) stress component for an axially infinite hollow cylinder with infinite external radius and a far-field temperature T_2 different from the internal temperature, T_1:

$$\sigma_\theta = -\frac{E_B}{1-\nu_B}\alpha_T(T_2 - T_1) \tag{12.99}$$

When the inner temperature is smaller than the external temperature, a positive (tensile) hoop stress is generated at the wall of the inner cylinder.

Let us consider now an infinite cylindrical borehole in a porous rock and assume that some fluid is injected at pressure P_1, and a temperature T_1 smaller than that of the ambient rock mass, T_2, away from the borehole. The hoop stress perturbation at the borehole wall associated with this radial injection process may be determined by summing equations (12.99) and (12.81):

$$\sigma_\theta = 2P_2 - P_1 - 2\eta_B(P_2 - P_1) - \frac{E_B}{1-\nu_B}\alpha_T(T_2 - T_1) \tag{12.100}$$

where P_2 is the far-field pressure.

12.3 Mechanical consequences of water–rock physicochemical interactions

We introduced in section 7.2.3 a mechanical consequence of water–rock chemical interaction when we were discussing stress corrosion mechanisms and we outlined how a very slow corrosion process may lead progressively to unstable crack growth.

Fig. 12.8 Example of stylolites, which result from dissolution (i.e. pressure solution) effects under high normal stress (reproduced from Schmittbuhl *et al.*, 2004, with permission from the American Physical Society).

Stress corrosion is just one of the various mechanical consequences of water–rock physicochemical interaction. *Pressure solution* is another significant phenomenon, which has particularly strong consequences for the deformation of sedimentary formations and is also important within fault zones (Gratier *et al.*, 1999).

Pressure solution was mentioned already in the nineteenth century, for example by Sorby when he observed impressed pebbles in limestone rocks in the molasse conglomerates of the Alpine foreland. Pressure solution markers are identified as pits, of various sizes from millimeters to centimeters, due to stress-enhanced dissolution at the contacts between the pebbles.

Stylolites constitute another feature indicating the presence of pressure solution, particularly in limestone (fig. 12.8). The formation and growth of stylolites has been fairly extensively discussed (e.g. Lehner, 1995; Fowler and Yang, 1999), with particular attention to its consequences for the creeping behavior of sedimentary rocks. Indeed, for such rocks, pressure solution results in deformations that may reach more than 20% of the original dimension in the direction parallel to the applied stress component. This mechanism was considered by Cornet and Röckel (2012) to be the source of the observed deviatoric stress in some limestone layers of the Paris Basin.

Also associated with pressure solution are some of the striations observed on fault slip planes. Indeed, whilst some of these striations result from the wearing of the surface during unstable slip (leading to an earthquake), others result from a very slow pressure solution process that leads to aseismic slip motion (e.g. Means, 1987).

Another significant mechanical consequence of physicochemical fluid–rock interactions is the swelling of clay upon its contact with water, as well as its reduction in strength. These issues are particularly important during drilling operations but also, for example, for the long-term isolation of nuclear waste.

These various physicochemical interactions involve processes at the nanometer scale, which are beyond the scope of this text book. Recently, Gratier *et al.* (2013) proposed a review of the role of pressure-solution creep in the ductility of the upper crust. They discussed models that take into account the various mechanisms involved in this deformation process. We also mention the paper by Chen *et al.* (2003) and the book by Fjaer *et al.* (2008) for discussions on the significance of the interaction between water and clay or shale during drilling operations.

12.4 Hydraulically induced fracturing processes

We explore now the various fracturing processes that may occur when the loading conditions associated with a fluid-pressure variation reach beyond the elastic limit of the rock mass.

Let us consider a rock mass under static equilibrium consisting of a poroelastic geomaterial in which a fracture set S has been identified. We assume here that the triaxial compressive strength of the material is characterized by a Mohr–Coulomb envelope and that a tensile strength σ^T may be defined for this porous rock. Further, the slip conditions for the fracture set are presumed to be characterized by some relationship between the resolved shear stress and the effective normal stress components supported by the fracture planes of the fracture set (e.g. a linear relationship for Byerlee's friction law but a nonlinear relationship, such as Barton's stability criterion, for low effective-normal-stress conditions; see section 10.1.2). Terzaghi's effective stress tensor acting within any small volume of the rock mass may be represented on a Mohr diagram, as well as the effective stress vector supported by any of the fracture planes that make up the fracture set S that exists within the same volume (we assume here that the stress field is uniform throughout the volume under consideration).

If the pore pressure is progressively increased, the three effective principal stress components are progressively equally decreased, according to Terzghi's effective stress concept, until one of the following three possibilities occurs (fig. 12.9):

1. the effective minimum principal stress, $\sigma_3 - P_p$, becomes equal to the tensile strength of the material, σ^T, and a mode I fracture develops;
2. the Mohr–Coulomb failure criterion is reached and failure occurs through the development of a fresh shear surface in the poroelastic material;
3. the slip condition for some fractures in the fracture set S is met, and slip occurs on these planes.

The rupture process observed when a pure mode I fracture develops because of a high fluid pressure is called hydraulic fracturing. We introduce first some factors that control the development of hydraulic fractures. Then, hydraulically induced nonreversible shear motions are discussed, with particular reference to correlative induced seismicity as observed in various engineering and natural contexts. The section ends with a discussion of the relationships between large-scale failure processes and fluid migration.

12.4.1 Hydraulic fracturing

Hydraulic fracturing is commonly used in the oil industry for improving hydrocarbon production. After a fracture has been initiated by increasing the pressure in a borehole, it propagates some distance while some so-called proppant (well-calibrated granular material) is added to the fracturing fluid. When injection stops, the fracture closes back on the proppant that has been injected, leaving in place an efficient hydraulically conductive

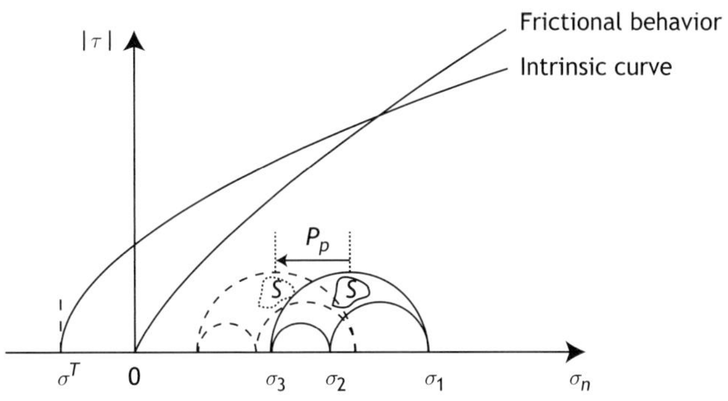

Fig. 12.9 The stability of a small volume in a saturated rock mass in which the pore pressure increases.

drain. The design of such operations requires an appraisal of the variation of fracture opening with the distance from the injection well. Indeed, fracture opening depends on many factors, such as the fluid viscosity, the fluid loss through the fracture wall and also, among other factors, the far-field stress state and its spatial variations.

Hydraulic fracturing also occurs spontaneously in nature, as for example with the volcanic dykes generated by high magmatic pressures. For such fracturing processes, dyke development depends on the viscosity and the gas content of the magma, both of which depend in turn on the pressure and temperature and also on the regional stress field (e.g. Cayol and Cornet, 1998; Dahm, 2000; Gudmundsson, 2002), as do industrial hydraulic fractures.

We introduce below the basic mechanisms that control the initiation and propagation of hydraulic fractures, with some attention to the fracture propagation velocity because of its consequences for microseismic activity.

Hydraulic fracture initiation

Consider first the problem of initiating a hydraulic fracture in a borehole of radius R located in an infinite homogeneous isotropic solid in which the far-field stress (i.e. the stress field more than five radii away from the borehole) is $\tilde{\sigma}^\infty$ (fig. 12.10).

The problem of hydraulic fracture development in a porous material, and more particularly the effect of pore pressure on the initiation of fracture, will be discussed in section 13.2 in the context of a description of stress determination by hydraulic fracturing. Here we consider an impervious formation for which pore pressure effects may be neglected.

The borehole axis is taken as parallel to one of the principal stress directions at infinity. The two other far-field principal stress components are normal to the borehole axis. We call σ_M^∞ the larger of them and σ_m^∞ the smaller.

Because the cylinder is infinite, we assume plane strain conditions and evaluate the hoop stress $\sigma_{\theta\theta}$ at the borehole wall. Assuming linear elasticity, we apply the superposition principle and add equations (5.94) and (5.304). If θ is the azimuthal coordinate defined with respect to the σ_M^∞ direction, the hoop stress is (here compressions are reckoned positive)

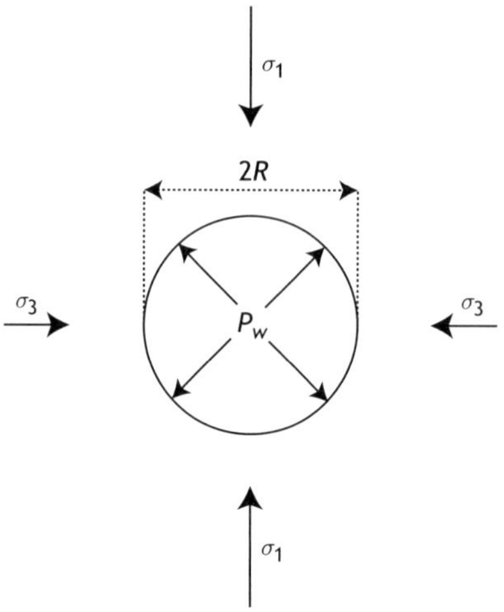

Fig. 12.10 Hydraulic fracturing with fluid pressure P_w applied in a cylindrical borehole parallel to one of the far-field principal stress directions. In this figure, we assume that $\sigma_M^\infty = \sigma_1$ and $\sigma_m^\infty = \sigma_3$.

$$\sigma_{\theta\theta} = \sigma_M^\infty + \sigma_m^\infty) - 2(\sigma_M^\infty - \sigma_m^\infty)\cos 2\theta - P_w = \sigma_\theta \tag{12.101}$$

For these loading conditions (where the borehole axis is parallel to one of the far-field principal stress directions), the hoop stress is a principal component and its magnitude varies with the azimuthal coordinate θ. Its extremum values occur for

$$\frac{\partial \sigma_\theta}{\partial \theta} = 4(\sigma_M^\infty - \sigma_m^\infty)\sin 2\theta = 0 \tag{12.102}$$

Minimum values are observed for $\theta = 0 + k\pi$, $k = 0, 1$, which correspond to the σ_M^∞ direction, whilst maximum values are reached for $\theta = \pi/2 + k\pi$, $k = 0, 1$. Hence the minimum hoop stress magnitude is

$$\sigma_{\theta min} = -\sigma_M^\infty + 3\sigma_m^\infty - P_w \tag{12.103}$$

As P_w increases, $\sigma_{\theta min}$ becomes negative and when it reaches the *tensile strength* of the rock a hydraulic fracture is generated at the borehole wall in the σ_M^∞ direction. The borehole pressure at the inception of fracture, P_{if}, is called the fracture initiation pressure and is given by

$$P_{if} = -\sigma_M^\infty + 3\sigma_m^\infty - P_w - \sigma^T \tag{12.104}$$

However, as we saw in chapter 7, using the concept of maximum stress for characterizing the *tensile strength* of a rock is not valid because of the existence of microcracks, which generate local stress concentrations. This is discussed further in section 13.2 when we are discussing the application of hydraulic fracturing to the measurement of *in situ* stresses.

It should be mentioned here that, in the early work on hydraulic fracturing (Hubbert and Willis, 1957; Haimson and Fairhust, 1967), the borehole pressure for which a hydraulic fracture initiates was called the *breakdown pressure* and was considered to correspond to

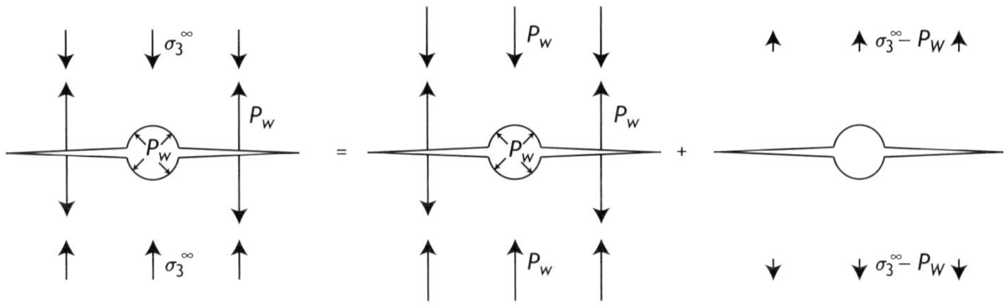

Fig. 12.11 Decomposition of the loading conditions for a hydraulic fracture propagating under pressure P_w in a rock under the uniaxial far-field stress component σ_3^∞.

a peak in the injection pressure (see fig. 13.3(b)). More recent work, however, has shown the necessity to differentiate between the initation pressure and the breakdown pressure, as explained after we discussed the parameters that control the stability of hydraulic fracture propagation in section 13.2.

Note that the difficulty with the tensile strength definition for rock is of importance when one is determining the borehole pressure for which initiation occurs, but it does not affect the azimuthal coordinate of fracture inception at the borehole wall. We conclude for now that the orientation of hydraulic fractures at the borehole wall is parallel to the direction of the far-field maximum principal stress component in the plane normal to the borehole axis, when this axis is parallel to one of the far-field principal stress directions.

Stability of the hydraulic fracturing process

As a hydraulic fracture propagates away from the borehole, the stress perturbation associated with the borehole becomes less and less significant: once the fracture has extended from the borehole for a distance larger than five times the borehole radius, the stress field surrounding the fracture tip depends only on the far-field stress $\tilde{\sigma}^\infty$. We may analyze this fracture propagation process in a manner similar to that of the mode I fracture introduced in section 7.1.2. The loading conditions are decomposed into a purely hydrostatic pressure P_w and a far-field stress state equal to $\tilde{\sigma}^\infty - P_w\tilde{I}$. The stress concentration near the hydraulic fracture tip is found to be identical to that generated by a uniaxial stress field with component $\sigma_3^\infty - P_w$ (fig. 12.11), where σ_3^∞ is the minimum principal stress away from the borehole.

Hence, using equation (7.13), the stress intensity factor for a hydraulic fracture propagating due to the application of a uniform fluid pressure is

$$K_I = (\sigma_3^\infty - P_w)\sqrt{\pi a} \quad (12.105)$$

where a is the distance between the fracture tip and the center of the borehole.

We note that the stress intensity factor increases with the crack length a, so that the fracturing process accelerates as the fracture propagates. Some kinetic energy is generated, giving rise to a seismic signal.

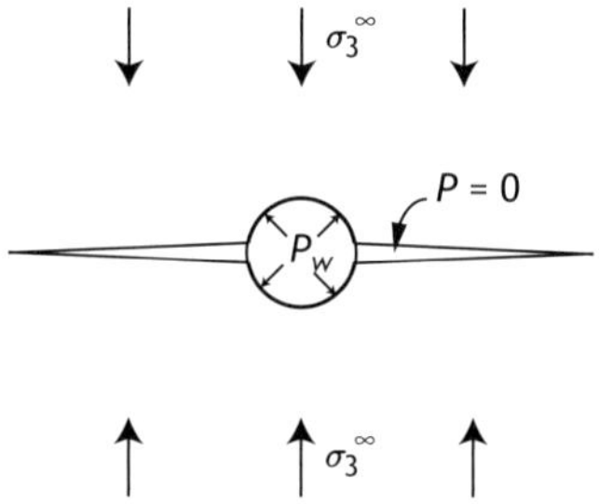

Fig. 12.12 Decomposition of the loading conditions for a hydraulic fracture propagating under pressure P_{pr} in a rock under a far-field stress component σ_3^∞.

As the fracture extends, the flow rate required to keep the pressure constant up to the fracture tip increases. However, in practice the injection flow rate is constant and controlled by the surface pumping system. Furthermore, as the fracture propagates away from the borehole, more and more fluid is lost by percolation through the pre-existing fracture network that intersects the walls of the fracture or through the pore space when this is significant. Hence, for hydraulic fracturing with incompressible fluids the pressure drops continuously along the fracture length.

In order to investigate how the stress intensity factor varies when the pressure distribution decreases along the fracture length, we consider the extreme case in which no fluid at all penetrates into the fracture (fig. 12.12).

When the pressure is applied only at the borehole wall, it may be shown that the stress intensity factor is inversely proportional to the square root of the distance a between the fracture tip and the borehole wall:

$$K_I \propto (\sigma_3^\infty - P_w)/\sqrt{\pi a} \tag{12.106}$$

Thus, K_T decreases as the fracture propagates (Zoback *et al.*, 1977). This implies that the fracturing process is stable and that the borehole pressure must be continuously increased to keep the fracture propagating. No seismic signal is generated by this quasistatic fracturing process.

In practice, a hydraulic fracture does propagate in a quasistatic manner. Indeed, the increase in fracture extension is faster than the fluid flow in the fracture, so that when fracture increase starts the pressure drops and the fracture stops. Then, however, because of the constant injection flow rate, the fluid flows into the fracture, progressively increasing the pressure and therefore the stress intensity factor at the fracture tip, until the fracture propagates again. This new fracture extension creates a new volume for the fluid to penetrate and therefore induces a pressure drop leading to a fracture stop, etc.

Clearly the dynamics of fracture propagation depends on the pressure distribution along the fracture and therefore on the flow conditions. In addition, because the surface energy of rocks is not uniform in large rock volumes, the fracture propagation becomes locally unstable along the short distances over which the surface energy drops. The fracture lengths for which the fracturing process is unstable control the wavelength of the seismic signals generated by these instabilities (see section 11.3). Experience has shown that the microseismic signals associated with hydraulic fracturing are in the kHz range and are associated with fracture jumps smaller than 10 meters (e.g. Sileny *et al.*, 2009).

Initiation pressure, breakdown pressure, propagation pressure

In the previous discussion we introduced the idea that the stability of a hydraulic fracturing process depends in particular on the flow conditions into the fracture and also on the surface energy of the rock.

The flow conditions depend on the fluid's intrinsic properties, i.e. its viscosity and compressibility as well as its density, and also on the conditions that describe the flux of fluid through the boundaries of the fracture, i.e. the flow conditions through the injection borehole and also the flow conditions through the walls of the fracture. When the inception of fracture is unstable, a sudden drop in pressure is observed in the borehole and this produces a clear breakdown pressure in the pressure–time record. When the inception of fracture is stable, the pressure peak observed in the pressure–time record reflects only the balance between the injection of fluid into the borehole and its flux through the hydraulic fracture.

For viscous fluids that do not penetrate up to the crack tip, the viscous drag gets larger and larger as the fracture propagates and the injection pressure must therefore be continuously increased to insure fracture propagation. Hence, after a certain breakdown pressure an increase in the propagation pressure may be observed when the injection flow rate is maintained constant.

If the material is completely impervious, all the injected fluid volume is used to increase the hydraulic fracture volume. Generally, however, some fluid is lost through the walls of the fracture, the more so the longer the fracture. Ultimately, in permeable rocks, fracture propagation stops when all the injected fluid is lost through the walls of the fracture.

An abundant literature has been devoted to the various issues raised by hydraulic fracture inception and propagation, particularly in view of the many applications of this process to fluid extraction from deep reservoirs (e.g. Adachi *et al.*, 2007; Detournay and Carbonell, 1997; Lecampion and Detournay, 2007; Prioul and Karpfinger, 2011).

Regional stress field and hydraulic fracture geometry

As a hydraulic fracture becomes larger and larger, the far-field stress tensor in the vicinity of the fracture tip can no longer be considered uniform, and these spatial far-field stress variations influence the hydraulic fracture geometry.

Let us first consider the growth of a vertical fracture in a homogeneous rock mass for which we assume that the minimum principal stress remains horizontal throughout the volume in which the fracture propagates (fig. 12.13). We consider a fracture that has extended by a height Δh symmetrically with respect to its initiation point at the borehole.

The stress intensity factor at the upper fracture tip is

$$K_{I_{up}} = \left[(\sigma_3^{\infty} - \alpha_{33}\Delta h) - (P_w - \varrho_F g \Delta h)\right]\sqrt{\pi a} \qquad (12.107)$$

where α_{33} is the vertical gradient of the minimum horizontal principal stress component (see section 13.2.2) whilst that at the lower fracture tip is

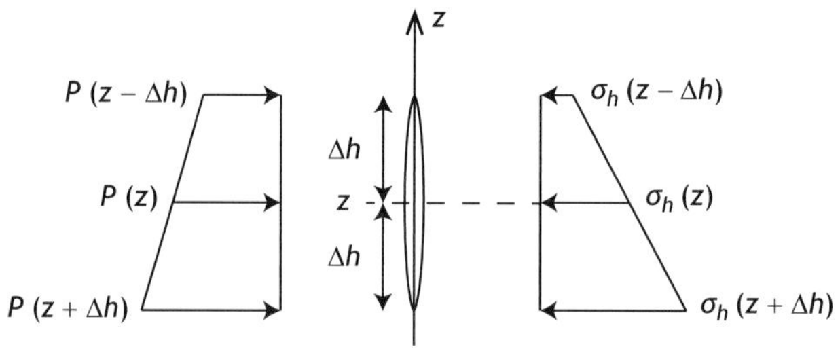

Fig. 12.13 Growth of a vertical hydraulic fracture when the vertical component of the far-field minimum principal stress component is α_{33} and the fracking fluid density is ϱ_F.

$$K_{I_{down}} = \left[(\sigma_3^{\infty} + \alpha_{33}\Delta h) - (P_w + \varrho_F g \Delta h)\right]\sqrt{\pi a} \tag{12.108}$$

If the rock mass density is ϱ_B, it will be argued in chapter 14 that the vertical component of the minimum horizontal principal stress vertical gradient is generally larger than $[\nu_B/(1-\nu_B)]\,\varrho_B g$, where ν_B is the drained Poisson's ratio for the rock mass. Hence, at the top of the fracture the fracturing-fluid pressure is larger than the minimum horizontal principal stress but the reverse occurs at the bottom of the fracture, so that the hydraulic fracture propagates upward rather than downward.

This applies for example to volcanic dyke propagation. Given that melted rocks are lighter than solid rocks, dykes have a tendency to propagate upward initially. However, as they get closer to the ground surface they remain oriented perpendicularly to the local least minimum principal stress, and this often results in a change of orientation (e.g. Cayol and Cornet, 1998; White *et al.*, 2011). In particular, if the minimum principal stress becomes the vertical component then the hydraulic fracture (or the magmatic intrusion) extends perpendicularly to the vertical direction, which leads to the formation of horizontal fractures, called *sills* when one is referring to magmatic intrusions.

Let us consider now the propagation of a vertical hydraulic fracture in a horizontally layered sedimentary rock formation. It will be shown in chapter 14 that the minimum horizontal principal stress is larger in soft sediments such as clay or shale and smaller in stiff formations such as sandstone or limestone (as already mentioned in chapter 2). Hence, in layered formations a hydraulic fracture grows mostly horizontally within the stiffest layer (i.e. that which sustains the largest differential stress $\sigma_1 - \sigma_3$) because its vertical extension is prevented by the softer layers (fig. 12.14).

These observations guided the early models proposed for evaluating the geometry of hydraulic fracture (Perkins and de Kern, 1961; Geertsma and de Klerk, 1969). These models, known as PKN or KGD respectively, based on closed-form solutions were intended to provide in real time an estimate of a hydraulic fracture aperture at various stages of the fracture extension.

Hydraulic fracture development from inclined boreholes

Now let us consider the case in which the borehole axis is not parallel to any of the far-field principal stress directions. For such loading conditions, the stress tensor at any point of the borehole wall was derived in section 5.6.4.

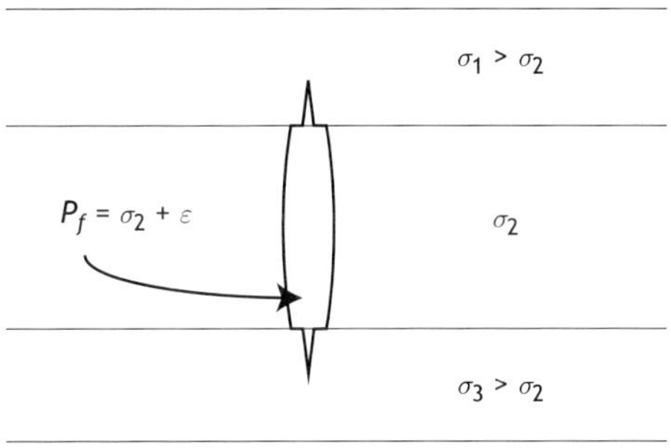

Fig. 12.14 Growth of a vertical hydraulic fracture in a horizontally layered formation. The vertical fracture extension is limited by the minimum horizontal principal stress component supported by the softest layer, so that the fracture extends mostly horizontally.

We will superpose this solution with that resulting from a uniform pressure P_{br} in the borehole in order to obtain the stress field at the borehole wall when fracture initiates:

$$\sigma_{\rho\rho} = P_{br} = \sigma_\rho \tag{12.109}$$

$$\sigma_{\theta\theta} = \sigma_{11} + \sigma_{22} - 4\left(\frac{\sigma_{11} - \sigma_{22}}{2} \cos 2\theta + \sigma_{12} \sin 2\theta\right) - P_{br} \tag{12.110}$$

$$\sigma_{zz} = \sigma_{33} - 4\nu\left(\frac{\sigma_{11} - \sigma_{22}}{2} \cos 2\theta + \sigma_{12} \sin 2\theta\right) \tag{12.111}$$

$$\sigma_{\theta z} = 2(\sigma_{23} \cos\theta - \sigma_{31} \sin\theta), \quad \sigma_{\rho\theta} = \sigma_{z\rho} = 0 \tag{12.112}$$

It may be noted that if the component $\sigma_{\theta z} = 2(\sigma_{23} \cos\theta - \sigma_{31} \sin\theta)$ is nonzero, the two components $\sigma_{\theta\theta}$ and σ_{zz} are not principal components; but hydraulic fractures are always normal to the minimum principal stress so that they must be inclined, at the borehole wall, to the borehole axis. While σ_ρ is one of the principal stress components, the two other principal components, σ_M and σ_m ($\sigma_M > \sigma_m$), at the borehole wall are

$$\left.\begin{matrix}\sigma_M \\ \sigma_m\end{matrix}\right\} = \tfrac{1}{2}(\sigma_{\theta\theta} + \sigma_{zz}) \pm \sqrt{(\sigma_{\theta\theta} - \sigma_{zz})^2 + 4\sigma_{z\theta}^2}$$

The azimuthal coordinate for which the smallest principal stress component σ_m reaches its lowest value at the borehole wall is given by $\partial\sigma_m/\partial\theta = 0$. Solutions of this equation generally yield two azimuths, although Peska and Zoback (1995) showed that, for some specific values of the far-field principal stress components and borehole orientations, four extrema or no extrema at all may exist. When two solutions exist, for both these azimuths a set of fractures inclined to the borehole axis is generated. They are referred to as *en échelon* fractures (fig. 12.15). Their inclination to the borehole axis depends on the magnitude of both σ_M and σ_m and therefore on the magnitude of the various far-field stress components. This may be used to determine the complete stress tensor in a rock mass, as will be discussed in section 13.2.

Let us note that the more general case of a hydraulic fracture in an inclined borehole intersecting a transversely isotropic medium (for which there are five elastic constants) was discussed by Abousleiman and Ekbote (2005).

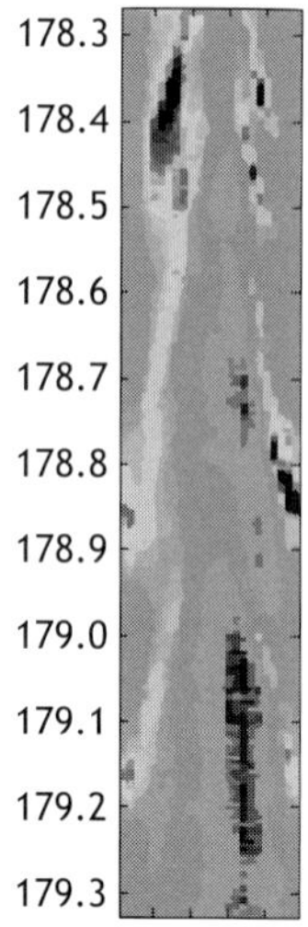

Fig. 12.15 Electrical images of *en échelon* fractures generated by hydraulic fracturing in a borehole inclined with respect to the far-field principal stress directions.

12.4.2 Hydraulically induced shear motions and related seismic activity

Since the catastrophic failure of the Malpasset dam, which occurred in 1956 in southern France, it has been well known that changes in pore pressure along faults may lead to their destabilization (Bernaix, 1967). As illustrated by fig. 12.9, an increase in pore pressure within a fault zone reduces the normal stress supported by the fault, leaving the shear stress unchanged. For large enough pore pressure increments, the decrease in effective normal stress leads to destabilization of the fault.

This mechanism is generally proposed to explain the microseismic activity observed upon the filling of large dams, such as that during the magnitude-6.4 earthquake that occurred in 1967 near the city of Koyna in India (Gupta *et al.*, 1969).

However, the filling of a dam not only leads to a pore pressure increase, it also modifies the local stress because of the weight of the water in the reservoir. This was discussed by Simpson *et al.* (1988), who identified two different types of induced seismicity associated with dam impoundments. They suggested that the microseismicity observed during the first filling of a dam results from the load increase associated with the changes in water level in the reservoir. This loading process may be described by a linearly poroelastic model that assumes undrained or partially drained conditions. It requires some understanding of the rock-mass hydraulic diffusivity in order to evaluate the correlated pore pressure variation. When the reservoir is completely full, the external load remains constant and only water diffusion can explain the observed microseismicity.

In order to better understand the role of water diffusion below reservoirs, Roelloffs (1988) investigated the effects of seasonal water level variation on the microseismicity observed in the vicinity of the Lake Mead reservoir in Arizona, USA. Indeed, Garder (1945) had pointed out that during the few years that followed the filling of the dam, the induced seismicity was correlated with high water levels. But after she conducted a poroelastic analysis of this diffusion process, Roeloffs observed that the induced seismicity is more sensitive to the slip plane orientation than to the local pore pressure variation. In other words, depending on their location with respect to the reservoir, the normal stress increments supported by some slip planes are larger than the pore pressure variations induced by change in water level, so that slip is inhibited for these planes. But, for some other source mechanisms, the reverse is observed. Roeloffs concluded that, in the vicinity of a reservoir, the effective stress variation induced by water-level changes depends as much (if not more) on the variation of the total stress applied to the pre-existing fractures as on the variation in the local pore pressure. However, she also observed that, after a few years, the peaks of seismic activity are no longer correlated with high water levels; rather, the opposite occurs, with peaks of seismic activity becoming associated with low water levels. This demonstrates that linear poroelasticity alone cannot explain all the microseismicity observed near dams and that some additional deformation mechanisms must be taken into account.

Turning now to the Koyna dam earthquake in India, we observe that its magnitude reached 6.4 and that it occurred a few years after the reservoir had been filled. This magnitude implies a seismic rupture length of the order of 15 to 20 km (see section 11.3). The exact hypocenter depth was not sufficiently well defined for an evaluation of the total stress perturbation caused by the water load to be undertaken, but, due to the time lag observed between the filling of the dam and the occurrence of the earthquake, the latter cannot be simply related to an instantaneous linearly poroelastic response to the load increase. It is generally accepted that this lag was caused by a diffusion process, i.e. it was related to the time it took for the pressure to build up at the hypocenter location. Nevertheless, pore pressure variations evaluated using standard hydraulic diffusivity values for the crust suggest extremely small pressure perturbations in the neighborhood of the hypocenter depth.

More recently, it was proposed (Lei, 2011) that the magnitude-7.9 Wenchuan earthquake, which in 2008 affected the Longmenshan fault system in China, was triggered by the filling of the Zipingpu reservoir. Lei argued that the filling of the dam generated effective stress variations of the order of a few kPa at the focal depth, some 10 km below ground level. He suggested that these very small effective stress variations were in fact large enough for an earthquake to be initiated.

Indeed, it has been proposed (e.g. King *et al.*, 1994; Brodsky *et al.*, 2003) that the loading conditions encountered in some parts of the seismogenic crust are very close to failure, i.e. are under critical loading conditions, so that very small effective-stress perturbations could generate earthquakes. This would explain why the very low pore-pressure perturbations that have been mentioned above might indeed have triggered the observed earthquakes.

However, many large dams have been built without generating large earthquakes. Clearly, the change in loading conditions at depth for these various sites was not large enough to initiate a large earthquake. This implies that the concept of critical loading may be applicable at a few sites but is not ubiquitously valid. We will explore this proposition regarding critical loading conditions below, in the context of the microseismic activity observed during hydraulic simulations conducted for the geothermal reservoir development at Soultz-sous-forêts, already introduced in fig. 12.6.

The four levels of nonlinear hydromechanical coupling observed during hydraulic stimulations

Whilst hydraulic fracturing is commonly used in the oil industry for improving the declining production of oil wells, this technology cannot be applied for the stimulation of hot geothermal wells, since the injected proppant is likely to interact with the hot geothermal fluids.

In crystalline formations where preexisting fractures are numerous, the object of stimulation is to induce large shear motions by building up progressively the pore pressure within the preexisting fracture field. The shearing of fractures is presumed to be associated with fracture dilatancy (see section 10.1.3), which leads to a significant increase in hydraulic conductivity for the whole rock mass. Monitoring the induced microseismic activity helps to identify the zones where the stimulation has been the most efficient.

This is illustrated by fig. 12.16, which presents some of the main features observed during the initial GPK1 well stimulation conducted in granite at Soultz-sous-forêts. This 3600 m deep vertical well is cased down to 2850 m, and the stimulation was conducted in the open-hole 2850–3400 m depth range, after the well was filled with sand from its bottom up to 3400 m. This sanding was done to prevent reactivating a significant fault zone identified at 3500 m. Panel I of fig. 12.16 represents the variation with time (with the horizontal axis scaled in days) of the depths of microseismic events (the vertical axis is scaled in meters) whilst Panel II exhibits the variation with time of the injection flow rate and the well-head injection pressure. Panel III is a plot of the projections on to a horizontal plane of microseismic event locations observed either within the 2850–3000 m depth range (III-1) or below 3000 m (III-2).

At Soultz-sous-forêts, the local stress field is such that the vertical stress component σ_v is a principal component throughout the depth range of interest and is nearly equal to the maximum horizontal principal stress component σ_H (Cornet *et al.*, 2007). The minimum horizontal principal stress σ_h is sub-horizontal throughout the domain of interest and is oriented N80°E. Its magnitude is such that the ratio σ_h/σ_v is constant and equal to about 0.54. This stress evaluation, together with the borehole pressure measurements, helps one to understand the various effects of the stimulation.

Cornet (2012) distinguished four different effects of this hydraulic stimulation. The first, which is associated with event locations *A* on fig. 12.16, is somewhat analogous to the Kaiser effect described for acoustic emissions detected during laboratory triaxial tests, as discussed in section 8.1.4. Microseismic activity starts only when the maximum differential stress supported by the rock exceeds a certain level, even though the overall rock mass deformation remains elastic. This critical stress level corresponds to the largest differential stress that has been supported by the rock material in the past for the corresponding effective minimum principal stress magnitude. For this type of stimulation, the rate of growth of the seismic cloud provides a direct observation of the velocity of pore pressure diffusion in the rock mass.

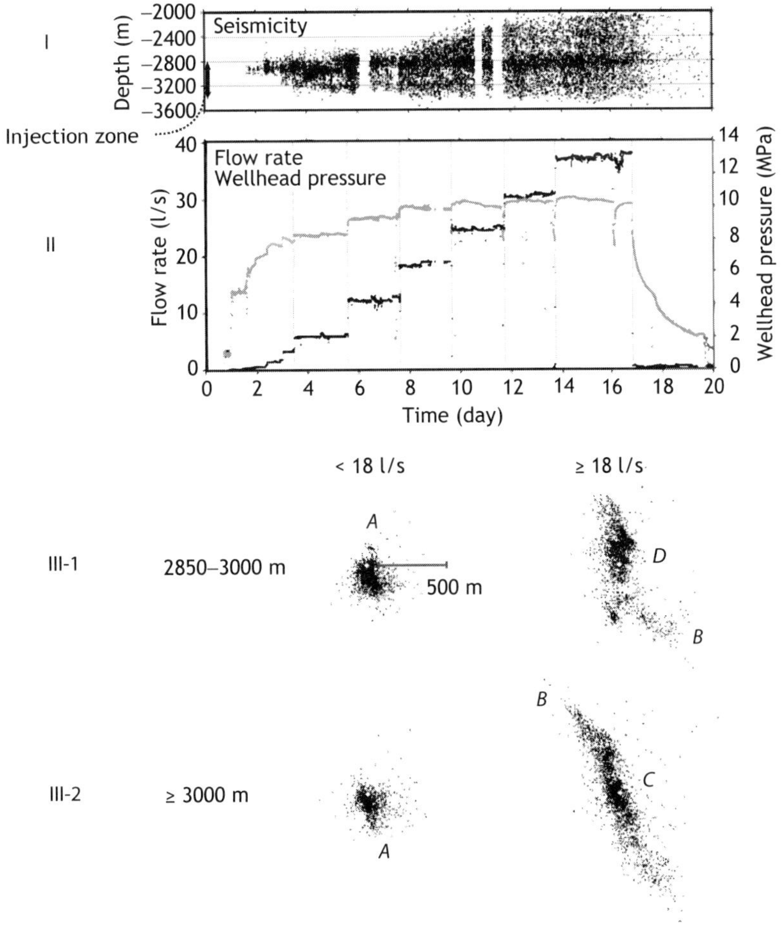

Fig. 12.16 Microseismic activity observed during the 1993 hydraulic stimulation conducted in well GPK1 at Soultz-sous-forêts, in eastern France. See the main text for an explanation. Note the 500 m scale shown in *A* in III-1 (reproduced from Cornet *et al.*, 2007, with permission from Elsevier).

Shapiro *et al.* (2002) used growth rate evaluations such as these for estimating the hydraulic diffusivity of rock masses. This supposes a uniformly distributed deformation process, which precludes the localization of failure within the rock mass. This process is equivalent to the loading phase *AB* shown in fig. 8.13, which describes acoustic emissions observed in the laboratory, but also includes the initial part of phase *BC*. It is observed before large-scale failure localization takes place. For these loading conditions the rock-mass behavior may be assumed to remain poroelastic, although some nonlinearity is required in order to take into account the volume variations associated with shear (see the discussion on dilatancy in section 8.1.3).

When fluid injection may be assimilated to a point source, Shapiro *et al.* proposed that the distance r at time t between this point source and the seismic event furthest away from the source is given by

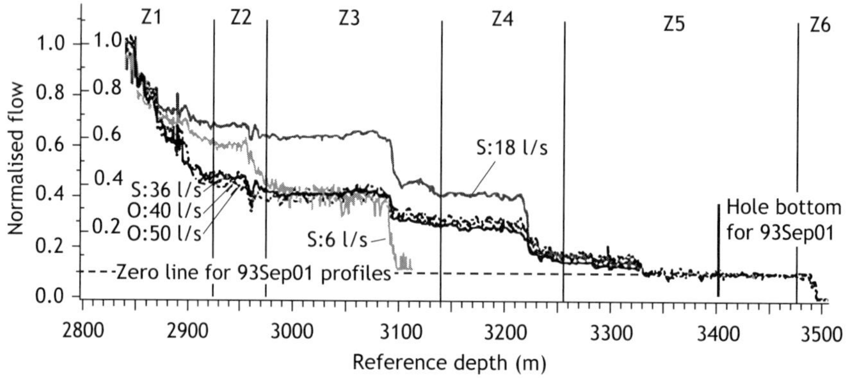

Fig. 12.17 Flow logs conducted in the well GPK1 during hydraulic stimulations. The distribution of flow with depth is plotted as a percentage of the total injected flow rate. For injection flow rates equal to 6 l/s, 18 l/s and 36 l/s these logs correspond to the stimulation illustrated in fig. 12.16, but the logs for flow rates 40 l/s and 50 l/s correspond to a different stimulation sequence, in which the sand was washed away from the bottom of the well (reproduced from Evans *et al.*, 2005, with permission from Wiley).

$$r = \sqrt{4\pi D t} \tag{12.113}$$

where D is the rock-mass hydraulic diffusivity, which is assumed to be isotropic (i.e. a scalar).

Whilst at Soultz-sous-forêts induced seismicity was detected for injection pressures as low as 3 MPa, the growth of the microseismic cloud and therefore of the fluid pressure diffusion process, remained axisymmetrical until the well-head injection pressure reached about 9.5 MPa. So, clearly, the natural effective loading conditions of the Soultz-sous-forêts granite are not critical for depths smaller than 5 km, i.e. they satisfy neither the rock-mass failure criterion (the Mohr envelope in fig. 12.9) nor the stability criterion for local faults and other *large* preexisting fractures. For the loading conditions at Soultz-sous-forêts, the rock-mass hydraulic diffusivity was evaluated by Shapiro *et al.* (2002) to be about $0.05\,\mathrm{m}^2/\mathrm{s}$.

The fact that some stimulation took place during the injection phase is apparent from the flow logs (plots of injected flow rate versus depth) conducted during the stimulation (fig. 12.17). We note that at low injection rates (6 l/s to 18 l/s), flow occurs in the rock mass through a very limited number of flow paths. As the injection pressure increases progressively along with the injection rate, proportionately more and more water flows down the localized flow paths below 3000 m (zones Z3 to Z6 in fig. 12.17). This outlines some effects of the water injection in these preexisting fault zones; yet no clear large-scale structure is detected in the map of microseismic activity for these flow rates (the left of panel III-2 in fig. 12.16), which suggests that no large-scale structure has been reactivated during this phase.

The three other stimulation effects observed at Soultz-sous-forêts concern failure localization, through slip along preexisting large-scale fractured zones (event locations B in fig. 12.16), or through the development of a fresh shear zone (event locations C in fig. 12.16) or through the formation of a hydraulic fracture (event locations D in fig. 12.16).

The development of a hydraulic fracture can be clearly detected once the well-head pressure reaches about 10 MPa. Indeed we note that, below this value, the injection pressure increases as the flow rate increases. But, after it reaches 10 MPa, the pressure remains more or less constant even if the flow rate is still increasing. The flow logs indicate that this hydraulic fracture developed in the upper 70 meters of the uncased portion of the well, i.e. where the horizontal minimum principal stress is the smallest (fig. 12.17, zone $Z1$). Indeed, when the injection flow rate is equal to 18 l/s, some 70% of the flow reaches 3000 m, but when the flow rate equals 36 l/s about 60% of the flow never reaches this depth for it has been lost through the hydraulic fracture. In addition, we note in panel I of fig. 12.16 that once the hydraulic fracture starts propagating, the front of the microseismic cloud moves upward; this is consistent with our discussion on the growth direction of hydraulic fractures in homogeneous rocks. Furthermore, this part of the microseismic cloud is oriented roughly in the N–S direction, i.e. perpendicularly to the minimum horizontal principal stress direction. Let us mention here that, as will be shown next, the microseismic sources are not related to the hydraulic fracture tip but to the shearing of preexisting fractures that intersect the main flow zone at an angle to the principal stress directions.

We have seen that hydraulic fracturing occurs when the fluid pressure is larger than the local minimum principal stress magnitude. Given that the granite density is close to 2.6 g/cm^3, the stress measurements of Cornet *et al.* (2007) indicate that the vertical gradient of the minimum horizontal principal stress magnitude is about 0.54×0.026 MPa/m or 0.014 MPa/m. The borehole pressure magnitude, which is controlled by the hydraulic fracturing process that occurs between 2920 m and 2850 m, remains smaller than the minimum principal stress magnitude at greater depths than these, so that hydraulic fracturing cannot occur in the lower parts of the well. This is the domain where the two other large-scale failure processes may develop: slip along preexisting large-scale structures or the development of a fresh shear zone.

Stimulation through the reactivation of preexisting large-scale structures is clearly illustrated by the mapping of induced microseismicity shown in zones B in fig. 12.16. These zones join the domain where hydraulic fracturing has occurred but they are associated with fluid pressures smaller than the local minimum principal stress component. They correspond to the reactivation of weak planar structures (100 m in scale and more). The sizes of the sources of seismic events located in these zones may be evaluated from their corner frequency. The results indicate frequencies in the 10–100 Hz range, i.e. source sizes in the 5–40 m range (see section 11.2.4), which are found to be much more energetic than those associated with hydraulic fracture propagation.

Whether fresh shear zones develop when the pore pressure in a rock mass has been questioned, for it is often considered that preexisting faults and fractures constitute weaker zones, i.e. they are mobile even before a fresh shear zone is generated. The results from Soultz-sous-forêts indicate the role of preexisting fracture sets in the development of fresh shear zones. Their influence is somewhat similar to that of microcracks in the development of failure in rock samples tested in the laboratory (section 8.1.3). For zone C in fig. 12.16, each microseismic event represents some slip on a preexisting fracture. The seismic moments for these various events show that the amount of slip remains very small (a few tens of microns) but all these events occur within a progressively decreasing volume,

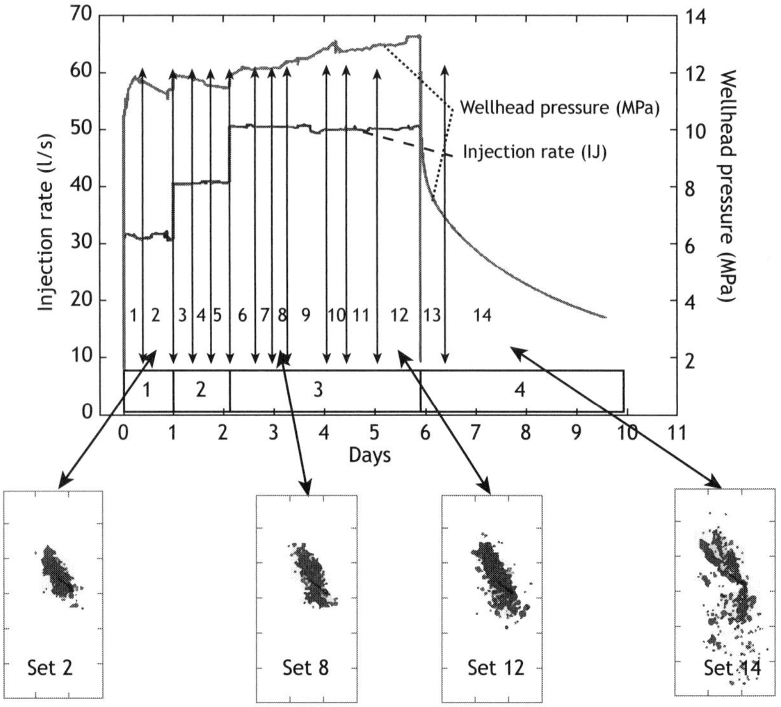

Fig. 12.18 Development of a fresh shear zone during the GPK2 stimulation at Soultz-sous-forêts. The induced seismicity is localized during the stimulation but becomes more diffuse when stimulation stops (after Calo *et al.*, 2011, with permission from Oxford University Press).

the geometry of which is governed by the regional effective stress field. At Soultz, the maximum horizontal principal stress is oriented N170°E, as shown in fig. 13.8 (Cornet *et al.*, 2007; Valley and Evans, 2007). As the density of events increases, the equivalent material gets more and more compliant and its hydraulic diffusivity increases. This takes place until deformation localizes, following a scheme somewhat similar to that proposed by Rudnicki and Rice (1975) for the localization of failure in intact rocks (section 9.2.2).

Calo *et al.* (2011) analyzed the microseismicity induced during the hydraulic stimulation conducted in the GPK2 well below 4500 m, some 600 m away from the GPK1 well on the Soultz-sous-forêts site. They provided a very clear description of the development of a fresh shear zone. During this stimulation, which did not reach the hydraulic fracturing stage, all the microseismic activity remained localized within a sub-planar zone, the orientation of which is similar to that of zone C identified for the GPK1 stimulation below 3000 m. It is inclined at some 25° to the maximum horizontal principal stress direction (fig. 12.18).

However, when injection ceases, so that the injection pressure drops, microseismic activity develops in a domain much broader than that observed during stimulation. It involves structures with quite diverse orientations. After shut-in, the water diffusion is no longer confined within the fresh shear zone but extends to a much broader volume because of the correlated variation in Biot's effective stresses. Let us note that a rigorous

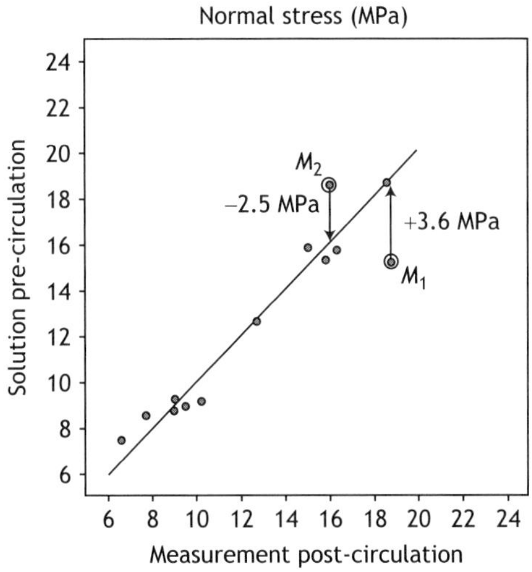

Fig. 12.19 Plot of the normal stress supported by preexisting fractures as computed from pre-stimulation measurements (vertical axis) and as computed from post-stimulation measurements (horizontal axis). After Scotti and Cornet (1994), with permission from Elsevier.

computation of these effective stresses would require taking into account the nonlinearity of the stress–strain relationship.

Seismic and aseismic motions induced by pore pressure variations

The analysis of the various effects associated with large-scale hydraulic stimulation presented above rests on a discussion of the development of induced seismicity. It assumes that both the regional stress field and the borehole injection pressure are known.

However, Scotti and Cornet (1994) suggested that, in addition to these seismic slips, water injections may also generate large-scale nonseismic (aseismic) motions that in turn induce large-scale effective stress variations. First we introduce Scotti's and Cornet's indirect observations of aseismic slips at the Le Mayet de Montagne test site in central France, before presenting direct evidence for such aseismic motions at the Soultz-sous-forêts site.

As a preliminary to the Soultz experimental program, a set of large-scale injections was conducted in two 800 m deep boreholes drilled in the granite of Le Mayet de Montagne (e.g. Cornet and Yin, 1995). Detailed stress measurements were conducted before and after these large-scale injections.

The normal stress components supported by preexisting fractures intersected by one of the boreholes and identified with an electrical imaging tool (see section 1.3.2) were measured using the HTPF method (see section 13.2) before and after hydraulic stimulation. The stress field, as determined from pre-stimulations measurements, was used to work out the expected values for the post-stimulation measurements. The results are presented in fig. 12.19 and show good agreement between the calculated and measured values for most post-stimulation results, except for two fracture planes. Both these planes were located close to a zone where some induced seismic activity had been observed.

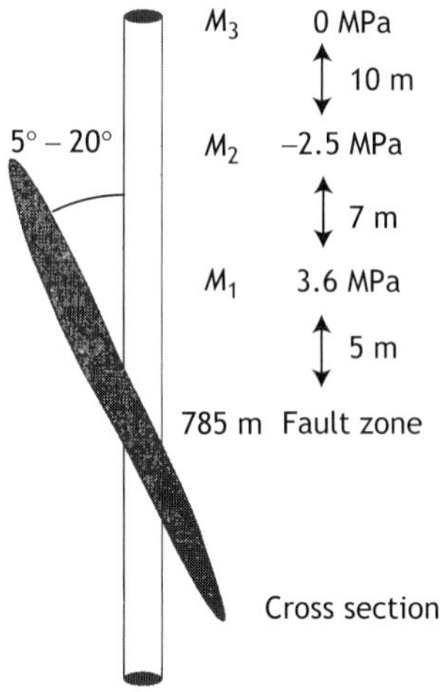

Fig. 12.20 Model for evaluating the stress perturbation generated by slip along a preexisting fracture that intersects the injection borehole. Measurements were conducted at points M_1, M_2 and M_3 of fig. 12.19 (after Scotti and Cornet, 1994, with permission from Elsevier).

A numerical model based on the displacement discontinuity method (see section 7.3.2) was developed for evaluating the stress perturbations caused by the progressive relaxation of the shear stress within the zone where induced seismicity was observed (fig. 12.20). The results show that the observations could be modeled provided that the sheared zone was about 20 m long and the stress relaxation was complete. Had such a stress relaxation occurred during an unstable slip of the fault, the correlative seismic magnitude would have been larger than 3, but no such event was observed. Given the relatively small number of seismic events detected in this zone, Scotti and Cornet concluded that the shear motion was mostly aseismic.

This modeling exercise was pursued further to characterize the stress perturbation throughout the volume surrounding the sheared zone (fig. 12.21). Only variations in the relative magnitudes of the principal stress components were considered and, furthermore, one principal stress direction was taken as vertical, σ_v, whilst the two other principal stress components, σ_H and σ_h, were horizontal ($\sigma_H > \sigma_h$). For this analysis, the *ellipticity* of the stress tensor is characterized by the ratio $\omega = (\sigma_H - \sigma_h)/(\sigma_v - \sigma_h)$. Hence normal faulting systems correspond to $\omega < 1$ and strike-slip faulting systems correspond to $\omega > 1$. Figure 12.21 illustrates the stress perturbation for an initial ellipticity ratio $\omega = 1.2$ and a friction coefficient equal to 0.2 for the slip zone. Once slip has occurred, the ellipticity ratio drops below 1 in some areas so that locally the faulting regime changes from strike-slip to normal. Observation of such variations in the focal mechanism during large-scale injections confirmed Scotti and Cornet in their conclusion on the existence of large aseismic slips during large-scale fluid injections.

Evidence for aseismic slips was also given by Cornet *et al.* (1997) and by Evans *et al.* (2005), after compiling borehole televiewer data gathered before and after the GPK1 stimulation at Soultz-sous-forêts described by fig. 12.16. They reported that this hydraulic

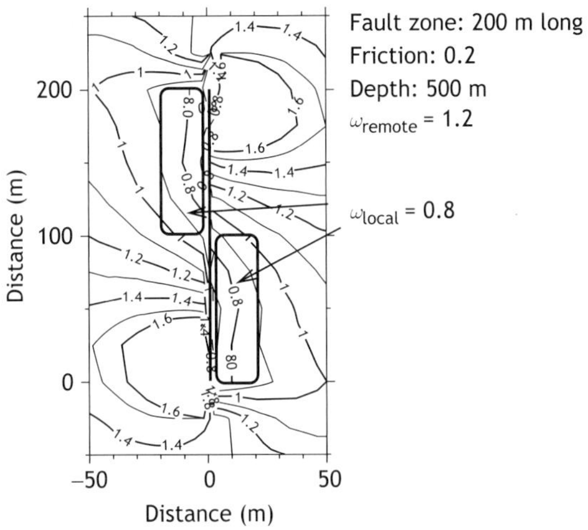

Fig. 12.21 Plot of the stress perturbation observed when the initial shear stress supported by a fault plane has completely relaxed. The stress field is characterized by its ellipticity ratio ω, as defined in the text (after Scotti and Cornet, 1994, with permission from Elsevier).

stimulation generated measurable displacements along some of the most significant flow zones and reached up to two to four centimeters. Had these motions been associated with one single seismic event, however, the corresponding magnitude would have been larger than 3, but no such event was observed.

Independently, Bourouis and Bernard (2007) reprocessed the microseismic data recorded during this stimulation using a double-difference relative relocation algorithm (see section 11.2.3). They observed that many of the sources are multiplets, some of which are located on the same fault planes.

One of these faults (fig. 12.22) intersects the GPK1 borehole at precisely the depth where Cornet *et al.* (1997) identified a 4 cm displacement. Further, not all the multiplets appear at once along the fault plane; the deeper ones start to be activated some 4 to 5 days after the first multiplets have been detected close to the borehole. Bourouis and Bernard concluded that a slow motion with variable velocity along the fault was activated by the water injection. Further, they evaluated the cumulated shear displacement for some multiplets. The values they obtained are within the same order of magnitude as the shear motions measured in the borehole with an acoustic televiewer by Cornet *et al.*

Calo *et al.* (2011) also concluded that aseismic shear motion existed, after they conducted a four-dimensional (three spatial coordinates plus time) P wave velocity (C_P) tomography, from the microseismic activity monitored during the GPK2 stimulation mentioned above (fig. 12.18). Tomography images were produced, for different periods, adjusted to the stimulation injection scheme. They detected a change in seismic velocity within a kilometer-scale volume precisely when the injection rate was incremented with a correlative 2 MPa injection pressure increase (set 3 of fig. 12.18). Another such coincidence between the pore pressure and the C_P velocity variation was detected at the next flow-rate increase (set 6). They attributed these seismic velocity variations to an effective stress

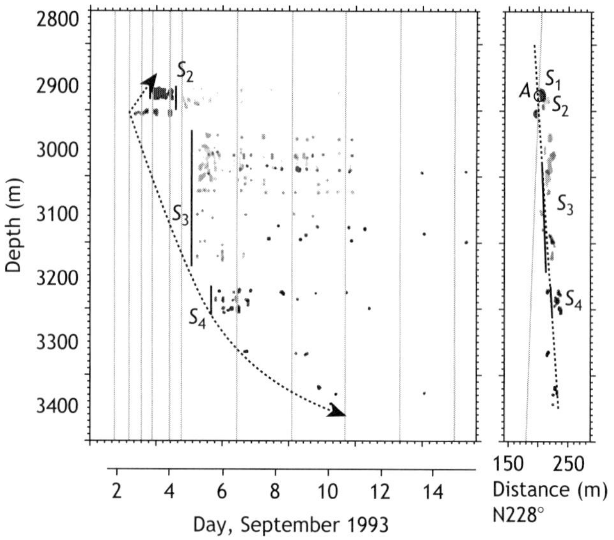

Fig. 12.22 Location of some multiplets activated during the 1993 GPK1 stimulation at Soultz-sous-forêts. The vertical axis gives the depth in m and the horizontal axis gives the time in days (reproduced from Bourouis and Bernard, 2007, with permission from Oxford University Press).

variation generated by undetected slip motions that occurred within the large seismogenic structure shown on fig. 12.18.

This shows that pore-fluid pressure variations may generate a variety of shear motions. Some are seismic but others correspond to nonseismic slip over a large area. Further, two different types of seismic source must be considered: sources due to slip on preexisting discontinuities within an elastic continuous material and sources due to slip along asperities embedded in a much larger discontinuity which itself is undergoing some irregularly distributed slow motion. Observation of these diverse shear motions requires seismic monitoring systems covering very different frequency ranges, and this cannot be handled with a unique sensor type. Whilst high-frequency signals generated by asperities (with a frequency spectrum reaching kHz) are attenuated over fairly short distances and can be recorded only with specific sensors located close to the source, slow-motion events are more easily detectable at some distance from the source but require broadband sensors.

The above remark applies to the monitoring of natural deformation processes as well. The necessity of covering large frequency domains for the monitoring of natural deformation processes is now well established, as illustrated for example by the magma flow monitoring on the Merapi volcano in Indonesia (e.g. Beauducel *et al.*, 2000). However, it is not commonly applied yet for the monitoring of man-made large-scale pore pressure perturbations, although Bame and Fehler (1986) did report the detection of so-called *long-period events* during a large-scale hydraulic stimulation conducted at the Fenton Hill geothermal field in New Mexico, USA.

A better understanding of the relation between seismic and aseismic slips is a prerequisite for many geomechanics problems. For example, large fluid injections in rock masses must be designed in such a way as to avoid generating unacceptable seismic disorders.

This is still an active domain of research, and the interested reader is referred to the excellent review on the interactions between slips of various sizes and various slip velocities by Ben-Zion (2008).

12.4.3 Large-scale shear-failure processes and fluid migration in the seismogenic crust

The C_P tomographic images produced by Calo *et al.* (2011) for the GPK2 hydraulic stimulation cited above exemplify the fact that the water content in some rock masses depends strongly on the stress field applied to the rock mass, especially when these loads exceed the rock-mass elastic limit. Stress variation patterns generated by slip along planar surfaces are illustrated in fig. 12.21. They are found to either increase or decrease the spherical stress component, depending on the location with respect to the slip motion. This implies that shear motions generate pore pressure increase or pore pressure decrease at both ends of a slipping surface, resulting in correlative local fluid flow.

The influence of shear motions on the fluid content of rocks, as observed in various tectonic environments, has often been described (e.g. Sibson, 1987; Muirwood and King, 1993; Bourouis and Cornet, 2009). For example, Sibson discussed how the extension zones associated with shear imply a local pore pressure decrease and therefore an effective minimum principal stress increase for the the arrest of earthquakes. Muirwood and King proposed also that such a mechanism could explain the variation in natural-spring discharge, observed after large earthquakes, depending on the spring's location with respect to the zone of fault arrest. Bourouis and Cornet showed how the recurrent seismic-swarms activity observed near the city of Aigion, in the Corinth Rift (western Greece) may be explained by a fluid diffusion process. In particular they identified regions of upward and downward fluid migrations depending on the proximity of the regions to extremities of the seismically activated domain. These upward and downward growth patterns seem to be consistent with a large-scale slow shear motion in the volume where the swarm initially developed.

Now we will follow up a discussion by Pine and Batchelor (1984) on the factors controlling the upward or downward migration of shearing in jointed rock masses during a large hydraulic stimulation (where the injected volume reaches a few tens of thousands of cubic meters). We concentrate on the destabilization of preexisting fractures (effect B of hydraulic stimulation; see fig. 12.16) and consider a homogeneous isotropic material with fracture set S inclined at an angle θ to the maximum principal stress direction. We consider the case when the maximum principal stress is horizontal, $\sigma_1 = \sigma_H$, as well as the minimum principal stress $\sigma_3 = \sigma_h$, so that the vertical stress component σ_v is the intermediate principal stress component (fig. 12.23). Further, all these components are presumed to vary linearly with depth. For simplicity, the strengths of all fractures of this set are assumed to satisfy the Coulomb criterion with no cohesion (also called Byerlee's friction law), which may be expressed in terms of Terzaghi's effective stresses as

$$|\tau| = (\sigma_n - P_p)\tan\phi_S = \sigma_n' \tan\phi_S \tag{12.114}$$

where ϕ_S is the friction angle for all fractures of set S (compressions are taken positive).

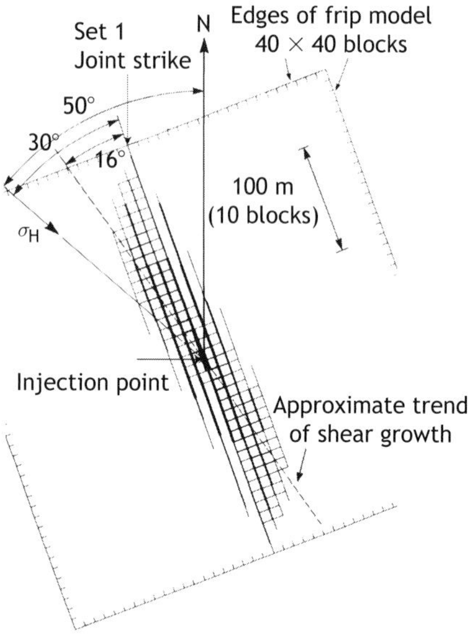

Fig. 12.23 Stability of a rock mass with two orthogonal fracture sets, one of them being inclined 30° to the maximum principal stress direction. The stress field is assumed to be linearly dependent on depth and the fracture system is saturated by a fluid under a pressure larger than hydrostatic (after Pine and Batchelor, 1984, with permission from Elsevier).

The system is assumed to be saturated by a fluid with density ϱ_w, which is pressurized above the hydrostatic level by a value P_0. The loading conditions at depth z are

$$\sigma_1 = \sigma_H = \alpha_H z + c_H \tag{12.115}$$

$$\sigma_2 = \sigma_v = \alpha_v z \tag{12.116}$$

$$\sigma_3 = \sigma_h = \alpha_h z + c_h \tag{12.117}$$

$$P_p = \varrho_w \, g \, z + P_0 \tag{12.118}$$

$$\tag{12.119}$$

or, in terms of Terzaghi's effective stresses,

$$\sigma'_H = (\alpha_H - \varrho_w g) z + (c_H - P_0) = \alpha'_H z + (c_H - P_0) \tag{12.120}$$

$$\sigma'_v = (\alpha_v - \rho_w g) z - P_0 = \alpha'_v z - P_0 \tag{12.121}$$

$$\sigma'_h = (\alpha_h - \varrho_w g) z + (c_h - P_0) = \alpha'_h z + (c_h - P_0) \tag{12.122}$$

$$\tag{12.123}$$

The shear and effective normal stresses supported by any surface of fracture set S are

$$\sigma'_n = \frac{\sigma'_H + \sigma'_h}{2} + \frac{\sigma'_H - \sigma'_h}{2} \cos\theta \tag{12.124}$$

$$\tau = \frac{\sigma'_H - \sigma'_h}{2} \sin\theta \tag{12.125}$$

At depth z, the critical loading for which a plane from fracture set S starts slipping is reached when (see section 9.2.2)

$$\frac{\sigma'_H}{\sigma'_h} = R_S = \frac{1+\omega_S}{1-\omega_S} \tag{12.126}$$

with

$$\omega_S = \frac{\tan\phi_S}{\sin 2\theta + \tan\phi_S \cos 2\theta} \tag{12.127}$$

where R_S is the limiting value of the ratio of the effective maximum and minimum principal stress components when slip occurs on a fracture of set S.

For rock masses with two or more fracture sets, the angles ϕ and θ refer to the most critically loaded fracture set (fig. 12.16). Equation (12.127) also applies to the development of a fresh shear zone in a rock mass with ubiquitous randomly oriented preexisting fractures. At failure, the shear plane orientation is such that $\theta = (\pi/2 - \phi_i)/2$, where ϕ_i is the intrinsic friction angle of the equivalent material. In this case, the limit of the effective principal stress ratio is

$$R_M = \frac{1+\sin\phi_i}{1-\sin\phi_i} \tag{12.128}$$

Equation (12.128) places limits on the vertical gradient of the horizontal principal stress components in a rock mass. This limit depends on both the fluid density and the eventual overpressure that may exist at a given depth. For example, considering a rock mass filled with a fluid of density ϱ_w under hydrostatic conditions ($P_0 = 0$), the maximum depth z_c for which the vertical gradient of the effective horizontal principal components α'_H and α'_h remains compatible with stability is given by

$$z_c = \frac{R_M c_h - c_H}{\alpha'_H - R_M \alpha'_h} \tag{12.129}$$

Further, for a given stress field, the critical pore pressure value P_{cr} required to destabilize a plane of fracture set S is such that $\tau = [\sigma_n - (g\varrho_w z + P_0)] \tan\phi_S$, so that

$$P_{cr} = \frac{\sigma_H + \sigma_h}{2} - \left(\frac{\sigma_H - \sigma_h}{2}\right)\frac{\sin 2\theta + \cos 2\theta \tan\phi_S}{\tan\phi_S} \tag{12.130}$$

$$P_{cr} = \frac{\sigma_H + \sigma_h}{2} - \frac{1}{\omega_S}\left(\frac{\sigma_H - \sigma_h}{2}\right) \tag{12.131}$$

If the pore pressure required to destabilize the rock mass decreases with depth (i.e. we have negative values for $\partial P_{cr}/\partial z = P_{cr,z}$) then shear failure grows downward when fluid is injected to keep up the pressure, so that the related induced seismicity develops downward as well. Such a downward development of induced seismicity upon fluid injection has been reported by Pine and Batchelor for the stimulation they conducted at a 2 km depth in Cornwall (western UK). The seismicity extended down to 3 km, a feature quite different from that described for the hydraulic stimulation conducted at Soultz-sous-forêts (fig. 12.16).

Negative values for $P_{cr,z}$ require that

$$P_{cr,z} = \frac{\sigma_{H,z} + \sigma_{h,z}}{2} - \frac{1}{\omega_S}\left(\frac{\sigma_{H,z} - \sigma_{h,z}}{2}\right) < 0 \qquad (12.132)$$

which implies that, during fluid injection, downward seismic activity migration is observed when

$$\frac{\sigma_{H,z}}{\sigma_{h,z}} > R_S \qquad (12.133)$$

12.5 Thermal fracturing processes

Anybody who has poured boiling water into a glass made of silica crystal knows that the glass may break. Because silica crystal exhibits very poor heat conduction together with high thermal-dilatation characteristics, large temperature gradients develop upon the pouring of boiling water and generate large deformation gradients, which in turn generate high thermal tensile stresses. Thermal fracturing processes depend on both temperature gradients and dilatation characteristics.

We address here the problem of the thermal fracturing processes that may be observed in boreholes because of the cooling of the rock caused by the fluid involved in the drilling operation. We assume that the rock mass is under uniform far-field stress conditions and that the borehole is parallel to one of the far-field principal stress directions. We call σ_M and σ_m the two principal stress components perpendicular to the borehole axis.

Axial thermal fractures

As shown in section 12.2.2, when the temperature at the wall of a borehole, T_b, differs from that in the unperturbed rock formation, T_r, an axisymmetrical hoop stress develops, which depends on the material's linear thermal expansion coefficient α_T:

$$\sigma_\theta = -\frac{E_B}{1 - \nu_B}\alpha_T(T_r - T_b) \qquad (12.134)$$

Here, we assume compressions to be positive. This axisymmetrical thermal hoop stress is superposed on the hoop stress associated with the far-field stress state. Hence, just as for hydraulic fracturing, the resulting hoop stress magnitude varies with the azimuthal coordinate θ and is a minimum in the direction of the maximum principal stress σ_M. When the temperature difference $T_r - T_b$ is positive and large enough, a tensile fracture develops parallel to the borehole axis in the σ_M direction.

The stress intensity factor that characterizes the stress singularity close to the fracture tip is somewhat analogous to that observed for hydraulic fracturing when no fluid penetrates into the fracture. If we apply a fictitious pressure P_{fic}, where

$$P_{fic} \sim \frac{E_B}{1 - \nu_B} \alpha_T (T_r - T_b) \tag{12.135}$$

at the borehole wall, it generates the same stress perturbation near the borehole wall as that generated by thermal stresses. The stress intensity factor associated with thermal cooling is

$$K_I \propto \frac{\sigma_3^{\infty} - P_{fic}}{\sqrt{\pi a}} \tag{12.136}$$

which decreases as the fracture extends. It may be concluded that thermal fractures in boreholes are stable and therefore have only a limited radial extension.

When the borehole is not parallel to any of the far-field principal stress directions, just as for hydraulic fracturing the two components $\sigma_{\theta\theta}$ and σ_{zz} are not principal components, so that thermal fractures appear *en échelon*, in diametrically opposed locations at the borehole wall.

Influence of fabric heterogeneity

The stress analysis given above assumes the material to be homogeneous. However, for materials such as granite that are made of different crystals with very diverse thermal expansion coefficients, even uniform temperature fields generate thermal stresses because of the differences in thermal expansion for the various constituents (e.g. Simmons and Cooper, 1978; Heard and Page, 1982; Darot *et al.*, 1992). Consequently, when the temperature is progressively decreased in a borehole in such a material, a zone under tensile stress develops at the borehole wall in two diametrically opposed locations, oriented according to the far-field stress. Because of the different thermal expansion coefficients of the constitutive crystalline grains, locally high intergrain tensile stresses may appear within both these zones, leading to a progressive disaggregation of the rock rather than to the formation of a unique planar tensile fracture. This mechanism was proposed by Morin *et al.* (1990) and Bérard and Cornet (2003) for interpreting borehole failure processes concomitant to high, but slow, temperature perturbations. It has also been considered as a mechanism for permeability enhancement during geothermal reservoir developments (Elsworth, 1989).

12.6 Further reading

Bai, M., and Elsworth, D., 2000. *Coupled Processes in Subsurface Deformation, Flow and Transport.* ASCE Press, 354 pp.

Berest, P., and Weber, Ph., 1988. *La thermomécanique des roches.* Éditions BRGM, 327 pp.

Carlslaw, H. S., and Jaeger, J. C., 1986. *Conduction of Heat in Solids.* Clarendon Press, 510 pp.

Coussy, O., 1991. *Mécanique des milieux poreux.* Éditions Technip, 437 pp.

Economides, M. J., and Nolte, K. G. (eds.), 2000. *Reservoir Stimulation, Third Edition.* Wiley, 620 pp.

Handy, M. P., Hirth, G., and Hovius, N. (eds.), 2005. *Tectonic Faults, Agents of Change on a Dynamic Earth.* MIT Press, 446 pp.

Stephansson, O., Jing, L. and Tsang C.-F. (eds.), 1997. *Coupled Thermo-Hydro-Mechanical Processes of Fractured Media.* Elsevier, 487 pp.

Wang, H. F., 2000. *Theory of Linear Poroelasticity.* Princeton University Press, 287 pp.

13 Methods for stress field evaluation from *in situ* observations

Stress is a continuum mechanics concept that supposes the existence of a representative elementary volume (REV) for the geomaterials of concern. Generally, the REV is small enough that the stress may be assumed to be constant within this volume.

Rock masses involve fractures and faults as well as continuous geomaterials, and the geometrical characteristics of fractures and faults are described by power laws which are not compatible with the definition of an REV. However, these power laws are valid within bounded domains.

The stress determination methods in geomechanics apply only to stress fields defined in continuous geomaterials and therefore it is assumed that no discontinuity exists within the volume of the measurement.

Geomaterials are very often elastoviscoplastic materials, i.e. they retain some memory of past loads for a while (chapter 2); efforts have been made to determine stresses from the monitoring of deformations that affect rock samples recently removed from their natural environment (e.g. Amadei and Stephansson, 1997, chapter 7). This supposes that such data depend only on the loading conditions that applied *in situ* to the samples before their extraction. However, in addition to the effect of stress boundary conditions, various sources of deformation exist, for example temperature variations or physicochemical interactions such as the swelling of clay or the subcritical growth of microcracks. Further, it has long been recognized that stresses exist within large blocks of rock even if these do not support any external load (e.g. Friedman and Logan, 1970); these stresses have been described as *residual stresses*. In order to avoid these difficulties, efforts have been concentrated on evaluating stresses from *in situ* measurements.

In this chapter we introduce various methods for stress determination from *in situ* measurements and in chapter 14 we present practical examples of applications. These examples illustrate how results obtained with different methods at the same site can provide confidence that the results are correct. They also shed some light on the rheology of rock masses at various scales, including the influence of faults and fractures.

First we discuss methods used by the mining industry, which assume an underground access to the location of interest. These include the flatjack and overcoring methods. Then we address methods that are adapted to deep boreholes, either through direct hydraulic testing or through borehole failure analysis. Finally we introduce methods that rely on seismic-wave propagation, whether natural (for which a focal mechanism interpretation is used) or artificial (for which a shear wave splitting analysis is used).

The International Society for Rock Mechanics (ISRM) has published recommendations for overcoring and hydraulic testing procedures, as well as for the strategy and quality

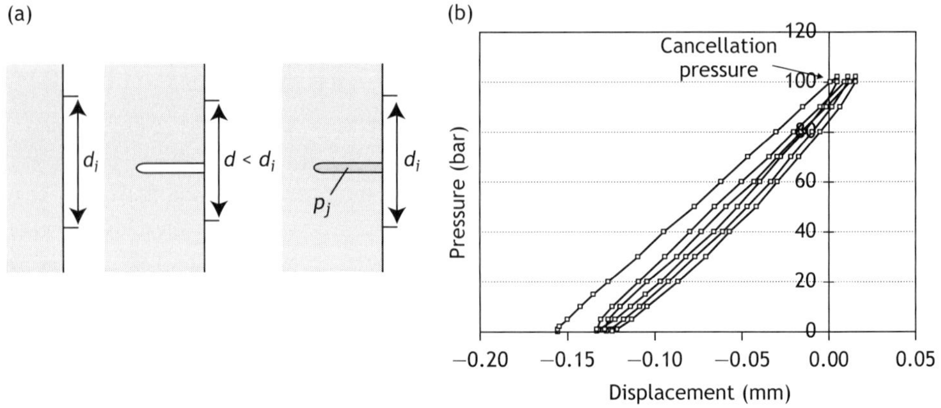

Fig. 13.1 Normal stress measurement with the flatjack technique. (a) Stress relief and restoration by the flatjack technique. (b) Example of load–unload cycles conducted for normal stress determination (the horizontal axis corresponds to displacement and the vertical axis corresponds to pressure).

control for stress measurement campaigns (Hudson and Cornet, 2003). The present chapter is not intended just to repeat these recommendations but rather to present a panorama of the various approaches that may be followed, depending on the context of the stress determination.

13.1 Stress measurements from underground access

The problem of stress estimates in geomechanics was first encountered in the context of the optimum design of mines, and different techniques have been developed depending on the mining context. These methods all take advantage of the direct access to a free rock face in the mine openings.

The issue of determining the effect of a mine opening on the local stress field is tackled through numerical modeling and is beyond the scope of this chapter. We address here only the two types of measurement technique normally used in underground mines: the flatjack and the overcoring techniques.

13.1.1 The flatjack method

Some of the very first *in situ* stress measurements were conducted in iron mines in order to evaluate the uniaxial stress supported by the pillars left in place for roof support (Tincelin, 1952). In this method, a set of pins is placed on both sides of a straight line that extends about 50 cm in length on the free face of interest. The pin location geometry is such that the alignment between some pins is normal to the straight line (fig. 13.1(a)).

The original distance d_i between these pins is measured. Then a slot is cut along a straight line perpendicular to the rock face, so as to relieve the stress vector components (normal and shear) that were acting on this surface. Some relative displacements occur between the various pins during this stress relaxation. A flat hydraulic jack is inserted into the slot and pressure is applied by the jack on the rock until the normal displacement observed during the stress relief has been canceled. An example of such a pressure–displacement curve is given in fig. 13.1(b). It shows a small drift to the right during the successive load–unload cycles, which shows that the rock–flatjack system is not linearly elastic. A common practice is to load the system three times so as to cancel, each time, the observed normal displacement. The mean pressure value of these three measurements is taken as equal to the normal stress component existing before the saw cut was made.

This method has been generalized by conducting similar tests on the walls of tunnels or mining drifts. For circular-shaped openings with radius R, the six equations (5.303) through (5.308) provide, in cylindrical coordinates, the relationship between the six stress tensor components at the wall of the opening, $\tilde{\sigma}|_{\rho=R}$, and the stress far from the opening, $\tilde{\sigma}^{\infty}$, for linearly elastic and isotropic rocks. More generally, for irregularly shaped openings in an elastic rock mass, a linear relationship exists between the stress at point $\boldsymbol{X}$ of the measurement and the far-field stress:

$$\tilde{\sigma}(\boldsymbol{X}) = [A]\tilde{\sigma}^{\infty} \tag{13.1}$$

where $[A]$ is a 6×6 matrix, the coefficients of which depend on the geometry of the local underground opening, the location and orientation of the flatjack and the elastic parameters of the rock mass. The matrix coefficients are determined using a numerical procedure after the elastic constants have been measured (e.g. Cornet, 1996).

If the jth flatjack orientation is characterized by that of its normal $\boldsymbol{n}^j$, the flatjack measurement, P^j, yields the following result:

$$P^j = \left[\tilde{\sigma}(\boldsymbol{X}^j)\boldsymbol{n}^j\right] \cdot \boldsymbol{n}^j = \left([A]\tilde{\sigma}^{\infty}\boldsymbol{n}^j\right) \cdot \boldsymbol{n}^j \tag{13.2}$$

Determining the complete natural-stress tensor requires a minimum of six measurements in six different directions. In practice, though, errors affect the measurements so that it takes a larger number of tests for the stress determination. This produces a set of N equations ($j = 1, \ldots, N$, $N > 6$), with six unknowns (the six stress tensor components), which may be solved by a least squares method (see the next section). Let us observe that, in this procedure, it is assumed that the far-field stress tensor remains uniform, i.e. there is no stress gradient along the tunnel axis. This implies that the REV volume is cylindrical and coaxial with the tunnel, with its radius equalling 4 to 5 times the tunnel radius.

The validity of the method rests on the assumption that the rock mass remains linearly elastic and homogeneous within the REV volume. Very often, however, the tunnel excavation process involves some blasting, so that an altered zone surrounds the tunnel. Clearly this altered zone will not exhibit the same material properties as the intact rock mass and therefore equation (13.1) will not be valid.

In order to avoid difficulties with the partially stress-relieved domain that surrounds the underground cavity where measurements are conducted, techniques have been developed to reach the unperturbed virgin rock mass.

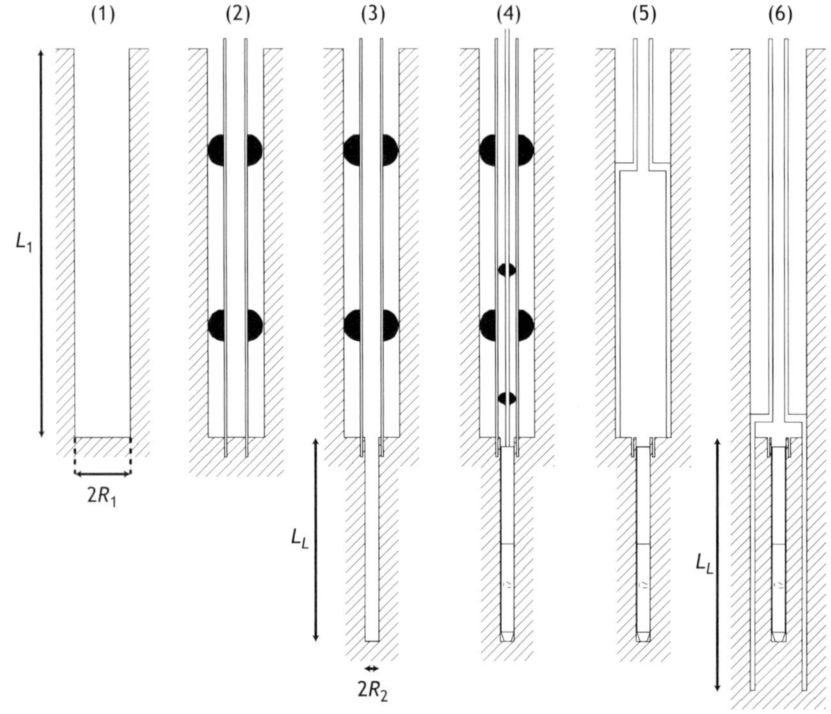

Fig. 13.2 Principle of an overcoring stress measurement technique. The measuring cell can be seen in position in the pilot hole, of diameter $2R_2$, in (4)–(6).

13.1.2 Stress relief methods

Stress determination methods based on stress relief assume an elastic response of the rock. The stress that exists in a given volume is completely relaxed and the corresponding strain is measured. When the elastic parameters of the material are known, displacement measurements provide data for an evaluation of the stress that has been relieved, which is assumed to correspond to the initial stress.

Principle of overcoring techniques

Most stress relief methods involve overcoring techniques (fig. 13.2). First a borehole with diameter $2R_1$ is drilled deep enough to reach domains, far from the free face that are assumed to be loaded only by the natural stress field ((1) in fig. 13.2). Once it has reached this critical length L_1, the borehole is lengthened by a coaxial pilot hole of length L_L with a smaller diameter, $2R_2$ ((2), (3) in fig. 13.2). A cell that measures displacements in various directions is inserted into the pilot hole and coupled securely to the rock ((4) in fig. 13.2). Then the first borehole is deepened, leaving in place a cylindrical piece of rock around the displacement-measuring cell, and the displacements associated with this stress relief operation are continuously recorded ((5), (6) in fig. 13.2).

At the end of the overcoring operation (when no further displacement is detected whilst overcoring proceeds), the cylindrical piece of rock, with the measuring cell still set into its inner cylindrical hole, is removed from the large borehole after its orientation has been properly recorded. It is pressurized on its outer surface and the corresponding displacements at the walls of the inner hole are measured with the cell in order to determine the elastic properties of the rock.

Interpretation of the results in terms of the *in situ* stress field rests on the assumption of a linear elastic response of the rock core. The measured displacements are converted to stress components using the measured rock-core elastic parameters. Because the drilling operation may be accompanied by temperature variations, the rock temperature is often continuously recorded in order to correct for thermal expansion (or retraction) effects.

Overcoring cells

Various displacement-measuring systems have been developed for overcoring operations (e.g. Amadei and Stephansson, 1997, chapter 5). Amongst the first of these we may cite the US Bureau of Mines (USBM) cell, which measures only radial displacements (Obert *et al.*, 1962). But, as may be observed from equation (5.309), radial displacements depend only on the far-field stress components applied in the plane normal to the cylinder axis, σ_{11}^{∞}, σ_{22}^{∞}, σ_{12}^{∞}. Hence, with the USBM cell, measurements must be conducted in boreholes with three different orientations for a complete stress determination and this implies relatively large REVs.

In order to avoid this three-borehole difficulty, various cells have been designed for measuring, within one unique borehole, displacements in the required diversity of orientations. For example, in some cells strain gauge rosettes are glued directly onto the borehole wall in various azimuths. With such cells, much less time is required by the measurements and furthermore the size of the REV remains fairly small. It corresponds roughly to the volume of the cylindrical core and is of the order of a few dm^3. In fact this decrease of size of the displacement-measuring sensors in comparison with that of the REV may become an issue for some cells.

Unfortunately, experience has shown that, for large differential stresses in the plane normal to the borehole axis, the core has a tendency to break with a regular pattern known as core *disking*. The core is split by regularly spaced fractures normal to the core axis (e.g. Li and Schmitt, 1998). Both the density of these fractures and their shape depend on the far-field stress state and also on the pressure applied to the drill bit during coring. This suggests that some microcracks may develop in the core even though disking has not occurred, so that the elastic properties of the core may differ from those of the intact rock.

Sugawara and Obara (1999) developed the compact conical-ended borehole overcoring (CCBO) technique in order to limit difficulties due to core disking and microcracking. When the first, large-diameter, borehole has reached the required length, its end is given a conical shape and strain gauges are glued to this conical surface. Then overcoring proceeds in such a way as to relieve the stresses acting on the volumes surrounding the cone-shaped surface and the correlative strains are continuously recorded. This conical shape modifies the stress concentration at the end of the large-diameter borehole and limits the chances of microcracking during overcoring.

Data reduction

Displacement or strain measurements are always affected by some error, which may be defined as the difference between the measured value and an expected value. Here, the expected value is that which would have been observed for a perfect system (a homogeneous isotropic linearly elastic rock with perfectly known elastic parameters and a perfect measuring procedure). The distribution of errors is supposed to satisfy a Gaussian law and a solution is sought that minimizes the sum of the squares of errors (this is the least squares method).

The volume for which all measurements can be assumed to sample the same stress field defines the size of the REV. Within this REV, because of the assumption of linear elasticity, a linear relationship exists between the expected value for the ith measured displacement and the relieved stress components:

$$u_i^e = A_{ij}S_j, \qquad j = 1, \ldots, 6, \qquad \text{or} \qquad \boldsymbol{u}^e = [A]\boldsymbol{S} \tag{13.3}$$

where u_i^e is the expected value for the ith measured displacement, $i = 1, \ldots, N$, $N \geq 6$; $S_1 = \sigma_{11}^{\infty}, \ldots, S_4 = \sigma_{12}^{\infty}, \ldots$ are the six far-field stress components. The A_{ij} coefficients characterize the linear relationship that exists between the expected value for the corresponding measured displacement and the far-field stress components (if one far-field stress component has no influence on the corresponding displacement component then its influence coefficient is set to zero).

For cylindrical-shaped pilot holes, these coefficients are derived from equations (5.309)–(5.311). For noncylindrical geometries, numerical solutions must be obtained. If the volume where measurements have been conducted is too large for stress variations to be neglected, equation (13.3) must be modified accordingly. However, we assume in what follows that there is no stress variation within the sampled volume.

Let $\boldsymbol{u}^M$ be the N-component column vector, the ith component of which is the ith displacement measurement; $N \geq 6$. The objective is to find the six far-field stress components S_j that minimize the square of the norm E of the error vector $\boldsymbol{e}$, defined by $\boldsymbol{e} = \boldsymbol{u}^M - \boldsymbol{u}^e$. The components of $\boldsymbol{e}$ are

$$e_i = u_i^M - u_i^e = u_i^M - A_{ij}S_j, \qquad i = 1, \ldots, N, \quad j = 1, \ldots, 6 \tag{13.4}$$

The square of the norm E of $\boldsymbol{e}$ is

$$E^2 = \boldsymbol{e}^T\boldsymbol{e} = \left(u_i^M - A_{ij}S_j\right)^T \left(u_i^M - A_{ij'}S_{j'}\right) \tag{13.5}$$

$$= (u_i^M)^T u_i^M - (A_{ij}S_j)^T u_i^M - (u_i^M)^T A_{ij'}S_{j'} + (A_{ij}S_j)^T A_{ij'}S_{j'} \tag{13.6}$$

and the objective is to find the S_j, $j = 1, \ldots, 6$, values that minimize E^2. The solution is such that $\partial E^2/\partial S_k = 0$, $k = 1, \ldots, 6$:

$$\frac{\partial E^2}{\partial S_k} = -\delta_{jk}A_{ji}u_i^M - (u_i^M)^T A_{ij'}S_{j'} + \delta_{jk}A_{ji}A_{ij'}S_{j'} + (A_{ij}S_j)^T A_{ij'}\delta_{j'k} = 0 \tag{13.7}$$

which implies that

$$2A_{ki}u_i^M = 2A_{ki}A_{ij'}S_{j'}, \qquad i = 1,\ldots,N, \quad k,j' = 1,\ldots,6 \tag{13.8}$$

$$= 2B_{kj'}S_{j'}, \qquad k,j' = 1,\ldots,6 \tag{13.9}$$

with $B_{kj'} = A_{ki}A_{ij'}$. The solution is thus

$$S_{k'} = B^{-1}_{k'k}A_{ki}u_i^M, \qquad i = 1,\ldots,N, \quad k,k' = 1,\ldots,6 \tag{13.10}$$

However, as discussed in chapters 8 and 9, rocks are not linearly elastic, and the lower the applied stresses the stronger the nonlinearity. Also, stress measurements by overcoring require that all the stress components applied to the rock volume under consideration drop to zero, which is precisely the domain where nonlinearity becomes most significant. Further, most overcoring techniques apply only in boreholes less than 100 m long, although some overcoring results have been obtained in boreholes nearly 1000 m deep.

In order to mitigate these various shortcomings, alternative stress determination methods have been developed.

13.2 Stress determination from hydraulic tests in boreholes

Hydraulic tests in boreholes have proven efficient for determining stresses at great depths. Brudy *et al.* (1997) reported measurements down to a depth of 8 km, which is the deepest direct stress determination ever conducted.

Two different testing methods may be applied: hydraulic fracturing (HF) or hydraulic tests on preexisting fractures (HTPF). It will be shown in section 13.2.3 that the best results are obtained when both methods are combined.

13.2.1 Hydraulic fracturing (HF) method

When a borehole is parallel to a principal stress direction, we saw in section 12.4.1 that the minimum value for the circumferential (hoop) stress component at the borehole wall is observed in the direction of the maximum principal stress component in the plane normal to the borehole axis. This observation prompted Scheidegger (1960) and later Kehle (1964) to propose exploiting this property to determine tectonic stresses from hydraulic fracturing operations conducted in the oil industry. The concept was then further investigated in the laboratory (Haimson and Fairhurst, 1970) before being applied in the field (Haimson, 1978).

The method is presented schematically in fig. 13.3. A portion of a borehole is isolated with a straddle packer and the pressure is progressively raised in the isolated interval until a hydraulic fracture develops, at the so-called breakdown pressure. Then the fracture is extended until it reaches zones outside the domain of influence of the borehole. After the fracture stops, the hydraulic injection system is kept shut so as to monitor the subsequent

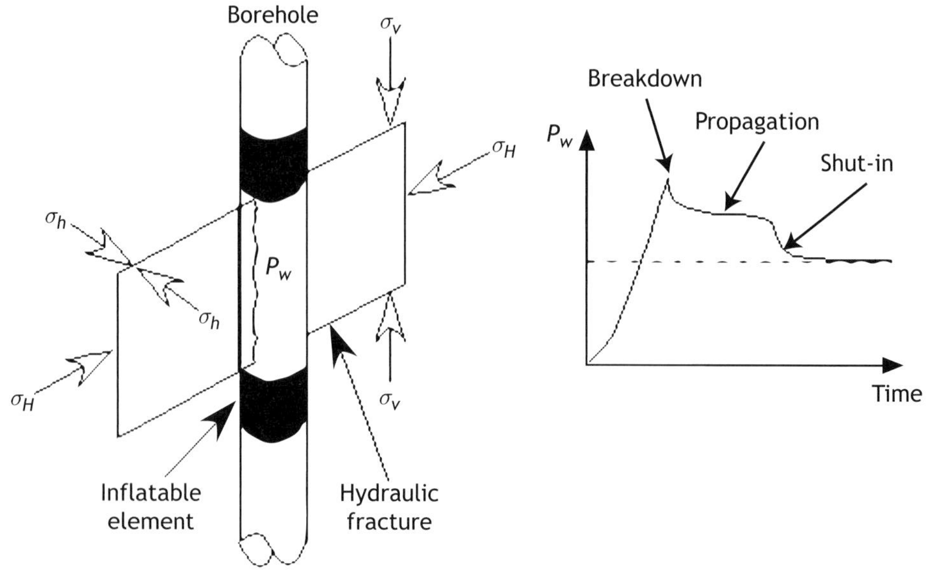

Fig. 13.3 Stress measurement by hydraulic fracturing. Left, principle of the method; right, typical pressure–time record.

pressure decay for a few minutes. This testing period is called the *shut-in* period. At the end of shut-in, the system is rapidly bled off (this takes a few seconds) and the subsequent pressure build up is observed for a few minutes. This part of the test is called flow-back and ends the first testing cycle. The testing cycle is reproduced at least twice and sometimes more than that, if the results show some drift from one cycle to the next (fig. 13.4).

On the basis of elasticity theory, analysis of the pressure records and of the hydraulic fracture orientation yields information on the regional stress field. We review first the main features of this elastic analysis (compressions are reckoned positive) assuming no porosity for the rock and therefore the absence of pore pressure.

Elastic analysis of the stress field associated with hydraulic fracturing

The stress field that prevails during hydraulic fracturing with a straddle packer is analyzed by application of the superposition principle (fig. 13.5). The packers are assumed to be far enough apart from each other that the central part of the packed-off interval may be assimilated to an infinite circular cylinder. The corresponding loading conditions are decomposed into three components.

Component I corresponds to the stress field associated with injection of water at pressure P_w and temperature T_1 into the borehole whilst the interstitial fluid away from the borehole is under a pressure P_0 and temperature T_2. The hoop stress at the borehole wall associated with this loading process is given by equation (12.100). Component II corresponds to the resulting stress when the far-field stress is $\tilde{\sigma}^{\infty}$ whilst the borehole wall is free of any loading. The corresponding stress components in the material are given by equations (5.303)–(5.308). Finally, component III results from the loading imposed by the straddle packer.

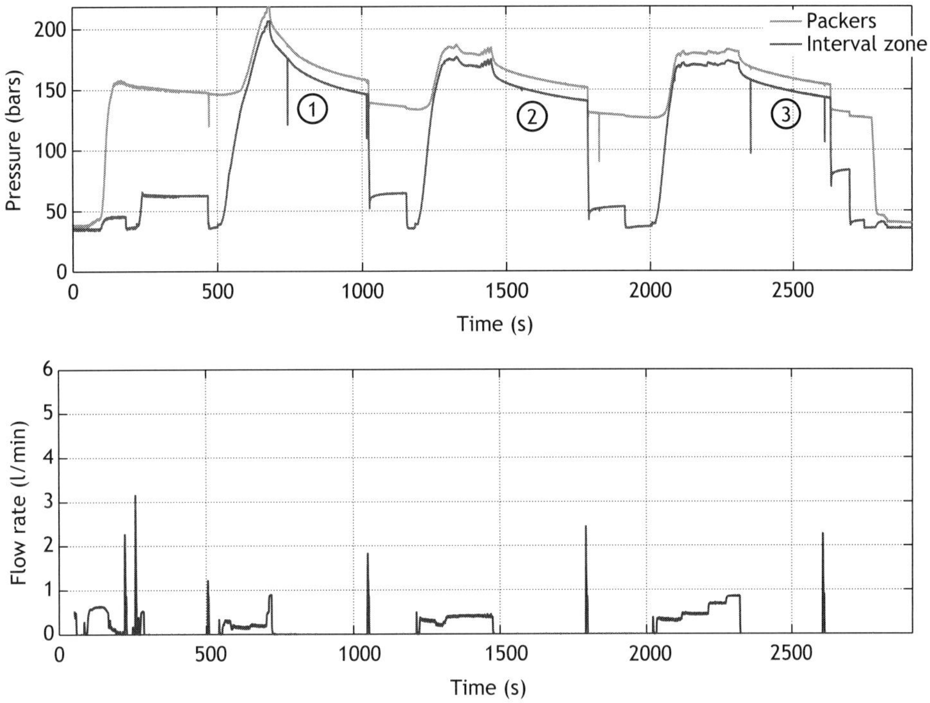

Fig. 13.4 Example of straddle packer and interval pressure as well as the flow-rate record for a hydraulic fracturing test. The short peaks of flow rate observed at the end of each cycle correspond to flow-back tests; they help to validate the assumption that the testing system is leakproof (courtesy of D. Ask).

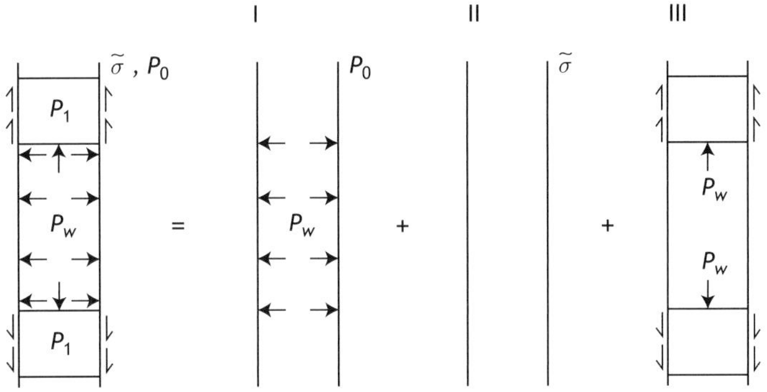

Fig. 13.5 Decomposition of the stress field generated during borehole pressurization.

Kehle (1964) considered the case in which the axial forces exerted on the packers by the fluid pressure in the borehole are balanced by uniformly distributed shear stresses along the contact between the borehole and the packers (figs. 13.6(a), (b)). He determined the axial and hoop stress components at the borehole wall, σ_{zz} and $\sigma_{\theta\theta}$, when a uniform pressure is applied along a borehole interval equal to five times the borehole diameter whilst a uniform shear stress is applied to both sides of the pressurized interval along the packer–borehole contacts.

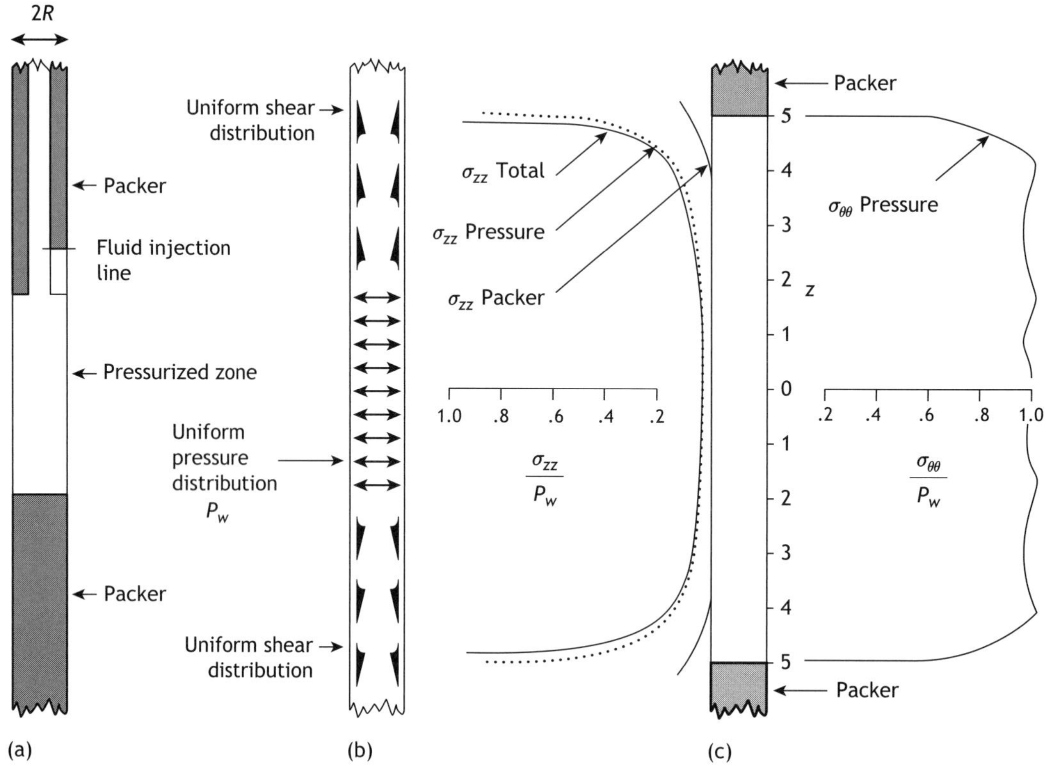

Fig. 13.6 Stress field generated by a straddle packer during a hydraulic fracturing test; only the absolute values of the stress magnitudes are shown (reproduced from Kehle, 1964, with pemission from Wiley).

The results are shown in fig. 13.6(c). They indicate that, at the fluid–packer interfaces, the axial stress $\sigma_{zz}|_{\rho=R}$ at the borehole wall is negative and has the same absolute value as the applied pressure P_w; but this axial tensile component decreases with the distance to the interface and becomes nearly zero at distances larger than three borehole diameters from the interface. The reverse is observed for the hoop stress $\sigma_{\theta\theta}|_{\rho=R}$, which is negligible close to the fluid–packer interface and is close to $-P_w$ in the center of the packed-off interval.

In summary, for a borehole of radius R inclined to the far-field principal stress directions, the stress state at the borehole wall in the center of the packed-off interval is given by

$$\sigma_{\rho\rho}|_{\rho=R} = P_w \tag{13.11}$$

$$\begin{aligned}\sigma_{\theta\theta}|_{\rho=R} &= (\sigma_{11}+\sigma_{22}) - 4\left(\frac{\sigma_{11}-\sigma_{22}}{2}\cos 2\theta + \sigma_{12}\sin 2\theta\right)\\ &\quad + \left[2P_0 - P_w - 2\eta_B(P_0-P_w) - \frac{E_B}{1-\nu_B}\alpha_T(T_2-T_1)\right]\end{aligned} \tag{13.12}$$

$$\sigma_{zz}|_{\rho=R} = \sigma_{33} - 4\nu\left(\frac{\sigma_{11}-\sigma_{22}}{2}\cos 2\theta + \sigma_{12}\sin 2\theta\right) - 2\eta_B(P_0-P_w) \tag{13.13}$$

$$\sigma_{\rho\theta}|_{\rho=R} = 0 \quad (13.14)$$

$$\sigma_{\theta z}|_{\rho=R} = 2\,(\sigma_{23}\cos\theta - \sigma_{31}\sin\theta) \quad (13.15)$$

$$\sigma_{z\rho}|_{\rho=R} = 0 \quad (13.16)$$

with

$$\eta_B = \frac{G_B}{\lambda_B + 2G_B}\alpha = \frac{1 - 2\nu_B}{2(1 - \nu_B)}\alpha$$

where σ_{ij}; $i,j = 1, 2, 3$, are the far-field stress components defined in the $\hat{\boldsymbol{I}}_i; i = 1, 2, 3$, cartesian frame of reference, z is the borehole axis parallel to the $\tilde{\boldsymbol{I}}_3$ direction and θ is the angular coordinate defined with respect to the $\hat{\boldsymbol{I}}_1$ direction.

However, following Kehle's analysis, close to the two fluid-packer interfaces (at the top and bottom of the packed-off interval) the borehole pressure P_w does not affect the hoop stress but decreases the axial stress $\sigma_{zz}|_{\rho=R}$ by equal amounts.

When the borehole is parallel to one of the principal stress directions, the component $\sigma_{\theta z}$ is zero so that $\sigma_{\rho\rho}|_{\rho=R}$, $\sigma_{\theta\theta}|_{\rho=R}$, $\sigma_{zz}|_{\rho=R}$ are the principal stress components, which we now call $\sigma_{\rho}|_{\rho=R}$, $\sigma_{\theta}|_{\rho=R}$, $\sigma_{z}|_{\rho=R}$. When the pressure is raised progressively in the packed-off section, both $\sigma_{\theta}|_{\rho=R}$ in the center of the interval and $\sigma_{z}|_{\rho=R}$ at the fluid–packer interface are decreased to become progressively tensile, until a fracture forms normal to the minimum principal stress. Whether $\sigma_{\theta}|_{\rho=R}$ or $\sigma_{z}|_{\rho=R}$ reaches this tensile strength value first depends on the relative magnitudes of the far-field principal stress components, which thus control the orientation of the hydraulic fracture observed at the borehole.

When the borehole is inclined to all the far-field principal directions, only $\sigma_{\rho}|_{\rho=R}$ is a principal direction and *en échelon* fractures inclined to the borehole axis are formed.

Identification of principal stress directions

Observation of a symmetrical axial fracture at the borehole wall implies that one of the principal stress directions is parallel to the borehole axis and that the least principal stress away from the hole is normal to this axis. In this instance, the principal stress directions away from the well are identified from the fracture orientation.

Observing that a fracture plane is normal to the borehole direction implies that the minimum principal stress is parallel to the borehole axis; therefore the two other principal stress directions are normal to the borehole axis but their orientations in this plane often remain undetermined. However, in some instances the fracture initiates parallel to the borehole axis and then rotates away from the borehole so as to become oriented normal to the far-field least, i.e. minimum, principal stress. Sometimes the fracture portion perpendicular to the borehole axis propagates back to the borehole, so that both a symmetrical axial fracture and a fracture normal to the borehole axis are observed. In such instances, all the three principal stress directions can be identified.

When *en échelon* fractures are observed, this demonstrates that none of the far-field principal stress directions is parallel to the borehole. Experience has shown that *en échelon* fractures form when the borehole axis is inclined by more than 20° to any of the far-field principal stress directions.

Formerly, fracture orientations were determined with an inflatable impression packer, a device that takes an oriented print of the borehole wall within the packed-off interval. Nowadays, however, standard geophysical imaging procedures, whether acoustic or electrical (section 1.3.2), are often used to obtain images of the complete borehole interval involved in the testing. This includes the packed-off zone and the two borehole sections where the straddle packers have been applied. Such geophysical images have revealed that often hydraulic fractures develop partly within the rock-packer contact zones, a feature of great importance for interpretation of the breakdown pressure, as will be discussed later.

Determination of the least (i.e. minimum) principal stress magnitude from shut-in or from leak-off tests (LOTs)

During fracture propagation the injection pressure is larger than the stress component normal to the fracture plane at the fracture tip, i.e. the minimum far-field principal stress if the fracture tip is far enough from the borehole. When injection stops, the pressure drops and the fracture closes when the pressure becomes lower than the normal stress magnitude. The pressure value that corresponds to the normal stress magnitude is called the *instantaneous shut-in pressure*, P_{isip}. Very robust methods have been proposed for measuring P_{isip}.

Aamodt and Kuriyagawa (1983) observed that, when the fracture is completely closed, the pressure P_w decays according to Darcy's law so that it is characterized by a negative exponential:

$$P_w = \exp(-at + b) + P_0 \tag{13.17}$$

where P_0 is the far-field pore pressure and a and b are two parameters that depend on the rock-mass hydraulic conductivity. Aamodt and Kuriyagawa plotted the logarithm of the recorded pressure versus time. When the fracture is closed the plot corresponds to a straight line, and the time at which the logarithm of the pressure decay starts becoming linear corresponds to the first time that the fracture is completely closed. The corresponding borehole pressure value yields a lower bound to the instantaneous shut-in pressure.

Hayashi and Haimson (1991) found an upper bound to the instantaneous shut-in pressure by analyzing the pressure–time record immediately after the end of injection. They noted that during fracture propagation the pressure is larger in the injection well than at the fracture tip. When injection stops, the pressure becomes uniformly distributed within the fracture, but the fracture keeps propagating. Then, because of fluid losses through the walls of the fracture, the pressure decreases uniformly and the rate of pressure decay is controlled by both fluid losses through the walls of the fracture and the decrease in fracture aperture (phase I). During phase I there is no contact between the two sides of the fracture, and this phase ends with the onset of fracture closure (contact) at the fracture tip. Phase II corresponds to the closing process of the fracture, i.e. to the period during which one side of the fracture becomes progressively in contact with the opposite side. Phase III of Hayashi and Haimson corresponds to the period when the pressure decay is controlled by Darcy flow, as had already been described by Aamodt and Kuriyagawa. Hayashi and

Haimson called P_1 the borehole pressure at the end of phase I and showed that the pressure decay observed during this phase I may be approximated as follows:

$$P_w - P_1 \simeq A\,(dt/dP_w) + C \tag{13.18}$$

where A and C are parameters which depend on the problem's geometry and on the material properties. This shows that, during phase I, the inverse of the pressure decay varies linearly with the pressure. The pressure value after which this linearity is no longer satisfied corresponds to the pressure P_1, which yields an upper bound to the instantaneous shut-in pressure.

After a fracture has been created during the first pressurization cycle, when the pressure is raised again at the beginning of the second cycle, the flow rate is sometimes decreased significantly so as to allow fluid penetration in the fracture. When the fluid pressure equals the normal stress supported by the fracture, the fracture opens and the rate of the borehole pressure increase slows down. For injection flow rates slow enough that pressure losses along the fracture may be neglected, the pressure value for which the fracture opens yields an estimate of the minimum principal stress magnitude. It is associated with the departure from linearity on the pressure–time record when the injection flow rate is constant. This testing procedure is called a leak-off test (LOT) in the oil industry (e.g. Zoback *et al.*, 2003).

An example of a large-scale leak-off test is shown in fig. 12.16. In this case no packer was used but the pressure was raised progressively during various steps that each lasted two days. At the end of the later steps, the borehole pressure was nearly equal to the pore pressure at distances larger than three to four borehole diameters (say about 40 cm), so that the borehole pressure yielded a small overestimate of the stress normal to the fracture planes intersected by the borehole. Only fractures that supported the lowest normal stress were involved, i.e. the borehole pressure yielded a small overestimate of the minimum principal stress magnitude for an interval at 2850–2900 m depth.

Interpretation of the breakdown pressure

In the best case, i.e. when the borehole is perpendicular to the minimum principal stress direction, hydraulic fracturing yields a good estimate for four of the six far-field stress components, namely the three Euler angles that characterize the three principal stress directions plus the minimum principal stress magnitude. In this case the fracture is parallel to the borehole axis and the breakdown pressure may be used to determine the magnitude of the maximum principal stress in the plane normal to the borehole axis, σ_M.

For isotropic linearly elastic rocks, when temperature effects are negligible, if σ^T is the rock's tensile strength and P_{br} is the breakdown pressure, equation (13.12) provides an expression for the hoop stress at the inception of hydraulic fracturing:

$$\sigma_\theta|_{\rho=R} = -\sigma_M + 3\sigma_m + [2P_0 - P_{br} - 2\eta_B(P_0 - P_{br})] = \sigma^T \tag{13.19}$$

where P_0 is the far-field pore pressure (see Detournay and Cheng, 1988, for a more complete discussion); σ_m is the minimum principal stress away from the borehole and has been shown to be equal to $\sigma|_{isip}$. When pore pressure effects are negligible, equation (13.19) simplifies to

$$\sigma_\theta|_{\rho=R} = -\sigma_M + 3\sigma_m - P_{br} = \sigma^T \tag{13.20}$$

So, for impervious materials, only the tensile strength of the material is required for an evaluation of the maximum principal stress normal to the borehole axis. For porous materials, the drained elastic parameters of the material must be known as well as Biot's α coefficient.

Equation (13.19) is given on the assumption that the various elastic coefficients are independent of stress, but experience has shown that when rocks are under tensile stress they become much softer than when they are under compression, and this introduces a source of error that depends very much on the size of the heterogeneities within the REV.

Also, we saw in chapter 7 that the so-called tensile strength of rocks depends on the test geometry and is strongly size and time dependent (e.g. Yamashita *et al.*, 2010).

Furthermore, identifying the time of fracture inception from the pressure–time record in order to determine the breakdown pressure has been shown to depend on the system's compliance (Ito *et al.*, 1999).

Finally, since the packer pressure must be larger than the interval pressure in order to keep the packed-off interval tight, fracture may occur at the fluid–packer interface because of fluid penetration. In fact this is generally observed, when the fracture imaging does indeed provide information on the complete interval length involved in the testing. This shows that for such tests the real breakdown pressure is not the interval fluid pressure at the time of fracture inception but that of the packer.

Because of all these various sources of uncertainty, alternative methods have been proposed for determining the missing stress components from complementary hydraulic tests.

13.2.2 Hydraulic tests on preexisting fractures (HTPF) method

When testing hydraulic fracturing in the laboratory, Cornet and Valette (1984) were able to reopen fractures normal to the applied maximum principal stress direction, provided that the injection flow rate was slow enough. The larger the applied normal stress, the lower the flow rate required to reopen the preexisting fracture without generating a fresh fracture perpendicular to the minimum principal stress direction.

Cornet and Valette proposed to generalize this testing procedure to measuring the normal stress supported by preexisting fractures with any orientation; hence the method is known as the hydraulic testing of preexisting fractures (HTPF). A straddle packer is located over a preexisting fracture and the packed-off interval pressure is very slowly raised until the fracture opens. Generally the fracture is inclined to the principal stress directions, so that the opening of a sufficiently large fracture results in a change in all the stress components except the normal stress supported by the fracture. For such fractures the instantaneous shut-in pressure yields the magnitude of the normal stress supported by the fracture away from the borehole.

An example of an HTPF record is shown in fig. 13.7. When the fracture is inclined to the borehole axis, it should be pointed out that leak-off tests are not appropriate for a normal

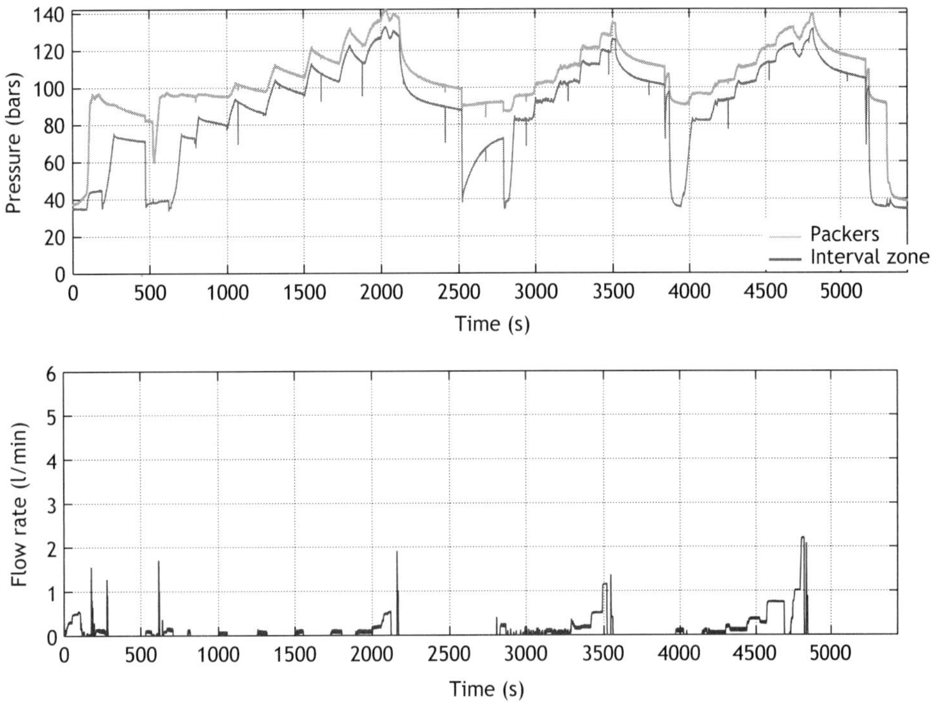

Fig. 13.7 Example of pressure and flow-rate records for a hydraulic test on a preexisting fracture (HTPF) (courtesy of D. Ask).

stress measurement. Indeed, as shown by Cornet *et al.* (2003), the normal stress supported by an inclined fracture at the borehole wall depends on the azimuth, so that initially the fracture opening is not uniformly distributed. Only when the opened part of the fracture has reached outside the domain of stress perturbation associated with the borehole does the pressure in the fracture yield a measurement of the far-field normal stress.

By measuring the normal stress supported by preexisting fractures at six different orientations, all six components of the local stress tensor may be evaluated. The HTPF method may be viewed as a generalization of the flatjack technique. If the kth preexisting fracture's orientation is characterized by that of its normal $\boldsymbol{n}^k$, the corresponding isip measurement, P^k_{isip}, yields

$$P^k_{isip} = \sigma_n = \tilde{\sigma}\boldsymbol{n}^k \cdot \boldsymbol{n}^k = \sigma_{ij} n^k_j n^k_i \tag{13.21}$$

When N ($N > 6$) fractures with six different orientations have been tested, the system of N equations may be solved by a least squares technique to determine the six stress components σ_{ij}.

Often, however, the distance between the various test sites is of the order of a few meters, if not tens of meters, so that the stress components σ_{ij} cannot be considered to be constant within the volume where measurements have been conducted. As a first approximation, the natural stress variations may be approximated by linear functions of the spatial coordinates:

$$\tilde{\sigma}(\boldsymbol{x}) = \tilde{\sigma}(\boldsymbol{x}^0) + (x_1 - x_1^0)\tilde{\alpha}^1 + (x_2 - x_2^0)\tilde{\alpha}^2 + (x_3 - x_3^0)\tilde{\alpha}^3 \tag{13.22}$$

where $\tilde{\alpha}^1$, $\tilde{\alpha}^2$, $\tilde{\alpha}^3$ are the stress gradients in the x_1, x_2, x_3 directions and $\tilde{\sigma}(\boldsymbol{x}^0)$ is the stress tensor at the point $\boldsymbol{x}^0$. Because the rock mass is at rest, the stress field $\tilde{\sigma}$ must satisfy the equilibrium condition

$$\nabla \cdot \tilde{\sigma} + \varrho \boldsymbol{g} = \boldsymbol{0} \tag{13.23}$$

where $\boldsymbol{g}$ is the acceleration due to gravity and ϱ is the local rock-mass density.

Equation (13.22) implies 4×6 unknowns whilst equation (13.23) introduces three equations but adds one more unknown, ϱ. So, when HTPF tests are conducted in a large real three-dimensional volume (where the distance between the furthest apart tests is larger than 50 m), a minimum of 30 tests is required for a reasonable evaluation of these 22 unknown parameters.

In practice measurements are usually conducted along one straight portion of a borehole, so that only the gradient along the borehole direction z is needed. Equation (13.22) then reduces to

$$\tilde{\sigma}(\boldsymbol{x}) = \tilde{\sigma}(\boldsymbol{x}^0) + (x_z - x_z^0)\tilde{\alpha}^z \tag{13.24}$$

where x_z is the axial coordinate and $\tilde{\alpha}^z$ is the stress gradient along the borehole axis.

Equation (13.24) involves 12 parameters so that, when the equilibrium conditions are introduced, only 10 unknown parameters are left to be determined. In the absence of topography and when the borehole is vertical, the vertical direction may be assumed to be a principal direction for both the stress and the stress gradient, so that now only eight unknown parameters must be determined: the magnitude of the three principal stress components at the point $\boldsymbol{x}^0$ and the azimuth of the maximum horizontal principal stress, the magnitude of the three principal components of the vertical stress gradient and the azimuth of the maximum horizontal principal component of the vertical stress gradient.

This may be further reduced to six unknowns when the density of the materials is known, so that the vertical principal stress magnitude and its vertical gradient can be deduced. Finally, if it is further assumed that there is no rotation of the horizontal principal stress directions within the depth range where measurements have been obtained, only five unknown parameters are left.

The HTPF method has been shown to be very efficient for stress evaluations in crystalline homogeneous formations, for these always exhibit a large diversity of preexisting fracture orientations. But it has been found much less efficient in sedimentary formations, where the rheological characteristics of the geomaterials vary rapidly so that spatial stress variations do not remain linear for long distances. Further, in sedimentary rocks, preexisting fractures have a tendency to heal partially so that injection tests fail to reopen these preexisting fractures but create new ones.

Combining the HF and HTPF methods exploits the advantage of one method to compensate the shortcomings of the other, as we now explain.

13.2.3 Integrating the HF and HTPF methods for complete stress determination

We saw that, even in the best case (when a borehole is parallel to a principal stress direction and normal to the least principal stress direction), the HF method yields only four of the six stress components. In this optimum configuration, the breakdown pressure may yield some information on the maximum principal stress component normal to the borehole axis but with an uncertainty that is difficult to evaluate. We also saw that the HTPF method is much less dependent on the material properties but that it assumes the continuity of the stress field within the domain where measurements are conducted. This may raise difficulties, in particular in sedimentary formations with rapid facies variations.

When the borehole is parallel to either the maximum or the intermediate principal stress direction, combining one HF test and two HTPFs conducted on two planes with different orientations yields a complete stress determination: the HF test provides all three principal stress directions and the minimum principal stress magnitude, σ_3, whilst the two HTPFs provide two linear combinations of the three principal stress components, σ_3, σ_2 and σ_1. If two to three HF tests and four to five HTPFs are conducted in such a way that the distance between the furthest apart test sites is less than 50 m, then the spatial stress variations may be neglected, and this combination of six to seven tests yields a complete stress field determination. Experience shows that, for crystalline rocks, an accuracy of the order of 10% is achieved for the principal stress magnitudes and a resolution of 10°–15° may be reached for orientation determinations. However for sedimentary rocks such a combination of tests is rarely possible, and alternative solutions have been proposed.

One possibility is to conduct HF tests in two different boreholes. The first borehole must be parallel to a principal stress direction and the other must be inclined by about 30° to any of the principal stress directions, so as to create *en échelon* fractures. Indeed, when four of the stress components are known, equations (13.12), (13.13) and (13.15), together with the discussion in section 12.4.1 may be used to interpret the azimuths of the two generatrices along which *en échelon* fractures are formed (section 12.4.1). Further, the *en échelon* fractures are normal to the minimum principal stress at the borehole wall. Their angle with the borehole axis, ω_f, yields a second equation that involves the stress magnitudes at the borehole wall (see equation (3.66)):

$$\tan \omega_f = \frac{2\sigma_{\theta z}|_{\rho=R}}{\sigma_{zz}|_{\rho=R} - \sigma_{\theta\theta}|_{\rho=R}} \tag{13.25}$$

Equation (13.25) thus yields another constraint on the far-field principal stress magnitudes.

When this method is applied to a sedimentary formation, only data acquired within the same sedimentary layer may be integrated in order to avoid difficulties with nonlinear spatial variations. In such formations the HTPF method may be used for measuring the stress component normal to the sedimentary layers.

13.3 Borehole-failure analysis for stress field characterization

As shown by equation (13.12), the hoop stress at the wall of a cylindrical borehole depends on the far-field principal stress field, on the difference between the temperature of the fluid in the borehole and the temperature of the rock, and finally on the difference between the borehole fluid pressure and the undisturbed pore pressure.

When a hole is being drilled, the far-field stresses increase with depth and the circulating drilling fluid cools down the rock. Further, both the density of the drilling fluid (mud) and the motion of the drill-string assembly in the hole generate overpressures in comparison with the original pore pressure. These overpressures and the cooling of the well generate negative axisymmetrical hoop stresses.

The variation of far-field stress components with depth results in hoop stress variations that depend on the azimuthal coordinate with respect to the direction of the maximum stress component normal to the borehole axis. In particular, when the borehole is parallel to a principal stress direction, the hoop stress at the borehole wall in the direction of the minimum principal stress component, σ_m, increases as $\sigma_{\theta,z} = 3\alpha_M^z - \alpha_m^z$, where α_M^z and α_m^z are the principal components in the direction normal to the borehole axis of the axial gradient of the far-field stress ($\alpha_M^z \geq \alpha_m^z$).

Failure occurs when the loading conditions reach a critical level. When the borehole is subparallel to a principal stress direction, the azimuthal coordinates of the failure zones coincide with those of the far-field principal stress directions.

When failure occurs in compression, it generates so-called *borehole breakouts* oriented in the direction of maximum compression at the borehole wall, which is that of the far-field minimum principal stress direction.

Bell and Gough (1979) were the first to take advantage of this systematic orientation of borehole breakouts, which they could detect with simple four-arm caliper tools. Today, with the development of borehole wall-imaging tools, the analysis of failure processes in deep boreholes has become routine and this information is systematically exploited by automatic log analysis for natural stress evaluations (e.g. Zoback *et al.*, 2003). However, great care must be taken to filter out imaging artifacts such as missing or double echoes and to differentiate borehole breakouts from drilling artifacts such as key seats (Etchecopar *et al.*, 2013).

When failure occurs in compression, it generates so-called borehole breakouts located where the loop stress is a maximum, i.e. breakouts occur in the direction of the far-field minimum principal stress. However, Morin *et al.* (1990) and Bérard and Cornet (2003) reported borehole elongations also in the direction of the maximum horizontal stress, which they attribute to thermal effects on a somewhat heterogeneous rock matrix.

Acoustic imaging tools provide accurate descriptions of the geometry of boreholes, so that both the width and depth of breakouts can be measured (fig. 13.8), and various propositions have been made to exploit these data (Zheng *et al.*, 1989; Chang and Haimson, 2000; Zoback *et al.*, 2003). However, as discussed in chapters 8 and 12, rock failure depends on time as well as on physicochemical rock–fluid interactions. For example, Wileveau *et al.* (2007) reported borehole breakouts in argillite (a shale-type material), when drilling was conducted with water-based mud, but an absence of breakouts when drilling was conducted with oil-based mud. Also, the drill string assembly usually includes some reamers (rotary cutting tools) for keeping the borehole in-gauge and these may affect the measured width of breakouts.

In order to avoid these difficulties Etchecopar *et al.* (2013) proposed to exploit the geometry of breakouts observed in wells with various orientations. Indeed, in the same way as *en échelon* tensile fractures are observed when boreholes are inclined to the principal stress

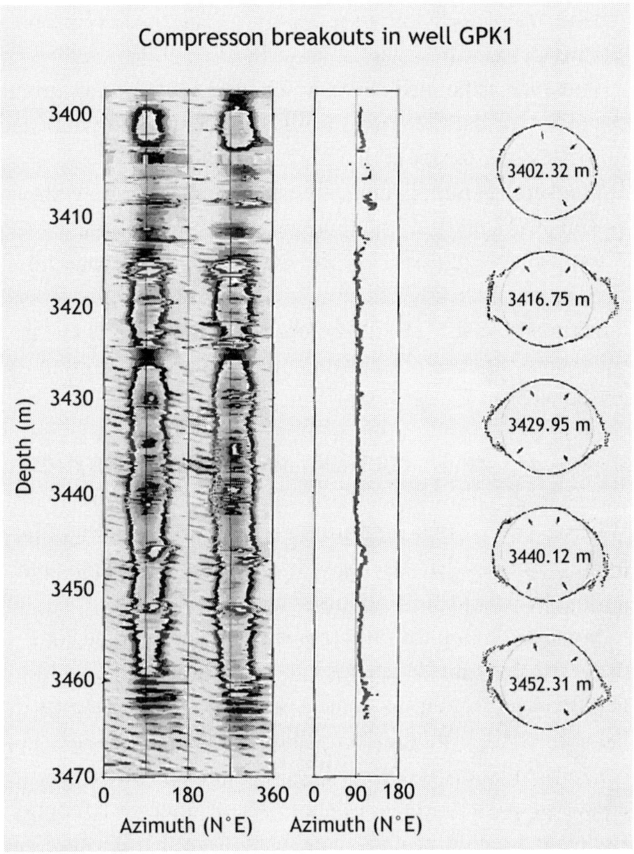

Fig. 13.8 Example of a borehole breakout observed in the Soultz-sous-forêts GPK2 well around 3400 m depth (reproduced from Cornet *et al.*, 2007, with permission from Elsevier).

directions, *en échelon* breakouts are observed in such inclined boreholes. The location of the *en échelon* breakouts at the borehole wall depends on both the far-field stress state and the relative orientation of the well. Etchecopar *et al.* showed how the geometry of the breakouts observed in inclined boreholes may be used to determine the magnitude of the maximum horizontal principal stress when these data are integrated with results from real HF tests conducted in vertical boreholes, but the integration of data requires that all observations pertain to the same sedimentary layer because of the nonlinearity of stress variation with depth in sedimentary rocks.

13.4 Stress field characteristics derived from focal plane solutions

All the above-mentioned stress determination methods require direct access to the location where the stresses are being investigated. In order to obtain some order of magnitude values for the stress fields at greater depths, geologists have proposed various approaches derived from fault analysis.

However, as pointed out by Mackenzie (1969), such structural analysis must differentiate the stress fields that exist during fault formation from those which prevail during fault reactivation. Célérier *et al.* (2012) gave a review of these various methods including those applied in structural geology. Often structural geology approaches tend to confuse stresses, as defined in chapter 3, with large-scale deformation.

We concentrate here on the evaluation of the natural stress field characteristics that may be retrieved from the analysis of earthquake focal mechanisms. This corresponds to the stress field that existed before the occurrence of the earthquake sequence considered. This stress field is not to be confused with the stress drops associated with the various earthquakes in a sequence.

13.4.1 From focal plane solutions to stress characterization

We assume first that the seismic sources are small and far enough from the locations of seismic sensors that they may be assimilated to points and, second, that they all exhibit a strong double couple component (see section 11.2).

We also assume that all seismic events considered for a stress inversion are sufficiently distant from previous events (even those not involved in the stress evaluation), so that the stress relaxation associated with any of these events does not affect the stress at the location of the events used for the stress determination. For this purpose we assume that, before the first event occurred, the volume of interest was under uniform stress. Then, once an event occurs, we draw a sphere whose radius is a function of the magnitude of the event to define a region of exclusion. Any event that occurs later within this sphere cannot be used for the stress inversion. For such events, however, we also define an exclusion sphere. Hence, as time passes the unperturbed volume decreases and sometimes may vanish nearly completely. This implies that any later event cannot be used to evaluate the original stress field.

Finally, we assume that, for each event, the shear displacement (or slip vector) at the source is parallel to the shear stress component supported by the fault (or fracture) plane before the event occurred, an assumption first proposed by Bott (1959).

Let $\boldsymbol{s}$ be a unit vector parallel and in the same direction as the seismic displacement $\boldsymbol{S}$, let $\boldsymbol{\tau}$ be the shear stress supported by the fault plane before the seismic rupture occurred and finally let $\tilde{\sigma}$ be the uniform original stress tensor in the source volume and $\boldsymbol{n}$ the unit normal to the fault plane. The resolved shear stress $\boldsymbol{\tau}$ supported by either of the nodal planes, before failure, is given by either of the following two expressions:

$$\boldsymbol{\tau} = \boldsymbol{n} \wedge [(\tilde{\sigma}\boldsymbol{n}) \wedge \boldsymbol{n}] = \tilde{\sigma}\boldsymbol{n} - (\tilde{\sigma}\boldsymbol{n} \cdot \boldsymbol{n})\boldsymbol{n} \tag{13.26}$$

so that Bott's assumption may be written as

$$\boldsymbol{s} \wedge \boldsymbol{\tau} = \boldsymbol{s} \wedge \boldsymbol{n} \wedge [(\tilde{\sigma}\boldsymbol{n}) \wedge \boldsymbol{n}] = 0 \tag{13.27}$$

as proposed by Mackenzie (1969), or also

$$\boldsymbol{s} \cdot (\boldsymbol{\tau}/|\boldsymbol{\tau}|) = 1 \tag{13.28}$$

Both expressions in (13.26) are independent of the seismic displacement amplitude and of the magnitude of the initial resolved shear stress supported by the fault plane. McKenzie showed that, for one single focal mechanism, the only restriction that exists on the orientation of the principal stress directions is that they must be within the quadrant containing the P axes. Our objective is to combine diverse focal plane solutions in order to constrain the natural stress tensor.

The stress tensor $\tilde{\sigma}$, assumed to be uniform within the source volume, may be decomposed as follows:

$$\tilde{\sigma} = \sigma_1 \tilde{\boldsymbol{I}} + (\sigma_3 - \sigma_1)\tilde{T} \tag{13.29}$$

where σ_i; $i = 1, 2, 3$, are the principal stress components, with $\sigma_1 \geq \sigma_2 \geq \sigma_3$, and

$$T_{ij} = \begin{pmatrix} 0 & 0 & 0 \\ 0 & R & 0 \\ 0 & 0 & 1 \end{pmatrix} \tag{13.30}$$

with $i, j = 1, 2, 3$ and $R = (\sigma_2 - \sigma_1)/(\sigma_3 - \sigma_1)$, $0 \leq R \leq 1$ (compressions are reckoned as positive).

The objective is to identify the three Euler angles, ψ, ϕ, θ that characterize the principal directions of $\tilde{T}$, and therefore of $\tilde{\sigma}$, as well as the parameter R, which is sometimes called the shape factor of the stress ellipsoid.

Theoretically it takes only four events, with four different focal plane solutions, to solve for these four unknowns. In practice, however, the method is generally applied to as large a set of events as possible in order to correct for the various sources of error, in particular those associated with stress heterogeneity. Thus the volume where events have occurred can be quite large, so that the uniformity hypothesis may not be valid. However, if the stress varies linearly with depth as in equation (13.24) and if the term $\tilde{\sigma}(\boldsymbol{x}^0)$ is negligible in comparison with the term $(x_z - x_z^0)\tilde{\alpha}^z$, then the stress inversion yields the principal directions and the shape factor of the vertical stress gradient $\tilde{\alpha}^z$. When the volume where events have occurred is too large for either of these two assumptions (uniformity or proportionality with depth) to be valid then the volume is subdivided into subregions where these assumptions are valid.

Focal plane solutions yield two planes, the actual fault plane and the auxiliary nodal plane perpendicular to it. The normal to one nodal plane is the slip vector in the other nodal plane, so that it takes three angles to characterize a focal mechanism as represented by the so-called beach balls (see section 11.2.1), e.g. the azimuth a and dip d of the fault plane and the angle i between the horizontal line and the slip vector in the fault plane (fig. 13.9).

However, as can be seen in fig. 11.9, the uncertainty associated with the determination of both nodal plane orientations is linked to the location of the seismic stations with respect to the radiation diagram of the source. Very often one plane is better constrained than the other, so that a focal plane solution yields six data with different associated uncertainties. One plane is usually well identified but the definition of the second plane is much less clear; thus a larger uncertainty affects the rake angle definition compared with those for the dip and dip direction of the well-defined nodal plane. The reverse applies to the other nodal plane. Stress characterization from focal plane inversions must take into account these differences in uncertainties.

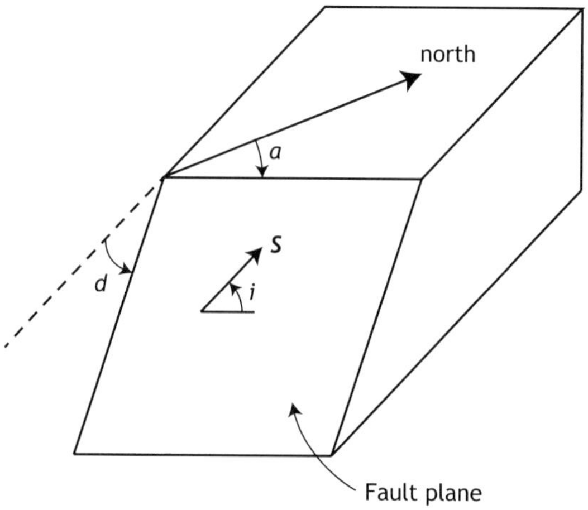

Fig. 13.9 Definition of the angles used to characterize the slip vector **S** in a fault plane. The fault is characterized by its azimuth *a* and its dip *d*. The slip vector describes the motion of the hanging wall with respect to the footwall, which is assumed to be fixed.

Sometimes (e.g. Julien and Cornet, 1989) the uncertainties are assumed to obey a Gaussian law as a first approximation. In practice, the most extreme positions for the nodal planes are identified and the solid angle between these extremes, for each nodal plane, is taken as four times the standard deviation associated with the determination (equivalent to a 95% confidence level). Hence, for each of the N focal plane solutions, the data include the six angles a_1, d_1, i_1, a_2, d_2, i_2 and the associated uncertainties ϵa_1, ϵd_1, ϵi_1, ϵa_2, ϵd_2, ϵi_2.

Identification of which of the two nodal planes is the actual fault plane remains the main difficulty for such stress inversions, and various approaches have been proposed for answering this question.

For example Angelier (2002) assumed that slip occurs parallel to the maximum shear stress in the rock mass, i.e. all seismic events are presumed to correspond to fresh ruptures that obey the Tresca criterion (see sections 8.3.1 and 9.2.1). But we saw in section 10.1.2 that, when preexisting fracture surfaces slip, the stability criterion is very similar to the Coulomb failure criterion and involves the effective normal stress supported by the fracture surfaces. Hence, depending on the local pore pressure value, the orientation of a slipping surface may be quite different from that which supports the maximum shear stress.

Michael (1984) assumed that all slipping planes fail for the same resolved shear stress magnitude. We saw in section 10.1.4, however, that the shear strength of fractures may be expressed as

$$\tau_p = \mu(\sigma_n - P_0) + C_f \tag{13.31}$$

where P_0 is the pore pressure in the fracture and C_f is the cohesion of the fracture. In many situations, e.g. when we are considering the aftershocks of an earthquake, the pore pressure varies both in time and space. Thus, the determination of the initial stress components from considerations of the shear stress magnitude as expressed in equation (13.31) requires some description of the pore pressure variation both in time and space, as well as an assumption on the value of C_f. In addition, because often the uncertainty associated with one nodal plane is very different from that associated with the other nodal plane, the domains of uncertainty associated with the shear stress magnitudes supported by the two nodal planes overlap.

Gephart and Forsyth (1984) developed a method that only requires Bott's hypothesis in order to identify the fault plane out of the two nodal planes and does not involve the magnitude of the shear stress components. Further it, also takes into account the uncertainties in the nodal plane orientation determinations. Below, we describe their method, step by step.

13.4.2 The stress determination method of Gephart and Forsyth

As already stated, the method of Gephart and Forsyth rests on three basic assumptions:

1. the seismic slip is parallel to the resolved shear stress supported by a fracture (or fault) plane before slip occurs;
2. the stress field is uniform throughout the volume within which all events included in the inversion have occurred;
3. the events are independent of each other, i.e. the volume within which the stress is being relaxed by a given event has not been perturbed by any previous event and therefore the stress before the event corresponds to the original natural stress.

The objective of inversion is to determine the three Euler angles that characterize the principal stress directions and the ratio R, $0 \leq R \leq 1$, introduced in equation (13.30).

Principle of the method

The unit vectors $\boldsymbol{n}$ and $\boldsymbol{s}$ are respectively normal to the fault plane and parallel to the slip vector in this fault plane. Their coordinates in the frame of reference B_G, with base vectors $\hat{\boldsymbol{I}}'_i; i = 1, 2, 3$, such that $\hat{\boldsymbol{I}}'_1$ is oriented to the north, $\hat{\boldsymbol{I}}'_2$ is oriented to the east and $\hat{\boldsymbol{I}}'_3$ is vertical positive downward, are respectively

$$\boldsymbol{n} = \begin{pmatrix} -\sin a \sin d \\ \cos a \cos d \\ -\cos d \end{pmatrix} \quad \text{and} \quad \boldsymbol{s} = \begin{pmatrix} \cos a \cos i + \sin a \cos d \sin i \\ \sin a \cos i - \cos a \cos d \sin i \\ -\sin d \sin i \end{pmatrix} \tag{13.32}$$

Let B_T be the base defined by the three eigenvectors $\boldsymbol{e}_i; i = 1, 2, 3$, of the stress tensor $\tilde{\sigma}$ and define a cartesian frame of reference B_F with its three orthogonal unit vectors $\hat{\boldsymbol{I}}_i; i = 1, 2, 3$, associated with the fault that slipped in such a way that $\hat{\boldsymbol{I}}_1$ is parallel to $\boldsymbol{n}$, $\hat{\boldsymbol{I}}_2$ is parallel to $\boldsymbol{n} \wedge \boldsymbol{s}$ and $\hat{\boldsymbol{I}}_3$ is parallel to $\boldsymbol{s}$. Let Q_{ij} be the direction cosines of the unit vectors $\hat{\boldsymbol{I}}_i; i = 1, 2, 3$, of the base B_F with respect to the base B_T.

Since the slip vector in the fault plane is assumed to be parallel to the resolved shear stress supported by the fault, the component σ'_{12} of the stress tensor $\tilde{\sigma}$ expressed in the B_F frame of reference is zero,

$$\sigma'_{12} = Q_{11}Q_{21}\sigma_1 + Q_{12}Q_{22}\sigma_2 + Q_{13}Q_{23}\sigma_1 = 0 \tag{13.33}$$

but the two vectors $\hat{\boldsymbol{I}}_1$ and $\hat{\boldsymbol{I}}_2$ are perpendicular to each other, so that

$$Q_{11}Q_{21} + Q_{12}Q_{22} + Q_{13}Q_{23} = 0 \tag{13.34}$$

from which we conclude that

$$R = \frac{\sigma_2 - \sigma_1}{\sigma_3 - \sigma_1} = -\frac{Q_{13}Q_{23}}{Q_{12}Q_{22}} \tag{13.35}$$

Equation (13.35) expresses the relationship that must exist between the principal stress magnitudes for slip to occur parallel to the observed slip vector. In addition, as mentioned above, the slip vector in the fault plane must have the same direction as the resolved shear stress $\boldsymbol{\tau}$ in this plane.

Definition of the misfit and identification of the fault plane

Consider any stress tensor $\tilde{T}^m$ and the slip vector s^n observed in the nth fault plane. The resolved shear stress associated with $\tilde{T}^m$ for that fault plane and the slip vector are not parallel unless $\tilde{T}^m$ itself is precisely the solution. Angelier and Goguel (1979) proposed defining the misfit between a stress tensor and the slip vector in a fault plane by the rotation angle around the normal to the fault plane that brings into coincidence the slip vector and the resolved shear stress.

However, Gephart and Forsyth (1984) pointed out that, first, some uncertainty is associated with the fault plane orientation determination and, second, that other rotation axes should be considered as well. Hence they proposed as the definition of the misfit the smallest rotation of the fault plane that brings into coincidence the observed slip vector in the fault with the resolved shear stress supported by the plane. For a given fault plane solution (with two nodal planes) and a given stress tensor, the fault plane is selected to be the nodal plane that requires the smallest rotation angle.

We may point out here that, nowadays, it is recognized that many induced microseismic events are multiplets (see section 11.2.3), and this fact may be used for effective direct fault plane identification (e.g. Gaucher *et al.*, 1998).

Normative measure of misfit

Gephart and Forsyth (1984) pointed out that the choice of fault plane out of the two nodal planes is either right or wrong. The associated error does not satisfy a Gaussian law, and therefore minimization of the least squares sum of misfits (the L^2 norm) is not appropriate since it gives too much weight to outliers. They proposed to consider instead the L^1 norm, i.e. the sum of the absolute value of errors, as a measure of the misfit between a given tensor and the set of focal plane solutions.

Identifying the best solution

The solution requires the determination of the factor R as well as the three Euler angles that characterize the orientation of the principal stress directions. The three variables ψ, ϕ, and θ vary between 0 and π whilst R varies between 0 and 1. The domain of all possible solutions remains therefore limited and lends itself to a systematic grid search approach. Gephart and Forsyth proposed to proceed in two steps. First an approximate solution is searched for, in order to identify the fault plane for each focal plane solution. This is achieved by considering, for each stress tensor generated with the grid search and for each focal mechanism, rotations with respect to $\boldsymbol{n}$, to $\boldsymbol{n} \wedge \boldsymbol{s}$ and to $\boldsymbol{s}$. The nodal plane that yields the smallest of all six rotations is selected as the fault plane for the corresponding stress tensor $\tilde{T}$. Once the set of fault planes is identified and the domain of possible solutions has been significantly reduced, a best solution is identified by exploring systematically the restricted domain of solutions through a grid search.

Only the approximate method is described below. We determine the rotation angles with respect to $\boldsymbol{n}$ (angle χ_1), to $\boldsymbol{s} \wedge \boldsymbol{n}$ (angle χ_2) and to $\boldsymbol{s}$ (angle χ_3) that bring $\boldsymbol{s}$ antiparallel to $\boldsymbol{\tau}_0$, the shear stress in the fault plane associated with the tensor $\tilde{T}$. Indeed, the factor $\sigma_3 - \sigma_1$ in equation (13.29) is negative, so that the shear stress component $\boldsymbol{\tau}_0$ is in the opposite direction to the shear stress component $\boldsymbol{\tau}$ associated with $\tilde{\sigma}$. The code developed by Gephart and Forsyth for this approximate solution, as well as that for the exact solution, which is applied after the approximate solution has been identified, is available on the internet at http://www.pages-perso-bernard-celerier.univ-montp2.fr/software/fsweb.software.html.

Rotation of angle χ_1 with respect to *n*. With this rotation, the B_F base, i.e. $(\boldsymbol{n}, \boldsymbol{s} \wedge \boldsymbol{n}, \boldsymbol{s})$, becomes the B_{χ_1} base, i.e. $(\boldsymbol{n}, [\boldsymbol{s} \wedge \boldsymbol{n}]_{\chi_1}, \boldsymbol{s}_{\chi_1})$, after application of the rotation matrix $[\alpha_1]$ defined by

$$[\alpha_1] = \begin{pmatrix} 1 & 0 & 0 \\ 0 & \cos\chi_1 & -\sin\chi_1 \\ 0 & \sin\chi_1 & \cos\chi_1 \end{pmatrix} \tag{13.36}$$

If $[Q]$ is the rotation matrix that transforms the base $B_{\tilde{T}}$ of tensor $\tilde{T}$, into the fault-plane-related base B_F then a vector $\boldsymbol{v}$ in these bases may be written as follows:

$$\boldsymbol{v}_{B_F} = [\alpha_1]\boldsymbol{v}_{\chi_1} \tag{13.37}$$

$$\boldsymbol{v}_{B_T} = [Q]^T \boldsymbol{v}_{B_F} \tag{13.38}$$

so that

$$\boldsymbol{v}_{B_T} = [Q]^T[\alpha_1]\boldsymbol{v}_{\chi_1} = [Q']\boldsymbol{v}_{\chi_1} \tag{13.39}$$

with $[Q'] = [Q]^T[\alpha_1]$.

Hence the compatibility condition (13.35), which expresses that the slip vector and the resolved shear stress are parallel after rotation, becomes

$$R = -\frac{Q'_{13}Q'_{23}}{Q'_{12}Q'_{22}} \tag{13.40}$$

However, we have

$$Q'_{13} = Q_{13} \tag{13.41}$$

$$Q'_{23} = \cos\chi_1 Q_{23} + \sin\chi_1 Q_{33} \tag{13.42}$$

$$Q'_{12} = Q_{12} \tag{13.43}$$
$$Q'_{22} = \cos\chi_1 Q_{22} + \sin\chi_1 Q_{32} \tag{13.44}$$

so that

$$-RQ_{12}(\cos\chi_1 Q_{22} + \sin\chi_1 Q_{32}) = Q_{13}(\cos\chi_1 Q_{23} + \sin\chi_1 Q_{33}) \tag{13.45}$$

which yields

$$\chi_1 = -\arctan\left(\frac{RQ_{12}Q_{22} + Q_{13}Q_{23}}{RQ_{12}Q_{32} + Q_{13}Q_{23}}\right) + k\pi, \qquad k = 0, 1, 2, \ldots \tag{13.46}$$

Rotation of angle χ_2 with respect to $\boldsymbol{s} \wedge \boldsymbol{n}$. The rotation matrix is

$$[\alpha_2] = \begin{bmatrix} \cos\chi_2 & 0 & \sin\chi_2 \\ 0 & 1 & 0 \\ -\sin\chi_2 & 0 & \cos\chi_2 \end{bmatrix} \tag{13.47}$$

so, if $[Q''] \equiv [\alpha_2]^T Q$, with

$$Q''_{13} = \cos\chi_2 Q_{13} - \sin\chi_2 Q_{33} \tag{13.48}$$
$$Q''_{23} = Q_{23} \tag{13.49}$$
$$Q''_{12} = \cos\chi_2 Q_{12} - \sin\chi_2 Q_{32} \tag{13.50}$$
$$Q''_{22} = Q_{22} \tag{13.51}$$

we find

$$\chi_2 = \arctan\left(\frac{RQ_{12}Q_{22} + Q_{23}Q_{13}}{RQ_{22}Q_{32} + Q_{23}Q_{33}}\right) + k\pi, \qquad k = 0, 1, 2, \ldots \tag{13.52}$$

Rotation of angle χ_3 with respect to $\boldsymbol{s}$. The rotation matrix is

$$[\alpha_3] = \begin{bmatrix} \cos\chi_3 & -\sin\chi_3 & 0 \\ \sin\chi_3 & \cos\chi_3 & 0 \\ 0 & 0 & 1 \end{bmatrix} \tag{13.53}$$

so if $[Q'''] \equiv [\alpha_3]^T [Q]$, with

$$Q'''_{13} = \cos\chi_3 Q_{13} - \sin\chi_3 Q_{23} \tag{13.54}$$
$$Q'''_{23} = -\sin\chi_3 Q_{13} + \cos\chi_3 Q_{23} \tag{13.55}$$
$$Q'''_{12} = \cos\chi_3 Q_{12} - \sin\chi_3 Q_{22} \tag{13.56}$$
$$Q'''_{22} = -\sin\chi_3 Q_{12} - \cos\chi_3 Q_{22} \tag{13.57}$$

the compatibility condition becomes

$$(\cos\chi_3 Q_{13} + \sin\chi_3 Q_{23})(-\sin\chi_3 Q_{13} + \cos\chi_3 Q_{23}) \tag{13.58}$$
$$= -R(\cos\chi_3 Q_{12} + \sin\chi_3 Q_{22})(-\sin\chi_3 Q_{12} + \cos\chi_3 Q_{22}) \tag{13.59}$$

which yields

$$\chi_3 = \frac{1}{2}\arctan\frac{2}{A} \tag{13.60}$$

with

$$A = \frac{R(Q_{12}^2 - Q_{22}^2) + Q_{13}^2 - Q_{23}^2}{RQ_{12}Q_{22} + Q_{13}Q_{23}} + k\frac{\pi}{2}, \qquad k = 0, 1, 2, \ldots \tag{13.61}$$

Characterizing the uncertainty of the solution

Once a best solution has been identified, its confidence limits must be evaluated. Gephart and Forsyth adopted the characterization of the uncertainty by Parker and McNutt (1980); this was defined by the quantity m:

$$m = \sum_{j=1}^{n} \frac{|x_j|}{\sigma_j} \tag{13.62}$$

where x_j is the misfit of the jth focal mechanism and σ_j is the standard deviation, which is unknown. Gephart and Forsyth proposed that σ_j is the same for all models and evaluated it from the best model, i.e. the model that minimizes $\sum_{j=1}^{n} |x_j|$. From Parker and McNutt (1980) they concluded that the standard deviation, σ_{est}, is given by

$$\sigma_{est} = \frac{\sum_{min}}{(2/\pi)^{1/2}(n-k)} \tag{13.63}$$

where k is the number of parameters that characterize the model (here $k = 4$) and $\sum_{min}$ is the value of $\sum_{j=1}^{n} |x_j|$ for the best model. This estimated standard deviation may be used (see Revets, 2010) to characterize the uncertainty level according to Fisher's statistics for the L^1 norm (Fisher *et al.*, 1987) .

13.4.3 Integrating focal plane solutions with results from hydraulic tests in boreholes for pore pressure mapping

As shown in fig. 12.16, the injection of water under pressure into rock masses generates some induced seismicity and various authors (e.g. Cornet and Julien, 1989; Dorbath *et al.*, 2010) have attempted to invert focal plane solutions from such induced seismicity in order to determine the reduced stress tensor within the volume where microseismic activity has been observed. But flow occurs along preexisting channels that sometimes constitute zones of local stress heterogeneity (Cornet *et al.*, 2007) and results from these stress evaluations have often yielded results very different from those produced by hydraulic tests or borehole breakout-orientation interpretations. However, these various data have been found to be consistent with one another provided that heterogeneous focal mechanisms are eliminated from the data collection.

A way of overcoming the difficulty of identifying zones of local stress heterogeneity was developed by Yin and Cornet (1994), who integrated in one single inversion procedure the results from hydraulic tests and from focal mechanisms. Indeed, one difficulty raised by the determination of a reduced stress tensor from a collection of focal plane solutions is the selection of the fault plane from the two possible nodal planes. But, as proposed by Gephart and Forsyth (1984), this difficulty disappears if the fault plane is considered to be the nodal plane that requires the smaller rotation to bring into coincidence the resolved shear stress supported by the plane and the observed slip vector. During the inversion procedure, when such rotations are found to be larger than the values permitted by the uncertainty in the nodal plane orientation, the corresponding mechanism is deemed "heterogeneous" (see the start of section 14.1.2) with the corresponding stress tensor. That mechanism is then eliminated from the data set for this specific solution. Correspondingly the best solution is identified as that which minimizes the misfit measure and which is consistent with the largest number of data.

The inversion proceeds in successive steps. First the stress field is approximated by the linear function introduced in equation (13.24), which involves only five unknown parameters when the vertical stress component is a principal component equal to the weight of overburden:

$$\tilde{\sigma}(\boldsymbol{x}) = \tilde{\sigma}(\boldsymbol{x}^0) + (x_z - x_z^0)\tilde{\alpha}^z$$

More parameters may be used to describe the stress field, depending on the *a priori* information available for the site.

We call model m the set of unknown parameters that characterizes the stress field. A nondimensional misfit function φ_m is associated with model m so as to characterize the quality of its fit with the complete data set (M focal mechanisms plus N results from HTPFs). This misfit function is made up of two parts, one for the results from HTPFs, φ_h, and one for the focal mechanisms, φ_f.

For HF tests the values obtained in the tests (the magnitude and orientation of the minimum principal stress component) are used to define the domain of *a priori* values to be investigated for the various models.

The misfit function for HTPFs is

$$\varphi_h = \sum_{i=1}^{N} \frac{|\sigma_{no}^i - \sigma_{nc}^i|}{\delta_h^i + \delta_r^i} \tag{13.64}$$

where σ_{no}^i is the ith observed normal stress component and σ_{nc}^i is the computed value for model m, whilst δ_h^i is the uncertainty associated with the normal stress determination (taken as equal to 2 or 3 standard deviations) and δ_r^i is the normal stress component variation associated with the maximum rotation of the plane compatible with the uncertainty limits of the orientation determination.

For focal mechanisms, the misfit function is

$$\varphi_f = \sum_{i=1}^{M} \frac{|\chi_i|}{\delta_f^i} \tag{13.65}$$

where χ_i is the rotation angle required to bring into conformity the resolved shear stress direction and the observed slip vector for the ith focal plane solution, whilst δ_f^i is the standard deviation associated with the evaluation of the nodal plane geometry.

Because in general many more focal plane solutions are available in the data set than HTPFs, weight factors must be introduced into the global misfit function. First the two data sets are inverted separately so as to identify, for each data set, the largest number of data (N_c^{max} and M_c^{max}) consistent with a stress model and the corresponding minimum values of the misfit (φ_h^{min} and φ_f^{min}), as defined by equations (13.64) and (13.65). Then we define the misfit φ_m that characterizes model m by

$$\varphi_m = \frac{1}{2}\frac{N_c^{max}}{N_c}\frac{\varphi_h}{\varphi_h^{min}} + \frac{1}{2}\frac{M_c^{max}}{M_c}\frac{\varphi_f}{\varphi_f^{min}} \tag{13.66}$$

where N_c and M_c are respectively the number of HTPFs and the number of focal plane solutions consistent with model m.

The solution is the model which minimizes φ_m. The solution may be solved by a Monte-Carlo technique (using intensive random sampling of the model space) or an improved Monte-Carlo method such as the genetic algorithm proposed by Yin and Cornet (1994).

Once the solution has been identified the stress field is now known at the location of all focal mechanisms consistent with this stress field and the fault plane orientation is identified. Consequently the shear stress supported by the various fault planes may be computed. Then the pore pressure required to induce slip on these planes at these locations may be estimated from Coulomb's effective stress criterion for a range of reasonable friction coefficients (e.g. 0.7–0.9). A plot of these various pore pressure values may be used to characterize the percolation process in the rock mass (Cornet and Yin, 1995).

13.5 Stress fields and seismic wave velocity anisotropy

We discussed in section 9.1.1 how microcracks may influence the elastic response of rocks and in section 10.1.1 how the compliance of fractures depends on the ambient stress field. We concluded that the elastic response of rock masses is generally nonlinear and that the lower the effective stresses, the stronger the nonlinearity. But we observed that, for effective mean stress components larger than a few hundred MPa, this nonlinearity becomes negligible.

Hence a lot of attention has been devoted during the last 20 years to the relationships between seismic wave velocities and stress field characteristics for rock masses in the upper 10 km of the crust. The idea is to exploit the fact that fractures and microcracks normal to the maximum principal stress direction are more *closed* than those normal to the minimum principal stress direction. In other words, when a triaxial stress is applied to a rock sample in which microcracks are randomly oriented the elastic response of the material becomes anisotropic, with the direction of fastest wave propagation parallel to the maximum principal stress direction.

In particular, we saw in section 5.3.5 that for isotropic materials the shear wave velocity depends on the shear modulus of the material. For anisotropic materials in which the shear modulus exhibits different values depending on direction, different shear waves appear depending on their direction of polarization. This effect is called the splitting of shear waves. When the rock anisotropy is only dependent on the applied stress field, the splitting of shear waves, i.e. the detection of different shear wave arrivals depending on their polarization direction, may be used to identify the orientation of the maximum principal stress direction in the rock mass where the recording station is located.

Understanding the effect of stress on the elastic response of a material requires a description of its nonlinear stress–strain relationship (e.g. Johnson and Rasolofosaon, 1996). Following the same reasoning as in section 5.4.2, we now keep the third-order term in the expression for the strain energy density w:

$$w = \tfrac{1}{2} C_{ijkl} E_{ij} E_{kl} + \tfrac{1}{6} C_{ijklmn} E_{ij} E_{kl} E_{mn}, \qquad i,j,k,l,m,n = 1,2,3 \tag{13.67}$$

where the strain component E_{ij} refers to the finite-strain tensor component,

$$E_{ij} = u_{i,j} + u_{j,i} + u_{k,i} u_{k,j}$$

defined in equation (4.41), with u_i the ith component of the displacement.

After analyzing previously published experimental results, Johnson and Rasalofosaon concluded that the third-order constants, C_{ijklmn}, are much larger than the second-order constants, C_{ijkl}. As a consequence the stress-induced anisotropy can become large. These properties have been exploited for determining the stress components from sonic log observations (e.g. Lei *et al.*, 2012).

Nonlinear elasticity is beyond the scope of this textbook, and the reader interested in anisotropy-related stress is referred to the publications by Johnson and Rasolofosaon (1996) and Lei *et al.* (2012).

13.6 Further reading

Amadei, B., and Stephansson, 0., 1997. *Rock Stress and Its Measurements.* Chapman & Hall, 490 pp.

Zoback, M. D., 2007. *Reservoir Geomechanics.* Cambridge University Press, 449 pp.

14 Elements of stress fields and crustal rheology

An evaluation of the relative significance of fractures and faults for the deformation processes of rock masses is provided by an examination of the stress field within the volume of interest.

When stress perturbations associated with these surfaces of weakness are only local, so that the stress field may be modeled by continuous functions defined within the whole volume of concern, it may be concluded that continuum mechanics provides a sound paradigm for understanding geomechanics issues. However, when the definition of continuous functions for the description of the regional stress field becomes too approximate, a different approach, such as for example the discrete element method (Cundall, 1988; Hart *et al.*, 1988; Shi and Goodman, 1988), may be found more appropriate.

We discuss below four examples of the stress field evaluations that have been conducted at different length scales. The objective is to explore whether the concept of stress as defined in continuum mechanics has been shown to be helpful for these examples and how stress field modeling may shed some light on loading, boundary conditions and rheological behavior.

We present, first, the results from a stress determination project conducted for the design of an underground hydraulic power station in northern Portugal (Figueiredo *et al.*, 2014). The scale is in the km^3 range, the material is a homogeneous granite and the region is mountainous.

Then we consider a stress determination conducted in the sedimentary Paris Basin (Wileveau *et al.*, 2007; Cornet and Röckel, 2012). The volume of interest is about 15 km $\times$ 15 km $\times$ 0.8 km and involves a series of limestones, shales and clays.

This discussion is followed by a presentation of results gathered in the upper Rhine graben, in the vicinity of the city of Basel in Switzerland, where data are available for a volume approximately equal to 50 km $\times$ 50 km $\times$ 15 km (Cornet and Burlet, 1992; Valley and Evans, 2009; Maury *et al.*, 2013). These results illustrate how the stress field in the upper kilometer of the crust is decoupled from that which prevails in the basement crystalline rock and how the data from a deep borehole are consistent with those of a focal mechanism inversion.

Finally we discuss a stress estimate conducted in a 350 km $\times$ 400 km $\times$ 250 km lithospheric domain located to the northwest of the European Alps (Maury *et al.*, 2014). It includes the Basel area, where a 6.5 ± 0.5 M_w-magnitude earthquake occurred in 1356 AD, but where no present-day horizontal motion of tectonic origin has been detected from GPS observations conducted over the last three decades (Nocquet, 2012).

These various examples illustrate stress determination methods and provide support for introducing ongoing debates on the interpretation of regional stress fields.

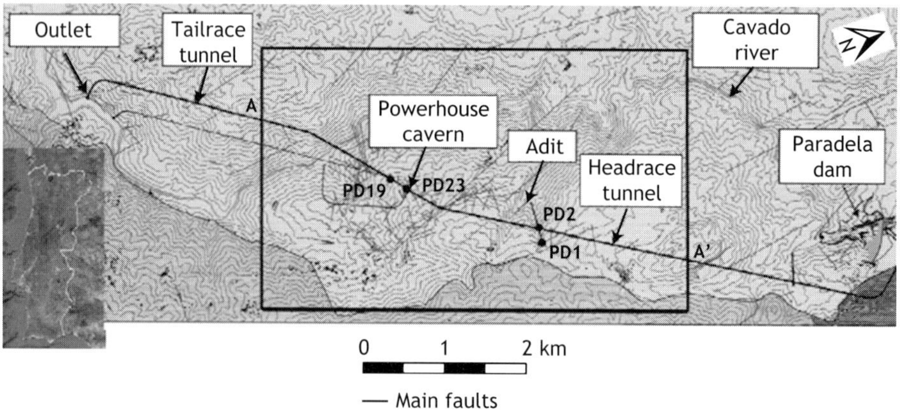

Fig. 14.1 Layout of the Paradela II hydroelectric scheme (courtesy of Energy of Portugal). The rectangular frame corresponds to the vertical limits of the numerical model shown in fig. 14.9. The mapped faults are indicated.

14.1 A stress field evaluation in a mountainous granite massif of northern Portugal

The repowering scheme of the Paradela II hydroelectric infrastructure developed on the Cávado river, in northern Portugal, has required an evaluation of the regional stress field (Figueiredo *et al.*, 2014). First, an *in situ* measurement campaign was designed to produce complementary data sets from different measurement methods. Then these various data sets have been integrated into a simple stress model for evaluating the stress at the future excavation locations. We present first the various stress measurements conducted on site and then discuss the integration method. Finally we conclude by discussing the rheology of this granite rock mass.

14.1.1 Results from the stress determination program

Different techniques have been used for evaluating the stress field: hydraulic tests were conducted in two 500 m deep boreholes located some 100 m apart from each other (PD19 and PD23 in fig. 14.1), whilst both overcoring and flatjack tests were conducted in a horizontal adit located some 1.7 km east of these vertical wells (boreholes PD1 and PD2 for overcoring and sites SFJ1, SFJ2 and SFJ3 for flatjack tests in figs. 14.1 and 14.2).

Hydraulic tests provided an opportunity to obtain precise data on the variation of stress with depth, whilst overcoring and flatjack tests explored the possibility of extrapolating results from the vertical boreholes to horizontal distances reaching a few kilometers.

The hydraulic tests included both hydraulic fracturing (HF) and hydraulic tests on preexisting fractures (HTPF). The results are presented in figs. 14.3 and 14.4.

In fig. 14.5 we plot the results from the instantaneous shut-in pressure measurements versus depth for all tests. For the HF tests these results correspond to the minimum principal stress magnitude. For HTPFs they correspond to the normal stress supported by the tested fractures, the orientations of which were identified after testing (see section 13.2).

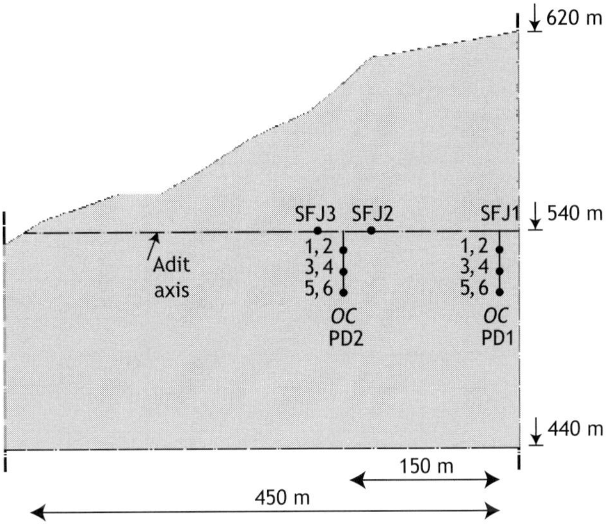

Fig. 14.2 Location of overcoring and flatjack tests run in the horizontal adit shown in fig. 14.1 (courtesy of B. Figueiredo).

Test	Type	Depth		Azimuth		Dip		Normal stress	
		z (m)	δz (m)	ϕ (°)	$\delta\phi$ (°)	θ (°)	$\delta\theta$ (°)	σ_n (MPa)	$\delta\sigma_n$ (MPa)
1	HF	471.8	0.5	108	4	87	2	10.3	0.4
2	HF	455.5	0.5	18	7	90	2	9.0	0.2
3	HF	455.5	0.5	126	5	90	2	7.8	0.2
	HTPF	450.4	0.5	281	6	60	2		
4	HF	442.1	0.5	133	5	90	2	9.0	0.2
	HTPF	442.2	0.5	303	5	42	2		
5	HTPF	436.3	0.5	108	6	32	2	8.9	0.2
6	HF	414.9	0.5	133	4	90	2	7.1	0.1
7	HTPF	393.4	0.5	270	4	66	2	7.3	0.3
		393.9	0.5	284	3	79	2		
		394.1	0.5	277	5	52	2		
		394.3	0.5	50	5	38	2		
8	HTPF	379.3	0.5	88	3	57	2	5.6	0.2
9	HF	335.6	0.5	119	3	61	2	6.7	0.1
10	HTPF	293.1	0.5	14	2	44	3	7.5	0.3
11	HF	279.8	0.5	22	4	90	2	5.8	0.1
12	HF	164.6	0.5	320	7	86	2	2.6	0.3

Fig. 14.3 The results of the hydraulic tests in borehole PD19.

Test	Type	Depth		Azimuth		Dip		Normal stress	
		z (m)	δz (m)	ϕ (°)	$\delta\phi$ (°)	θ (°)	$\delta\theta$ (°)	σ_n (MPa)	$\delta\sigma_n$ (MPa)
1	HF	490.7	0.5	199	5	90	2	20.2	0.5
2	HF	421.8	0.5	19	5	90	2	9.9	0.2
	HTPF	420.9		33	5	42	2		
3	HF	402.4	0.5	348	5	68	2	8.9	0.2
	HTPF	402.5		252	5	34	3		
4	HTPF	377.4*	0.5	215*	5*	81*	3*	9.7	0.2
		377.8*		106*	5*	12*	3*		
		377.8*		46*	5*	20*	3*		
		378.4		2	5	35	2		
		377.9		137	5	22	2		
		377.9		59	5	24	2		
		377.7		358	10	37	2		
5	HF	364.7*	0.5	325*	5*	81*	3*	7.0	0.2
	HTPF	364.6		252	5	35	2		
6	HI	356.8	0.5	269	4	90	2	5.9	0.1
	HTPF	357.0*		33*	5*	6*	3*		
7	HF	176.6	0-5	243	3	77	2	3.2	0.2

Fig. 14.4 Results from hydraulic tests in borehole PD23.

The mean standard deviation of the normal stress measurements is equal to 0.22 MPa, so that the 99% confidence level spans about 1.3 MPa, (i.e. $\pm 3 \times 0.22$ MPa) for these normal stress evaluations.

The orientations of hydraulic fractures are shown in fig. 14.6 and correspond to the local minimum principal stress orientations. All 14 hydraulic fractures except for two were within 15° of the vertical direction. The two nonvertical fractures were inclined at respectively 22° and 29° to the vertical direction and were in the vicinity of large preexisting fractures that apparently had locally altered the principal stress directions. Hence it may be concluded that the vertical direction is close to being a principal stress direction within the depth-range of testing.

Further, we note that whilst the principal stress magnitudes are very similar for tests conducted within the same depth-range (say within 50 m), the orientations of fractures vary by as much as 90°. This suggests that the two horizontal principal stress components are sub-equal (recall that when two eigenvalues are equal, any vector in the plane normal to the eigenvector associated with the third eigenvalue is an eigenvector).

The results of the stress determination from the overcoring operations are presented in fig. 14.7, whilst the results from the flatjack tests are presented in fig. 14.8. These two types of stress measurement assume that the rock behaves elastically and, further, for

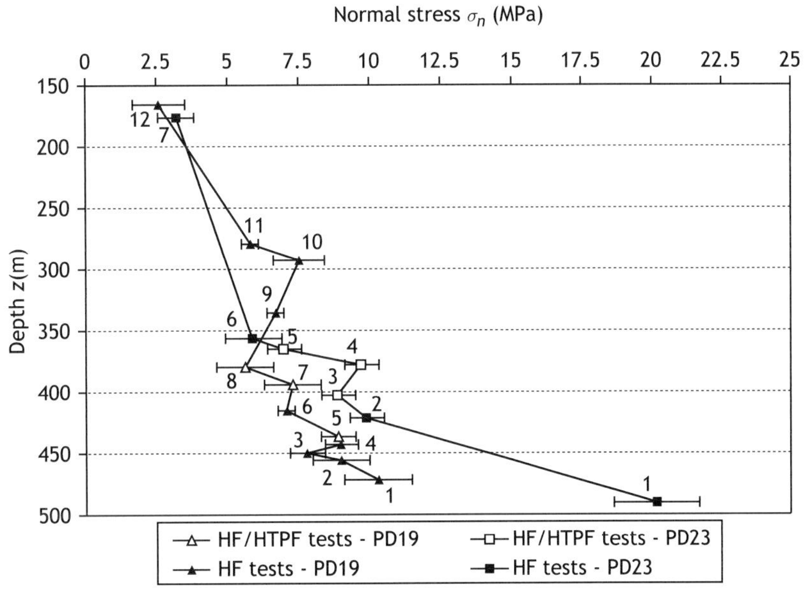

Fig. 14.5 Plot of normal stress magnitudes versus depth as measured by hydraulic testing (courtesy of B. Figueiredo).

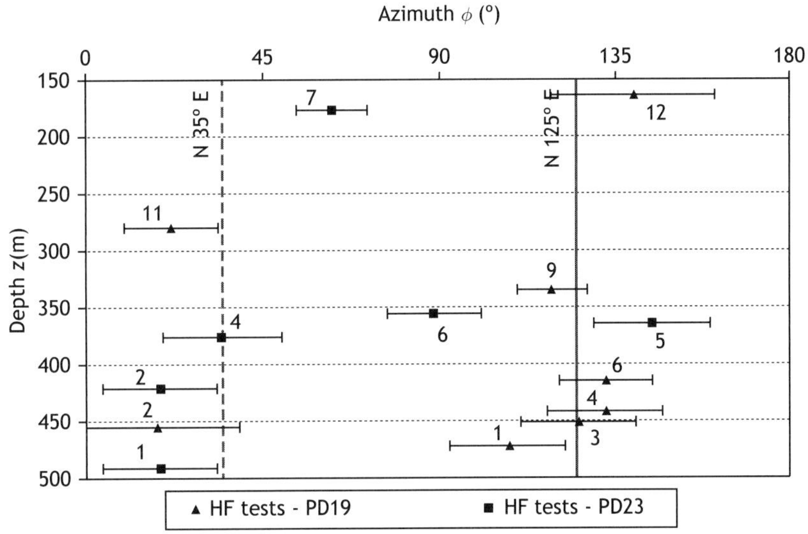

Fig. 14.6 Horizontal directions of hydraulic fractures (courtesy of B. Figueiredo).

the overcoring tests the results depend on the accuracy of the elastic constant determination. Results derived from the pressurization of the cores collected during the overcoring operations have yielded the following values.

Borehole PD1: Young's modulus $E = 60.8$ GPa; $\delta E = 12.9$ GPa; Poisson's ratio $\nu = 0.30$; $\delta \nu = 0.05$.

Borehole	Test	Depth z (m)	Principal stress magnitudes (MPa)			Principal stress orientations (°)		
			σ_I	σ_{II}	σ_{III}	σ_I	σ_{II}	σ_{III}
PD1	1	202.8	6.4	5.6	4.7	325/66	85/12	179/20
	2	206.5	7.0	3.4	3.2	249/86	147/1	56/4
	3	221.5	8.7	3.4	3.3	354/86	231/2	141/4
	4	222.2	9.3	6.9	5.9	36/64	188/23	283/11
	5	250.9	10.6	8.2	7.0	94/54	200/11	298/34
	6	251.7	11.3	6.5	6.0	26/73	150/10	242/14
PD2	1	162.5	11.9	6.9	5.9	94/79	260/11	350/3
	2	164.1	8.0	6.3	4.3	328/75	229/2	139/15
	3	183.5	6.2	2.8	2.0	324/79	177/10	86/6
	4	184.4	5.1	4.6	4.3	285/63	24/5	116/26
	5	201.5	−1.9	−3.6	−4.0	267/77	69/12	160/4
	6	202.9	6.0	4.2	3.4	289/39	147/45	36/20

Fig. 14.7 Results from overcoring tests in boreholes PD1 and PD2. For the principal stress direction determination, the angle before the slash corresponds to the azimuth whilst the angle after the slash refers to the dip angle (the angle made with horizontal plane).

Location	Test	d (m)	ϕ (°)	θ (°)	σ_n (MPa)
SFJ1	1	447	0	0	8.9
	2	448	110	45	6.4
	3	437	110	90	9.4
	4	446	290	45	3.7
SFJ2	5	332	0	0	9.9
	6	333	290	45	2.6
	7	331	290	90	2.0
	8	332.5	110	45	6.4
SFJ3	9	278	0	0	9.9
	10	275	290	45	3.2
	11	279	290	90	4.1
	12	280	110	45	3.0

Fig. 14.8 Results from flatjack tests run in the adit at the locations shown in fig. 14.2.

Borehole PD2: Young's modulus $E = 57.7$ GPa; $\delta E = 9.5$ GPa; Poisson's ratio $\nu = 0.35$; $\delta\nu = 0.11$.

However, the results from uniaxial compression tests run on cores collected at the adit wall have yielded:

Young's modulus $E = 44.6$ GPa; $\delta E = 9.9$ GPa; Poisson's ratio $\nu = 0.25$; $\delta\nu = 0.06$.

Hence, whilst the results from the cores collected during overcoring operations (which produces hollow cylinders, see fig. 13.2) are consistent with each other, the results from full cylindrical cores collected at the walls of the adit suggest significantly lower values for Poisson's ratio. This is consistent with our proposition of section 13.1.2 that overcoring generates some microcracking, which in turn influences the elastic response of the material.

Ascertaining the uncertainty of overcoring results involves the identification of upper bounds both to the explicit errors on the displacement measurements and also to the implicit errors associated with the elasticity hypothesis. Defining a precise procedure for ascertaining these uncertainties remains an open question. On the basis of the reproducibility of the results, we assume for the present discussion that the uncertainties are somewhere between 10% and 15% of the measurements, and we will ignore the results from the test run at 201.5 m in PD2 because of its abnormal values.

From overcoring tests conducted at distances larger than 20 m from the adit, we observe that the maximum principal stress direction is inclined at some 23° to the vertical direction in well PD1 and 26° in well PD2. Further, the mean difference between the intermediate and the minimum principal stress magnitudes is equal to 0.7 MPa, which indicates that both values are sub-equal. This sub-equality is confirmed by the large dispersion in the sub-horizontal principal stress direction determinations.

This dispersion is consistent with that already observed for the orientation of hydraulic fractures. So, it may be concluded that the intermediate and the minimum principal stress directions are sub-horizontal and that their magnitudes are sub-equal throughout the volume where measurements were conducted. However, this does not apply close to the adit, which locally alters the stress field.

Our objective is first to identify a stress model that fits the measurements conducted away from the adit. Then the consistency of this stress model with the measurements conducted close to the adit will be explored, with particular attention to the results from the flatjack tests.

14.1.2 Integration of hydraulic and overcoring test results for an optimum evaluation of the natural stress field

Our objective here is to determine the model for the stress field that fits best the various data that have been obtained. For this purpose we need to define a procedure for evaluating the fit between a given stress model and the data. The model concept is very general and includes both explicit parameters and implicit hypotheses (chapter 1 in Tarantola, 1987).

It may be observed that if a model includes more explicit parameters than the constraining data, it is generally possible to propose a set of values for the parameters that will fit the observations exactly. Such models are often developed for exploring various hypotheses and imply a complementary process for selecting values for some parameters. For our purpose here, a satisfactory model is one that involves fewer parameters than the constraining data. We could even propose that, for two models that fit the same data set equally well, the better model is that which requires the smaller number of parameters.

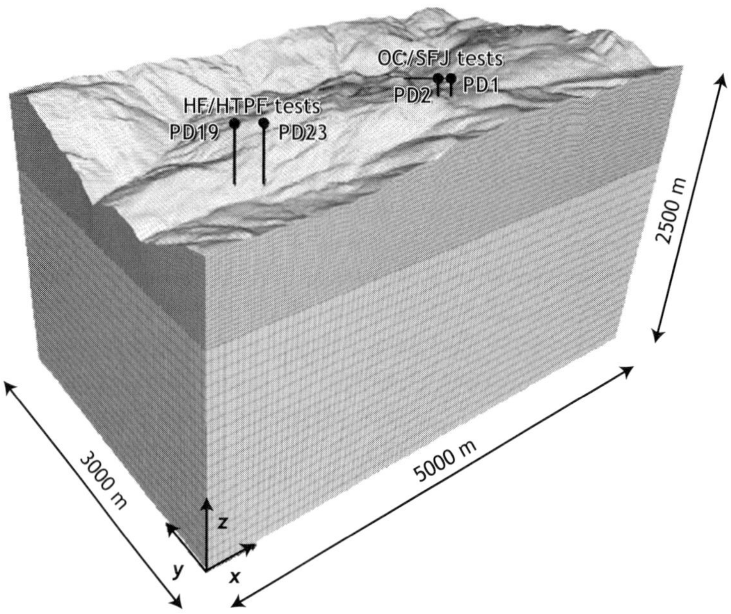

Fig. 14.9 Grid used for evaluating the stress field throughout the volume of interest (courtesy of B. Figueiredo).

Another important consideration is the concept of heterogeneous data. When the misfit between a given model and some given data is larger than the uncertainty on the data, it is customary to consider that such data are inconsistent with the corresponding model; an ideal model must fit all points of the data set. Often, however, part of the data has been obtained in domains where the implicit hypotheses of the model are not met, so that the misfit between the measured and predicted values is larger than the uncertainty on the data. Such data are called *heterogeneous* and are removed from the data set when the misfit measure associated with the corresponding model is being identified. This implies that a significant characteristic of a given model is the amount of heterogeneous data associated with it.

For the Paradela project, the volume of interest (fig. 14.9) is assumed to be filled with an equivalent homogeneous geomaterial, the material properties of which are part of the model definition. We consider the geometry of this volume to be defined without uncertainty, i.e. the topography and the geometry of the adit are well known and are not considered to be sources of error. Furthermore, the uncertainties on the deep boreholes dips and azimuths are neglected, so that the uncertainties in the locations of the hydraulic tests (which are smaller than one meter) are also neglected.

The vertical boundaries for the Paradela project were placed far enough from the points of interest to avoid influencing the stress field at these locations, whilst the horizontal base was located some 2.5 km below sea level. The boundary conditions imposed on the vertical boundaries are part of the model definition, but a condition of no vertical displacement is imposed on the horizontal basal boundary.

The commercial software FLAC3D (Itasca, 2009) was used to solve the set of differential equations that describes the system. The mesh constructed with this software is composed of 600 000 elements, with cubic 25 m elements for volumes above sea level. Below sea level, the elements were 50 m × 50 m × 100 m (see fig. 14.9).

Four different models were considered by Figueiredo *et al.* (2014). In the first the geomaterial is assumed to be isotropic and linearly elastic, with its two elastic constants taken as equal to the values measured for the full cylindrical cores. The volume is under gravity loading only, i.e. we impose a condition of no displacement normal to the vertical boundaries. For this first model the adit was ignored and only the results from hydraulic tests were considered for evaluating the quality of the model.

The second model is similar to the first but introduces horizontal tectonic stresses, the components of which are to be optimized to fit both the hydraulic and overcoring results.

The third model is again similar to the first (with gravity loading only) but the Poisson's ratio for the rock mass is optimized to fit both the hydraulic and overcoring data.

The fourth model was dedicated to analysis of the stress concentration associated with the adit, in order to exploit the results from flatjack tests and from the overcoring tests located closest to the adit. It is discussed in section 14.1.3.

Definition and measurement of the misfit

The difference between the *i*th hydraulic normal stress measurement, $\sigma^i_{n,mes}$, and the value predicted for this value by a model, $\sigma^i_{n,comp}$, is used to define the nondimensional misfit ψ_i^{HT}:

$$\psi_i^{HT} = \frac{|\sigma^i_{n,mes} - \sigma^i_{n,comp}|}{\delta^i_n + \delta^i_f} \tag{14.1}$$

where δ^i_n is the uncertainty in the measurement and δ^i_f is the uncertainty in the normal stress computation due to the uncertainty in the fracture plane orientation.

For an ambiguous test with one or more observed fracture planes, the fracture plane that fits the observations best is selected as the result of the test and used for the corresponding model. Uncertainties in the hydraulic measurements and in the fracture orientation determinations are generally chosen to correspond to the 99% confidence intervals.

For the overcoring results a simplified definition for the misfit, ψ_j^{OC}, was adopted:

$$\psi_j^{OC} = \frac{|\sigma^j_{mes} - \sigma^j_{comp}|}{\delta^j_\sigma + \delta^j_{loc}} \tag{14.2}$$

where σ^j_{mes} and σ^j_{comp} are respectively the measured and modeled *j*th principal stress components at the location of the *k*th overcoring test (with $j = 3k - 2 + p$; $p = 0, 1, 2$); δ^j_σ is the uncertainty in the *j*th principal stress determination, which was taken as equal to the maximum difference between the values obtained from tests run at the same location (i.e. within 10 m); δ^j_{loc} is the uncertainty in the value computed for the corresponding model due to the uncertainty in the geometry (the test location and cell orientation).

Hence the fit between any model q, with $q = 1, 2, \ldots, Q$, may be defined by the overall misfit measure, ψ_q^{HTOC}:

$$\psi_q^{HTOC} = \frac{1}{N+M}\left(\sum_{i=1}^{M} \omega_i^{HT}\psi_i^{HT} + \sum_{J=1}^{N} \omega_j^{OC}\psi_j^{OC}\right) \tag{14.3}$$

where N and M are respectively the total number of hydraulic and overcoring tests included in the model evaluation; ω_i^{HT} and ω_j^{OC} are weight functions that may be adapted for correcting specific bias. The bias may come from the number of tests (if there were many more tests of one kind than of the other kind) or more simply from the nature of the measurement. For example, overcoring operations involve displacement measurements conducted with strain gauges that are 1 to 2 cm long, whilst hydraulic tests generally involve fracture areas of the order of 1 m^2. Also, weight factors are usually the same for all tests of the same kind, but they may be adapted to correct for bias introduced by tests of a specific type. For the present investigation, the weight factors selected by Figueiredo *et al.* (2014) gave slightly more weight to the results from hydraulic tests.

In equation (14.3) the misfit measure involves the sum of absolute values of the misfits for the various measurements. It is often referred to as the L^1 norm, whilst in section 13.1.2 we introduced the L^2 norm, which corresponds to the least squares method. We have adopted the L^1 norm here, since some errors do not follow a Gaussian probability density function. This is particularly true for the effect of microcracking on the determination of the elastic parameters from the overcored samples.

For the L^1 norm, once a global minimum ψ_{min}^{HTOC} has been identified, the limits of the 90% confidence level may be estimated as (Parker and McNutt, 1980)

$$\psi_{90\%}^{HTOC} = \frac{1.645(\pi/2-1)^{1/2}(M+N)^{1/2} + M + N}{M+N-W} - \psi_{min}^{HTOC} \tag{14.4}$$

where W is the number of unknown parameters of the model whilst M and N are as before the total numbers of hydraulic and overcoring tests.

Best fit model

Model 1, i.e. a model with gravity loading only and elastic coefficients of the full cylindrical cores applied to the complete rock mass volume, was not found consistent with any of the hydraulic tests. For model 2, which assumed the same elastic constants as those measured on the cores but with horizontal tectonic stress components, a solution was identified that implied equality between the horizontal stress components; yet this solution was consistent with only 40% of the hydraulic tests.

For model 3, which assumed gravity loading only, an optimum Poisson's ratio of 0.47 was identified for the rock mass, which is a very high value as discussed below. This solution was found to be consistent with 75% of the results from both hydraulic and overcoring tests, i.e. the differences between the observed and measured values were smaller than 0.7 MPa for 75% of the measurements, as shown in figs. 14.10 and 14.11.

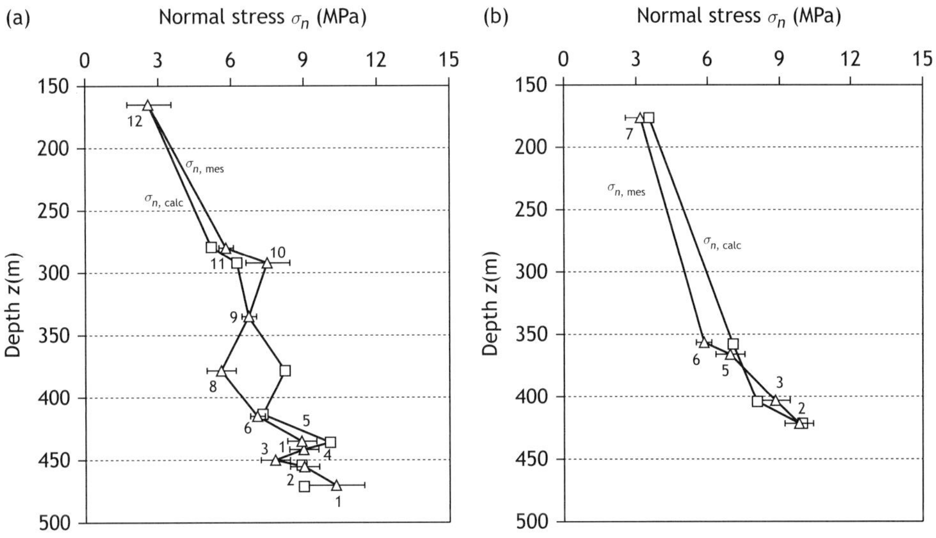

Fig. 14.10 The measured and computed normal stress for the fractures of hydraulic tests. The model (model 3) assumes a value of 0.47 for Poisson's ratio and gravity loading alone (courtesy of B. Figueiredo).

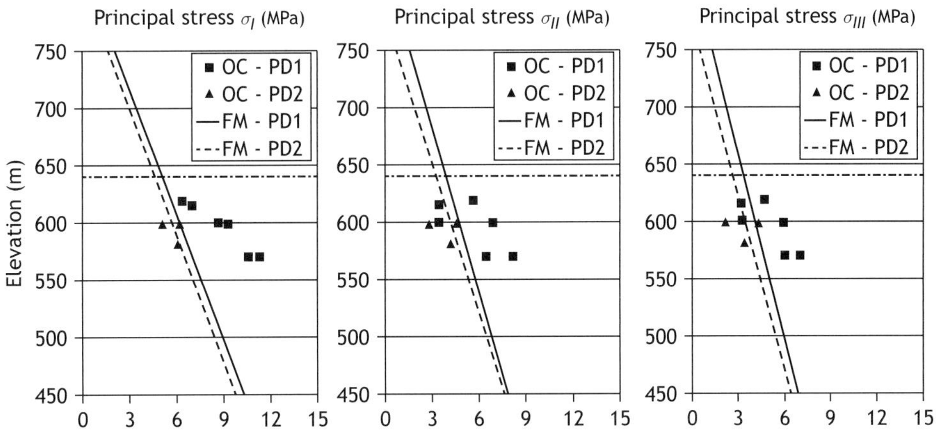

Fig. 14.11 The principal stress components as measured by overcoring and as computed with FLAC3D for a value 0.47 of Poisson's ratio and gravity loading alone (model 3)(courtesy of B. Figueiredo).

However, we should point out that the values for the vertical stress component determined with the overcoring procedure are systematically larger than those computed with the model. This overestimate of the stress component parallel to the borehole axis is a recurrent feature of the overcoring method, and various reasons have been proposed for this, including the role of microcracking during the overcoring operation.

In order to verify whether this solution is also compatible with results from the flatjack tests, a local analysis of the stress concentrations associated with the adit was undertaken.

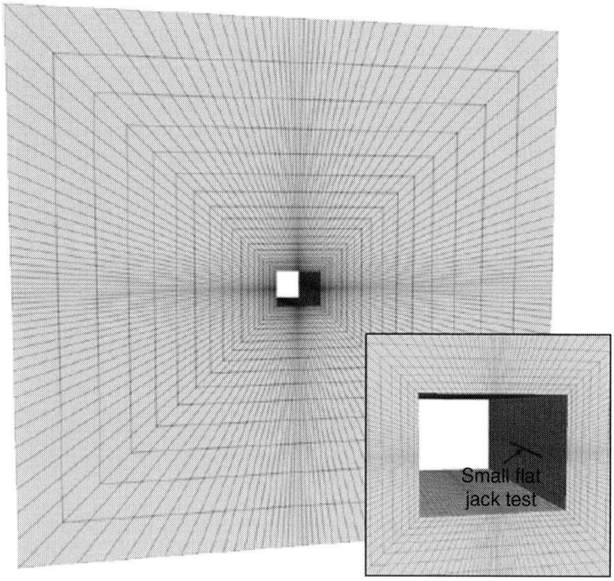

Fig. 14.12 The grid used for investigating local stress concentrations associated with the adit (model 4)(courtesy of B. Figueiredo).

14.1.3 Discussion of adit influence on the stress field and conclusions for the rock mass rheology

As shown in fig. 5.11, an infinite square opening in an elastic medium generates significant stress concentrations close to the free surface of the opening, and these stress concentrations depend strongly on the orientation of the far-field principal stress directions. However, the cross section of the adit where flatjack measurements were conducted is 2.4 m $\times$ 2.0 m and, more importantly, we may anticipate some effects of the local topography on the principal stress directions.

Hence the plane strain hypothesis implied by the analytical solution derived in section 5.6.4 for fig. 5.11 is not strictly applicable to the adit, and the FLAC3D code was used for determining the stress at the location of flatjack tests and overcoring tests conducted close to the adit (see fig. 14.12). Poisson's ratio for the geomaterial surrounding the adit was taken as equal to 0.25, i.e. the value measured on the full cylindrical cores collected locally.

A no-vertical-displacement condition was imposed on the horizontal basal surface and the lower left corner was fixed. The stresses computed with model 3, discussed above (gravity loading only, with a Poisson's ratio equal to 0.47), were applied on the vertical boundaries and on the upper horizontal boundary.

The measured and computed values for the flatjacks are shown for comparison in fig. 14.13. This stress modeling showed that the stress concentration associated with the adit becomes negligible at distances larger than five times the adit height, so that the results from overcoring tests are unaffected by the adit. The principal stress magnitudes

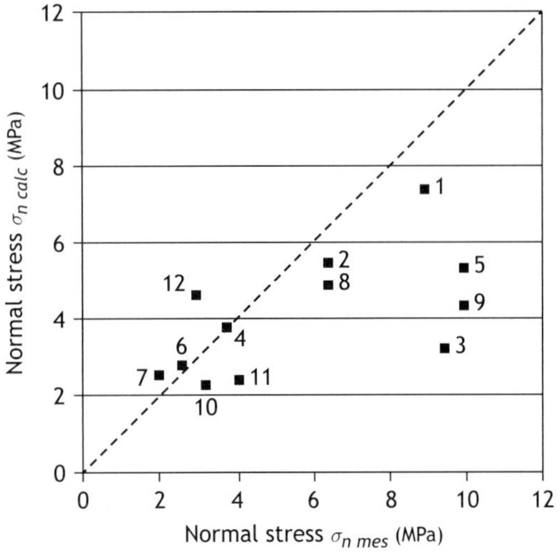

Fig. 14.13 Measured (horizontal axis) and computed (vertical axis) normal stress values for the flatjack tests. The model assumes a 0.25 Poisson's ratio around the adit and the far-field stress state is that determined with model 3 (courtesy of B. Figueiredo).

Borehole	Test	Elevation (m)	Method	Principal stress magnitudes (MPa)			Principal stress orientations (°)		
				σ_I	σ_{II}	σ_{III}	σ_I	σ_{II}	σ_{III}
PD1	1	619.5	OC	6.4	5.6	4.7	325/66	85/12	179/20
			FM	6.0	4.5	3.8	225/71	127/3	36/19
	2	617.9	OC	7.0	3.4	3.2	249/86	147/1	56/4
			FM	6.1	4.6	3.9	225/70	127/3	36/20
PD2	1	618.2	OC	11.9	6.9	5.9	94/79	260/11	350/3
			FM	5.2	3.8	3.0	193/58	293/6	27/31
	2	614.5	OC	8.0	6.3	4.3	328/75	229/2	139/15
			FM	5.2	3.8	3.1	193/60	293/6	26/29

Fig. 14.14 The principal stress components as determined from overcoring tests and as computed with the FLAC3D code. The model assumes a 0.25 Poisson's ratio around the adit and the far-field stress state is that determined with model 3 (gravity loading only and a value 0.47 for Poisson's ratio for the rock mass surrounding the adit's grid volume).

and directions derived from the overcoring tests and those computed with the FLAC3D code are shown in fig. 14.14.

We note that the flatjack tests 3, 5 and 9 and the overcoring tests 1 and 2 run in PD2 were not consistent with this solution but that all other tests were consistent with it.

Conclusion on the rock mass rheology

Results from this numerical modeling show that about 75% of all the stress measurements may be fitted with an isotropic linearly elastic rock mass with a very large Poisson's ratio

(0.47). At this scale the rock mass includes only a few faults but with very variable orientations, as shown on fig. 14.1. These results suggest that the material filling up these faults has crept over time, thus relaxing nearly completely the shear stresses supported by these fault surfaces.

Yet the stress field in the vicinity of the adit is satisfactorily modeled with an isotropic linearly elastic material with a 0.25 Poisson's ratio. This shows that the shear stress supported by the many fractures that affect the rock mass at this scale has not been relaxed over the existence of the adit (50 years). Furthermore, local stress anomalies have been detected that may be explained by simple models of slip along faults in a manner somewhat similar to that discussed in section 12.4.2 and illustrated by fig. 12.21. The very existence of these stress anomalies demonstrates that the long-term behavior of the geomaterial in which the faults have developed is not that of a very viscous fluid.

We conclude that the soft homogeneous linearly elastic model, to which we have assimilated the rock mass, provides a convenient means for interpolating the local stress measurements to other points of interest. However, the real rock mass behavior is that of an elastic material in which there exist a few faults that have relaxed only locally the shear stress components transmitted through these surfaces. This is of particular importance for evaluating the minimum principal stress component that may be encountered along the planned pressure tunnel. Indeed it is this minimum principal stress value that controls potential leakage problems, when the tunnel is filled by water under pressure, rather than the expected value predicted by the model. Nevertheless it is the expected value associated with an apparent 0.47 Poisson's ratio that is of use for designing the support of the large caverns to be constructed for turbine installation.

14.2 A stress field characterization in the sedimentary Paris Basin

A long-term nuclear waste repository is being planned in the Callovo-Oxfordian argillite (clayish shale) of the eastern Paris Basin. The design of this repository has required a detailed characterization of the stress field for a volume that extends 15 km × 15 km × 0.8 km. Measurements have proceeded in two phases. First, detailed measurements were conducted in the vicinity of the underground laboratory developed for investigating *in situ* various issues related to the long-term behavior of the material (Wileveau *et al.*, 2007). Then, additional measurements were conducted in vertical boreholes drilled some distance from the underground laboratory for the purpose of evaluating the horizontal consistency of the results (Cornet and Röckel, 2012).

The target zone of the repository is the Bure argillite formation, which is a mixture of 45%–50% mica clays, 25%–30% calcite and 20% quartz. It is about 150 m thick and lies above the Jurassic Dogger limestone and below the Oxfordian limestone formations. On site, these sedimentary layers are sub-horizontal, dipping some 2° to 3° to the northwest (fig. 14.15).

We discuss first the methodology that has been adopted for constraining all the six stress components within a depth interval that extends from the lower part of the dogger to the

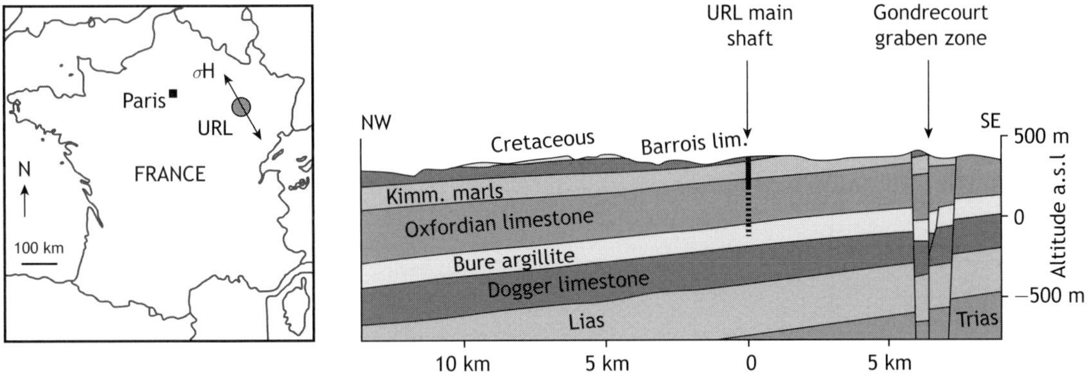

Fig. 14.15 Sedimentary series at the site of the stress measurements. The dip shown in the figure is somewhat exaggerated; it is actually 2° to 3° (redrawn from Gunzburger and Cornet, 2007, with permission from Oxford University Press).

upper part of the Oxfordian limestone. Then we introduce various hypotheses that have been proposed for explaining the observed stress field.

14.2.1 Results from the stress determination program

The development of the underground research laboratory (URL) installed prior to the construction of the repository required an assessment of the local stress field. A stress determination program was designed so as to take advantage of the many boreholes drilled on site with various dips and azimuths (fig. 14.16). First, a vertical reconnaissance well was drilled down to the top of the Dogger. It was used for conducting a series of true hydraulic fracturing stress-measurement tests. Also, a number of additional vertical reconnaissance wells were drilled with water-based mud and, for all these wells, some borehole breakouts were observed in the argillite layers. Hence the conjunction of vertical hydraulic fractures together with borehole breakouts in vertical wells provided a very good constraint on the principal stress directions, both in the argillite and in the overlying Oxfordian limestone (the mean orientation was N151°E, with a local variation of 10° to 15° from one layer to another in the maximum horizontal stress direction). Furthermore, instantaneous shut-in tests run during the hydraulic fracturing operations (see section 13.2) provided a good constraint on the minimum principal stress magnitude at various depths in these formations.

After hydraulic fracturing tests were complete, a vertical 6 m diameter shaft was sunk over the well. When the shaft reached a depth of 471 m in the argillite, where a vertical hydraulic fracture had been observed during the stress measurement program, a horizontal fracture was detected some distance from the vertical well. This demonstrated that, at this depth, the vertical stress component is in fact the minimum principal stress component, or possibly is sub-equal to the minimum horizontal principal stress, so the anisotropy of the argillite tensile strength is in fact large enough to control fracture propagation.

Very few preexisting fractures were observed, and those that were identified were not appropriate for proper HTPF testing (no single fracture could be isolated within the length of the pressurized straddled interval). However, numerous horizontal weakness planes were

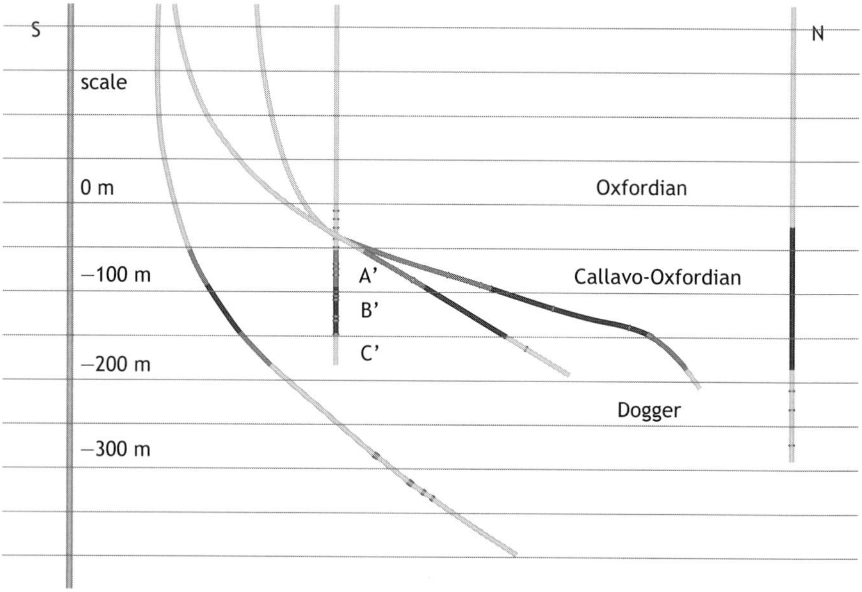

Fig. 14.16 Geometry of the various reconnaissance boreholes available for the stress reconnaissance program (redrawn from Wileveau *et al.*, 2007, with permission from Elsevier).

observed and these were used to ascertain the local vertical stress component magnitude through hydraulic testing at very slow injection flow rates.

The principal stress directions and the minimum horizontal principal stress magnitude in the Dogger limestones were later established through vertical wells drilled with oil-based mud. Interestingly, no breakouts were observed in the argillite for any of the wells drilled with oil-based mud. This showed that the failure criterion for the argillite is strongly dependent on environmental conditions, and this has prevented the obtaining of precise stress estimates from the width of breakouts (see the discussion in section 13.3).

The evaluation of the maximum horizontal principal stress magnitude was conducted by three different methods.

For the Oxfordian limestone, above the argillite, the breakdown pressure observed during hydraulic fracturing conducted at large flow rates was used to obtain upper and lower bounds to the stress magnitude. An upper bound is provided by neglecting completely the effect of the pore pressure on the tensile rupture of the limestone and by considering that fracture initiation occurs at the peak injection pressure. A lower bound is obtained by assuming that the tensile strength is controlled by Terzaghi's effective stress (see chapter 12).

For the argillite, a lower bound was obtained by assuming that the maximum tangential stress was barely lower than the dry uniaxial compressive strength of the material (no breakout was observed in wells drilled with oil-based mud). The upper bound was defined from the breakdown pressure observed during hydraulic fracture tests, assuming that there is no pore pressure effect and assuming a tensile strength equal to values obtained from Brazilian tests.

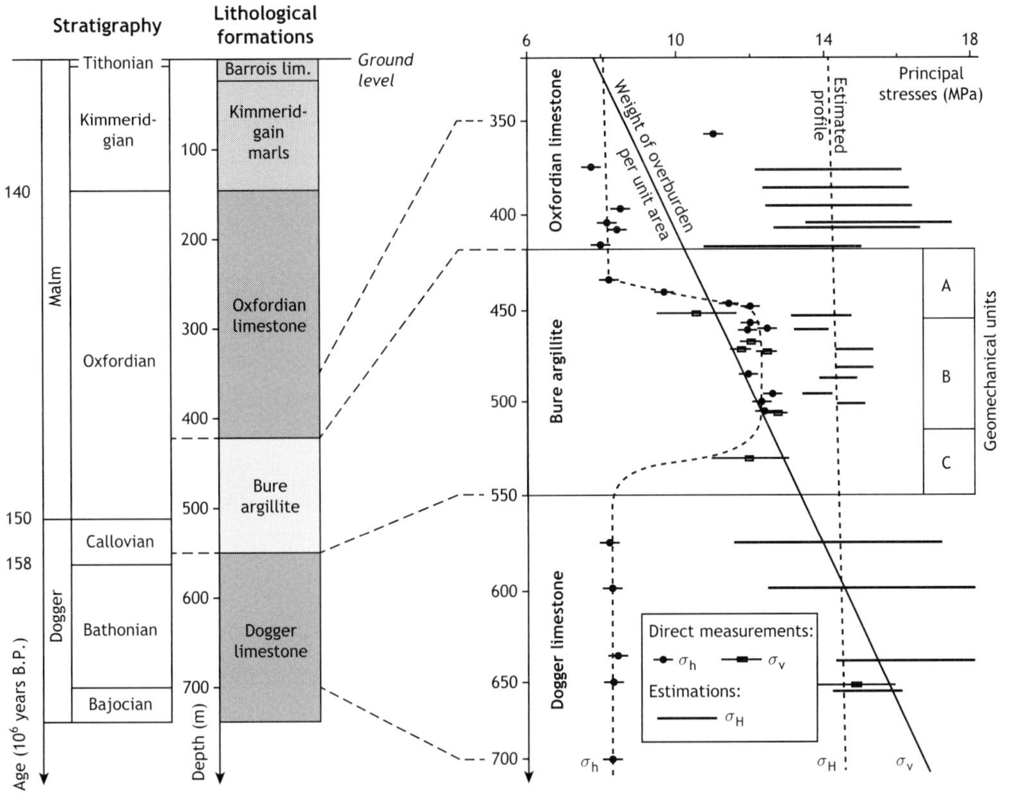

Fig. 14.17 Vertical stress profile obtained in the vicinity of the underground research laboratory in the Bure argillite (reproduced from Gunzburger and Cornet, 2007, with permission from Oxford University Press).

For the Dogger limestone, inclined boreholes were used for conducting hydraulic fracture tests, and the geometry of the observed *en echelon* fractures provided a very good control on the maximum principal stress magnitude since all other five stress components were known for this formation (see section 13.2.3).

A compilation of all the results obtained at the URL location is summarized in fig. 14.17.

The horizontal consistency of these results was established by conducting mostly hydraulic fracturing and one hydraulic test on a preexisting fracture in vertical wells drilled some 15 km to the northeast and 15 km to the northwest of the URL site (Cornet and Röckel, 2012). The fact that, in all wells drilled with water-based mud, breakouts have been observed in the argillite demonstrate that in this formation the maximum horizontal principal stress is larger than the minimum horizontal principal stress.

14.2.2 Discussion on the origin of the local stress field

These stress measurements illustrate the well-known dependence of the stress on the material properties of sedimentary formations: the softer the material, the smaller the deviatoric stress that may be supported by this layer.

Many questions arise, however, concerning the mechanisms that control the magnitude of the maximum horizontal principal stress above the argillite formation.

Indeed, in fig. 14.15 it may be observed that the Oxfordian limestones are outcropping to the east of the measurement locations. Hence, if some strain rate is applied in the maximum horizontal principal stress direction observed in the underlying Dogger limestones, the softer argillite will deform. It will carry with it the overlying limestone without placing any load on it since the Oxfordian limestone is outcropping and can move horizontally without any resistance to its motion.

In addition, no present-day strain rate has been detected in this area; no regional micro-seismic activity has ever been detected, and repeated GPS measurements have not detected any measurable horizontal motion (Nocquet, 2012).

Finally, the laboratory creep and relaxation tests conducted on this argillite suggest that it cannot support any deviatoric stress lasting more than one thousand years (Gasc-Barbier *et al.*, 2004; Zhang and Rothfuchs, 2004). If these results are valid then they would imply that loads applied during the last tectonic phase, which took place in the late Miocene, or about 6 million years ago, according to Gunzburger and Magnenet (2014) should have relaxed by now.

Cornet and Röckel (2012) observed that the maximum horizontal principal stress direction coincides with that of most fractures mapped in both the Dogger and the Oxfordian limestones, and they proposed that a link exists between these two features. Assuming that the fracture pattern coincides with a direction of preferential regional flow, they proposed that the dissolution associated with this hydric flux removes some matter from the walls of the fractures, thereby decreasing the stress in the normal direction. In other words, it was suggested that limestone–water physicochemical interactions through pressure solution effects (section 12.3) alter the stress field in a preferential direction. This would imply that the limestone is "shrinking" today, preferentially perpendicularly to the flow direction, and hence applying a load to the argillite. The detailed modeling of this effect has not been conducted yet.

It may be noted that this presently active deformation process seems also to be consistent with the still unexplained pore pressure profile that has been documented in the argillite (fig. 14.18) (Toussaint and Cornet, 2011). The stess relaxation in the limestones may imply some straining of the argilite, which would be accompanied by a pore pressure decrease close to the argillite–limestone boundary. The pore pressure would remain higher in the center of the formation because of its very small intrinsic permeability (10^{-19}–10^{-20} m^2).

This proposition, which remains to be properly documented and modeled, has not been validated yet. The above discussion was given just to illustrate the possible effects of the rheology and fluid–rock physicochemical interaction on the present-day stress field of sedimentary formations.

14.3 A stress field investigation in the upper Rhine graben

The objective of the previous example was to illustrate nonlinear variations of stress with depth and to raise some questions on mechanisms that must be considered for modeling

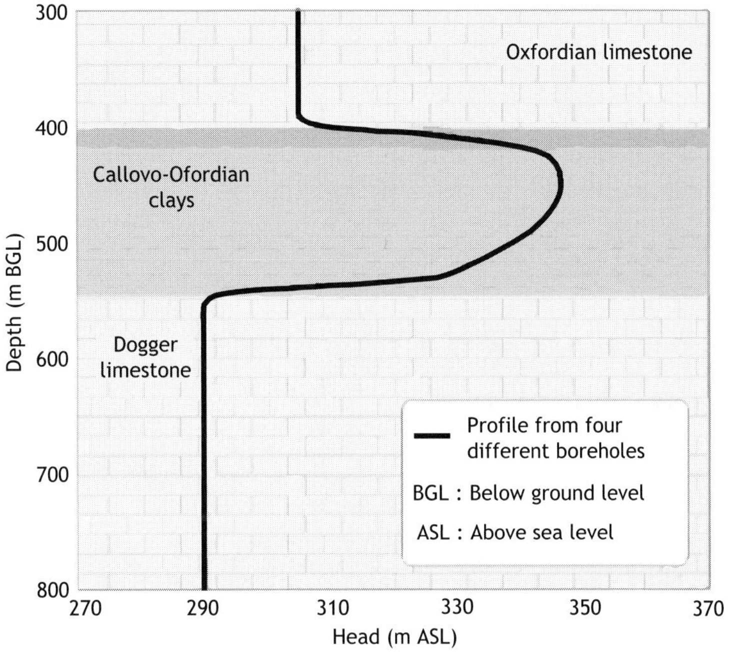

Fig. 14.18 Vertical pore pressure profile measured in the Bure argillite (reproduced from Delay *et al.*, 2007, with permission from Elsevier).

these nonlinear variations properly. In this previous example, however, no systematic rotation of the principal stress direction with depth was observed.

We present now a well-documented example from the upper Rhine graben in which the principal stress directions were found to be rotated by about 45° from measurements conducted in a potash mine at depths ranging from 600 m to 1000 m and breakout observations gathered in a 5 km deep borehole. Further, this deep stress orientation is consistent with results from a focal mechanism inversion that involves microseismic events deeper than 10 km.

Stress evaluation in the potash deposit of the Wittelsheim area, near Mulhouse, France

In the mid 1980s, the MDPA company (Mines De Potasse d'Alsace) undertook some detailed *in situ* stress measurements in order to investigate the possibility of developing solution mining. The ore body is made up of two main layers two to three meters thick, interbedded between salty deposits (KCl and NaCl) with variable shale and anhydrite content and with a larger concentration of anhydrite in the shaley layers. The deposit is made up of sub-horizontal panels, displaced vertically with respect to one another through two sets of faults oriented N10°E and N140°E.

First, MDPA conducted a small-scale hydraulic fracture experiment at a depth of 550 m during which a few m^3 of slurry were injected. Then a drift was mined through the fracture

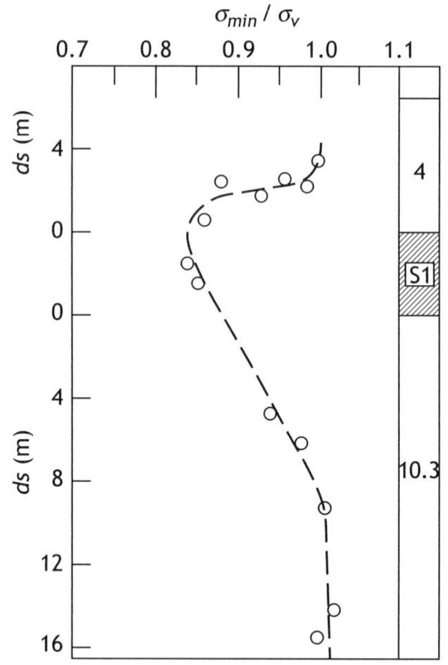

Fig. 14.19 Variation of the ratio of the minimum horizontal principal stress and the vertical principal stress magnitudes as a function of anhydrite content (reproduced from Cornet and Burlet, 1992, with permission from Wiley).

created during the experiment, and this allowed a precise mapping of the fracture for about 50 m. The fracture was found to be vertical and striking N7°E.

Then a set of stress measurements was conducted by hydraulic fracturing from the bottom of the mine, at a depth of around 1000 m, so as to determine the ratio of the minimum horizontal principal stress magnitude and the vertical stress magnitude. The minimum principal stress magnitude σ_h was provided by the shut-in measurements conducted during hydraulic tests, whilst the magnitude of the vertical component σ_v was estimated from the density distribution of the overlying formation. The test site was located at a distance larger than 500 m away from mining operations. The results are presented in fig. 14.19. Anhydrite is a very stiff material and is found to act as a stiffener for the much softer salt and potash. For sufficiently thick salt beds free of anhydrite the stress state is sub-hydrostatic (i.e. there is no deviatoric stress component).

The salt formation is at the origin of a still active diapir about 10 km long, oriented N10°E and about 20 km to the northeast of the mine. It may be assimilated to a large-scale hydraulic fracture, so that its orientation yields the regional principal stress directions.

Hence it may be concluded that the regional maximum horizontal principal stress direction, for depths shallower than 1000 m, is oriented about N10°E near Wittelsheim.

Principal stress direction determination in the city of Basel, Switzerland

The city of Basel is located about 50 km to the east of the MDPA mine. A 5 km deep borehole was drilled in this city in order to exploit the local geothermal energy. The objective was to follow the same development procedure as that adopted for the Soultz site development, located some 200 km to the north of Basel; this was discussed in section 12.4.2.

For this purpose two vertical wells 1 km distant from each other were drilled. The 2755 m deep vertical exploration well OT2 provided information on the sedimentary cover until a granite interface was encountered at 2649 m. Then a 5 km deep vertical well BS1 was drilled, which encountered the granite basement at a depth of 2426 m.

Continuous ultrasonic borehole images have been obtained from 2550 m to 2753 m in the exploration well OT2 and from 2569 m to 4992 m in the well BS1. These images provide an excellent vertical profile of the maximum horizontal principal stress orientations SH_{max}, both in the granite and at the granite–sediment interface (Valley and Evans, 2009).

In the granite, the mean SH_{max} orientation determined from drilling-induced tensile fractures (DITF, see section 13.3) was found to be $N151°E \pm 13°$, whilst the mean breakout orientation yields $N144°E \pm 14°$.

Hence these mean orientations are more than 40° to the west of the orientations determined above 1000 m in the Wittelsheim area. But the orientation of SH_{max} determined from the borehole images in well OT2 yields an orientation $N115°E \pm 12°$ for the sedimentary cover just above the granite.

No model has yet been proposed for this variability of SH_{max} orientation, but clearly, in this region, these results demonstrate that the principal stress directions observed in the sediments cannot be used for estimating the principal stress directions in the basement rock: a stress decoupling is observed between the sedimentary cover and the granite basement.

Focal plane inversion for the Sierentz 1980 seismic crisis

The Sierentz locality is situated in between Basel and the Wittelsheim area, in the southern upper Rhine graben. A magnitude-4.8 earthquake occurred close to this locality on July 15, 1980 (Rouland *et al.*, 1980).

Twelve hours after the main shock occurred, a temporary seismic network composed of 17 stations was deployed (Maury *et al.*, 2013). This provided the means to locate 86 events, out of which 39 fault plane solutions were determined from the first-motion polarities. The main event, which was located at a depth of 13.5 km, was a nearly pure strike-slip event with sub-vertical nodal planes striking respectively N120°E and N210°E.

After having eliminated all events for which the local stress field may have been influenced by a previous event (see section 13.4), a collection of 26 events, assumed to sample the original unperturbed stress field were included in the stress determination. The smallest magnitude was 1.7, which was large enough to smooth out local small-scale stress heterogeneities along slip surfaces (Cornet *et al.*, 2007).

Maury *et al.* (2013) used this data set to compare the various focal plane inversion methods discussed in section 13.4. The method of Gephard and Forsyth is consistent with the largest number of focal mechanisms (which amounted to 90% of them). It yielded two possible solutions (fig. 14.20) with similar values for the misfit function.

The first solution (shaded in gray) fits all but two of the focal mechanisms whilst the other solution is not consistent with three focal mechanisms. Furthermore, with this second

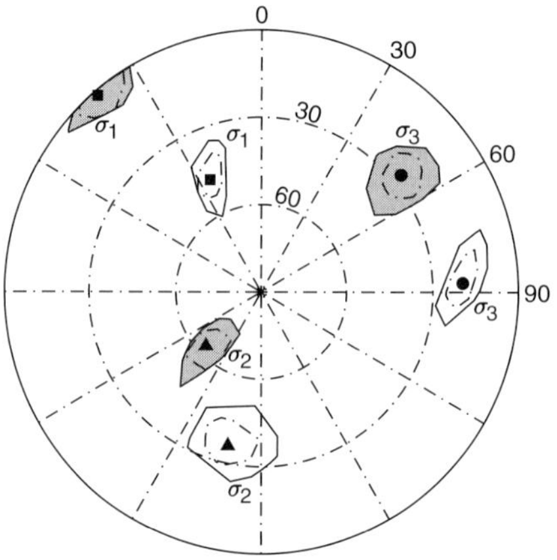

Fig. 14.20 Principal stress directions from an inversion of the focal plane solutions computed for the aftershocks of the Sierentz earthquake. The best solution is shaded in gray (reproduced from Maury *et al.*, 2013, with permission from the Geological Society of France).

solution, 62% of the fault planes selected correspond to the nodal plane determined with the largest uncertainty. For the best solution, the minimum principal stress orientation is found to strike N51°E with a ±16° uncertainty at the 90% confidence level. This value compares very well with the value $N54°E \pm 14°$ determined for the SH_{max} orientation by Valley's and Evans's analysis of the deep Basel borehole failure pattern.

14.4 An evaluation of the stress field in the north-central European lithosphere

The city of Basel was the site of a 6.5±0.5 magnitude earthquake in 1356 AD (Fäh *et al.*, 2009). But, as already pointed out, no horizontal motion has been detected by the continuous GPS measurements that have been conducted over the last 30 years in north central Europe (Nocquet, 2012). So, questions arise about the loading mechanism that led to the occurrence of this earthquake and whether a magnitude-7 or larger earthquake could occur in this area.

Before investigating this question, we must understand the loading mechanism that controls the stress field observed today in the regional lithosphere. Further, let us note that a magnitude-7 or larger earthquakes implies the existence of a lithospheric-scale seismically active structure. So, the question arises whether there is indeed a seismogenic lithospheric structure in this area.

We have just demonstrated that the stress field observed in sedimentary materials cannot be used for understanding the stress field in the underlying materials, especially in the Basel area. However, we also noted that the inversion of focal mechanisms yields

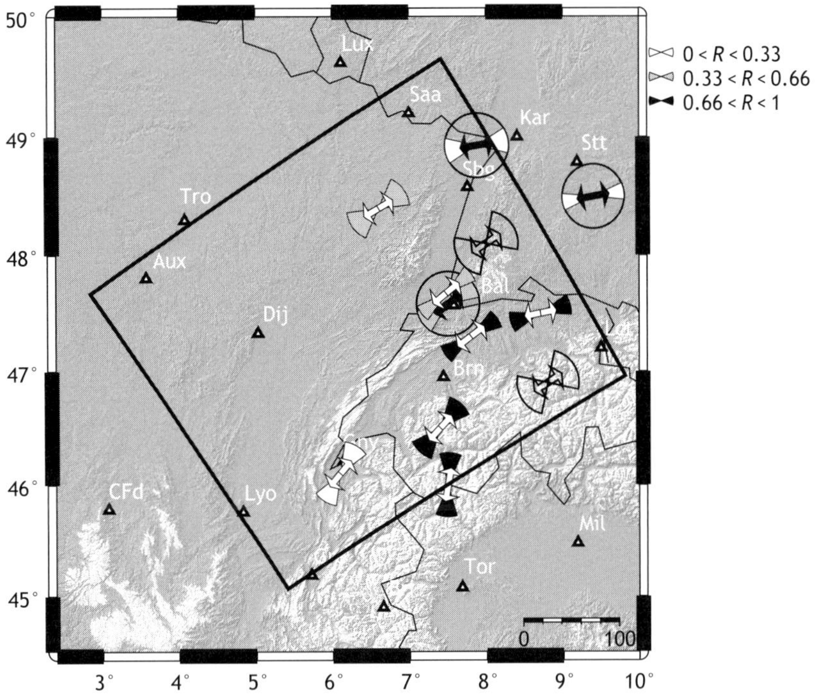

Fig. 14.21 Stress data observed in the seismogenic crust and the horizontal extent of the model (courtesy of J. Maury).

reliable information on the orientation of principal stress directions within seismically active volumes.

In this section we discuss the modeling of the stress field in a volume 400 km long, 360 km wide and 230 km deep that includes the southern Rhine graben area, with the cities of Mulhouse and Basel (fig. 14.21). It extends vertically below the lithosphere–asthenosphere boundary (LAB). The objective is to try to reproduce the stress field where it is documented in the seismogenic crust. This discussion is reproduced from a paper by Maury *et al.* (2014).

First we review the various data that have been considered for constraining the numerical model. Then the main features of the model are presented and we show how different choices of rheological characteristics may lead to very different conclusions concerning the loading mechanisms at the origin of the present-day stress field.

14.4.1 Constraints from the stress data and from the mapping of seismic activity

Before developing the numerical model, we first identify stress field data that may help to constrain the model as well as the lithospheric structures that may affect this stress field.

Data that constrain the stress field in the regional seismogenic crust

The inversion of focal plane solutions yields the orientation of the three principal stress directions as well as the value of a parameter R that characterizes the ratio of the principal stress magnitude differences, $R = (\sigma_2 - \sigma_1)/(\sigma_3 - \sigma_1)$.

We show in fig. 14.21 the locations where sets of focal mechanisms were inverted, so that both the principal stress directions and the R values are available. For all those sites, one principal stress direction is close to vertical, so only the maximum horizontal principal stress direction is shown. The uncertainty in the direction is also indicated as well as the ranges of values for R. Indeed, most of the inversions provided a poor resolution of the R values and only three classes of values could be identified: $0 \leq R \leq 0.33$; $0.33 \leq R \leq 0.66$; $0.66 \leq R \leq 1$.

For three of these sites (identified by circles over the SH_{max} direction), the maximum horizontal principal stress directions were corroborated by borehole failure analysis (Soultz, shown north of Strasbourg (Sbg), Basel (Bas) and the Urach site shown south of Stutgart (Stt)). Indeed, for all these sites the development of geothermal energy exploitation has led to the drilling of wells reaching depths greater than 4 km in the granite.

Identification of lithospheric-scale seismogenic structures

Combining the seismic catalogue of the Rhine graben area with that of Switzerland, historical and instrumental observations indicate a mainly low to moderate seismicity, with moment magnitudes lower or equal to 5, for this region. The largest earthquake of the area was the 1356 Basel earthquake, with a few magnitude-6 earthquakes that occurred during the two last centuries in the Valais region of southern Switzerland (Grünthal and Wahhlström, 2012).

In order to better characterize this seismicity, maps of the accumulated radiated seismic energy for 5×5 km^2 areas are drawn. The radiated seismic energy E is evaluated from the moment magnitude M_w, according to the Gutenberg–Richter relation:

$$\log E = 1.5M_w + 11.8 \tag{14.5}$$

where the radiated seismic energy is expressed in erg (dyne cm).

The intensities I of historical earthquakes were converted to magnitudes following the procedure of Grünthal and Wahhlström (2012), which assumes empirical relationships between the peak ground acceleration and intensities on the one hand and the peak ground acceleration and magnitude on the other hand (e.g. Gutenberg and Richter, 1956); see section 11.3.

The domain where the cumulated radiated seismic energy is the largest is interpreted as the region where present-day deformation is the most active. The catalogue covers a period of 1000 years. The results are shown in fig. 14.22.

The northeast part of the area investigated with the geomechanical model is found to be the most active, with very little activity observed in the western part of the model. The only lithospheric structure identified by the accumulated radiated seismic energy is the Valais zone, where Sion city is located (SIO on the map), whilst the Basel earthquake appears as an isolated event.

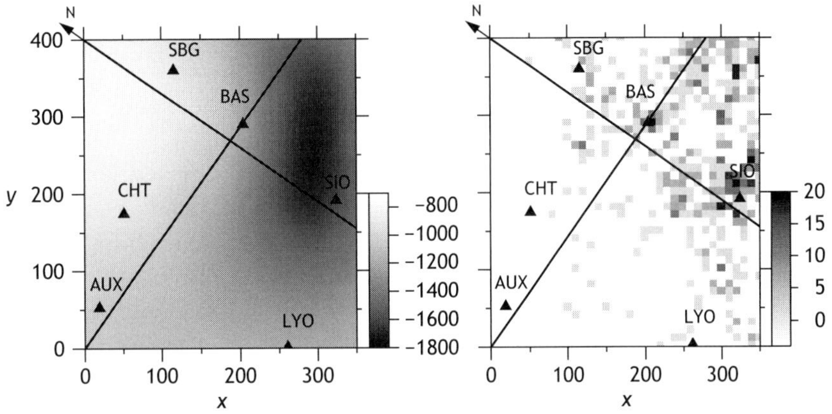

Fig. 14.22 Maps of cumulated radiated seismic energy (right) and plastic deformation (left); see the discussion of fig. 14.25 (courtesy of J. Maury).

14.4.2 Numerical modeling investigation

The first objective here is to develop a three-dimensional model that takes into account the actual three-dimensional geometry of the structural elements that are of significance at the scale of the region of interest. Then this model is used to discuss how the stress field mapped in the upper 15 km of the seismogenic crust sheds some light on the rheology of the deeper geomaterials as well as on the loading processes that generate this stress field.

Definition of the model

Only the main lithospheric structural elements are included. A structural element is considered of importance when one of its dimensions is in the 100 km range.

Structural elements

Two main tectonic events have marked the regional structures: the Alpine orogeny and the Cenozoïc rifting.

The subduction of the European plate under Adria, the Molasse basin and the Jura fold-and-thrust belt system are direct consequences of the Alpine orogeny. The Molasse Basin is no more than 6 km deep (Sissingh, 1998), whilst the Jura corresponds to an approximately 2 km thick decollement. Hence the thickness of these structures is negligible compared with that of the subducting lithosphere (see fig. 14.23). Consequently, Maury *et al.* considered only the various layers that constitute the lithosphere with the three-dimensional geometry of the various interfaces: the Moho discontinuity at the base of the crust and the lithosphere–asthenosphere boundary (LAB) at the base of the lithosphere. The lateral variations of crustal density in the Alps are of the order of 0.2 g/cm^3 (Masson *et al.*, 2002), whereas the density differences between the crust and the upper lithosphere are equal to 0.6 g/cm^3, which led Maury *et al.* to neglect the lateral density variations in the crust.

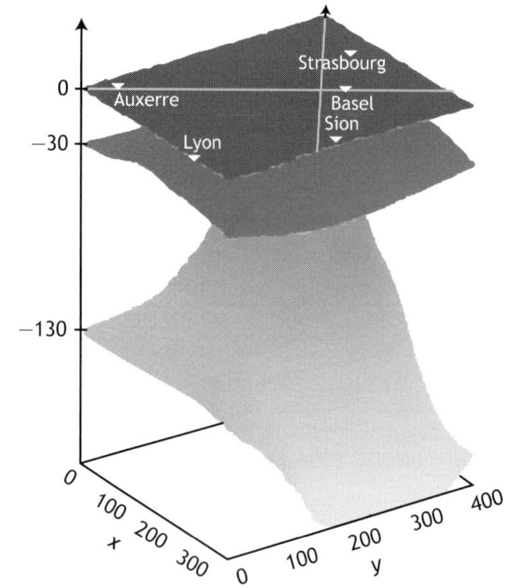

Fig. 14.23 Three-dimensional geometry of the ground surface topography, of the Moho and of the LAB, taken into account in this investigation (courtesy of J. Maury).

The Cenozoïc rifting generated the upper Rhine graben and the Bresse graben. These grabens are marked by the deposit of up to 5 km of sediments, a value that remains small compared with the thickness of the crust (which is around 34 km) and so has been neglected. They also correspond to a fractured basement that may be assumed to extend down to the Moho, and the effect of which is discussed along with the model. Further, these grabens correspond to a small upwelling of the Moho.

The three-dimensional geometry of the Moho and of the LAB were taken from the literature (Eaton *et al.*, 2009; Plomerova and Babuska, 2010; Tesauro *et al.*, 2009; Wagner *et al.*, 2012). The geometry of the ground surface topography, of the Moho and of the LAB are shown in fig. 14.23, which indicates that the only significant lateral depth variations for these various surfaces are those observed at the LAB.

Rheology of the various geomaterials

The mechanical behavior of the asthenosphere is that of an elastoviscous Maxwell fluid, i.e. in the absence of motion the asthenosphere supports only hydrostatic pressures (section 2.3). Its constitutive equation may be represented by

$$\dot{\varepsilon}_{ij} = \frac{1+\nu_A}{E_A}\dot{\sigma}_{ij} - \frac{\nu_A}{E_A}\dot{\sigma}_{kk}\delta_{ij} + \frac{1}{2\eta_L}\left(\sigma_{ij} - \frac{\sigma_{kk}}{3}\delta_{ij}\right) \tag{14.6}$$

where E_A and ν_A are respectively the Young's modulus and Poisson's ratio of the asthenosphere and η_A is its viscosity. It is assumed that there is no long-term motion in the asthenosphere. The duration of the relaxation after a load has been applied is of no consequence for the purpose of the discussion below, and so it was chosen so that no shear stress exists 5000 years after a load perturbation has been applied.

The lithosphere is an elastoviscoplastic solid, with time constants much longer than those of the asthenosphere. At the time scale of a few thousand years it can be assimilated to an elastoplastic solid:

$$\varepsilon_{ij} = \frac{1+\nu_L}{E_L}\sigma_{ij} - \frac{\nu_L}{E_L}\sigma_{kk}\delta_{ij} \tag{14.7}$$

where E_L and ν_L are respectively the Young's modulus and Poisson's ratio of the lithosphere. The plastic criterion has been chosen to be the von Mises criterion, for the plasticity of the lithosphere is essentially linked to the motion of dislocations and this does not depend on the hydrostatic stress component (see section 9.2):

$$(\sigma_1 - \sigma_2)^2 + (\sigma_2 - \sigma_3)^2 + (\sigma_3 - \sigma_1)^2 = 2\sigma_0^2 \tag{14.8}$$

where σ_0 is the yield stress.

The lithosphere is often considered to comprise three layers, the upper seismogenic crust, the lower crust and the lithospheric mantle (e.g. Burov and Watts, 2006; Burov, 2009). In this study we consider only two layers but we have investigated the effects of the coupling between the crust and the lithospheric mantle on the stress field in the seismogenic crust. The physical parameters chosen for the three layers are:

Young's modulus, equal to 75 GPa for the crust and 100 GPa for both the lithosphere and the asthenosphere;
Poisson's ratio, equal to 0.25 for all three layers;

Density, equal to 2.7 g/cm^3 for the crust and 3.3 g/cm^3 for the lithospheric mantle and the asthenosphere.

Various values for the yield criterion have been considered and the consequences will be discussed after the numerical model has been described.

The numerical model

The objective of the model is to take into account the influence of the real three-dimensional geometry of the main structural elements on the key rheological parameters when only gravity loading is considered. The model is also intended to embrace the possible existence of a regional horizontal tectonic stress in this area.

In order to limit the effect of boundary locations on the stress field within the volume of interest, the modeled region was increased in size until further increments of the discretized volume did not affect the results at locations where data are available. A model with an area nine times larger than that of the domain of interest was finally defined, but the results presented in the figures are just those for the domain of interest.

The domain of interest was discretized into elements 20 km × 20 km in the horizontal plane and around 7 km in the vertical direction. The geometry of the interfaces between the various layers is shown in fig. 14.23. The X axis of the model, parallel to the smallest dimension of the rectangular area of interest, is oriented N55°W.

The FLAC3D code, already mentioned in section 14.1 was used for solving the set of differential equations that describe the problem. Indeed, it is well adapted to solving the various constitutive equations of interest and can take into consideration large deformations. The equations are solved explicitly with a step-by-step procedure appropriate for modeling path-dependent rheologies. Although FLAC3D is designed to consider plastic localization, this feature has not been applied in the case of the Valais region.

Conclusions from the crustal stress field analysis

First, an elastic solution was computed for the lithosphere when gravity loading only is considered. For this elastic solution, all shear stresses in the asthenosphere relaxed so that only hydrostatic pressures are applied along the LAB. In a second step, the shear stresses were relaxed in the lithosphere wherever the local stress conditions were larger than those implied by the yield criterion. The plastic deformations were computed until the chosen plastic criterion was satisfied everywhere.

The rheological characteristics of lithospheric geomaterials

When the lithosphere is assumed to remain elastic, if gravity loading alone is considered then we note a regional rotation of the maximum horizontal principal stress direction, the wavelength of which is governed by that of the flexure of the LAB. However, the computed principal stress directions do not fit the measured values (fig. 14.24). We call this model R_1, for later reference.

In the second step, we let the plastic deformation develop in such a way as to satisfy the von Mises criterion. First, a uniform yield stress equal to 60 MPa is applied throughout the lithosphere (model R_4, shown in fig. 14.25). In order to map the locations where the plastic deformation is largest, the vertical displacement amplitude associated with plastic deformation is taken from fig. 14.22. For this evaluation, the amplitude of the displacement relative to the base of the lithosphere obtained with the purely elastic model R_1 is subtracted from the amplitude computed with the elastoplastic model.

It may be observed that, with this elastoplastic model, the computed principal stress directions are satisfactorily consistent with observation and that the domain where the plastic deformation is largest coincides with that where the seismic activity is the most significant, but the computed *R* ratio does not fit the observations in quite a few places.

Maury *et al.* (2014) then investigated the effect of decoupling the crustal material and the lithospheric mantle by assuming a 100 MPa yield stress in the crust and two possible yield stress magnitudes (10 MPa and 60 MPa) for the mantle. The best results are obtained when the crust and the mantle are fully coupled (i.e. the yield stress in the crust and in the mantle is the same). This suggestion of a thick elastic layer is somewhat consistent with the conclusions reached by Milne *et al.* (2004), who proposed a 120 km thick elastic layer for Fennoscandia as a result of their analysis of the postglacial rebound observed in this area. However, their model does not address the limit-of-plasticity issue.

The effects of rifting in the cenozoïc have been tested by introducing some anisotropy in the elastic constants of the local crustal material (a Young's modulus smaller in the direction normal to the rift than in the direction parallel to the rift; different values of Poisson's ratio and the shear moduli in the planes parallel and normal to the rift). These

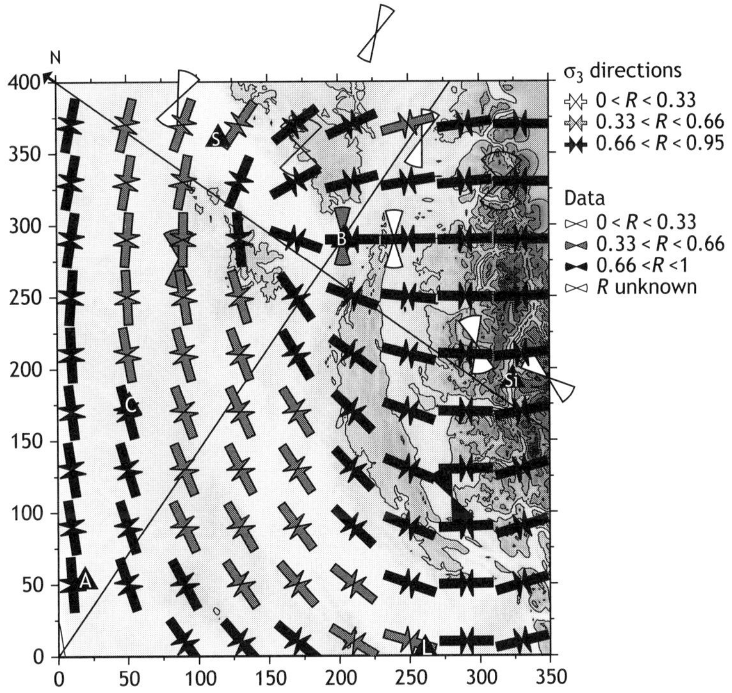

Fig. 14.24 The computed and measured minimum horizontal principal stress directions at a depth of 15 km for an elastic lithosphere loaded by gravity only (model R_1 of Maury *et al.*, 2014). The site locations, shown by the dark triangles, are at Auxere (A), Basel (B), Sion (Si) and Soultz (S) (courtesy of J. Maury).

perturbations induce variations in the principal stress directions with wavelengths related to those of the rift. However, such perturbations are not in practice observed and therefore the effects of rifting on the crustal-material elastic properties, if they exist, do not seem to be strong.

Our objective here is not to conduct a detailed sensitivity analysis but rather to indicate that the large-scale horizontal variations in the principal directions observed for the crustal stress field correlate well with the LAB geometry. Furthermore, these directions are consistent with a strong coupling between the crust and the mantellic lithosphere and a von Mises plastic criterion for this material. However, this model cannot provide any information on strain rates since no viscosity has been introduced for the lithospheric materials. It also requires further refinement in order to fit the stress amplitude ratios properly.

Loading conditions

It is often considered that the stress field north of the Alps is dominated by Alpine tectonics (e.g. Hieronymus *et al.*, 2008; Sue *et al.*, 1999). The influence of tectonic stresses may be investigated by superimposing on the effect of gravity a horizontal uniform uniaxial stress field oriented N145°E, so as to fit the principal stress directions observed in Basel if one assumes an elastic response for the whole lithosphere (model R_6). The results are shown in fig. 14.26.

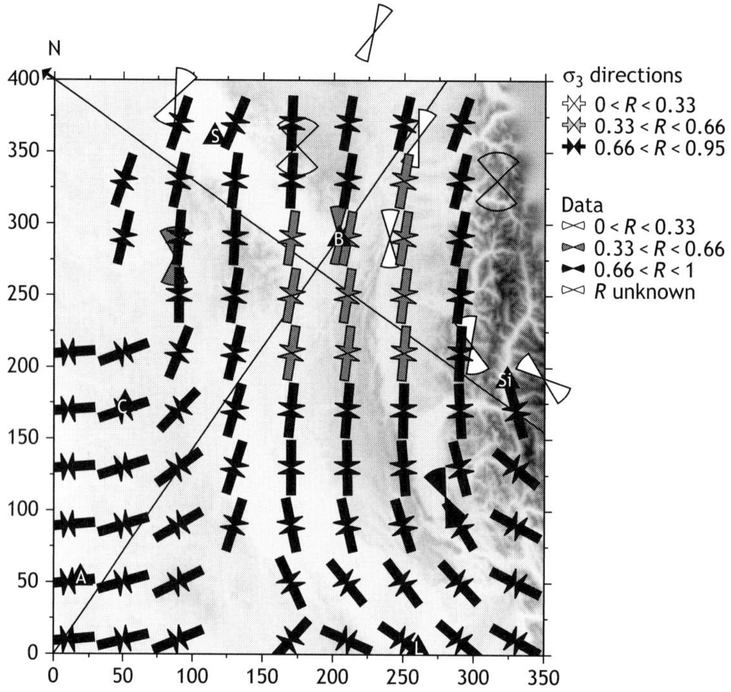

Fig. 14.25 The computed and measured minimum horizontal principal stress directions at a depth of 15 km for an elastoplastic lithosphere with a 60 MPa yield stress, loaded by gravity only (model R_4 of Maury *et al.*, 2014) (courtesy of J. Maury).

Figure 14.26 shows that, indeed, if the lithosphere is elastic, then superimposing a horizontal tectonic stress on the effect of gravity yields principal stress orientations that are consistent with observation, but then nowhere in the lithosphere is the plastic yield criterion satisfied. Furthermore, the computed horizontal displacement field is not consistent with the absence of motion implied from GPS monitoring. It may be argued that this tectonic stress implies a much slower strain rate than that which may be detected from GPS observations. Even so, for very slow strain rates some relaxation should occur, which should in turn invalidate the elasticity hypothesis.

Also, no work has yet been undertaken to explain the high seismicity observed in the Valais area, and this modeling would require that localization be introduced in the computed plastic deformation. Clearly this localization would result in the perturbation of the stress field surrounding this area. Interestingly, the Valais is the only location where the measured horizontal principal stress direction does not fit the model R_4 predictions (elastoplasticity without plastic localization).

These are but a few of the questions that must be answered when one is modeling the geomechanical behavior of this continental lithosphere. We conclude for now that the simple elastoplastic model that has been proposed fits the stress field data gathered in the seismogenic crust quite satisfactorily and would suggest that the Basel earthquake did not occurr on a lithospheric-scale seismogenic structure.

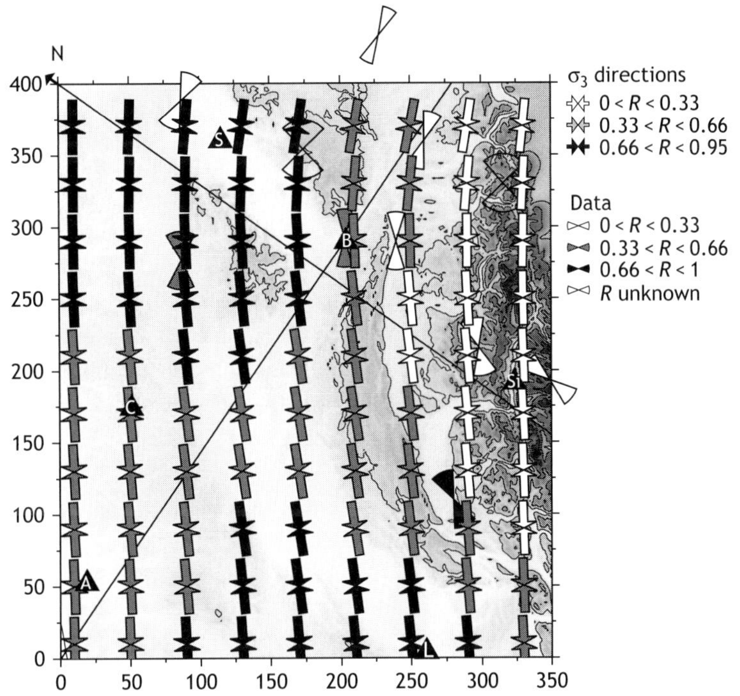

Fig. 14.26 The computed and measured minimum horizontal principal stress directions at a depth of 15 km for an elastic lithosphere loaded by gravity and a horizontal uniform uniaxial stress oriented N145°E (model R_6 of Maury *et al.*, 2014) (courtesy of J. Maury).

14.5 Elements of a conclusion

The four stress field examples that have been discussed illustrate various rheological behaviors of geomaterials that depend on both time and length scales.

For the Paradela example we concluded that the large-scale (i.e. kilometer-scale) stress field may be modeled by that observed in a very soft elastic material and that the smaller-scale (i.e. 10 meter scale) rheological behavior of the granite is that of an elastic brittle material. Clearly the fault network morphology is an essential part of the large-scale mechanical behavior of the massif.

For the sedimentary Paris Basin, we observed that the stress field at the kilometer scale is completely dominated by the rheology of the various sedimentary layers. A significant deviatoric stress has been measured in the limestone formations but a nearly total absence of deviatoric stress is indicated for the most clay-rich formations. This effect of rheology on the regional, km-scale, stress field is even more evident for the Basel area in which complete stress decoupling exists between formations above the salty deposits and the crystalline basement.

However, a larger-scale (10 km scale) homogeneity has been identified for the stress, in the basement that corresponds to the definition of the seismogenic crust. Understanding the stress field in the basement requires an understanding of its coupling with the mantellic lithosphere, and this has led us to conclude that north of the Alps the basement stress field is dominated by the LAB geometry provided that a von Mises plastic criterion is satisfied.

In summary, crustal geomaterials are elastoviscoplastic materials, the yield strength of which depends on many factors including temperature and velocity and also fluid–solid interactions. The constitutive elements of the seismogenic crust are affected by discontinuities that influence the mechanics of both the solids and the fluids.

A basic paradigm of continuum mechanics is that of mass conservation. However, mass may be transferred from the solid phase to the fluid phase and conversely. These mass exchanges are generally ignored in geomechanics. They may possibly explain the stress profile observed in sedimentary rocks and also the occurrence of some intraplate earthquakes.

It is to be hoped that future work will be devoted to a better understanding of fluid motion and the associated fluid–rock interactions at all scales.

14.6 Further reading on inverse-problem theory

Menke, W., 1984. *Geophysical Data Analysis: Discrete Inverse Theory.* Academic Press, 260 pp.

Tarantola, A., 2005. *Inverse Problem Theory and Methods for Model Parameter estimation.* Society for Industrial and Applied Mathematics, 342 pp.

Appendix Elements of tensors in rectangular coordinates

Tensor calculus is most useful when one is dealing with curvilinear coordinates. Although in this book only rectangular cartesian coordinates are used, the notation of tensor calculus is very useful for programming. This appendix presents the main elements of tensor calculus used throughout the book. Derivations not given in this appendix may be found e.g. in Malvern (1969) or Fung (1965).

A.1 Definitions

Linear algebra is the branch of mathematics that concerns vector spaces and linear mappings between such spaces. Linear algebra assigns vectors as the coordinates of points in a space so that operations on the vectors define operations on the points in the space.

For any three-dimensional vector space **V**, an orthogonal frame of reference is defined by three perpendicular unit vectors $\hat{\boldsymbol{I}}_i;\ i = 1, 2, 3$. Any point $\boldsymbol{X}$ of the space is characterized by its three coordinates, X_1, X_2, X_3 in the frame of reference $\hat{\boldsymbol{I}}_i;\ i = 1, 2, 3$.

Consider two points $\boldsymbol{A}$, with coordinates $A_i;\ i=1,2,3$, and $\boldsymbol{B}$, with coordinates $B_i;\ i=1,2,3$. The difference $\boldsymbol{A} - \boldsymbol{B}$ defines a vector $\boldsymbol{V}$ with components $V_i = A_i - B_i;\ i=1,2,3$, so that $\boldsymbol{V} = V_i\hat{\boldsymbol{I}}_i;\ i = 1, 2, 3$. By convention, when an index i is repeated once, summation over this index is implied.

The *scalar product* (also called the *dot product*) of two vectors $\boldsymbol{a}$ and $\boldsymbol{b}$ is defined by

$$\boldsymbol{a} \cdot \boldsymbol{b} = (a_i\hat{\boldsymbol{I}}_i) \cdot (b_j\hat{\boldsymbol{I}}_j) = a_i b_j\,(\hat{\boldsymbol{I}}_i \cdot \hat{\boldsymbol{I}}_j) \tag{A.1}$$

The scalar product of two basis vectors $\hat{\boldsymbol{I}}_i$ and $\hat{\boldsymbol{I}}_j$ is characterized by the Kronecker δ_{ij} symbol:

$$\hat{\boldsymbol{I}}_i \cdot \hat{\boldsymbol{I}}_j = \delta_{ij} = 0 \qquad \text{when } i \neq j \tag{A.2}$$

$$\hat{\boldsymbol{I}}_i \cdot \hat{\boldsymbol{I}}_j = \delta_{ij} = 1 \qquad \text{when } i = j \tag{A.3}$$

or, in matrix form,

$$[\delta_{ij}] = \begin{pmatrix} 1 & 0 & 0 \\ 0 & 1 & 0 \\ 0 & 0 & 1 \end{pmatrix} \tag{A.4}$$

so that

$$\boldsymbol{a} \cdot \boldsymbol{b} = a_i b_j \delta_{ij} = a_i b_i \tag{A.5}$$

The indices are repeated in equation (A.5) which implies summation over i and j. This summation leads to the term on the right. Note that when the summation convention is in use, $\delta_{ii} = 3$.

The *vector* or *cross* product, $\boldsymbol{a} \times \boldsymbol{b}$, also written as $\boldsymbol{a} \wedge \boldsymbol{b}$, is defined by

$$\boldsymbol{a} \times \boldsymbol{b} = \boldsymbol{a} \wedge \boldsymbol{b} = \varepsilon_{ijk} a_i b_j \hat{\boldsymbol{I}}_k \tag{A.6}$$

with summation on i, j, and k. The ε_{ijk} permutation symbol is defined by

$$\begin{aligned} \varepsilon_{ijk} &= +1 \quad \text{for } i = 1, j = 2, k = 3, \text{ or any circular permutation of indices,} \\ \varepsilon_{ijk} &= -1 \quad \text{for } i = 2, j = 1, k = 3, \text{ or any circular permutation of indices,} \\ \varepsilon_{ijk} &= 0 \quad \text{when any two indices are equal.} \end{aligned}$$

A.2 Second-order tensor

A second-order tensor is a linear vectorial transform operation defined by the two relations

$$\boldsymbol{t}(\boldsymbol{u} + \boldsymbol{v}) = \boldsymbol{t}(\boldsymbol{u}) + \boldsymbol{t}(\boldsymbol{v}), \qquad \forall \boldsymbol{u} \text{ and } \boldsymbol{v} \in \boldsymbol{V} \tag{A.7}$$

$$\boldsymbol{t}(\alpha \boldsymbol{u}) = \alpha \boldsymbol{t}(\boldsymbol{u}) \qquad \forall \alpha \text{ (where } \alpha \text{ is a scalar)} \tag{A.8}$$

Dyadic product

The dyadic product $\otimes$ is a linear vectorial transform defined by

$$(\boldsymbol{a} \otimes \boldsymbol{b})\boldsymbol{c} \equiv \boldsymbol{a}(\boldsymbol{b} \cdot \boldsymbol{c}) \tag{A.9}$$

A second-order tensor $\tilde{T}$ is defined by

$$\tilde{T} \equiv T_{ij} \hat{\boldsymbol{I}}_i \otimes \hat{\boldsymbol{I}}_j \tag{A.10}$$

Indeed, we have

$$\boldsymbol{t}(\boldsymbol{u}) = \tilde{T}\boldsymbol{u} = (T_{ij} \hat{\boldsymbol{I}}_i \otimes \hat{\boldsymbol{I}}_j) u_k \hat{\boldsymbol{I}}_k \tag{A.11}$$

$$= T_{ij} \hat{\boldsymbol{I}}_i (\hat{\boldsymbol{I}}_j \cdot u_k \hat{\boldsymbol{I}}_k) \tag{A.12}$$

$$= T_{ij} u_k \delta_{jk} \hat{\boldsymbol{I}}_i = T_{ik} u_k \hat{\boldsymbol{I}}_i \tag{A.13}$$

Hence

$$\boldsymbol{t}(\boldsymbol{u}) = \tilde{T}\boldsymbol{u} = T_{ij} u_j \hat{\boldsymbol{I}}_i \tag{A.14}$$

or

$$t_i(\boldsymbol{u}) = T_{ij} u_j \tag{A.15}$$

Note that in equation (A.15) summation is made over the second index of the tensorial component T_{ij} (the notation $\boldsymbol{u}\tilde{T}$ has not been defined), and we observe that

$$\hat{\boldsymbol{I}}_i \cdot (\tilde{T}\hat{\boldsymbol{I}}_j) = \hat{\boldsymbol{I}}_i \cdot (T_{kp}\hat{\boldsymbol{I}}_k \otimes \hat{\boldsymbol{I}}_p)\hat{\boldsymbol{I}}_j \tag{A.16}$$

$$= \hat{\boldsymbol{I}}_i \cdot (T_{kp}\delta_{pj})\hat{\boldsymbol{I}}_k \tag{A.17}$$

$$= \hat{\boldsymbol{I}}_i \cdot T_{kj}\hat{\boldsymbol{I}}_k = T_{kj}\delta_{ik} = T_{ij} \tag{A.18}$$

or, in summary,

$$\hat{\boldsymbol{I}}_i \cdot (\hat{T}\hat{\boldsymbol{I}}_j) = T_{ij} \tag{A.19}$$

The zero tensor $\tilde{0}$ transforms any vector $\boldsymbol{a}$ of $\mathbf{V}$ into the vector $\mathbf{0}$. The unit tensor, $\tilde{1} = (\hat{\boldsymbol{I}}_i \otimes \hat{\boldsymbol{I}}_i)$, with components δ_{ij}, $i, j = 1, 2, 3$, transforms any vector $\boldsymbol{a}$ into itself ($\boldsymbol{a}$).

A.3 Algebra of tensors

For second-order tensors $\tilde{T}$, $\tilde{S}$, $\tilde{R}$, a vector $\boldsymbol{u}$ and a scalar α we have

$$(\tilde{T} + \tilde{S})\boldsymbol{u} = \tilde{T}\boldsymbol{u} + \tilde{S}\boldsymbol{u}, \qquad (\tilde{T} + \tilde{S})_{ij} = T_{ij} + S_{ij} \tag{A.20}$$

$$\alpha(\tilde{T}\boldsymbol{u}) = (\alpha\tilde{T})\boldsymbol{u}, \qquad (\alpha\tilde{T})_{ij} = \alpha T_{ij} \tag{A.21}$$

$$(\tilde{T}\tilde{S})\boldsymbol{u} = \tilde{T}(\tilde{S}\boldsymbol{u}) \qquad \forall \boldsymbol{u} \in \mathbf{V} \tag{A.22}$$

$$(\tilde{T}\tilde{S})\tilde{R} = \tilde{T}(\tilde{S}\tilde{R}) \tag{A.23}$$

$$\tilde{T}(\tilde{S} + \tilde{R}) = \tilde{T}\tilde{S} + \tilde{T}\tilde{R} \tag{A.24}$$

$$\alpha(\tilde{T}\tilde{S}) = (\alpha\tilde{T})\tilde{S} = \tilde{T}(\alpha\tilde{S}) \tag{A.25}$$

$$(\tilde{T}\tilde{S})_{ij} = T_{ik}\,S_{kj} \tag{A.26}$$

The derivation of equation (A.26) makes use of equation (A.19):

$$(\tilde{T}\tilde{S})_{ij} = \hat{\boldsymbol{I}}_i \cdot [(\tilde{T}\tilde{S})\hat{\boldsymbol{I}}_j] \tag{A.27}$$

$$= \hat{\boldsymbol{I}}_i \cdot [\tilde{T}(\tilde{S}\hat{\boldsymbol{I}}_j)] \tag{A.28}$$

$$= \hat{\boldsymbol{I}}_i \cdot [\tilde{T}(S_{mn}\hat{\boldsymbol{I}}_m \otimes \hat{\boldsymbol{I}}_n)\hat{\boldsymbol{I}}_j] \tag{A.29}$$

$$= \hat{\boldsymbol{I}}_i \cdot [\tilde{T}S_{mn}\delta_{nj}\hat{\boldsymbol{I}}_m] = \hat{\boldsymbol{I}}_i \cdot (\tilde{T}S_{mj}\hat{\boldsymbol{I}}_m) \tag{A.30}$$

$$= S_{mj}(\hat{\boldsymbol{I}}_i \cdot \tilde{T}\hat{\boldsymbol{I}}_m) = S_{mj}T_{im} = T_{im}S_{mj} \tag{A.31}$$

Nonsingular tensor; inverse tensor

A tensor $\tilde{T}$ is said to be nonsingular if

$$\tilde{T}\boldsymbol{u} = 0 \qquad \Leftrightarrow \qquad \boldsymbol{u} = 0 \tag{A.32}$$

If the tensor $\tilde{T}$ is nonsingular, an inverse tensor $\tilde{T}^{-1}$ is defined by

$$\tilde{T}\boldsymbol{u} = \boldsymbol{v} \qquad \Rightarrow \qquad \boldsymbol{u} = \tilde{T}^{-1}\boldsymbol{v} \tag{A.33}$$

Let us show that $(\tilde{T}\tilde{S})^{-1} = \tilde{S}^{-1}\tilde{T}^{-1}$:

$$(\tilde{T}\tilde{S})\boldsymbol{u} = \boldsymbol{v} \qquad \Rightarrow \qquad \boldsymbol{u} = (\tilde{T}\tilde{S})^{-1}\boldsymbol{v} \tag{A.34}$$

but

$$(\tilde{T}\tilde{S})\boldsymbol{u} = \tilde{T}(\tilde{S}\boldsymbol{u}) = \boldsymbol{v} \qquad \Rightarrow \qquad \tilde{S}\boldsymbol{u} = \tilde{T}^{-1}\boldsymbol{v} \tag{A.35}$$

and

$$\boldsymbol{u} = \tilde{S}^{-1}\tilde{T}^{-1}\boldsymbol{v} = (\tilde{T}\tilde{S})^{-1}\boldsymbol{v} \tag{A.36}$$

Transpose of a tensor

The transpose $\tilde{T}^T$ of a tensor $\tilde{T}$ is defined by

$$\tilde{T}\boldsymbol{v}\cdot\boldsymbol{u} = \tilde{T}^T\boldsymbol{u}\cdot\boldsymbol{v} \qquad \forall\,\boldsymbol{u}, \forall\,\boldsymbol{v} \in \mathbf{V} \tag{A.37}$$

The components of the transpose tensor $\tilde{T}^T$ are

$$(\tilde{T}^T)_{ij} = \hat{\boldsymbol{I}}_i\cdot(\tilde{T}^T\hat{\boldsymbol{I}}_j) = \tilde{T}\hat{\boldsymbol{I}}_i\cdot\hat{\boldsymbol{I}}_j = T_{ji} \tag{A.38}$$

Symmetric and skew tensors

A tensor $\tilde{T}$ is symmetric if $\tilde{T} = \tilde{T}^T$. A symmetrical tensor, $\tilde{T} = \tilde{T}^T$, is said to be definite positive when

$$\boldsymbol{a}\cdot\tilde{T}\boldsymbol{a} > 0 \qquad \forall\,\boldsymbol{a} \neq 0 \tag{A.39}$$

A skew tensor $\tilde{T}$ is such that $\tilde{T} = -\tilde{T}^T$.

Orthogonal tensors

A tensor $\tilde{Q}$ is orthogonal if $\tilde{Q}^T = \tilde{Q}^{-1}$. Hence an orthogonal tensor $\tilde{Q}$ is such that $\tilde{Q}\tilde{Q}^T = \tilde{1}$.

Transformation by an orthogonal tensor does not affect the scalar product of two vectors. Indeed, if $\boldsymbol{a} = \tilde{Q}\boldsymbol{u}$ and $\boldsymbol{b} = \tilde{Q}\boldsymbol{v}$ then

$$\boldsymbol{a}\cdot\boldsymbol{b} = (\tilde{Q}\boldsymbol{u})\cdot(\tilde{Q}\boldsymbol{v}) = [(\tilde{Q}^T\tilde{Q})\boldsymbol{u}]\cdot\boldsymbol{v} = \boldsymbol{u}\cdot\boldsymbol{v} \tag{A.40}$$

A.4 Trace and determinant

Trace

The trace of a tensor, $\operatorname{tr}\tilde{T}$, is a scalar defined as follows:

$$\operatorname{tr}(\boldsymbol{a}\otimes\boldsymbol{b}) = \boldsymbol{a}\cdot\boldsymbol{b} \tag{A.41}$$

so that

$$\operatorname{tr}\tilde{T} = \operatorname{tr}[T_{ij}(\hat{\boldsymbol{I}}_i\otimes\hat{\boldsymbol{I}}_j)] = T_{ij}\operatorname{tr}(\hat{\boldsymbol{I}}_i\otimes\hat{\boldsymbol{I}}_j) = T_{ij}\delta_{ij} = T_{ii} \tag{A.42}$$

Thus the trace of a tensor is the sum of its diagonal components. Note that

$$\operatorname{tr}\tilde{T} = \operatorname{tr}\tilde{T}^T, \quad \operatorname{tr}\tilde{T}\tilde{S} = \operatorname{tr}\tilde{S}\tilde{T}, \quad \operatorname{tr}\tilde{T}\tilde{T}^T > 0 \qquad \forall\,\tilde{T} \neq \tilde{0} \tag{A.43}$$

The magnitude of a tensor, $|\tilde{T}|$, is given by

$$|\tilde{T}| = \sqrt{\mathrm{tr}\,\tilde{T}\tilde{T}^T} \tag{A.44}$$

Determinant

Let us consider three nonparallel vectors $\boldsymbol{e}_i$; $i = 1, 2, 3$. A tensor $\tilde{T}$ transforms these three vectors into three vectors $\boldsymbol{a}_i$ such that $\boldsymbol{a}_i = \tilde{T}\boldsymbol{e}_i$.

The determinant of tensor $\tilde{T}$, $\det\tilde{T}$, is a scalar defined as the ratio of the volume $V(\boldsymbol{a}_i)$ associated with the vectors $\boldsymbol{a}_i$ and the volume $v(\boldsymbol{e}_i)$ associated with the vectors $\boldsymbol{e}_i$:

$$\det\tilde{T} = \frac{V(\boldsymbol{a}_i)}{v(\boldsymbol{e}_i)} = \frac{\boldsymbol{a}_1 \wedge \boldsymbol{a}_2 \cdot \boldsymbol{a}_3}{\boldsymbol{e}_1 \wedge \boldsymbol{e}_2 \cdot \boldsymbol{e}_3} \tag{A.45}$$

Let $\boldsymbol{e}_i$; $i = 1, 2, 3$, be the unit vectors of an orthogonal base $\boldsymbol{e}_i = \hat{\boldsymbol{I}}_i$, so that $v(\hat{\boldsymbol{I}}_i) = 1$. Then

$$\boldsymbol{a}_i = \tilde{T}\hat{\boldsymbol{I}}_i = T_{kl}(\hat{\boldsymbol{I}}_k \otimes \hat{\boldsymbol{I}}_l)\hat{\boldsymbol{I}}_i = T_{ki}\hat{\boldsymbol{I}}_k \tag{A.46}$$

and

$$\boldsymbol{a}_1 \wedge \boldsymbol{a}_2 \cdot \boldsymbol{a}_3 = T_{k1}\hat{\boldsymbol{I}}_k \wedge T_{p2}\hat{\boldsymbol{I}}_p \cdot T_{l3}\hat{\boldsymbol{I}}_l \tag{A.47}$$

$$= T_{k1}T_{p2}T_{l3}(\hat{\boldsymbol{I}}_k \wedge \hat{\boldsymbol{I}}_p \cdot \hat{\boldsymbol{I}}_l) \tag{A.48}$$

$$= T_{k1}T_{p2}T_{l3}\varepsilon_{kpr}\hat{\boldsymbol{I}}_r \cdot \hat{\boldsymbol{I}}_l = T_{k1}T_{p2}T_{l3}\,\varepsilon_{kpr}\,\delta_{rl} \tag{A.49}$$

so that

$$\det\tilde{T} = \varepsilon_{kpl}\,T_{k1}T_{p2}T_{l3} \tag{A.50}$$

Note that

$$\varepsilon_{rst}\det\tilde{T} = \varepsilon_{rst}\,\varepsilon_{ijk}\,T_{i1}T_{j2}T_{k3} \tag{A.51}$$

$$= \varepsilon_{rst}\,\varepsilon_{ijk}\,T_{ir}\delta_{r1}T_{js}\delta_{s2}T_{kt}\delta_{t3} \tag{A.52}$$

$$= \varepsilon_{123}\,\varepsilon_{ijk}\,T_{ir}T_{js}T_{kt} \tag{A.53}$$

$$= \varepsilon_{ijk}\,T_{ir}T_{js}T_{kt} \tag{A.54}$$

since $\varepsilon_{123} = +1$. Note that

$$\det\tilde{T}\tilde{S} = \det\tilde{T}\det\tilde{S} \tag{A.55}$$

$$\det\left(\tilde{T}^{-1}\right) = \left(\det\tilde{T}\right)^{-1} \tag{A.56}$$

$$\det\tilde{T} = \det\tilde{T}^T \tag{A.57}$$

A.5 Change of orthogonal frame of reference

A.5.1 Vector components

Let us consider two orthogonal frames of reference $\hat{\boldsymbol{I}}_i$; $i = 1, 2, 3$, and $\hat{\boldsymbol{I}}'_i$; $i = 1, 2, 3$, such that $\hat{\boldsymbol{I}}'_i = \tilde{Q}\hat{\boldsymbol{I}}_i$ where $\tilde{Q}$ is an orthogonal tensor.

If a vector $\boldsymbol{u}$ has components u_i in the frame of reference $\hat{\boldsymbol{I}}_i$ and components u'_i in the frame of reference $\hat{\boldsymbol{I}}'_i$ ($\boldsymbol{u} = u_i\hat{\boldsymbol{I}}_i = u'\hat{\boldsymbol{I}}'_i$), we may write

$$\boldsymbol{u} \cdot \hat{\boldsymbol{I}}'_j = u_i\hat{\boldsymbol{I}}_i \cdot \hat{\boldsymbol{I}}'j = u'_i\hat{\boldsymbol{I}}'_i \cdot \hat{\boldsymbol{I}}'_j = u'_i\delta_{ij} = u'_j \tag{A.58}$$

so that

$$u'_j = u_i\hat{\boldsymbol{I}}_i \cdot (\tilde{Q}\hat{\boldsymbol{I}}_j) = u_iQ_{ij} = Q^T_{ji}u_i \tag{A.59}$$

Note that $Q_{ij} = \hat{\boldsymbol{I}}_i \cdot \tilde{Q}\hat{\boldsymbol{I}}_j = \hat{\boldsymbol{I}}_i \cdot \hat{\boldsymbol{I}}'_j = \cos(\hat{\boldsymbol{I}}_i, \hat{\boldsymbol{I}}'_j)$. In equation (A.59) the tensor $\tilde{Q}$ is transposed, in order to satisfy a convention that summation must involve the second index of the second-order tensor component, and we conclude that

$$u'_j = Q^T_{ji}u_i \tag{A.60}$$

A.5.2 Second-order tensor components

Let us consider a tensor $\tilde{T}$ as expressed in the basis $\hat{\boldsymbol{I}}_i$ and in the basis $\hat{\boldsymbol{I}}'_i$:

$$\tilde{T} = T_{ij}(\hat{\boldsymbol{I}}_i \otimes \hat{\boldsymbol{I}}_j) = T'_{ij}(\hat{\boldsymbol{I}}'_i \otimes \hat{\boldsymbol{I}}'_j) \tag{A.61}$$

from which we observe that

$$\tilde{T}\hat{\boldsymbol{I}}'_p = T_{ij}(\tilde{\boldsymbol{I}}_i \otimes \hat{\boldsymbol{I}}_j)\hat{\boldsymbol{I}}'_p = T'_{ij}(\hat{\boldsymbol{I}}'_i \otimes \hat{\boldsymbol{I}}'_j)\hat{\boldsymbol{I}}'_p = T'_{ip}\hat{\boldsymbol{I}}'_i \tag{A.62}$$

but

$$T_{ij}(\tilde{I}_i \otimes \hat{\boldsymbol{I}}_j)\hat{\boldsymbol{I}}'_p = T_{ij}\hat{\boldsymbol{I}}_i(\hat{\boldsymbol{I}}_j \cdot \hat{\boldsymbol{I}}'_p) = T_{ij}\hat{\boldsymbol{I}}_iQ_{jp} \tag{A.63}$$

so that

$$T_{ij}Q_{jp}(\hat{\boldsymbol{I}}_i \cdot \hat{\boldsymbol{I}}'_m) = T'_{ip}\hat{\boldsymbol{I}}'_i \cdot \hat{\boldsymbol{I}}'_m = T'_{ip}\delta_{im} = T'_{mp} \tag{A.64}$$

and

$$T_{ij}Q_{im}Q_{jp} = T'_{mp} \tag{A.65}$$

or

$$T'_{mp} = Q^T_{mi}T_{ij}Q_{jp} \tag{A.66}$$

A.6 Eigenvalues and eigenvectors of a second-order tensor

For a given tensor $\tilde{T}$, $\boldsymbol{e}$ is an eigenvector of $\tilde{T}$ if

$$\tilde{T}\boldsymbol{e} = t\tilde{T} \quad \text{or} \quad (\tilde{T} - t\tilde{1})\boldsymbol{e} = \boldsymbol{0} \tag{A.67}$$

where t is a scalar. Any eigenvector $\boldsymbol{e}$ of $\tilde{T}$ can be transformed into a collinear vector. The tensor $\tilde{T} - t\tilde{1}$ must be singular if $\boldsymbol{e} \neq \boldsymbol{0}$, which implies that

$$\det(\tilde{T} - t\tilde{1}) = 0 \tag{A.68}$$

or

$$-t^3 + I_{\tilde{T}}t^2 - II_{\tilde{T}}t + III_{\tilde{T}} = 0 \tag{A.69}$$

Equation (A.69) is called the characteristic equation. It is a cubic equation with roots t_1, t_2 and t_3 such that

$$I_{\tilde{T}} = t_1 + t_2 + t_3 \tag{A.70}$$

$$II_{\tilde{T}} = t_1t_2 + t_2t_3 + t_3t_1 \tag{A.71}$$

$$III_{\tilde{T}} = t_1t_2t_3 \tag{A.72}$$

Proposition

If $\tilde{T}$ is symmetric, $\tilde{T} = \tilde{T}^T$, then the three roots are real.

Let $t = \lambda + i\mu$, where λ is the real part of t and μ the imaginary part. Associated with t we have the eigenvector $\boldsymbol{e} = \boldsymbol{u} + i\boldsymbol{v}$, which satisfies equation (A.67), so that

$$(\tilde{T} - t\tilde{\boldsymbol{I}})\boldsymbol{e} = [\tilde{T} - (\lambda + i\mu)\tilde{1}](\boldsymbol{u} + i\boldsymbol{v}) = \boldsymbol{0} \tag{A.73}$$

This has a real part and an imaginary part:

$$\tilde{T}\boldsymbol{u} - \lambda\boldsymbol{u} + \mu\boldsymbol{v} = \boldsymbol{0} \tag{A.74}$$

$$\tilde{T}\boldsymbol{v} - \lambda\boldsymbol{v} - \mu\boldsymbol{u} = \boldsymbol{0} \tag{A.75}$$

Dot-multiplying equation (A.74) by $\boldsymbol{v}$ and equation (A.75) by $\boldsymbol{u}$, we obtain

$$(\tilde{T}\boldsymbol{u}) \cdot \boldsymbol{v} - \lambda\boldsymbol{u} \cdot \boldsymbol{v} + \mu\boldsymbol{v} \cdot \boldsymbol{v} = \boldsymbol{0} \tag{A.76}$$

$$(\tilde{T}\boldsymbol{v}) \cdot \boldsymbol{u} - \lambda\boldsymbol{v} \cdot \boldsymbol{u} - \mu\boldsymbol{u} \cdot \boldsymbol{u} = \boldsymbol{0} \tag{A.77}$$

The tensor $\tilde{T}$ is symmetrical, so that $(\tilde{T}\boldsymbol{u}) \cdot \boldsymbol{v} = (\tilde{T}^T\boldsymbol{v}) \cdot \boldsymbol{u} = (\tilde{T}\boldsymbol{v}) \cdot \boldsymbol{u}$. Subtracting (A.76) from (A.77) we obtain

$$\mu(|\boldsymbol{u}|^2 + |\boldsymbol{v}|^2) = 0 \tag{A.78}$$

which implies that μ is null and so the eigenvalue $t = \lambda$ is real.

It may be shown that to each eigenvalue there corresponds an eigenvector and that the three eigenvectors are orthogonal and form an orthogonal frame of reference. In this eigenvector frame of reference the components of the tensor $\tilde{T}$ are

$$(\tilde{T})_{ij} = \begin{pmatrix} \lambda_1 & 0 & 0 \\ 0 & \lambda_2 & 0 \\ 0 & 0 & \lambda_3 \end{pmatrix} \qquad \text{for } i,j = 1,2,3 \tag{A.79}$$

so that

$$I_{\tilde{T}} = \lambda_1 + \lambda_2 + \lambda_3 = \operatorname{tr} \tilde{T} \tag{A.80}$$

$$II_{\tilde{T}} = \lambda_1\lambda_2 + \lambda_2\lambda_3 + \lambda_3\lambda_1 = \tfrac{1}{2}[(\operatorname{tr} \tilde{T})^2 - \operatorname{tr} \tilde{T}^2] \tag{A.81}$$

$$III_{\tilde{T}} = \lambda_1\lambda_2\lambda_3 = \det \tilde{T} \tag{A.82}$$

(A.83)

Invariants of a tensor

For all second-order symmetrical tensors, the three quantities $I_{\tilde{T}}$ $II_{\tilde{T}}$ and $III_{\tilde{T}}$ are independent of the orthogonal frame of reference. They are known as the invariants of $\tilde{T}$ and are such that

$$I_{\tilde{T}} = \operatorname{tr} \tilde{T}, \quad II_{\tilde{T}} = \tfrac{1}{2}[(\operatorname{tr} \tilde{T})^2 - \operatorname{tr} \tilde{T}^2], \quad III_{\tilde{T}} = \det \tilde{T} \tag{A.84}$$

Geometric characterization

Let $\tilde{T}$ be a positive definite tensor, and consider the scalar function ϕ of a vector $\boldsymbol{u}$ defined by

$$\phi(\boldsymbol{u}) \equiv \boldsymbol{u} \cdot (\tilde{T}\boldsymbol{u}) \tag{A.85}$$

i.e. we consider the projection of $\tilde{T}\boldsymbol{u}$ onto $\boldsymbol{u}$ for all possible orientations of $\boldsymbol{u}$.

Let $\boldsymbol{e}_i$, $i = 1, 2, 3$, be the eigenvectors associated with $\tilde{T}$ and $\hat{u}_i$ be the components of $\boldsymbol{u}$ in the eigenvector frame of reference; then

$$\phi(\boldsymbol{u}) = \lambda_1\hat{u}_1^2 + \lambda_2\hat{u}_2^2 + \lambda_3\hat{u}_3^2 \tag{A.86}$$

where $\lambda_1, \lambda_2, \lambda_3$ are the eigenvalues of $\tilde{T}$.

Equation (A.86) defines an ellipsoid, the axes of which are the eigenvectors of $\tilde{T}$. The lengths of the semi-axes of the ellipsoid are directly related to the eigenvalues. When all three eigenvalues are equal, the ellipsoid becomes a sphere and any vector in the vector space is an eigenvector. When any two eigenvalues are equal, the ellipsoid exhibits an axis of symmetry which corresponds to the eigenvector associated with the third eigenvalue, $\boldsymbol{e}_3$. Any vector in the plane normal to $\boldsymbol{e}_3$ is then an eigenvector.

A.7 Polar decomposition of a tensor

Proposition

Any second-order tensor $\tilde{F}$ may be decomposed into the product of a symmetric tensor and an orthogonal tensor

$$\tilde{F} = \tilde{R}\tilde{U} = \tilde{V}\tilde{R} \tag{A.87}$$

where $\tilde{U}$ and $\tilde{V}$ are positive definite symmetric tensors and $\tilde{R}$ is an orthogonal tensor.

Demonstration

We observe first that $\tilde{F}\tilde{F}^T$ is a positive definite tensor. Indeed, $(\tilde{F}\tilde{F}^T)^T = \tilde{F}\tilde{F}^T$ and

$$\boldsymbol{a} \cdot (\tilde{F}\tilde{F}^T)\boldsymbol{a} = \tilde{F}^T\boldsymbol{a} \cdot \tilde{F}^T\boldsymbol{a} > 0 \qquad \forall \boldsymbol{a} \neq 0 \tag{A.88}$$

Let define $\tilde{U} = \sqrt{\tilde{F}\tilde{F}^T}$ such that the eigenvalues of $\tilde{U}$ are equal to the square roots of the eigenvalues of $\tilde{F}\tilde{F}^T$ and the eigenvectors of $\tilde{U}$ are the same as the eigenvectors of $\tilde{F}\tilde{F}^T$. Then $\tilde{U} = \tilde{U}^T$ is positive definite.

Define $\tilde{R} \equiv \tilde{F}\tilde{U}^{-1}$. We note that

$$\tilde{R}^T = (\tilde{U}^{-1})^T\tilde{F}^T = (\tilde{U}^T)^{-1}\tilde{F}^T = \tilde{U}^{-1}\tilde{F}^T \tag{A.89}$$

so that

$$\tilde{R}^T\tilde{R} = \tilde{U}^{-1}(\tilde{F}^T\tilde{F})\tilde{U}^{-1} = \tilde{U}^{-1}(\tilde{U}\tilde{U})\tilde{U}^{-1} = \tilde{1} \tag{A.90}$$

which demonstrates that $\tilde{R}$ is orthogonal and such that

$$\tilde{R}\tilde{U} = \tilde{F}\tilde{U}^{-1}\tilde{U} = \tilde{F} \tag{A.91}$$

Let us now define $\tilde{V} \equiv \tilde{R}\tilde{U}\tilde{R}^T$; we note that

$$\tilde{V}^T = \tilde{R}\tilde{U}\tilde{R}^T = \tilde{V} \tag{A.92}$$

and

$$\tilde{V}\tilde{R} = \tilde{R}\tilde{U}\tilde{R}^T\tilde{R} = \tilde{R}\tilde{U} = \tilde{F}\tilde{U}^{-1}\tilde{U} = \tilde{F} \tag{A.93}$$

Hence any tensor may be decomposed into the product of a symmetric tensor and an orthogonal tensor.

References

Aamodt, R. L., and Kuriyagawa, M., 1983. Measurement of instantantaneous shut-in pressure, pp. 139–142 of Zoback, M., and Haimson, B. (eds.), *Hydraulic Fracturing Stress Measurements*, National Academy Press, Washington D.C.

Abercrombie, R., McGarr, A., Di Toro, G., and Kanamori, H., 2006. *Earthquakes: Radiated Energy and the Physics of Faulting*. American Geophysical Union.

Abousleiman, Y., and Ekbote, S., 2005. Solutions for the inclined porothermoelastic transversely isotropic medium. *J. Appl. Mech.*, **72**, 102–114.

Adachi, J., Siebrits, E., Peirce, A., and Desroches, J., 2007. Computer simulation of hydraulic fractures. *Int. J. Rock Mech. Min. Sci.*, **44**, 739–757.

Adelinet, M., Dorbath, C., Le Ravalec, M., Fortin, J., and Guéguen, Y., 2011. Deriving microstructure and fluid state within the Icelandic crust from the inversion of tomographic data. *Geophys. Res. Lett.*, **38**, L03305.

Adler, P., and Thovert, J.-F., 1999. *Fractures and Fracture Networks*. Kluwer Academic.

Aki, K., and Richards, P. G., 1980. *Quantitative Seismology: Theory and Methods, Vol. 1*. W. H. Freeman.

Amadei, B., and Stephansson, O., 1997. *Rock Stress and its Measurement*. Chapman and Hall.

Anderson, E. M., 1905. The dynamics of faulting. *Trans. Edinburgh Geol. Soc.*, **8**, 387–402.

Anderson, T. L., 1995. *Fracture Mechanics: Fundamentals and Applications*. CRC Press.

Angelier, J., 2002. Inversion of earthquake focal mechanisms to obtain seismotectonic stress, IV – a new method free of choices among nodal planes. *Geophys. J. Int.*, **150**, 588–609.

Angelier, J., and Goguel, J., 1979. Sur une méthode simple de détermination des axes principaux de contraintes pour une population de failles. *C. R. Acad. Sci. Paris*, **288**, 307–310.

Atkinson, B. K., 1987. *Fracture Mechanics of Rock*. Academic Press.

Atkinson, B. K., and Meredith, P. G., 1987a. The theory of subcritical crack growth with application to minerals and rocks, pp. 111–162 of Atkinson, B. K. (ed.), *Fracture Mechanics of Rock*, Academic Press.

Atkinson, B. K., and Meredith, P. G., 1987b. Experimental fracture mechanics data for rocks and minerals, pp. 477–427 of Atkinson, B. K. (ed.), *Fracture Mechanics of Rock*, Academic Press.

Baer, M., and Kradolfer, U., 1987. An automatic phase picker for local and teleseismic events. *Bull. Seism. Soc. Am.*, **77**, 1437–1445.

Bame, D., and Fehler, M., 1986. Observation of long period earthquakes accompanying hydraulic fracturing. *Geophys. Res. Lett.*, **13**, 149–152.

Bandis, S. C., Lumsden, A. C., and Barton N. R., 1981. Experimental studies of scale effects on the shear behaviour of rock joints. *Int. J. Rock Mech. Min. Sci. Geomech. Abs.*, **18**, 1–21.

Bandis, S. C., Lumsden, A. C., and Barton N. R., 1983. Fundamentals of rock joint deformation. *Int. J. Rock Mech. Min. Sci. Geomech. Abs.*, **20**, 249–269.

Bandis, S. C., 1990. Mechanical properties of rock joints, pp. 125–140 in Barton, N., and Stephansson, O. (eds.) *Rock Joints, Proc. Int. Symp. on Rock Joints*, A. A. Balkema (CRC Press).

Barton, N., 1976. The shear strength of rock and rock joints. *Int. J. Rock Mech. Min. Sci. Geomech. Abs.*, **13**, 255–279.

Barton, N., Bandis, S. C., and Bakhtar, K., 1985. Strength, deformation and conductivity coupling of rock joints. *Int. J. Rock Mech. Min. Sci. Geomech. Abs.*, **22**, 121–140.

Barton, N., and Stephansson, O. (eds.), 1990. *Rock Joints Proc. Int. Symp. on Rock Joints*, A. A. Balkema (CRC Press).

Beauducel, F., and Cornet, F. H., 1999. Collections and three-dimensional modeling of GPS and tilt data at Merapi volcano, Java. *J. Geophys. Res.*, **104**, 725–736.

Beauducel, F., Cornet, F. H., Suhanto, E., Duquesnoy, T., and Kasser, M., 2000. Constraints on magma flux from displacements data at Merapi volcano, Java, Indonesia. *J. Geophys. Res.*, **105**, 8193–8203.

Bell, J. S., and Gough, D. I., 1979. Northeast–southwest compressive stress in Alberta: evidence from oil wells. *Earth Planet. Sci. Lett.*, **45**, 475–482.

Ben-Zion, Y., and Sammis, C., 2003. Scaling of fracture systems in geological media. *Rev. Geophys.*, **39**, 347–383.

Ben-Zion, Y., 2008. Collective behavior of earthquakes and faults: continuum–discrete transitions, progresssive evolutionary changes and different dynamic regimes. *Rev. Geophys.*, **46**, RG4006.

Bérard, T., and Cornet, F. H., 2003. Evidence of thermally induced borehole elongation: a case study at Soultz, France. *Int. J. Rock Mech. Min. Sci.*, **40**, 1121–1140.

Berest, P., and Weber, Ph., 1988. *La thermomécanique des roches*. Éditions BRGM.

Bernaix, J., 1967. *Étude géotechnique de la roche de Malpasset; contribution à l'étude de la stabilité des appuis des barages-voûtes*, Dunod.

Bernard, P., Boudin, F., Sacks, S., Linde, A., Blum, P. A., Courteille, C., *et al.* 2004. Continuous and tilt monitoring on the Trizonia Island, Rift of Corinth, Greece. *C.R. Geosci.*, **336**, 313–325.

Berry, D. S., 1960. Some kinetic considerations on the Griffith criterion of fracture. *J. Mech. Phys. Solids*, **8**, 194–216.

Berryman, J. G., 1992. Effective stress for transport properties of inhomogeneous porous rock. *J. Geophys. Res.*, **78**, 17409–17424.

Bésuelle, P., and Rudnicki, J. W., 2004. Localization: shear bands and compaction bands, pp. 219–321 of Guéguen Y., and Boutéca, M. (eds.), *Mechanics of Fluid Saturated Rocks*, Elsevier.

Biot, M., 1941. General theory of three dimensional consolidation. *J. Appl. Phys.*, **12**, 155–164.

Biot, M., 1956a. General solutions of the equations of elasticity and consolidation for a porous material. *J. Appl. Mech.*, **78**, 91–96.

Biot, M., 1956b. Theory of propagation of elastic waves in a fluid saturated porous solid. Low-frequency range. *J. Acoust. Soc. Am.*, **28**, 168–178.

Biot, M., 1962. Mechanics of deformation and acoustic propagation in porous media. *J. Appl. Phys.*, **33**, 1482–1498.

Biot, M., and Willis, D. G., 1957. The elastic coefficients of the theory of consolidation. *J. Appl. Mech.*, **24**, 594–601.

Bonnet, E., Bour, O., Odling, N. E., Davy, P., Main, I., Cowie, P., *et al.*, 2001. Scaling of fracture systems in geological media. *Rev. Geophys.*, **39**, 347–383.

Bott, M. H. P., 1959. The mechanics of oblique slip faulting. *Geol. Mag.*, **96**, 109–117.

Boullier, A. M., Fujimoto, K., Ohtani, T., Roman-Ross, G., Lewin, E., Ito, H., *et al.*, 2004. Textural evidence for recent co-seismic circulation of fluids in the Nojima fault zone, Awaji island, Japan. *Tectonophys.*, **378**, 165–181.

Bourbié, T., Coussy, O., and Zinszner, B., 1986. *Acoustique des milieux poreux*. Editions Technip, 337 pp.

Bourouis, P., and Bernard, P., 2007. Evidence of coupled seismic and aseismic fault slip during water injection in the geothermal site of Soultz (France), and implications for seismogenic transients. *Geophys. J. Int.*, **169**, 723–732.

Bourouis, S., and Cornet, F. H., 2009. Microseismic activity and fluid fault interactions: some results from the Corinth Rift Laboratory. *Geophys. J. Int.*, **178**, 561–580.

Brace, W. F., and Bombolakis, E. G., 1963. A note on brittle crack growth in compression. *J. Geophys. Res.*, **68**, 3709–3713.

Brace, W. F., Paulding, E. T., and Scholtz, C., 1966. Dilatancy in the fracture of crystalline rocks. *J. Geophys. Res.*, **71**, 3939–3953.

Brace, W. F., and Martin, R. J., 1968. A test of the law of effective stress for crystalline rocks of low porosity. *Int. J. Rock Mech. Min. Sci.*, **5**, 415–426.

Brodsky, E. E., Roeloffs, E., Woodcock, D., Gall, I., and Manga, M., 2003. A mechanism for sustained groundwater pressure changes induced by distant earthquakes. *J. Geophys. Res.*, **108**, 2390. doi:10.1029/2002JB002321.

Brown, R. J. S., and Korringa, J., 1975. On the dependence of the elastic properties of a porous rock on the compressibility of the pore fluid. *Geophysics*, **40**, 608–616.

Brown, E. T. (ed.), 1981. *ISRM Suggested Methods.* Pergamon Press (Elsevier).

Brown, S. R., and Scholz, C. H., 1985. Broad bandwidth study of the topography of natural rock surfaces. *J. Geophys. Res.*, **90**, 12575–12582.

Brudy, M., Zoback, M. D., Fuchs, K., Rummel, F., and Baumgärtner, J., 1997. Estimation of the complete stress tensor to 8 km depth in the KTB scientific drill holes: implication for crustal strength. *J. Geophys. Res.*, **102**, 18453–18475.

Brune, J. N., 1970. Tectonic stress and the spectra of seismic shear waves from earthquakes. *J. Geophys. Res.*, **75**, 4997–5009.

Brune, J. N., 1991. Seismic source dynamics, radiation and stress. *Rev. Geophys.*, **29**, 688–699.

Burov, E., and Diament, M., 1995. The effective elastic thickness (Te) of continental lithosphere: what does it really mean? *J. Geophys. Res.*, **100**, 3905–3927.

Burov, E., and Watts, A., 2006. The long-term strength of continental lithosphere: "jelly sandwich" or "crème brulée"? *GSA Today*, **16**, 4.

Burov, E. B., 2009. Plate rheology and mechanics, pp. 99–151 of Watts, A. B. (ed.), *Crust and Lithosphere Dynamics. Vol. 6, Treatise of Geophysics.* Elsevier.

Burov, E., 2010. The equivalent elastic thickness (Te), seismicity and the long-term rheology of continental lithosphere: time to burn-out "crème brûlée"? Insights from large-scale geodynamic modeling. *Tectonophys.*, **484**, 4–26.

Burridge, R., and Knopoff, L., 1964. Body force equivalents for seismic dislocation. *Bull. Seism. Soc. Am.* **54**, 1875–1888.

Byerlee, J., 1978. Friction of rocks. *Pure Appl. Geophys.*, **116**, 615–626.

Byerlee, J., and Brace, W. F., 1968. Stick-slip, stable sliding, and earthquakes – Effect of rock type, pressure, strain rate and stiffness. *J. Geophys. Res.*, **73**, 6031–6037.

Callen, H. B., 1985. *Thermodynamics and an Introduction to Thermostatistics*. John Wiley & Sons.

Calo, M., Dorbath, C., Cornet, F. H., and Cuenot, N., 2011. Large scale aseismic motion identified through 4D P-wave tomography. *Geophys. J. Int.*, **186**, 1295–1314.

Carslaw, H. S., and Jaeger, J. C., 1986. *Conduction of Heat in Solids*. Clarendon Press.

Cauchy, A. L., 1829. Memoire sur l'équilibre et le mouvement des corps élastique. *Mém. Acad Paris*, **8**.

Cauchy, A. L., 1864. Recherches sur l'équilibre et le mouvement intérieur des corps solides ou fluides, élastiques ou non élastiques. *Bull. Soc. Phil. Math.*, **2**, 300–304.

Cayol, V., and Cornet, F. H., 1998. 3D Modeling of the 1983–1984 dike intrusion at Piton de la Fournaise Volcano, Reunion Island, Indian Ocean. *J. Geophys. Res.*, **103**, 18025–18037.

Célérier, B., Etchecopar, A., Bergerat, F., Vergely, P., Arthaud, F., and Laurent, P., 2012. Inferring stress from faulting: from early concepts to inverse methods. *Tectononophys.*, **581**, 206–219.

Chang, C., and Haimson, B. C., 2000. True triaxial strength and deformability of the German continental deep drilling program (KTB) deep hole amphibolite. *J. Geophys. Res.*, **105**, 18999–19013.

Chen, G., Chenevert, M. E., Sharma, M. M., and Yu, M., 2003. A study of wellbore stability in shales including poroelastic, chemical and thermal effects. *J. Petrol. Sci. Eng.*, **38**, 167–176.

Colmenares, L. B., and Zoback, M. D., 2002. A statistical evaluation of intact rock failure criteria constrained by polyaxial test data for five different rocks. *Int. J. Rock Mech. Min. Sci.*, **39**, 695–729.

Comninou, M., and Dundurs, J., 1979. A closed crack tip terminating at an interface. *J. Appl. Mech.*, **46**, 97–100.

Cook, N. G. W., 1965. The failure of rock. *Int. J. Rock. Mech. Min. Sci.*, **2**, 389–403.

Cook, N. G. W., and Hodgson, K., 1965. Some detailed stress strain surves for rock. *J. Geophys. Res.*, **70**, 2883–2888.

Cook, N. G. W., 1993. Jaeger Memorial Dedication Lecture. Natural joints in rock: mechanical, hydraulic and seismic behaviour and properties under normal stress. *Int. J. Rock Mech. Min. Sci. Geomech. Abs.*, **29**, 198–223.

Cornet, F. H., and Fairhurst, C., 1974. Influence of pore pressure on the deformation behavior of saturated rocks, pp. 638–644 of Fairhurst, C. (ed.), *Proc. 3rd Congress on Rock Mechanics, Denver*, US National Academy of Sciences.

Cornet, F. H., 1979. Comparative analysis by the displacement discontinuity method of two energy criteria of fracture. *J. Appl. Mech.*, **46**, 349–355.

Cornet, F. H., and Valette, B., 1984. In-situ stress determination from hydraulic injection test data. *J. Geophys. Res.*, **89**, 11527–11537.

Cornet, F. H., and Julien, P., 1989. Stress determination from hydraulic test data and focal mechanisms of induced seismicity. *J. Geophys. Res.*, **89**, 11527–11537.

Cornet, F. H., and Burlet, D., 1992. Stress field determinations in France by hydraulic tests in boreholes. *J. Geophys. Res.*, **97**, 11829–11849.

Cornet, F. H., and Yin, J. M., 1995. Analysis of induced seismicity for stress field determination and pore pressure mapping. *Pure Appl. Geophys.*, **145**, 677–700.

Cornet, F. H., 1996. A complete 3D stress determination for the design of an underground power station, pp. 755–762 of Aubertin, Hassani and Mitri (eds.), *Rock Mechanics Tools and Techniques, Proc. 2nd North Am. Rock Mechanics Symp.*, A. A. Balkema.

Cornet, F. H., Helm, J., Poitrenaud, H., and Etchecopar, A., 1997. Seismic and aseismic slips induced by large scale fluid injetions. *Pure Appl. Geophys.*, **150**, 563–583.

Cornet, F. H., Li, L., Hulin, J. P., Ippolito, I., and Kurowski, P., 2003. The hydromechanical behaviour of a single fracture: an *in situ* experimental case study. *Int. J. Rock Mech. Min. Sci.*, **40**, 1257–1270.

Cornet, F. H., Doan M. L., Moretti I., and Borm, G., 2004. Drilling through the active Aigion Fault: the AIG10 well observatory. *C.R. Geosci.*, **336**, 395–406.

Cornet, F. H., Bérard, Th., and Bourouis, S., 2007. How close to failure is a natural granite rock mass at 5 km depth? *Int. J. Rock Mech. Min. Sci.*, **40**, 47–66.

Cornet, F. H., and Röckel, T., 2012. Vertical stress profiles and the significance of stress decoupling. *Tectonophys.*, **581**, 193–205.

Cornet, F. H., 2012. The relationship between seismic and aseismic motions induced by forced fluid injections. *Hydrogeology J.*, **20**, 1463–1466.

Coussy., O., 1991. *Mécanique des milieux poreux*. Éditions Technip.

Craig, R. F., 1987. *Soil Mechanics*. Van Nostrand Reinhold.

Crawford, F. H., and Curran, J. H., 1982. Higher-order functional variation displacement discontinuity elements. *Int. J. Rock Mech. Min. Sci.*, **19**, 143–148.

Crouch, S. L., 1970. Volumetric changes during rock loading. *Int. J. Rock Mech. Min. Sci.*, **7**, 589–603.

Cundall, P. A., 1988. Formulation of a three-dimensional distinct element model – Part I. A scheme to detect and represent contacts in a system composed of many polyhedral blocks. *Int. J. Rock Mech. Min. Sci. Geomech. Abs.*, **25**, 107–116.

Cundall, P. A., and Lemos, J. V., 1990. Numerical simulation of fault instabilities with a continuously yielding joint model, pp 147–152 of Fairhurst, C. (ed.) *Proc. of Symp. on Rockburst and Seismicity in Mines*, A. A. Balkema.

Dahm, T., 2000. Numerical simulations of the propagation and arrest of fluid filled structures in the Earth. *Geophys. J. Int.*, **141**, 623–638.

Darot, M., Guéguen,Y., and Baratin, M. L., 1992. Permeability of thermally cracked granite. *Geophys. Res. Lett.*, **19**, 869–872.

Davis, R. O., and Selvadurai, A. P. S., 2002. *Plasticity and Geomechanics*. Cambridge University Press.

Davis, K., Burbank, D. W., Fisher, D., Wallace, S., and Nobes, D., 2005. Thrust fault growth and segment linkage in the active Ostler fault zone, New Zealand. *J. Structural Geol.* **27**, 1528–1546.

Deere, D. U., 1963. Technical description of rock cores for engineering purposes. *Rock Mech. Eng. Geol.*, **1**, 18.

Delay, J. D., Distinguin, M., and Dewonck, S., 2007. Characterization of a clay-rich rock through development and installation of specific hydrogeological and diffusion test equipment in deep boreholes. *Phys. Chem. Earth*, **32**, 393–407.

de Marsily, G., 1981. *Hydrogéologie quantitative*. Masson.

Detournay, E., and Cheng, A. H.-D., 1988. Poroelastic response of a borehole in a non-hydrostatic stress field. *Int. J. Rock Mech. Min. Sci.*, **25**, 171–182.

Detournay, E., and Cheng, H. D., 1993. Fundamentals of poroelasticity. pp. 113–171 of Hudson, J. A. (ed.), *Comprehensive Rock Engineering, Vol. 2*, Pergamon Press.

Detournay, E., and Carbonell, R., 1997. Fracture-mechanics analysis of the breakdown process in minifracture of leakoff tests. SPE paper 28076. *Soc. Pet. Eng. Prod. Facil.*, **12**, 195–199.

Dieterich, J. M., 1978. Time dependent friction of rocks. *J. Geophys. Res.*, **77**, 3690–3697.

Dieterich, J. M., 1979. Modeling of rock friction: 1. Experimental results and constitutive equations. *J. Geophys. Res.*, **84**, 2161–2168.

Dorbath, L., Evans, K., Cuenot, N., Valley B., Charlety, J., and Frogneux, M., 2010. The stress field at Soultz-sous-forêts from focal mechanisms of induced seismic events: cases of the well GPK2 and GPK3. *C.R. Geosci.*, **342**, 600–606.

Drucker, D. C., and Pragger, W., 1952. Soil mechanics and plastic analysis of limit design. *Quart. J. Appl. Math.*, **10**, 157–165.

Duquesnoy, T. E., Barrier, E., Kasser, M., Aurelio, M., Gaulon R., and the French–Philippine Cooperation team, 1994. Detection of creep along the Philippine Fault: First results of geodetic measurements on Leyte Island, central Philippine. *Geophys. Res. Lett.*, **21**, 975–978.

Dusseault, M. B., and Fordham, C. J., 1993. Time dependent behavior of rock pp 119–149 of Hudson, J. A. (ed.), *Comprehensive Rock Engineering, Vol. 3*, Pergamon Press.

Eaton, D., Darbyshire, F., Evans, R., Grütter, H., Jones, A., and Yuan, X., 2009. The elusive lithosphere–asthenosphere boundary (LAB) beneath cratons. *Lithos*, **109**, 1–22.

Earle, P. I. S., and Shearer, P. M., 1994. Characterization of global seismograms using an automatic-picking algorithm. *Bull. Seism. Soc. Am.*, **84**, 366–376.

Einstein, H. H., and Baecher, G. B., 1983. Probabilistic and statisitcal methods in engineering geology. *Rock Mech. Rock Eng.*, **16**, 39–72.

Ellis, D. V., and Singer, J. M., 2008. *Well Logging for Earth Scientists*. Springer.

Egholm, D. L., Clausen, O. R., Sandiford, M., Kristensen, M. B., and Korstård, J. A., 2008. The mechanics of clay smearing along faults. *Geology*, **36**, 786–790.

Elsworth, D., 1989. Thermal permeability enhancement of blocky rocks: one dimensional flows. *Int. J. Rock Mech. Min. Sci. Geomech. Abs.*, **26**, 329–339.

Etchecopar, A., Yomoda, T., and Cheung, P., 2013. Borehole images for assessing present day stresses. *Bull. Soc. Geol. Fr.*, **184**, 307–318.

Evans, K. F., Genter, A., and Sause, J., 2005. Permeability creation and damage due to massive fluid injections into granite at 3.5 km at Soultz: 1. Borehole observations. *J. Geophys. Res.*, **110**, B0423.

Ewy, R., 1999. Wellbore stability predictions by use of a modified Lade criterion. *SPE Drill Completion*, **14**, 85–91.

Fäh, D., Gisler, M., Jaggi, B., Kästli, P., Lutz, T., Masciadri, V., *et al.*, 2009. The 1356 Basel earthquake: an interdisciplinary revision. *Geophys. J. Int.*, **178**, 351.

Figueiredo, B., Cornet, F. H., Lamas, L., and Muralha, J., 2014. Determination of the stress field in a mountainous granite rock mass. *Int. J. Rock Mech. Min. Sci.*, **72**, 37–68.

Fisher, N. I., Lewis, T., and Embleton, B. J. J., 1987. *Statistical Analysis of Spherical Data*. 329 pp. Cambridge Univeristy Press.

Fjaer., E., Holt, R. M., Horsrud, P., Raaen, A. M., and Risnes, R., 2008. *Petroleum Related Rock Mechanics, Second Edition*. Elsevier.

Fowler, A. C., and Yang, X., 1999. Pressure solution and viscous compaction in sedimentay basins. *J. Geophys. Res.*, **104**, 12989–12997.

Frechet, J, Martel, L., Nikolla, L., and Poupinet, G., 1989. Application of the cross-spectral moving-window technique (CSMWT) to the seismic monitoring of forced fluid migration in a rock mass. *Int. J. Rock Mech. Min. Sci. Geomech. Abs.*, **26**, 221–234.

Friedman, M., and Logan, J. M., 1970. The influence of elastic residual elastic strain on the orientation of experimental fractures in three quartzose sandstones. *J. Geophys. Res.*, **75**, 387–405.

Fung, Y. C., 1965. *Foundations of Solid Mechanics*. Prentice Hall.

Gabay, S., Shalev, E., Yeshieli, Y., Sagy, A., and Weibrod, A., 2014. The permeability of fault zones : a case study of the Dead Sea rift (Middle East). *Hydrogeol. J.*, **22**, 425–440.

Garder, D. S., 1945. Seismic investigations in the Boulder Dam area 1940–1944, and the infuence of reservoir loading on local earthquake activity. *Bull. Seism. Soc. Am.*, **35**, 175–192.

Gasc-Barbier, M., Chanchole, S., and Berest, P., 2004. Creep behavior of Bure Clayey rock. *Appl. Clay Sci.*, **26**, 449–458.

Gaucher, E., Cornet, F. H., and Bernard, P., 1998. Induced seismicity analysis for structure identification and stress field determination. SPE paper SPE-47324, pp. 545–554 of *Euroc-98*, Society of Petroleum Engineering.

Geertsma, J., 1957. The effects of fluid pressure decline on volumetric changes of porous rocks. *Petrol. Trans. Am. Inst. Min. Eng.*, **210**, 331–339.

Geertsma, J., and de Klerk, F., 1969. A rapid method of predicting width and extent of hydraulically induced fractures. *J. Petrol. Tech.*, **21**, 1571–1581.

Genter, A., Evans, K., Cuenot, N., Fritsch, D., and Sanjuan, B., 2010. Contribution to the exploration of deep crystalline fractures reservoir at Soultz to the knowledge of enhance geothermal systems (EGS). *C.R. Geosci.*, **342**, 502–516.

Gephart, J. W., and Forsyth, D. W., 1984. An improved method for determining the regional stress tensor using earthquake focal mechanism data: application to the San Fernando earthquake sequence. *J. Geophys. Res.*, **89**, 9305–9320.

Goodman, R. E., Taylor, R. L., and Brekke, T., 1968. A model for the mechanics of jointed rock. *J. Soil Mech. Proc. Am. Soc. Civ. Engrs.*, **94**, 637–659.

Goodman, R. E., 1989. *Introduction to Rock Mechanics*. John Wiley & Sons.

Gratier, J. P., Renard, F., and Labaume, P., 1999. How pressure solution, creep and fracturing processes interact in the upper crust to make it behave in both a brittle and viscous manner. *J. Struct. Geol.*, **21**, 1189–1197.

Gratier, J. P., Dysthe, D. K., and Renard, F., 2013. The role of pressure solution in the ductility of the Earth's upper crust. *Adv. Geophys.*, **54**, 47–178.

Grasseli, G., and Egger, P., 2003. Constitutive law for the shear strength of rock joint based on three-dimensional surface parameters. *Int. J. Rock Mech. Min. Sci.*, **40**, 25–40.

Griffith, A. A., 1920. The phenomena of rupture and flow in solids. *Phil. Trans. Roy. Soc. London A*, **221**, 163–198.

Griffith, A. A., 1924. The theory of rupture, pp. 55–63 of *Proc. Int. Cong. on Applied Mechanics, Delft*.

Griggs, D. T., Turner, F. J., and Heard, H. C., 1960. Deformation of rock at 500 to 800 °C. *Geol. Soc. Am. Mem.*, **79**, 39–104.

Grünthal, G., and Wahhlström, R., 2012. The European–Mediterranean earthquake catalogue for the last millennium. *J. Seism.*, **16**, 535–570.

Gudmundsson, A., 2002. Emplacement and arrest of sheets and dykes in central volcanoes. *J. Volc. Geotherm. Res.*, **116**, 279–298.

Guéguen, Y., and Palciauskas, V., 1992. *Introduction à la physique des roches*. Herrmann.

Guéguen, Y., and Palciauskas, V., 1994. *Introduction to the Physics of Rocks*, Princeton University Press.

Guéguen, Y., Dormieux, L., and Boutéca, M., 2003. Fundamentals of poromechanics, pp. 1–54 of Guéguen and Boutéca (eds.), *Mechanics of Fluid Saturated Rocks*, Elsevier, Academic Press.

Guillou-Frottier, L., Carré, C., Bourgine, B., Bouchot, V., and Genter, A., 2013. Structure and hydrothermal convection in the upper Rhine graben as inferred from corrected temperature data and basin-scale numerical models.. *J. Volc. Geotherm. Res.*, **256**, 29–49.

Gunzburger, Y., and Cornet, F. H., 2007. Rheological characterization of a sedimentary formation from a stress profile inversion. *Geophys. J. Int.*, **168**, 402–418.

Gunzburger, Y., and Magnenet, V., 2014. Stress inversion and basement-cover stress transmission across weak layers of the Paris Basin, France. *Tectonophys.*, **617**, 44–57.

Gupta, H., Narain, H., Rastogi, B. K., and Mohan, I., 1969. A study of the Koyna earthquake of December 10, 1967. *Bull. Seism. Soc. Am.*, **59**, 1149–1162.

Gutenberg, B., and Richter, C. F., 1942. Earthquake magnitude, intensity, energy and acceleration. *Bull. Seism. Soc. Am.*, **32**, 163–191.

Gutenberg, B., and Richter, C. F., 1956. Earthquake magnitude, intensity, energy and acceleration. *Bull. Seism. Soc. Am.* **46**, 105–145.

Guyon, E., Hulin, J. P., and Petit, L., 1991. *Hydrodynamique physique*. CNRS Éditions.
Hadley, K., 1975. Azimuthal variation of dilatancy. *J. Geophys. Res.*, **80**, 4845–4850.
Haimson, B. C., and Fairhurst, C., 1967. Initiation and extension of hydraulic fractures in rocks. *Soc. Petrol. Eng. J.*, **7**, 310–318.
Haimson, B. C., and Fairhurst, C., 1970. *In situ* stress determination at great depth by means of hydraulic fracturing, pp. 559–584 of Somerton, W. H. (ed.), *Rock Mechanics – Theory and Practice, Proc. 11th Symp. on Rock Mechanics*, American Institute of Metallurgical and Petroleum Engineering.
Haimson, B. C., and Kim, C. M., 1972. Mechanical behavior of rock under cyclic fatigue, pp. 845–863 of Cording (ed.), *Proc. 13th Symp.on Rock Mechanics*, American Society of Civil Engineering.
Haimson, B. C., 1978. The hydrofracturing stress measuring method and recent field results. *Int. J. Rock Mech. Min. Sci. Geomech. Abs.*, **15**, 167–178.
Hallbauer, D. K., Wagner, H., and Cook, N. G. W., 1973. Some observations concerning the microscopic and mechanical behaviour of quartzite specimens in stiff, triaxial compression tests. *Int. J. Rock Mech. Min. Sci. Geomech. Abs.*, **10**, 713–726.
Handin, J., Haeger, R. V., Friedman, M., and Feather, J. N., 1963. Experimental deformation of sedimentary rocks under confining pressure: pore pressure tests. *Bull. Am. Assoc. Petrol. Geol.*, **47**, 717.
Hanks, T. C., and Thatcher, W., 1972. A graphical representation of seismic source parameters. *J. Geophys. Res.*, **77**, 4393–4405.
Hart, R. D., Cundall, P. A., and Lemos, J.,1988. Formulation of a three-dimensional distinct element model – Part II. Mechanical calculations for motion and interaction of a system composed of many polyhedral blocks. *Int. J. Rock Mech. Min. Sci. Geomech. Abs.*, **25**, 117–126.
Haskell, N., 1964. Total energy and energy spectral density of elestic wave radiation from faults. *Bull. Seism. Soc. Am.*, **56**, 1811–1842.
Hayashi, K., and Haimson, B. C., 1991. Characteristics of shut-in curves in hydraulic fracturing stress measurements and the determination of the *in situ* minimum compressive stress. *J. Geophys. Res.*, **96**, 18311–18321.
Heard, H. C., and Page, L., 1982. Elastic moduli, thermal expansion, and inferred permeability of two granites to 350 °C and 55 MPa. *J. Geophys. Res.*, **87**, 9340–9348.
Hearst, J. R., 2000. *Well Logging for Physical Properties: A Handbook for Geophysicist, Geologists and Engineers*. John Wiley & Sons.
Hieronymus, C., Goes, S., Sargent, M., and Morra, G., 2008. A dynamical model for generating Eurasian lithospheric stress and strain rate fields: effects of rheology and cratons. *J. Geophys. Res.*, **113**, B07404.
Hill, R. A., 1958. A general theory of uniqueness and stability of elastic–plastic solids. *J. Mech. Phys. Solids*, **6**, 236–249.
Hiramatsu, Y, and Oka, Y., 1968. Determination of the stress in rock unaffected by boreholes or drifts, from measured strains or deformations. *Int. J. Rock Mech. Min. Sci.*, **5**, 337–353.
Hoek, E., and Bieniawski, Z. T., 1965. Brittle fracture propagation in rock under compression. *Int. J. Fracture Mech.*, **1**, 137–155.
Hoek, E., 1968. Brittle failure of rock, pp. 99–124 of Stagg, K. G., and Zienkiewicz, O. C. (eds), *Rock Mechanics in Engineering Practice*, John Wiley & Sons.
Hoek, E., and Brown, E. T., 1980. *Underground Excavations in Rock.* The Institute of Mining and Metallurgy, London.
Hoek, E., and Brown, E. T., 1988. The Hoek–Brown failure criterion – a 1988 update, pp. 31–38 of Curran, J. H. (ed.), *Rock Engineering for Underground Excavations, Proc. 15th Canadian Rock Mechanics Symp.*, University of Toronto.

Huang, X., Haimson, B. C., Plesha, M. E., and Qiu, X., 1993. An investigation of the mechanics of rock joints – Part I. Laboratory investigation. *Int. J. Rock Mech. Min. Sci.* **30**, 257–269.

Hubbert, M. K., and Willis, D. G., 1957. Mechanics of hydraulic fracturing. *Trans. Am. Inst. Min. Eng.*, **210**, 153–163.

Hudson, J. A., Brown, E. T., and Fairhurst, C., 1971. Optimizing the control of rock failure in servo-controlled laboratory test. *Rock Mech.*, **3**, 217–224.

Hudson, J. A., Crouch, S. L., and Fairhurst, C., 1972. Soft, stiff and servo-controlled testing machines: a review with reference to rock failure. *Eng. Geol.*, **6**, 155–189.

Hudson, J. A., and Harrison, J. P., 1997. *Engineering Rock Mechanics.* Pergamon, Elsevier.

Hudson, J. A., and Cornet, F. H., 2003. Rock stress estimation – special issue. *Int. J. Rock Mech. Min. Sci.*, **40**, 955–1276.

Hutson, R. W., and Dowding, C. H., 1990. Joint asperity degradation during cyclic shear. *Int. J. Rock Mech. Min. Sci.*, **27**, 109–119.

Inglis, C. E., 1913. Stresses in a plate due to the presence of cracks and sharp corners. *Trans. Inst. Naval Architects*, **55**, 219–241.

Ingraffea, A. R., 1987. Theory of crack initiation and propagation in rock, pp. 71–108 of Atkinson, B. K. (ed.), *Fracture Mechanics of Rock*, Academic Press.

Irwin, G. R., 1957. Analyses of stresses and strains near the end of a crack traversing a plate. *J. Appl. Mech.*, **24**, 361–364.

Itasca, 2009. FLAC3D, version 4.0. User's manual. Itasca Consulting Group, Minneapolis.

Ito, T., Evans, K., Kawai, K., and Hayashi, K., 1999. Hydraulic fracture reopening pressure and the estimation of the maximum horizontal principal stress. *Int. J. Rock Mech. Min. Sci.*, **36**, 811–826.

Jaeger, J. C., 1967. Brittle fracture of rocks, pp. 3–57 of Fairhurst, C. (ed.), *Failure and Breakage of Rock, Proc. 8th Symp. on Rock Mechanics.*

Jaeger, J. C., 1969. *Elasticity, Fracture and Flow, Third Edition.* Halsted Press, a division of John Wiley & Sons.

Jaeger, J. C., and Cook, N. G. W., 1979. *Fundamentals of Rock Mechanics, Third Edition.* Chapman and Hall.

Jaeger, J. C., Cook, N. G. W., and Zimmerman, R. W., 2007. *Fundamentals of Rock Mechanics, Fourth Edition.* Blackwell.

Johnson, P. A., and Rasolofosaon, P. N. J., 1996. Nonlinear elasticity and stress-induced anisotropy in rock. *J. Geophys. Res.*, **101**, 3113–3124.

Jolivet, L., and Faccenna, C., 2000. Mediterranean extension and the Africa–Eurasia collision. *Tectonics*, **19**, 1095–1106.

Julian, B. R., Miller, A. D., and Foulger, G. R., 1998. Non-double couple eartquakes – 1 Theory. *Rev. Geophys.*, **36**, 525–549.

Julien, P., and Cornet, F. H., 1989. Stress determination from aftershocks of the Campania-Lucania earthquake of November 23, 1980. *Ann. Geophys.*, **5B**, 289–300.

Kachanov, M., 1980. Continuum model of medium with cracks. *J. Eng. Mech. Div. Am. Soc. Civ. Eng.*, **106**, 1039–1051.

Kachanov, M., Prioul, R., and Jocker J., 2010. Incremental linear-elastic response of rocks containing multiple rough fractures: Similarities and differences with traction-free cracks. *Geophysics*, **75**, D1–D11.

Kanamori, H., 1977. The energy release of great earthquakes. *J. Geophys. Res.*, **82**, 2981–2987.

Kanamori, H., and Anderson, D. L., 1975. Theoretical basis for some empirical relations in seismology. *Bull. Seism. Soc. Am.* **65**, 1073–1095.

Kehle, R. 0., 1964. The determination of tectonic stresses through analysis of hydraulic well fracturing. *J. Geophys. Res.*, **69**, 259–273.

King, G. C. P., Stein, R. S., and Lin, J., 1994. Static stress changes and the triggering of earthquakes. *Bull. Seism. Soc. Am.*, **84**, 935–953.

Kozlovski, Y. A., 1984. *The Super Deep Well of the Kola Peninsula.* Springer

Labuz, J. F., Shah, S. P., and Dowding, C. H., 1985. Experimental analysis of crack propagation in granite. *Int. J. Rock Mech. Min. Sci. Geomech. Abs.*, **22**, 85–98.

Lade, P., 1977. Elasto-plastic stress–strain theory for cohesionless soil with curved yield surfaces. *Int. J. Solids Struct.*, **13**, 1019–1035.

Lecampion, B., and Detournay, E., 2007. An implicit algorithm for the propagation of a hydraulic fracture with a fluid lag. *Comput. Methods Appl. Mech. Eng.*, **196**, 4863–4880.

Lehner, F. K., 1995. A model for intergranular pressure solution in open systems. *Tectonophys.*, **245**, 153–170.

Lei, T., Sinha, B. K., and Sanders, M., 2012. Estimation of horizontal stress magnitudes and stress coefficients of velocities using borehole sonic data. *Geophysics*, **77**, WA181–WA196.

Lei, X., 2011. Possible role of the Zipingpu Reservoir in triggering the 2008 Wenchuan earthquake. *J. Asian Earth Sci.*, **40**, 844–854.

Li, F., and Schmitt, D. R., 1998. Drilling-induced core fractures and *in situ* stress. *J. Geophys. Res.*, **103**, 5225–5239.

Lockner, D. A., Byerlee, J. D., Kuksenko, V., Ponomarev, A., and Sidril, A., 1991. The quasi-static fault growth and shear fracture energy in granite. *Nature*, **350**, 39–42.

Lockner, D., 1993. The role of acoustic emission in the study of rock. *Int. J. Rock Mech. Geomech. Abs.*, **30**, 883–899.

Lockner, D. A., Tanaka, H., Ito, H., Ikeda, R., Omura, K., and Naka, H., 2009. Geometry of the Nojima Fault at Nojima-Hirabayashi, Japan – I. A simple damage structure inferred from borehole core permeability. *Pure Appl. Geophys.*, **166**, 1649–1667.

Mackenzie, D. P., 1969. The relation between fault plane solutions for earthquakes and the direction of the principal stresses. *Bull. Seism. Soc. Am.*, **59**, 591–601.

Madariaga, R., 1977. High-frequency radiation from crack (stress drop) models of earthquake faulting. *Geophys. J. Roy. Astron. Soc.* **51**, 625–651.

Malvern, L. E., 1969. *Introduction to the Mechanics of a Continuous Medium.* Prentice Hall.

Manighetti, I., King, G. C. P., Gaudemer, Y., Scholz, C. H., and Doubre, C., 2001. Slip accumulation and lateral propagation of active normal faults in Afar. *J. Geophys. Res.*, **106**, 13667–13696.

Masson, F., Gal, F., and Leloup, P. H., 2002. Une carte haute résolution du massif du Mont Blanc: implications structurales. *C.R. Geosci.*, **334**, 1011–1019.

Matsuki, K., Kimura, Y., Sakaguchi, K., Kizaki, A., and Giwelli, A. A., 2010. Effect of shear displacement on the hydraulic conductivity of a fracture. *Int. J. Rock Mech. Min. Sci..* **47**, 436–449.

Maury, J., Cornet, F. H., and Dorbath, L., 2013. A review of methods for determining stress fields from earthquakes focal mechanisms: application to the Sierentz 1980 seismic crisis (Upper Rhine graben). *Bull. Soc. Geol. Fr.*, **184**, 319–334.

Maury, J., Cornet, F. H., and Cara, M., 2014. Influence of the lithosphere–asthenosphere boundary on the stress field north-west of the Alps. *Geophys. J. Int.*, **199**, 1006–1017.

Means, W., 1987. A newly recognized type of slickenside striation. *J. Struct. Geology*, **9**, 585–590.

Meyer, J. P., and Labuz, J. F. W., 2011. Linear failure crireria with three principal stresses. *Int. J. Rock Mech.*, **60**, 180–187.

Micarelli, D. U., Moretti, I., Jaubert, M., Moulouel, H., 2006. Fracture analysis in the south-western Corinth rift (Greece) and implications on fault hydraulic behavior. *Tectonophys.*, **426**, 31–59.

Michael, A. J., 1984. Determination of stress from slip data: faults and folds. *J. Geophys. Res.*, **89**, 11517–11526.

Milne, G. A., Mitrovica, J. X., Scherneck, H. G., *et al.*, 2004. Continuous GPS measurements of post glacial adjustment in Fennoscandia. *J. Geophys. Res.*, **109**, B02412.

Milsom, J., and Eriksen, A., 2011. *Field Geophysics, Fourth Edition*. Wiley & Sons.

Mogi, K., 1967. Effect of the intermediate principal stress on rock failure. *J. Geophys. Res.*, **72**, 5117–5131.

Morin, R. H., Newmark, R. L., Barton, C. A., and Anderson, R. N., 1990. State of lithospheric stress and borehole stability at Deep Sea Drilling Project Site 504B, Eastern Equatorial Pacific. *J. Geophys. Res.*, **95**, 9293–9303.

Mosnier, J., 1982. Détection électrique des fractures naturelles ou artificielles dans un forage. *Ann. Geophys.*, **38**, 537–540.

Muirwood, R., and King, G. C. P., 1993. Hydrological signatures of earthquake strain. *J. Geophys. Res.*, **98**, 22035–22068.

Mussett, A. E., and Khan, M. A, 2000. *Looking Into the Earth*. Cambridge University Press.

Musson, R. M. W., and Cecic, I., 2011. Intensity and intensity scales, in *New Manual of Seismological Observatory Practice, MNSOP-2*, 38 pp.

Myer, L. R., Tsang, C. F., Cook, N. G. W., and Goodman, R. E. (eds.), 1995. *Fractured and Jointed Rock Masses*. A. A. Balkema (CRC Press).

Nadeau, R., and McEvilly, T. V., 1999. Fault slip rates at depth from recurrence intervals of repeating microearthquakes. *Science.*, **285**, 718–721.

Nadeau, R., Antolik, M., Johnson, P. A., Foxall, W., and McEvilly, T. V., 1994. Seismological studies at Parkfield. III. Microearthquake clusters in the study of fault zone dynamics. *Bull. Seism. Soc. Am.* **84**, 247–263.

Nemat-Nasser, S., and Horii, H., 1982. Compression induced non-planar crack extension with application to splitting, exfoliation and rockburst. *J. Geophys. Res.*, **87**, 6805–6821.

Neuville, A., Toussaint, R., and Schmittbuhl, J., 2011. Hydraulic transmissivity and heat exchange efficiency of rough fractures: a model based on lowpass filtered apertures. *Geophys. J. Int.*, **186**, 1064–1072.

Nocquet, J. M., 2012. Present-day kinematics of the Mediterranean: a comprehensive overview of GPS results. *Tectonophys.*, **579**, 220–242.

Obert, L., Merill, R. H., and Morgan, T. A., 1962. Borehole deformation gauge for determining the stress in mine rock. *US Bureau of Mine Rep. Invest.*, **5978**

Ouchterlony, F., 1988. International Society for Rock Mechanics suggested methods for determing the fracture toughness of rock. *Int J. Rock Mech. Min. Sci. Geomech. Abs.*, **25**, 71–98.

Paillet, F., and Cheng, M., 1991. *Acoustic Waves in Boreholes – The Theory and Application of Acoustic Full Waveform Logs*. CRC Press.

Palaniswamy, K., and Knauss, W. G., 1974. On the problem of crack extension in brittle solids under general loading. Report SM 74–8; Graduate Aeronautical Laboratories, California Institute of Technology.

Papadopoulos, I. S., and Cooper, H. H., 1967. Drawdown in a well of large diameter. *Water Resource Res.*, **3**, 241–244.

Parker, R. L., and McNutt, M. K., 1980. Statistics for one-norm misfit measure. *J. Geophys. Res.*, **85**, 4429–4430.

Patton, F. D., 1966. Multiple modes of shear failure in rock, pp. 509–513 of Rocha, M. (ed.), *Proc. 1st ISRM Congr.*

Perkins, T. K., and de Kern, L. R., 1961. Widths of hydraulic fractures. *J. Petr. Tech.*, **13**, 937–949.

Peska, P., and Zoback, M. D., 1995. Compressive and tensile failure of inclined well bores and determination of *in situ* stress and rock strength. *J. Geophys. Res.*, **100**, 12791–12811.

Pigott, A. R., and Elsworth, D., 1995. A comparison of methods of characterizing fracture surface roughness, pp. 471–477 of Mayer, Cook, Goodman and Tsang (eds.), *Proc. Symp. on Fractured and Jointed Rock Masses*, A. A. Balkema.

Pine, R. J., and Batchelor, A. S., 1984. Downward migration of shearing in jointed rock during hydraulic injections. *Int. J. Rock Mech. Min. Sci. Geomech. Abs.*, **13**, 937–949.

Plomerova, J., and Babuska, V., 2010. Long memory of mantle lithospheric fabric – European LAB constrained from seismic anisotropy. *Lithos*, **120**, 131–143.

Poupinet, G., Ellsworth, W. L., and Frechet, J., 1984. Monitoring velocity variations in the crust using erthquake doublets: an application to the Calaveras fault, Califormia. *J. Geophys. Res.*, **89**, 5719–5731.

Pribnow, D., and Schellschmidt, R., 2000. Thermal tracking of upper crustal fluid flow in the Rhine graben. *Geophys. Res. Lett.*, **105**, 13595–13612.

Price, D. G., 2009. *Engineering Geology, Principles and Practice*. Springer

Priest, S. D., and Hudson, J. A., 1981. Estimation of discontinuity spacing and trace length using scanline surveys. *Int. J. Rock Mech. Min. Sci. Geomech. Abs.*, **18**, 183–197.

Prioul, R., Cornet, F. H., Dorbath, C., Dorbath, L., Ogena M., and Ramos, E., 2000. An induced sesmicity experiment across a creeping segment of the Philippine Fault. *J. Geophys. Res.*, **27**, 1957–1960.

Prioul, R., and Karpfinger, F., 2011. Improving fracture initiation prediction on arbitrarily oriented wells in anisotropic shales. SPE 147 Society of Petroleum Engineers.

Pyrak-Nolte, L. J., and Morris, J. P., 2000. Single fractures under normal stress: the relation between fracture specific stiffness and fluid flow. *Int. J. Rock Mech. Min. Sci.*, **37**, 245–262.

Revets, S. A., 2010. http://arxiv.org/pdf/1008.0471

Rice, J. R., 1968. Mathematical analysis in the mechanics of fracture, pp. 191–311 of Liebowitz, (ed.), *Fracture, An Advanced Treatise, Vol. II*, Academic Press.

Rice, J. R., and Cleary, M. P., 1976. Some basic stress diffusion solutions for fluid-saturated elastic porous media with compressible constituents. *Rev. Geophys. Space Phys.*, **14**, 227–241.

Rietbrock, A., Tiberi, C., Scherbaum, F., and Lyon-Caen, H., 1996. Seismic slip on a low angle normal fault in the Gulf of Corinth: evidence from high-resolution cluster analysis of microearthquakes. *Geophys. Res. Lett.* **23**, 1817–1820.

Roelloffs, E. A., 1988. Fault stability changes induced beneath a reservoir with cyclic variations in water level. *J. Geophys. Res.*, **93**, 2104–2124.

Rouland, D., Haessler, H., Bonier, K. P., Gilg, B., Mayer-Rosa, D., and Pavoni, N., 1980. The Sierentz southern-Rhine graben earthquake of July 15, 1980. Preliminary results, in *Proc.. 17th Assembly Europ. Seism. Conf.*

Royden, L. H., 1993. The tectonic expression of slab pull at continental convergent boundaries. *Tectonics*, **12**, 303–325.

Rudnicki, J. W., and Rice, J. R., 1975. Conditions for the localization of deformation in pressure-sensitive dilatant materials. *J. Mech. Phys. Solids*, **23**, 371–394.

Ruina, A. L., 1983. Slip instability and state variable friction laws. *J. Geophys. Res.*, **88**, 10359–10370.

Rummel, F., and Fairhurst, C., 1971. Determination of the post failure behavior of brittle rock using a servo controlled testing machine. *Rock Mech.*, **12**, 189–204.

Rummel, F., 1974. Changes in the P wave velocity with increasing inelastic deformation in rock, pp. 517–523 of Fairhurst, C. (ed.), *Proc. 3rd Congr. Int. Soc. Rock Mechanics*, Balkema.

Salençon, J., 1974. *Théorie de la plasticité pour les applications à la mécanique des sols*. Eyrolles éditions, Paris.

Savage, W. Z., and Swolfs, H. S., 1986. Tectonic and gravitational stress in long symmetric ridges and valleys. *J. Geophys. Res.*, **91**, 3677–3685.

Scheidegger, A. E., 1960. On the connection between tectonic stresses and well fracturing data. *Geofis. Pura Appl.* **46**, 66–76.

Schmittbuhl, J., Schmitt, F., and Scholz, C. H., 1995. Scaling invariance of crack surfaces. *J. Geophys. Res.*, **100**, 5953–5973.

Schmittbuhl, J., Renard, F., Gratier, J. P., and Toussaint, R., 2004. Roughness of stylolites: implications of 3D high resolution topography measurements. *Phys. Rev. Lett.*, **93**, 238501.

Scholz, C. H., 1968. Microfracturing and the inelastic deformation of rock in compression, *J. Geophys. Res.*, **73**, 1417–1432.

Scholz, C. H., 1990. *The Mechanics of Earthquakes and Faulting*. Cambridge University press.

Scholz, C. H., and Saucier, F. J., 1993. What do the Cajon Pass stress measurements say about stress on the San Andreas Fault? *J. Geophys. Res.*, **98**, 17867.

Schubnel, A., Benson, B. M., Thompson, B. D., Hazard, J. F., and Young, R. P., 2006. Quantifying damage, saturation and anisotropy in cracked rocks by inverting elastic wave velocities. *Pure Appl. Geophys.*, **163**, 947–973.

Scotti, O., and Cornet, F. H., 1994. *In situ* evidence for fluid-induced aseismic slip events along fault zones. *Int. J. Rock Mech. Min. Sci. Geomech. Abs.*, **31**, 347–358.

Segall, P., 2010. *Earthquake and Volcano deformation.* Princeton University Press.

Shamir, G., and Zoback M. D., 1992. Stress orientation profile to 3.5 km depth near the San Andreas Fault at Cajon Pass, California. *J. Geophys. Res.*, **97**, 5059–5080.

Shapiro, S. A., Rothert, E., Rath, V., and Rindschwentner, J., 2002. Characterization of fluid transport properties of reservoirs using induced microseismicity. *Geophysics*, **67**, 212–220.

Sheriff, R. E., and Geldart, L. P., 1995. *Exploration Seismology*. Cambridge University Press.

Shi, G. H., and Goodman, R. E., 1988. Discontinuous deformation analysis – a new method for computing stress, strain and sliding of block systems, pp. 381–393 of Cundall, Sterling and Starfield (ed.), *Key Questions in Rock Mechanics, Proc. 29th US Symp. Rock Mechanics*, Balkema.

Sibson, R. H., 1977. Fault rocks and fault mechanisms. *J. Geol. Soc. London*, **133**, 191–213.

Sibson, R. H., 1987. Earthquake rupturing as a mineralizing agent in hydrothermal systems. *Geology*, **15**, 701–704.

Sih, G. C., and Liebowitz, H., 1968. Mathematical theories of brittle fracture, pp. 67–190 of Liebowitz H. (ed.), *Fracture, An Advanced Treatise, Vol. II*, Academic Press.

Sileny, J., and Milev, A., 2008. Source mechanisms of mining induced seismic events – resolution of double couple and non double couple models. *Tectonophys.*, **456**, 3–15.

Sileny, J., Hill, D. P., Eisner, L., and Cornet, F. H., 2009. Non double couple mechanisms of microearthquakes induced by hydraulic fracturing. *J. Geophys. Res.*, **114**, B08307.

Simons, G., and Brace, W. F., 1965. Comparison of static and dynamic measurement on compressibility of rock. *J. Geophys. Res.*, **70**, 5649–5656.

Simons, G., and Cooper, H. W., 1978. Thermal cycling cracks in three igneous rocks. *Int. J. Rock Mech. Min. Sci. Geomech. Abs.*, **15**, 145–148.

Simpson, D. W., Leith, W. S., and Scholz, C. H., 1988. Two types of reservoir-induced seismicity. *Bull. Seism. Soc. Am.*, **78**, 2025–2040.

Sissingh, W., 1998. Comparative tertiary stratigraphy of the Rhine Graben, Bress Graben and Molasse basin: correlation of alpine foreland events. *Tectonophys.*, **300**, 249–284.

Skempton, A. W., 1954. The pore pressure coefficients *A* and *B*. *Geotechnique*, **4**, A143–A147.

Smith, D. L., and Evans, B., 1984. Diffusional crack healing in quartz. *J. Geophys. Res.*, **89**, 4125–4135.

Stein, S., and Wysession, M., 2003. *An Introduction to Seismology, Earthquakes, and Earth Structure*. Blackwell.

Stephansson, O., Jing, L., and Tsang, C.-F. (ed.), 1996. *Coupled Thermo-Hydro Mechanical Processes of Fractured Media.* Elsevier.

Stuart, W. D., 1974. Diffusionless dilatancy model for earthquake precursors. *Geophys. Res. Lett.*, **1**, 261–264.

Sue, C., Thouvenot, F., Frechet, J., and Tricart, P., 1999. Widespread extension in the core of the western Alps revealed by earthquake analysis. *J. Geophys. Res.*, **104**, 25611–25622.

Sugawara, K., and Obara, Y., 1999. Draft ISRM suggested method for in situ stress measurement using the Compact Conical Ended Borehole overcoring (CCBO) technique. *Int. J. Rock Mech. Min. Sci.*, **36**, 307–321.

Sulem, J., Vardoulakis, I., Ouffroukh, H., Boulon, M., and Hans, J., 2004. Experimental characterization of the thermo-poro-mechanical properties of the Aigion Fault gouge. *C.R. Geosci.*, **336**, 455–466.

Sulem, J., 2007. Stress orientation evaluated from strain localisation analysis in Aigion Fault. *Tectonophys.*, **442**, 3–13.

Tarantola, A., 1987. *Inverse Problem Theory*. Elsevier.

Terzaghi, K. van, 1923. Die Berechnung der Durchlassigkeitsziffer des Tones aus dem Verlauf der hydrodynamishen Spannungserscheinungen. *Sber. Akad. Wiessenshaften Wien, Sitzungs. Mathnatur. Kl. Abt*, **132 3/4**, 125–138.

Terzaghi, K. van, 1943. *Theoretical Soil Mechanics*. Wiley.

Tesauro, M., Kaban, M. K. J., and Cloetingh, S. A. P. L., 2009. Eucrust-07: a new reference model for the European crust. *Geophys. Res. Lett.*, **35**, L05313.

Thatcher, W., and Hanks, T. C., 1973. Source parameters of Southern California earthquakes. *J. Geophys. Res.*, **78**, 8547–8556.

Thurber, C., and Eberhardt-Phillips, D., 1999. Local earthquake tomography with flexible gridding. *Comp. Geosci.* **25**, 809–818.

Thurber, C., Roecker, S., Roberts, K., Gold, M., Powell, L., and Rittger, K., 2003. Earthquake locations and three dimensional fault zone structure along a creeping section of the San Andreas fault near Parkfield CA: preparing for SAFOD. *Geophys. Res. Lett.*, **30**, 12–1.

Tincelin, M. E., 1952. Mesures de pressions de terrain dans les mines de fer de l'est. *Ann. Inst. Tech. Bat. Trav. Publ., Sols Fond.*, **58**, 972–990.

Toussaint, R., and Cornet, F. H., 2011. How chemistry can change stress in anisotropic fractured carbonates, in *Proc. Workshop on Seismic and Aseismic Deformation in Crustal Domains subject to very slow Deformation Rates*, http://workshopmontstodile.blogspot.com/p/abstract.html.

Tsang, Y. W., and Tsang, C. F., 1989. Flow channeling in a single fracture as a two-dimensional strongly heterogeneous permeable medium. *Water Resource Res.*, **16**, 1016–1024.

Turpeinen, H., Hamplel, A., Karow, T., and Maniatis, G., 2008. Effect of ice sheet growth and melting on the slip evolution of thrust faults. *Earth Planet. Sci. Lett.*, **269**, 230–241.

Twiss, R. J., and Moores, E. M., 1992. *Structural Geology*. W. H. Freeman.

Valley, B., and Evans, K. W., 2007. Stress state at Soultz-sous-forêts to 5 km depth from wellbore failure and hydraulic observations, in *Proc. 32nd Workshop on Geothermal Reservoir Engineering*, SGP-TR-183, Stanford.

Valley, B., and Evans, K. F., 2009. Stress orientation to 5 km depth in the basement below Basel (Switzerland) from borehole failure analysis. *Swiss J. Geosci.*, **102**, 467–480.

Vardoulakis, I., and Sulem, J., 1995. *Bifurcation in Geomechanics*. Taylor and Francis.

Vollmer, F. W., 1995. C programme for automatic contouring of spherical orientation data using a modified Kamb method. *Computers Geosci.*, **21**, 31–49.

von Karman, T., 1911. Festigkeitsversuche unter allseitigem Druck. *Z. Verein Deutscher Ing.*, **55**, 1749–1757.

Wagner, M., Kissling, E., and Husen, S., 2012. Combining controlled-source seismology and local earthquake tomography to derive a 3-D crustal model of the western Alpine region. *Geophys. J. Int.*, **191**, 789–802.

Walsh, J. B., 1965a. The effects of cracks on the compressibility of rocks. *J. Geophys. Res.*, **70**, 381–389.

Walsh, J. B., 1965b. The effects of cracks on the uniaxial elastic compression of rocks. *J. Geophys. Res.*, **70**, 399–411.

Walsh, J. B., 1965c. The effects of cracks in rocks on Poisson's ratio. *J. Geophys. Res.*, **70**, 5249–5257.

Wang, H. F., Bonner, B. P., Carlson, S. R., Kowallis, B. J., and Heard, H. C., 1989. Thermal stress cracking in granite. *J. Geophys. Res.*, **94**, 1745–1758.

Wang, H. F., 2000. *Theory of Linear Poroelasticity.* Princeton University Press.

Wawersik, W. R., 1968. Detailed analysis of rock failure in laboratory compression tests. Ph.d. thesis, University of Minnesota.

Wawersik, W. R., and Fairhurst, C., 1970. A study of brittle rock fracture in laboratory compression experiments. *Int. J. Rock Mech. Min. Sci.*, **7**, 561–575.

White, R. S., Drew, J., Martens, H. R., Key, J., Soosalu, H., and Jakobsdóttir, S. S., 2011. Dynamics of dyke intrusion in the mid-crust of Iceland. *Earth Planet. Sci. Lett.*, **304**, 300–312.

Wiebols, G. A., and Cook, N. G. W., 1968. An energy criterion for the strength of rocks in polyaxial compression. *Int. J. Rock Mech. Min. Sci.*, **5**, 529–549.

Wileveau, Y., Cornet, F. H., Desroches, J., and Blumling, P., 2007. Complete *in situ* stress determination in an argillite sedimentary formation. *Phys. Chem. Earth*, **32**, 866–878.

Witherspoon, P. A., Wang, J. S. W., Iwai, K., and Gale, J. E., 1980. Validity of cubic law for fluid flow in a deformable rock fracture. *Water Resource Res.*, **16**, 1016–1024.

Wong, T.-F., and Baud, P., 2012. The brittle-ductile transition in porous rock: a review. *J. Structural Geol.*, **44**, 25–53.

Xie, H., and Pariseau, W. G., 1995. Fractal estimation of joint roughness coefficients, pp. 125–131 of Myer, L. R., Tsang, C.-F., Cook, N. G., and Goodman, R. E. (eds.), *Proc. Symp. on Fractured and Jointed Rock Masses*, A. A. Balkema.

Yamashita, F., Mizoguchi, K., Fukuyama, E., and Omura, K., 2010. Reexamination of the present stress state of the Atera fault system, central Japan, based on the calibrated crustal stress data of hydraulic fracturing tests obtained by measuring the tensile strength of rocks. *J. Geophys. Res.*, **115**, B0449.

Yin, J. M., and Cornet, F. H., 1994. Integrated stress determination by joint inversion of hydraulic tests and focal mechanisms. *Geophys. Res. Lett.*, **21**, 2645–2648.

Zemanek, J., Glenn, E. E., *et al.*, 1970. Formation evaluation by inspection with the borehole televiewer. *Geophysics*, **35**, 254–269.

Zhang, C., and Rothfuchs, T., 2004. Experimental study of the hydromechanical behavior of the Callovo–Oxfordian argillite. *Applied Clay Sci.*, **26**, 325–336.

Zhang, H., and Thurber, C. H., 2003. Double difference tomography. The method and its application to the Hayward Fault, California. *Bull. Seism. Soc. Am.*, **93**, 1875–1889.

Zheng, Z., Kemeny, J., and Cook, N. G., 1989. Analysis of borehole breakouts. *J. Geophys. Res.*, **94**, 7171–7182.

Zhou, S., 1994. A program to model the initial shape and extent of borehole breakout. *Comp. Geosci.*, **20**, 1143–1160.

Zinszner, B., and Pellerin, F. M., 2007. *A Geoscientist's Guide to Petrophysics*. Editions Technip.

Zoback, M. D., Rummel, F., Jung, R., and Raleigh, C. B., 1977. Laboratory hydraulic fracturing experiments in intact and pre-fractured rock. *Int. J. Rock Mech. Min. Sci. Geomech. Abs.*, **14**, 49–58.

Zoback, M. D., Barton, C. A., Brudy, M., Castillo, D. A., Finkheiner, T., Grollimund, B. R., *et al.*, 2003. Determination of stress orientation and magnitude in deep wells. *Int. J. Rock Mech. Min. Sci.*, **40**, 1049–1076.

Zoback, M. D., 2007. *Reservoir Geomechanics.* Cambridge University Press.

Index